Multitstatic Passive Radar Target Detection

Other titles in this series:

Advances in Bistatic Radar Willis and Griffiths
Airborne Early Warning System Concepts, 3rd Edition Long
Bistatic Radar, 2nd Edition Willis
Design of Multi-Frequency CW Radars Jankiraman
Digital Techniques for Wideband Receivers, 2nd Edition Tsui
Electronic Warfare Pocket Guide Adamy
Foliage Penetration Radar: Detection and characterisation of objects under trees Davis
Fundamentals of Ground Radar for ATC Engineers and Technicians Bouwman
Fundamentals of Systems Engineering and Defense Systems Applications Jeffrey
Introduction to Electronic Warfare Modeling and Simulation Adamy
Introduction to Electronic Defense Systems Neri
Introduction to Sensors for Ranging and Imaging Brooker
Microwave Passive Direction Finding Lipsky
Microwave Receivers with Electronic Warfare Applications Tsui
Phased-Array Radar Design: Application of radar fundamentals Jeffrey
Pocket Radar Guide: Key facts, equations, and data Curry
Principles of Modern Radar, Volume 1: Basic principles Richards, Scheer and Holm
Principles of Modern Radar, Volume 2: Advanced techniques Melvin and Scheer
Principles of Modern Radar, Volume 3: Applications Scheer and Melvin
Principles of Waveform Diversity and Design Wicks *et al.*
Pulse Doppler Radar Alabaster
Radar Cross Section Measurements Knott
Radar Cross Section, 2nd Edition Knott *et al.*
Radar Design Principles: Signal processing and the environment, 2nd Edition Nathanson *et al.*
Radar Detection DiFranco and Rubin
Radar Essentials: A concise handbook for radar design and performance Curry
Radar Foundations for Imaging and Advanced Concepts Sullivan
Radar Principles for the Non-Specialist, 3rd Edition Toomay and Hannan
Test and Evaluation of Aircraft Avionics and Weapons Systems McShea
Understanding Radar Systems Kingsley and Quegan
Understanding Synthetic Aperture Radar Images Oliver and Quegan
Radar and Electronic Warfare Principles for the Non-specialist, 4th Edition Hannen
Inverse Synthetic Aperture Radar Imaging: Principles, algorithms and applications Chen and Martorella
Stimson's Introduction to Airborne Radar, 3rd Edition Baker, Griffiths and Adamy
Test and Evaluation of Avionics and Weapon Sytems, 2nd Edition McShea
Angle-of-Arrival Estimation Using Radar Interferometry: Methods and applications Holder
Biologically Inspired Radar and Sonar: Lessons from Nature Balleri, Griffiths and Baker
The Impact of Cognition on Radar Technology Farina, De Maio and Haykin
Novel Radar Techniques and Applications, Volume 1: Real Aperture Array Radar, Imaging Radar, and Passive and Multistatic Radar Klemm, Nickel, Gierull, Lombardo, Griffiths and Koch
Novel Radar Techniques and Applications, Volume 2: Waveform Diversity and Cognitive Radar, and Target Tracking and Data Fusion Klemm, Nickel, Gierull, Lombardo, Griffiths and Koch
Radar and Communication Spectrum Sharing Blunt and Perrins
Systems Engineering for Ethical Autonomous Systems Gillespie
Shadowing Function from Randomly Rough Surfaces: Derivation and applications Bourlier and Li
Photo for Radar Networks and Electronic Warfare Systems Bogoni, Laghezza and Ghelfi
Multidimensional Radar Imaging Martorella
Radar Waveform Design Based on Optimization Theory Cui, De Maio, Farina and Li
Micro-Doppler Radar and Its Applications Fioranelli, Griffiths, Ritchie and Balleri
Maritime Surveillance with Synthetic Aperture Radar Di Martino and Antonio Iodice
Electronic Scanned Array Design Williams
Advanced Sparsity-Driven Models and Methods for Radar Applications Li
Deep Neural Network Design for Radar Applications Gurbuz
New Methodologies for Understanding Radar Data Mishra and Brüggenwirth
Radar Countermeasures for Unmanned Aerial Vehicles Clemente, Fioranelli, Colone and Li
Holographic Staring Radar Oswald and Baker
Polarimetric Radar Signal Processing Aubry, De Maio and Farina

Multistatic Passive Radar Target Detection

A detection theory framework

Amir Zaimbashi and Mohammad Mahdi Nayebi

The Institution of Engineering and Technology

Published by SciTech Publishing, an imprint of The Institution of Engineering and Technology, London, United Kingdom

The Institution of Engineering and Technology is registered as a Charity in England & Wales (no. 211014) and Scotland (no. SC038698).

First published 2023

The Institution of Engineering and Technology
Futures Place
Kings Way, Stevenage
Herts, SG1 2UA, United Kingdom

www.theiet.org

British Library Cataloguing in Publication Data
A catalogue record for this product is available from the British Library

ISBN 978-1-83953-852-0 (hardback)
ISBN 978-1-83953-853-7 (PDF)

Typeset in India by Exeter Premedia Services Private Limited

Cover credit: Baac3nes/Moment via Getty Images

Contents

About the Authors xi
Preface xiii

PART I Passive radar basic principles and conventional target detection algorithms 1

1 An introduction to passive radar 3
1.1 Passive radar illuminators 5
1.1.1 FM radio signal characteristic 6
1.1.2 DVB-T signal characteristic 11
1.1.3 ATV signal characteristic 19
1.2 Geometry of passive radars 22
1.3 Power budget and passive radar coverage 31
1.4 Summary 35
References 36

2 Passive radar conventional target detection algorithms 39
2.1 Surveillance and reference signal models 40
2.2 Conventional interference removal approaches 44
2.2.1 Subspace-based interference removal approaches 44
2.2.2 Adaptive filter-based interference removal approaches 46
2.2.2.1 Least mean squares algorithms 47
2.2.2.2 Recursive least squares algorithms 48
2.3 Conventional PR target detection approaches 51
2.4 Performance results 52
2.4.1 Multipath removal capability: stationary multipath scatterers 53
2.4.2 Multipath removal capability: moving multipath scatterers 59
2.4.3 Detection performance evaluation 64
2.5 Summary 67
References 67

PART II Target detection in passive bistatic radar under high-SNR reference channel 71

3 Multitarget detection problem in single-band FM-based passive radar 73
3.1 Introduction 73
3.2 High-SNR RC-based problem formulation 74
3.2.1 GLRT-based detector: known σ^2 and when $K = 1$ 76
3.2.2 GLRT-based detector: unknown σ^2 and when $K = 1$ 80
3.2.3 Two-stage GLRT-based detector: general problem 81
3.2.4 First stage of 2S-GLRT-based detector 81
3.2.5 Second stage of 2S-GLRT-based detector 85
3.3 Analytical performance analysis 87
3.4 Simulation results 89
3.4.1 False alarm regulation evaluation 92
3.4.2 Detection performance evaluation 94
3.5 Summary 102
References 102
Appendix 3A: Derivation of detection performance 104

4 Multitarget detection in multiband FM-based passive bistatic radar: target detection quality improvement 107
4.1 Introduction 107
4.2 Signal modeling 108
4.3 Design of UMPI-based detector 111
4.3.1 First stage of UMPI-based detector 111
4.3.2 Second stage of UMPI-based detector 116
4.4 Analytical performance analysis 120
4.4.1 SW1-I target model 121
4.4.2 SW1-C target model 122
4.5 Performance results 123
4.6 Summary 132
References 132
Appendix 4A: Maximal invariant statistic distribution 134
Appendix 4B: Detection probability formulas for the SW1-I and SW1-C target models 135

5 Multitarget detection in multiband FM-based PBR: target range resolution improvement 137
5.1 Introduction 137
5.2 Signal modeling 138
5.3 Design of high-resolution UMPI test 142
5.3.1 First stage of UMPI-based detector 143
5.3.2 Second stage of UMPI-based detector 145
5.4 Analytical performance analysis 147

5.5 Analytical assessment of improved range resolution 148
5.6 Analytical performance analysis under amplitude mismatch 153
5.7 Simulation results 154
5.8 Summary 163
References 164
Appendix 5A: Maximal invariant derivation 166
Appendix 5B: Maximal invariant statistic pdfs 168

6 Broadband target detection algorithm in FM-based passive bistatic radar systems 171
6.1 Introduction 171
6.2 Signal modeling 172
6.3 Design of a broadband UMPI test 174
6.3.1 First stage of a UMPI-based detector 174
6.3.2 Second stage of a UMPI-based detector 177
6.4 Analytical performance evaluation 180
6.5 Enhancing range resolution with broadband detection 181
6.6 Enhancing target performance quality with broadband detection 185
6.7 Simulation results 187
6.8 Summary 194
References 195
Appendix 6A: MI statistic derivation 196

7 Multitarget detection in FM and digital TV-based passive radar: M-ary hypothesis testing framework 199
7.1 Signal modeling 199
7.2 M-ary hypothesis testing-based detection approach 200
7.2.1 BH-GLRT detector design 202
7.2.2 Forward and recursive implementation of BH-GLRT detector 205
7.2.3 Parallel implementation of FR-BH-GLR detector 206
7.2.4 FFT-based implementation of PI-FR-BH-GLR detector 207
7.3 Analytical performance evaluation 209
7.4 Simulation results 210
7.4.1 The importance of optimal detectors for passive radar 211
7.4.2 Performance of the proposed detection algorithm 212
7.4.3 A comparative evaluation of passive radar detection algorithms 217
7.5 Summary 222
References 222

8 Multitarget detection in analog TV-based passive radar systems 225
8.1 Introduction 225
8.2 Detection problem formulation 226

8.3 Multilayer GLR-based detection algorithm 228
8.3.1 Chirp z-transform implementation of MLa-GLR detection algorithm 229
8.3.2 Recursive implementation of the CZT-MLa-GLR detection algorithm 230
8.4 Robust S-CZT-MLa-GLR detection algorithm 233
8.5 Performance results 234
8.6 Summary 244
References 245

9 Multi-accelerating-target detection in passive radar systems 247
9.1 Introduction 247
9.2 Signal modeling 248
9.3 Detection problem formulation 252
9.4 Multilayer GLR-based detection algorithm 253
9.5 Chirp-FFT implementation of 3D-MLa-GLRT 255
9.6 Successive implementation of 3D-MCFFT-MLa-GLRT 256
9.7 Performance results 257
9.7.1 Detection performance of the 2D-SaR detection algorithm 257
9.7.2 Performance evaluation of the 3D-SaR detection algorithm 262
9.8 Summary 268
References 268

PART III Multistatic passive radar target detection under noisy reference channels 271

10 Noisy RC-based bistatic passive target detection 273
10.1 Introduction 273
10.2 Target detection problem formulation 274
10.3 LRT-based detector 275
10.4 Kernel-based detectors 278
10.4.1 Preliminaries 278
10.4.2 Kernelized LRT-based proposed detectors 280
10.5 Performance results 282
10.6 Summary 294
References 297

11 Noisy RC-based multistatic passive radar target detection 299
11.1 Introduction 299
11.2 Problem formulation 300
11.3 LRT-based proposed detectors 303
11.4 Performance results 306
11.5 Summary 312
References 312
Appendix 11A: General detection problems P1 313

PART IV Multistatic passive radar target detection without reference channels **315**

12 Multistatic passive radar target detection without direct-path interference **317**
12.1 Signal model and problem formulation 317
12.2 Detectors design 321
12.2.1 Proposed detectors for detection problem P_1 322
12.2.1.1 P_1-Rao detector 322
12.2.1.2 P_1-Vol detector 323
12.2.1.3 P_1-LRT-based proposed detectors 323
12.2.2 Proposed detectors for detection problem P_2 324
12.2.3 Proposed detector for detection problem P_3 325
12.3 CFAR analysis of proposed detectors 325
12.4 Performance results 326
12.4.1 False alarm probability evaluations 327
12.4.2 Detection performance evaluations 329
12.5 Summary 332
References 333
Appendix 12A: P_1-Rao detector 335
Appendix 12B: P_1-LRT detector 339
Appendix 12C: P_2-C2-LRT detector 340
Appendix 12D: P_3-LRT detector 343

13 MIMO passive radar target detection with direct-path interference **347**
13.1 Introduction 347
13.2 Detection problem formulation 347
13.3 Detectors design 350
13.4 Performance results 352
13.4.1 Performance of SIMO systems 353
13.4.2 Performance of MIMO systems 355
13.4.3 Rao test statistic for target localization 360
13.5 Summary 362
References 362
Appendix 13A: Unc-Rao test statistic derivation 363

Index **371**

About the Authors

Amir Zaimbashi is an associate professor and head of the Optical and RF Communication Systems Laboratory at the Department of Electrical Engineering, Shahid Bahonar University of Kerman, Iran. He was named best researcher in his engineering faculty in 2020. He is a senior member of IEEE since 2023. His current research interests include statistical signal processing, array signal processing, kernel theory, optimization theory, and their applications in active/passive radars and wireless communication systems.

Mohammad Mahdi Nayebi is a full professor in the Department of Engineering, Sharif University of Technology, Iran. He is a senior member of IEEE since 2005. He is a member of the editorial board of *IET Radar, Sonar, and Navigation* journal since 2018. He is the founder of the Iran Radar conference and the chairman of its organizing committee. His main research interests are radar signal processing and detection theory.

Preface

Radar sensing and wireless communications have been progressing in parallel for decades, with limited interactions. Regulators now mandate that certain frequency bands be shared with wireless communication systems, due to the increasing congestion of the electromagnetic spectrum. Therefore, it is imperative to develop advanced radar signal processing algorithms that ensure the acceptable performance of both radar and communication systems. Specifically, the radar waveform needs to be adaptively designed to maximize radar performance, such as signal-to-interference-and-noise ratio, and guarantee acceptable communication systems performance by ensuring a prescribed total throughput or sum capacity of communication systems. This approach is known as radar-centric communication technology. However, the main drawback of this strategy is the limited data rates provided to the wireless communication system, as the radar signaling remains largely unaltered to achieve near-optimal radar performance.

Passive radar, on the other hand, can be viewed as a form of Joint Communication and Radar Sensing (JCRS) system where a radar system is integrated with a communication system that transmits signals. The communication system is given high priority in the design, while the radar operation is considered an add-on with no priority. This approach may be known as a communication-centric radar system with low priority on the radar system. In this special case, the transmitted signals are not specifically designed for radar sensing but are received and processed by a passive radar receiver using optimal signal processing algorithms. Fortunately, modern digital transmitters for broadcast communications provide waveforms that are more suitable for radar use than those available in the past.

Passive Radar (PR) technology employs communication or broadcast signals not specifically designed for radar purposes to detect targets. Extracting useful information from these signals requires significant signal processing due to various factors that may impact the detection process, including interference from direct-path signals, multi-path signals, and strong targets that may mask weaker ones. To effectively design PR systems, these factors must be considered when modeling the received signal and developing target detection methods. Detecting multiple targets is another significant challenge that requires careful consideration in these systems. The book focuses on examining the multistatic PR target detection problem using the detection-theory framework, with the aim of presenting the latest research developments in this field. Early methods in this area were based on intuition and lacked optimality. However, more recent methods with a clear theoretical basis have emerged, grounded in detection theory. The book aims to offer timely and valuable information to researchers and PR system designers.

In conventional PR systems, a spatial and/or temporal filter is used to separate the direct-path signal and the target-path signal into two channels known as the reference channel (RC) and the surveillance channel (SC), respectively. The RC contains a delayed replica of the transmitted signal, which is used by PR for matching and processing the signal echo in the SC. However, using an RC is not the only approach to solving this problem. The authors believe that techniques for detecting PR targets can be classified into two categories based on whether an RC is included in formulating PR target detection problems or not. These categories are known as the two-channel and single-channel PR target detection problems. In the two-channel approach, both the surveillance and reference channels are utilized to formulate the PR target detection problem. In contrast, single-channel PR refers to the utilization of only the SC for formulating the target detection problem in PR.

This book is organized into four parts, with each part addressing a specific aspect of target detection in various radar systems. The first part, consisting of two chapters, covers the fundamentals of PR and traditional target detection algorithms. Part two comprises seven chapters and deals with the target detection issue in passive bistatic radar (PBR) with a reliable RC. Part three includes two chapters and focuses on the detection of targets in multistatic PR systems in the presence of noisy RC PR systems. Finally, part four, which consists of two chapters, discusses the target detection problem in multistatic and multiple-input multiple-output (MIMO) PRs when no reliable RC is available.

Chapter 1 discusses the fundamental principles of PR, which include its history, terminology, the characteristics of opportunity signals, bistatic geometry, power budget, and the coverage of PR.

Chapter 2 introduces a model for the received signals in the surveillance and reference channels, followed by an examination of detection theory-based techniques for PR target detection. These techniques are categorized into two groups based on whether an RC is used or not. The chapter also highlights the significance of using detection theory-based methods by comparing them to conventional PR signal processing approaches, where interference removal and target detection are performed separately. Although the latter strategies are acceptable for active radar design, they are not recommend for PR applications.

In Chapter 3, the focus is on solving the multi-target detection problem using a single channel for target detection. The chapter includes a detailed tutorial on this topic and presents analytical and simulation results to demonstrate the effectiveness of the proposed two-step generalized likelihood ratio test (GLRT) in removing interference and detecting targets. While Frequency Modulation (FM) signals are used to evaluate the detector's false alarm regulation and target detection performance, other opportunity signals can also be utilized.

FM-based PBRs that rely on a single broadcast channel are limited by their time-varying detection performance, which is influenced by changes in the program content of transmitted signals. This results in time-varying detection performance as well as poor target range resolution. To address this limitation, we propose using multiple broadcast channels from a single transmitter to improve target detection quality (TDQ), target range resolution (TRR), or both. In Chapter 4, we describe a multiband uniformly most powerful invariant (UMPI) test as an optimal invariant detector to enhance TDQ. We derive closed-form expressions for calculating false alarm probability and detection probability for the target models of Swerling 0 and Swerling 1 and present a comprehensive performance evaluation that demonstrates the robustness of the proposed multiband detector against the time-varying program content of FM radio channels. In addition to reliable detection performance, the proposed multiband PBR system offers non-coherent combined diversity gain and frequency diversity gain, thanks to the joint exploitation of multiple broadcasted channels for detection and independent returns received from one target over different frequency channels, respectively. Chapter 5 of the book aims to improve the target range resolution in FM-based PBRs systems by exploiting multiple FM radio broadcasted channels from the same transmitter. The problem is formulated as a composite hypothesis testing problem in the presence of various interferences. A multiband UMPI is developed to demonstrate how utilizing multiple broadcasted channels can enhance target range resolution. Closed-form expressions are derived for the false alarm probability and detection probability of the proposed detector. Simulation results show that a multiband PBR system provides coherently combined diversity gain and TRR improvement compared with a single-band PBR system. In Chapter 6, the authors present a target detection algorithm that leverages the techniques presented in Chapters 4 and 5 to improve target quality detection and range resolution simultaneously.

The use of non-cooperative illuminators of opportunity in PRs can lead to masking of target echoes by echoes from stronger targets in multi-target scenarios. To address this issue, the problem of multitarget detection is modeled as an M-ary hypothesis testing problem in Chapter 7. A GLRT-based detector has been proposed. It is implemented in a parallel and recursive manner, enabling sequential detection of targets. It treats previously detected targets as interferences to be removed, thus facilitating the detection of weaker ones. The false alarm and detection performance of the proposed sequential GLRT-based detector are studied analytically, and extensive simulation results for both FM and DVB-T-based PRs are provided to verify the analytical findings as well as to demonstrate the effectiveness of the proposed detection algorithm.

Chapter 8 focuses on the issue of target detection in analog TV (ATV)-based PRs. The chapter introduces a new sequential GLRT-based detection algorithm that uses the Chirp z-Transform (CZT). The algorithm enables simultaneous estimation of bistatic range-Doppler coordinates of targets, even in cases where the ATV signal has severe range-ambiguity.

Chapter 9 focuses on detecting high-speed and accelerating targets. In order to establish a PR system with a long range, one solution is to prolong integration times. However, this can pose difficulties for the detectors discussed in previous chapters. These detectors heavily depend on simplified target motion models, which consequently constrain their capacity to detect accelerating targets and enhance signal-to-noise ratios. To address this limitation, the authors propose a new three-dimensional (3D) sequential detection algorithm that formulates the target detection problem as an M-ary hypothesis testing problem and applies the GLRT principle.

This algorithm estimates the range, Doppler, and acceleration coordinates of desired targets and uses a modified version of the Chirp Fast Fourier Transform to implement the detector over the desired range-Doppler-acceleration map. The authors also derive a closed-form expression for false alarm probability to adjust the detection threshold. Simulation results show that the proposed 3D-detection algorithm outperforms classical target detection algorithms in FM-based PBR systems.

Part 3 of this work focuses on examining various methods for PR target detection in the presence of noisy RCs. The main aim is to account for the impact of thermal noise on the detection performance and false alarm regulation of the SC. Two different configurations are considered: single-input single-output (SISO) and single-input multiple-output (SIMO). In Chapter 10, the SISO target detection problem is discussed, and it is shown that any uncertainty in the direct-path signal-to-noise ratio of the RC (referred to as DNR_r) can lead to an excessive false alarm probability of the proposed detector, especially in low DNR_r regime. To address this issue, a new threshold-setting strategy is proposed to adjust the level of the detector, enabling efficient operation even under DNR_r uncertainty. Additionally, the kernel theory framework is applied to the target detection problem of both noisy and ideal RC PR scenarios, resulting in creating some new detection algorithms. In Chapter 11, the SIMO target detection problem is covered, and two detection methods are developed using the GLRT principle. One of the proposed detectors has a fixed level, while the other maintains a fixed size in the presence of uncertainty in DNR_r. The effectiveness of the proposed detection methods is illustrated and compared through numerous simulation results.

Chapter 12 of the book discusses the target detection problem in a passive multistatic radar that comprises one transmitter and multiple spatially separated receivers, specifically in an SIMO configuration, without utilizing the RC. The chapter presents several solutions for this problem within different frameworks, including Rao, LRT, and geometrical representation, and under different assumptions about the problem's parameter space, thus creating a unified framework for multistatic PR target detection. It is proven analytically that the proposed target detection methods exhibit constant false alarm rate (CFAR) behavior against noise variance uncertainties across different receivers. Finally, the chapter presents results from Monte Carlo simulations that demonstrate the detection capabilities of the proposed detectors.

In Chapter 13, the problem of detecting targets using a passive MIMO radar is examined. The radar system consists of multiple transmitters that do not overlap in frequency but share a common bandwidth and multiple receivers that are spatially separated. Direct-path interference is also considered. The study considers both calibrated and uncalibrated receiver scenarios. In the calibrated case, all receivers are assumed to have the same noise variance, while in the uncalibrated case, the noise variances differ across spatial and frequency receivers. The paper uses Rao's principle to develop closed-form tests for detection in both calibrated and uncalibrated scenarios. The performance of the proposed detectors is evaluated through extensive simulations.

The author assumes that the reader possesses basic knowledge of radar signal processing, as well as linear and matrix algebra, and a strong background in detection theory. However, the author also makes sure that the discussions are comprehensive enough to enable a clear understanding of the topic. Thus, the material presented would be useful not only for students but also for practicing engineers. The intended audience is those who are involved in the design and implementation of PR signal processing, with a focus on the development of detection algorithms that can be executed on a digital computer.

Finally, I express my deepest gratitude to my loved ones for their unwavering support and motivation throughout my journey. I am especially grateful to my family, who have demonstrated persistent patience and provided unconditional backing. I owe them a debt of gratitude for enabling me to conquer this demanding endeavor. I want to extend my heartfelt appreciation to my beloved wife, Malih, whose unwavering love, encouragement, and understanding have been my pillar of strength during the highs and lows of this creative process. Her presence and belief in me have been a constant source of inspiration.

Furthermore, I would like to express my sincere thanks to Prof. Mohammad Mahdi Nayebi, the second author of this book, whose unwavering encouragement inspired and motivated me to embark on this writing journey. Ultimately, I dedicate this book as a tribute to the memory of Alireza Afzalipour (March 26, 1909, to April 7, 1993), whose unwavering resolve paved the way for the founding of Shahid Bahonar University in Kerman.

Amir Zaimbashi
Shahid Bahonar University of Kerman
Kerman, Iran
Email: a.zaimbashi@uk.ac.ir

Part I

Passive radar basic principles and conventional target detection algorithms

The part consists of two chapters. The first chapter is dedicated to exploring the fundamental principles of passive radar, including its history, the characteristics of opportunity signals, bistatic geometry, power budget, and passive radar coverage. The aim of this chapter is to provide readers with a comprehensive understanding of the theoretical foundations of passive radar.

The second chapter delves into the modeling of the received signals in the surveillance and reference channels. It then introduces two main categories of passive radar target detection techniques. Within this chapter, conventional approaches to passive radar signal processing are examined, where interference removal and target detection are performed as separate steps. These conventional methods rely on intuition rather than intrinsic optimality. The author stresses the importance of adopting detection theory-based approaches to effectively formulate the passive radar target detection problem in the presence of interference signals, including multipath and interfering target signals. Since opportunistic signals are not designed for radar applications and are beyond the control of radar designers, it becomes crucial to concentrate on devising more sophisticated and optimal signal processing algorithms for passive radar receivers. Together, these two chapters comprehensively cover diverse aspects of passive radar systems and serve as a valuable resource for researchers, engineers, and students interested in passive radar technology.

Chapter 1

An introduction to passive radar

This book explores the problem of detecting targets using multistatic passive radar (PR) under the framework of detection theory. The term "multistatic" indicates that multiple transmitter–receiver pairs are used for target detection, with the receivers being spatially separated. In this type of radar, there is no dedicated transmitter, but instead, transmitters of opportunity, such as radio or television (TV) transmitters, are used.

PR systems can utilize a variety of illuminators, including analog frequency modulation (FM) radio, analog television (ATV), digital broadcasters like digital audio broadcast (DAB), Digital Video Broadcasting – Terrestrial (DVB-T), cell phone base stations, Wireless Fidelity (WiFi), Worldwide Interoperability for Microwave Access (WiMAX), and satellite-borne transmitters. These signals were not originally intended for radar target detection purposes, making it challenging to achieve optimal performance for PR applications. Since only the receiver is under the control and design of the radar designer, a suboptimal performance seems to be expected from these systems [1]. Therefore, this book aims to examine the PR target detection problem to better understand the challenges involved in this type of radar system and propose solutions to enhance its performance.

Throughout the following chapters, we will delve into how the performance of PR systems can be enhanced by taking into account the characteristics of the opportunity signals on the receiver side. In comparison with active radars that have been developed over many decades, PRs have generated significant interest due to the many advantages they offer.

First, PRs are typically priced lower and are smaller in size compared with their active counterparts. This makes them more accessible and easier to deploy in a variety of situations. Additionally, PRs are undetectable as they do not emit any signals or require transmitters, which is a considerable advantage in certain scenarios.

Second, PRs have the potential to detect stealth targets due to their bistatic or multistatic geometry and low operating frequency. This geometry allows them to detect targets that are not visible to active radars, which rely on emitted signals that can be detected by the target. PRs, on the other hand, can use existing signals in the environment to detect targets, making them less likely to be detected by the target itself.

Third, PRs often do not suffer from range and velocity ambiguities, which can occur in active radar systems. This ambiguity arises due to the radar signal bouncing off multiple objects, causing difficulties in determining the actual location and velocity of the target. PRs, however, do not emit any signals and rely on exploiting multiple opportunity signals, which eliminates this ambiguity.

Fourth, PRs do not require any specific frequency allocation, as they utilize existing broadcasting transmitters. This eliminates the need for any additional spectrum allocation, which can be a significant advantage in crowded frequency bands.

Fifth, PRs are capable of detecting low-flying objects that use environmental geography as a shield due to their bistatic geometry. This makes them useful in detecting targets that are difficult to detect with other radar systems.

Finally, PRs are more immune to anti-radiation missiles (ARMs) because they do not emit any signals, and therefore, there is no transmitter for the missile to target. This is a significant advantage in situations where the threat of ARMs is a concern.

In summary, PRs have many advantages over active radars, which make them attractive in various scenarios. In the upcoming chapters, we will discuss how these advantages can be leveraged to improve the performance of PR systems.

PR, also known as passive coherent location, passive covert radar, broadcast radar, noncooperative radar, parasitic radar, symbiotic radar, and hitchhiking, refer to a class of radar systems that operate without their own dedicated transmitter. Instead, they rely on other sources of electromagnetic radiation, such as TV or radio signals, to detect and track objects. The term “hitchhiking” is used specifically when an existing monostatic radar is used as the source of illumination.

In this book, the authors use the widely used term “passive radar” to refer to this class of radar systems. Over the last two decades, the field of PRs has seen many special issues, which have repeatedly discussed system concepts, experimental systems, and technology demonstrators, as evidenced by References 2–4.

Despite the various names used to describe PRs, the authors believe that the field has reached an unprecedented maturity, thanks to the research efforts carried out in universities and industrial and research centers. PRs have proven to be a valuable and effective alternative to traditional active radar systems, especially in situations where stealth or covert operations are required and where active radar emissions can reveal the position of the radar system to hostile forces.

PR technology has a long history that spans more than 90 years [1]. The concept of PR involves using existing electromagnetic radiation sources, such as TV or radio broadcast signals, as illumination sources for detecting and tracking targets. The first operational PR was the bistatic radar system Klein Heidelberg, which used the British Chain Home radar as an illumination source and became operational in 1943 [5].

In the 1960s, the first satellite-based passive bistatic radar with a randomly modulating continuous waveform signal was designed to detect ground vehicle targets [6]. In the same decade, the United States developed the over-the-horizon high-frequency passive bistatic radar [7].

In the 1980s, a hitchhiking-based passive bistatic radar using an air traffic control radar illumination source was developed at the University College London [8]. Later on, this system was updated to use ultra high frequency (UHF) TV transmitters. However, there were some difficulties working with 100% duty cycle signal and periodic features associated with the line scan of the TV signals. In 1990, Howland developed a passive bistatic radar based on the vision carrier of the ATV signal in order to measure Doppler frequencies of desired targets and track them [9].

In the mid-1990s, the Manastash Ridge Radar was developed at the University of Washington, which used FM broadcast transmitters to study turbulence in the ionosphere [10]. This system had a maximum range of 1,200 km. In the late 1990s and early 2000s, the US Lockheed Martin company developed an FM-based passive bistatic radar, known as Silent Sentry [11].

In 2005, Howland developed a very high-frequency (VHF) FM-based passive bistatic radar to detect and track aircraft targets with a maximum range of 100 km [12]. In recent years, many companies all over the world have developed and demonstrated PR systems. These include Homeland Alerter from THALES company in France [13], European Aeronautic Defense and Space, and AULOS from the SELEX company in Italy [14, 15]. For instance, the AULOS radar has two receiving systems connected to two uniform circular arrays, each with eight dipole elements, one array operating in the FM band and the other one in the digital TV band [16]. Each receiving system contains eight coherent receiving channels connected to the array dipoles. The radio frequency (RF) signal from each dipole is filtered, amplified, and then digitized. The system uses digital beamforming and multistatic techniques, exploiting one or more broadcasting stations simultaneously. The signal processor performs cochannel interference cancellation, generates the range–Doppler maps, applies constant false alarm rate (CFAR) detection, and sends plots to the tracker.

PR systems are advantageous over active radar systems because they do not require dedicated transmitters, which makes building and operational costs relatively lower. As a result, many university research groups all over the world have developed PR setups that exploit FM radio, digital radio and TV, cellphone base station, WiFi, WiMAX, and some satellite signals.

1.1 Passive radar illuminators

PRs are a type of radar that uses signals from existing RF radiation in the environment. These signals are typically produced by different communication systems, including FM, ATV, DVB-T, WiFi, WiMAX, and so forth. PRs are useful in situations where active radars cannot be used, such as in areas where radar emissions are restricted or when stealth operations are required.

The selection of the transmitter type for PRs is based on several factors, including spatial and time coverage, transmitter power, frequency band, signal bandwidth, and the shape of the ambiguity function. Transmitter power determines the maximum range of the PR, while the signal bandwidth dictates the achievable target range resolution (RR). The shape of the ambiguity function determines the target detection quality [17, 18].

In this book, the main focus is on PR target detection using FM, ATV, and DVB-T signals. These signals are of particular interest because they are readily available in the environment and can provide good range and velocity information about targets. FM signals have a narrow bandwidth, making them suitable for detecting small targets at long ranges. ATV signals have a wider bandwidth than FM signals, making them suitable for detecting larger targets at shorter ranges. DVB-T signals have a wide bandwidth and are suitable for detecting both small and large targets at a variety of ranges. Overall, PRs are a useful tool for detecting targets in situations where active radars cannot be used. The selection of the transmitter type is critical to achieving good performance, and FM, ATV, and DVB-T signals are all suitable for use in PR systems.

1.1.1 FM radio signal characteristic

This section provides a comprehensive overview of the characteristics of FM signals, highlighting key parameters that are commonly used to describe and analyze them. The first parameter discussed is the peak-to-sidelobe level ratio (PSLR), which is a measure of the ratio of the peak amplitude of the FM signal to the amplitude of its sidelobes. It represents the relative strength between the main peak of an FM signal and the unwanted sidelobes that occur in the output of detector statistics. This parameter is essential in evaluating the masking effect of strong targets on weak targets.

Another important parameter discussed is the integrated sidelobe level ratio (ISLR), which is a measure of the power contained in the sidelobes of the FM signal. This parameter is critical in determining the level of interference caused by the FM signal in adjacent rang-Doppler cells.

The section also covers target range resolution (RR), which is a measure of the ability of the FM signal to distinguish between closely spaced targets at different ranges. The FM radio, invented in the 1930s, is a widely used analog modulation scheme that operates in the 88–108 MHz portion of the VHF spectrum [19]. In most countries, each adjacent channel is separated by 200 kHz, and the frequency deviation is $\pm$75 kHz, resulting in a maximum bandwidth of 150 kHz for an FM channel [20, 21]. However, due to the low and varying bandwidth depending on transmitted program content, the FM radio transmission has limited RR.

The section also notes that FM transmitters are usually sited on tall towers or masts in high locations, where the radiation patterns are typically omnidirectional in azimuth. However, the vertical-plane radiation patterns of FM transmitters are tilted below the horizon to avoid wasting too much power above the horizontal. This tilt results in good airspace illumination for high altitude targets at medium to far ranges from the transmitter [22, 23].

The transmitting stations broadcast many radio programs on different FM channels, each with different power levels to meet various coverage requirements. The effective radiated power of FM transmitters typically ranges from 2 kW to 250 kW [21]. Additionally, owing to the high transmitted power of FM transmitters and the long wavelength, PRs seem to be the best candidate for a system with long-range detection capabilities [20, 22].

In general, FM transmitters operate by transmitting continuously, which means that the signal is broadcasted without interruption. This continuous signal transmission can be advantageous for radar applications because it allows for relatively long integration times, which in turn can yield very good Doppler resolution. Doppler resolution refers to the ability of the radar system to accurately distinguish between different moving targets by measuring their relative velocities. The continuous signal transmission of FM transmitters can also enable radar systems to accurately separate targets in frequency dimension, even if they are not separated in range dimension. This is because the frequency of the radar signal can change as it reflects off of different objects with varying velocities. By analyzing the changes in frequency, a radar system can accurately determine the relative velocities of different targets, which can be useful for identifying and tracking moving objects such as vehicles or aircraft.

Furthermore, most FM stations transmit multiple radio channels simultaneously. This means that there is a significant amount of energy being broadcasted over a wide range of frequencies, which can be potentially harnessed by radar designers to improve the target RR. By exploiting multiple FM radio channels, a radar system can effectively increase its available bandwidth, which can in turn improve its ability to accurately measure the range of different targets.

An FM radio signal is a type of radio transmission that uses FM to carry audio information. This type of transmission can be broadcasted in two ways, either mono or stereo modes. In the mono mode, the audio signals are routed through a single channel, whereas, in stereo FM signals, audio signals are routed through two or more channels to simulate depth and direction perception, just like in the real world. At present, there are a few FM stations around the world that still broadcast in mono. The primary reason for this is that it is less expensive for recording and reproduction. Moreover, stereo FM signals are more susceptible to noise and multipath distortion than mono FM signals. However, stereo sound has almost completely replaced mono because of the improved audio quality that it provides. To create a more natural listening experience, a stereo transmission involves separate left (L) and right (R) signals rather than a single signal containing all of the audio information.

There are two systems for the transmission of FM stereo defined by the International Telecommunication Union (ITU): the stereophonic multiplex signal system and the pilot tone system. In the pilot tone system, the left and right audio signal channels are multiplexed to create a monocompatible signal that is equal to the sum of the left and right channels ($L + R$). The difference of the left and right channels, say $L - R$, is modulated using suppressed-carrier amplitude modulation with a carrier frequency of 38 kHz. As such, the complex baseband stereo FM signal can be considered as follows:

$$s(t) = e^{j2\pi \Delta f \int_0^t m(\tau)d\tau} \tag{1.1}$$

where $m(\tau)$ is called the message signal which contains the data to be transmitted, and Δf represents the maximum shift away from carrier frequency in one direction, assuming $m(\tau)$ is limited to the range ± 1. The message signal $m(\tau)$ at 10% modulation can be given by

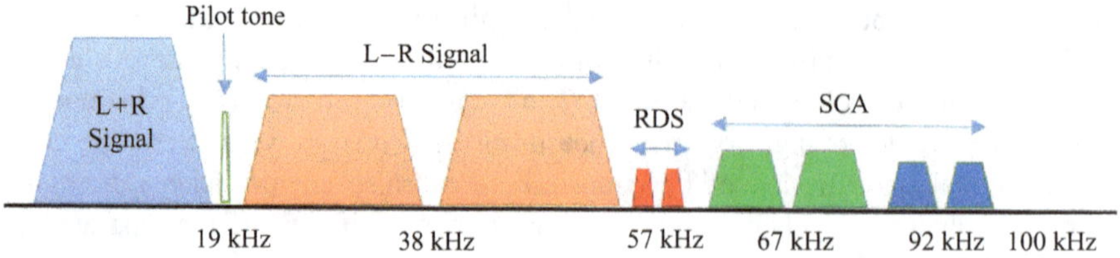

Figure 1.1 Spectrum for FM stereo modulating signal

$$m(\tau) = 0.9\left(\frac{L(\tau)+R(\tau)}{2} + \frac{L(\tau)-R(\tau)}{2}\sin(4\pi f_p\tau)\right) + 0.1\sin(2\pi f_p\tau) \quad (1.2)$$

The left and right audio signals after pre-emphasis are denoted as $L(\tau)$ and $R(\tau)$, respectively, in (1.2). A reference signal with a frequency of $f_p = 19$ kHz, also known as a pilot tone, is used. Optional auxiliary data transmission channels such as Radio Data System (RDS) and Subsidiary Carrier Authorization (SCA) can also be included, which are transmitted at higher frequencies beyond 53 kHz with lower power. RDS signal can carry weather information, FM radio channel information, or text-based information, and it is becoming increasingly common, while SCA is used on some transmissions. The baseband signal can extend up to 100 kHz if all the optional parts are used, as shown in Figure 1.1. It is important to note that stereo broadcasts are compatible with mono receivers, which use only the $L + R$ signal, and a listener will hear both channels through a single loudspeaker. Carson's rule can be used to approximate the bandwidth of an FM signal as:

$$B_T = 2\Delta f + W \quad (1.3)$$

where W is the highest frequency in the modulating signal. In FM radio broadcasting, the transmitted bandwidth is equal to $B_T = 165$ kHz, resulting from transmitting music and speech with up to a 15 kHz bandwidth. This bandwidth is achieved with a frequency deviation of $\Delta f = 75$ kHz. Figure 1.2 displays the radiophonic signal spectrum, while Figure 1.3 shows the complex envelope of a simulated stereo FM signal with duration of 1 second.

The PR faces a different challenge than active radar because it must deal with waveforms that are not specifically designed for radar purposes and can exhibit varying characteristics over time. This is particularly evident in FM radio signals, which contain music, speech, and silence that alternate on each channel with varying durations and timing. Consequently, the ambiguity function of the received FM signal has time-varying properties that affect both RR and sidelobe levels, which are often significant and comparable with the peak value. Figure 1.4 illustrates the changes in PSLR, ISLR, and RR values for a simulated stereo FM radio transmission that includes silence, music, and speech tracks over time. Each batch in the figure represents a 1-second integration time of the same FM signal with no overlap. In the simulated signals, the PSLR values range from 0 dB to 41 dB, the ISLR values range from –29 dB to –14 dB, and the RR values range from 3.5 km to 18 km.

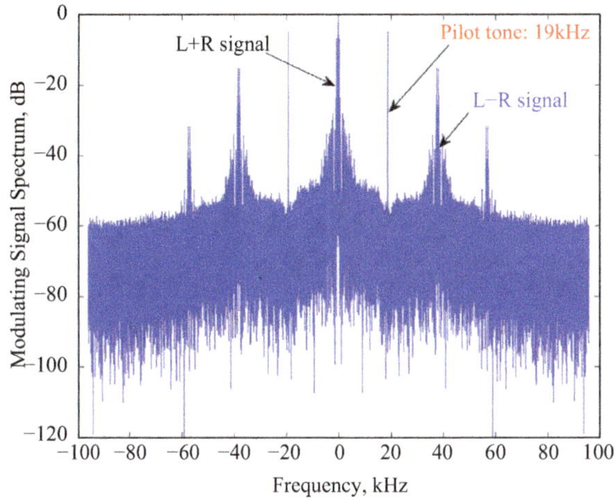

Figure 1.2 Normalized spectrum for simulated FM stereo modulating signal without optional signals

The effectiveness of a radar system in detecting targets depends on several factors, including RR, sidelobe level, and their interaction with the detection threshold. However, the time-varying characteristics of the transmitted FM waveform can significantly impact the reliability of PR system. For instance, a low PSLR can lead to a masking effect, where weak targets are obscured by stronger ones. Additionally, high ISLR can result in the capture effect, which makes it difficult for conventional CFAR detectors to detect targets. Poor RR of an FM waveform can also cause discrimination issues for close targets. Even when a "good channel signal" is chosen, which on average exhibits high RR and PSLR, the instantaneous PR performance may still experience significant degradation. Therefore, the use of a single-frequency channel does not always guarantee high-quality target detection in PR systems. Thus, it is essential to consider the

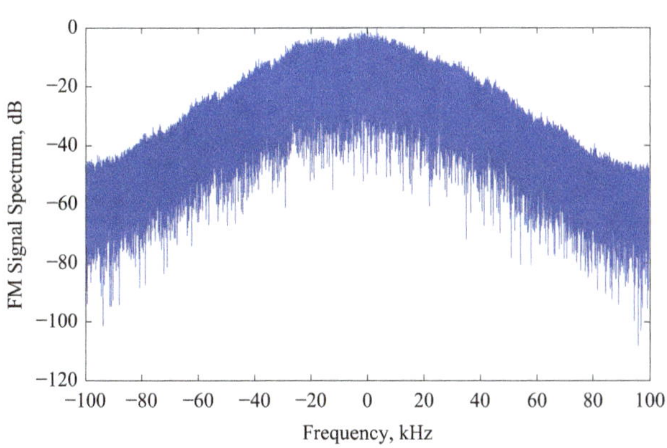

Figure 1.3 Normalized spectrum for simulated FM stereo signal

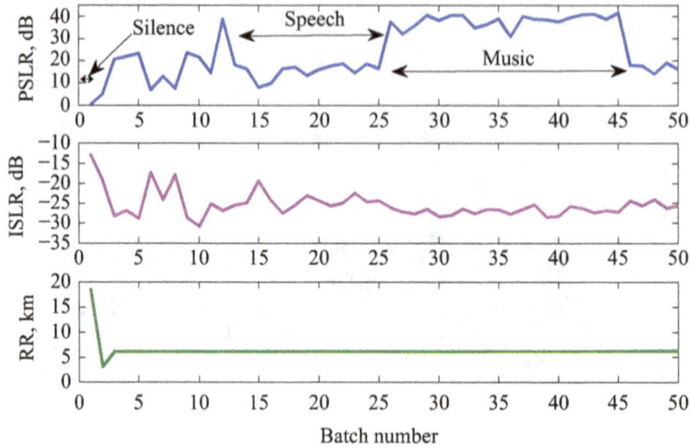

Figure 1.4 *Variation of PSLR, ISLR, and RR over time corresponding to a stereo FM radio signal contains silence, music, and voice tracks*

time-varying characteristics of the transmitted waveform when designing and operating a PR system to ensure reliable target detection.

The autoambiguity function (AAF) was used to examine the relationship between the FM transmission's ambiguity performance and the instantaneous modulation, which is dependent on the program content. The AAF was plotted as a function of delay time and velocity for three different signals: silence, speech, and jazz music from stereo FM radio signals. The results, shown in Figures 1.5–1.7, reveal that music with high spectral content yields the narrowest ambiguity function peak, resulting in the best RR. In contrast, for speech signals, the width of the peak of the ambiguity function decreases significantly during pauses between words, leading to poor RR. With silence, a periodic ambiguity function is observed due to the stereo pilot tone, resulting in a PSLR of 0 dB, as demonstrated in Figure 1.5.

In Part 2 of this book, we will delve deeper into the topic of multichannel detection to enhance target detection quality and target RR. This will lead to improved target

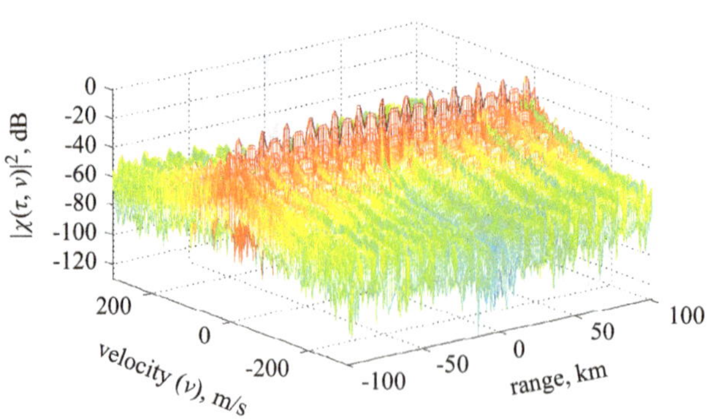

Figure 1.5 *AAF for stereo silence FM signal corresponding to the first batch of Figure 1.4 with integration time of 1 second*

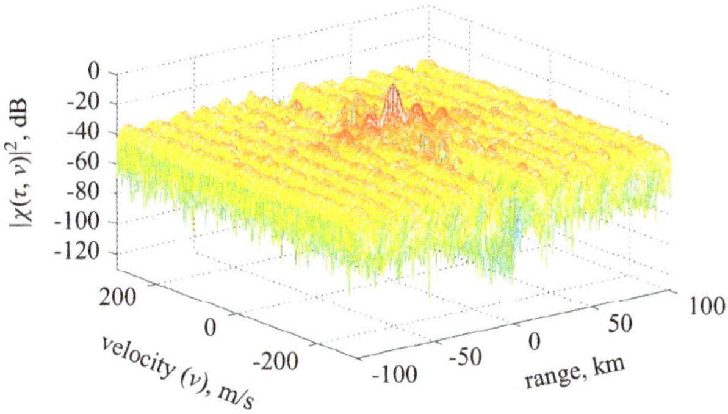

Figure 1.6 *AAF for stereo speech FM signal corresponding to the 20th batch of Figure 1.4 with integration time of 1 second*

detection quality and target range resolution. In other words, detection performance can be enhanced by utilizing multiple FM radio channels for detection. This approach achieves frequency diversity and robustness against bandwidth fluctuations that arise from the combination of the FM radio modulation technique and the time-varying program content. By leveraging the available broadcasted energy, we can enhance the resolution and accuracy of our target detection systems and better meet the demands of modern applications in various fields, including communications, defense, and security.

1.1.2 DVB-T signal characteristic

Digital TV standards vary across the world, with some of the most widely used standards including DVB-T, Digital Video Broadcasting – Second Generation Terrestrial (DVB-T2), Advanced Television Systems Committee (ATSC), Integrated Services Digital Broadcasting – Terrestrial (ISDB-T), and Digital Terrestrial Multimedia Broadcasting (DTMB). While these standards are not compatible with each other, some of them share similar signal characteristics that can be used for PR target

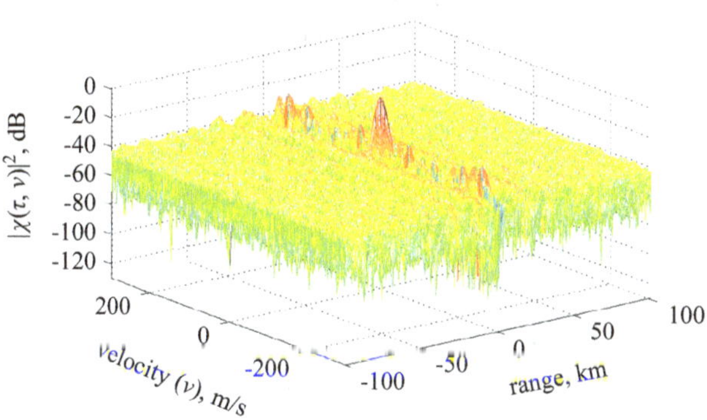

Figure 1.7 *AAF for stereo jazz music FM signal corresponding to the 40th batch of Figure 1.4 with integration time of 1 second*

detection. Among these standards, DVB-T is the most popular and is used in many regions around the world, including Asia, Australia, Africa, Europe, and South America.

DVB-T has been designed to operate within the existing VHF and UHF spectrum allocation, which covers the frequency band of 470–860 MHz. This standard has achieved maximum spectrum efficiency by utilizing large area single-frequency network operation, which enables different transmitters to transmit the same signal on the same carrier frequency. However, this can result in some ambiguities in the association of target detection results with the transmitter.

Despite these ambiguities, DVB-T is often used for PR target detection purposes due to its use of orthogonal frequency division multiplexing (OFDM) transmission with a noise-like nature in the time domain. This feature makes it easier to distinguish radar echoes from background noise, leading to improved target detection performance. However, it should be noted that the use of digital TV signals for PR target detection is still an emerging field, and further research is needed to fully understand its capabilities and limitation.

A simplified block diagram of this system is depicted in Figure 1.8, which is based on the description provided in [24]. At the input of the DVB-T transmission system, there is an Moving Pictures Experts Group (MPEG)-2 video data stream, which represents the content that is to be transmitted. To ensure that the transmitted signal is robust against channel impairments, such as noise and interference, the system employs a number of processing stages that are designed to mitigate the effects of these impairments.

The first stage in the processing chain is bit randomization, which is used to disperse the energy of the data stream across a wider bandwidth. This helps to reduce the impact of narrowband interference, such as that caused by adjacent channels or cochannel interference.

The next stage is channel coding, which involves the use of error-correcting codes to protect the data against errors that may occur during transmission. This stage consists of several substages, including outer coding, outer interleaving, inner coding, and inner interleaving. Outer coding is used to add redundancy to the data stream, while outer interleaving is used to spread any errors across a wider area of

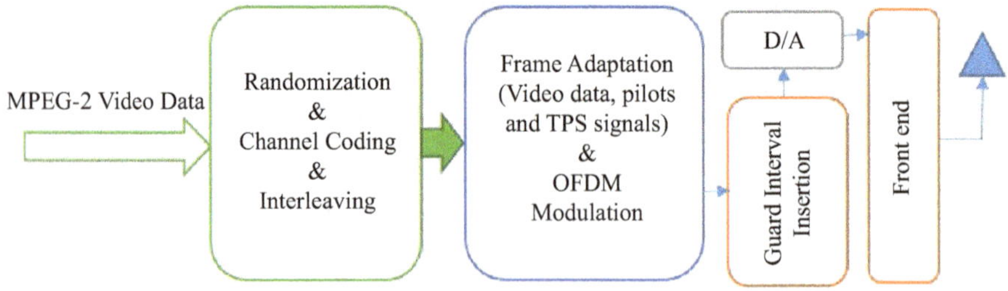

Figure 1.8 Functional block diagram of the DVB-T system, where D/A denotes digital-to-analog converter

the data stream. Inner coding and interleaving are used to protect against burst errors that may occur due to channel fading.

After channel coding, the data stream is subjected to frame adaption, which involves the addition of synchronization and control information to the data stream. This information is used to ensure that the receiver can correctly interpret the data stream.

The data stream is then modulated using OFDM, which is a technique that is used to transmit data over a wide bandwidth. OFDM is particularly well suited for use in broadcasting systems, as it is robust against multipath interference.

The next stage is guard interval insertion, which involves the addition of a guard interval between each OFDM symbol. The guard interval helps to mitigate the effects of multipath interference, by allowing the receiver to effectively filter out any delayed copies of the signal.

Finally, the modulated signal is converted to an analog signal using digital signal to analog conversion, and then transmitted over the airwaves to the receiver. Overall, the DVB-T transmission system is designed to provide robust and reliable transmission of digital content over terrestrial TV channels, and the processing stages, including bit randomization, channel coding (e.g., outer coding, outer interleaving, inner coding, and inner interleaving), frame adaption, OFDM modulation, guard interval insertion, and digital signal to analog conversion, are key to achieving this goal.

In the DVB-T standard, the transmitted signal is organized in frames. Each frame has a duration of T_F and consists of 68 OFDM symbols. Each OFDM symbol is constituted by a set of K_{8K} carriers in the $8K$ mode and K_{2K} carriers in the $2K$ mode and transmitted with a duration T_S, i.e., $T_F = 68T_S$. Each symbol itself contains two parts: (1) useful part with duration T_U and (2) guard interval with a duration Δ; thus, $T_S = T_U + \Delta$. In general, T_U is equal to $8{,}192T_B$ in the $8K$ mode, while it is $2{,}048T_B$ in the $2K$ one with T_B being the elementary period depending on the channel bandwidth B. The elementary period T_B is, respectively, equal to $\frac{7}{40}\mu$s, $\frac{7}{48}\mu$s, $\frac{1}{8}\mu$s, and $\frac{7}{64}\mu$s for nominal channel bandwidths of 5 MHz, 6 MHz, 7 MHz, and 8 MHz. The guard interval is used to prevent possible intersymbol interference (ISI), where it is created by copying the last part of the OFDM symbol. Depending upon the expected level of multipath, different values of Δ are considered in the standard, ranging from $\frac{T_U}{4}$ to $\frac{T_U}{16}$ [24]. The spacing between adjacent carriers is $\Delta f = \frac{1}{T_U}$. Let the spacing between the first and the end carriers be B_e, termed as effective channel bandwidth (ECB). Now, the number of subcarriers is equal to $B_e T_U$; thus, it is equal to $8{,}192B_e T_B$ in $8K$ mode and $2{,}048B_e T_B$ in $2K$ mode, depending upon nominal channel bandwidth B and system modes. In the DVB-T standards, there are 5 MHz, 6 MHz, 7 MHz, and 8 MHz channels with ECBs of 4.76 MHz, 5.71 MHz, 6.66 MHz, and 7.61 MHz, respectively. In Table 1.1, OFDM parameters for the $8K$ and $2K$ modes and nominal channel bandwidth B are summarized [24].

Besides the video data, the OFDM symbols carry the DVB-T transmission parameter signal, known as transmission parameter signaling (TPS), and pilots [25]. The TPS carriers convey information related to the transmission scheme, i.e., channel coding and modulation. The TPS carrier locations are fixed and are defined by the standard. The pilot symbols help the receiver in the acquisition, demodulation,

Table 1.1 *OFDM parameters for the 8K and 2K modes and different nominal channel bandwidths, where* T_B *is, respectively, equal to* $\frac{7}{40}\mu s$, $\frac{7}{48}\mu s$, $\frac{1}{8}\mu s$, *and* $\frac{7}{64}\mu s$ *for nominal channel bandwidths of 5 MHz, 6 MHz, 7 MHz, and 8 MHz*

Mode	**8*K* mode**				**2*K* mode**			
Guard interval ratio $\frac{\Delta}{T_U}$	1/4	1/8	1/16	1/32	1/4	1/8	1/16	1/32
T_U	$8{,}192T_B$				$2{,}048T_B$			

and decoding of the received signal. In the DVB-T standard, two types of pilots are used in an OFDM structure, named scattered and continual. The scattered pilots are uniformly distributed among the carriers in any given symbol, while the continual pilot signals occupy the same carrier from symbol to symbol. The location of all pilot symbol carriers is defined by the DVB-T standard. The TPS carriers are modulated according to differential binary phase shift keying. The pilot carriers are transmitted at a boosted power level of $\frac{16}{9}$, while the power level of video data and TPS pilots is normalized to 1. The carriers related to the video data are modulated using either Quadrature Phase Shift Keying (QPSK), 16-Quadrature Amplitude Modulation (QAM), 64-QAM, nonuniform 16-QAM, or nonuniform 64-QAM constellations. The constellations and the details of the Gray mapping are defined by the DVB-T standard [24].

The baseband structure of the OFDM signal can be represented as

$$x[n] = \sum_{m=0}^{\infty}\sum_{l=0}^{L-1}\sum_{k=0}^{K-1} c_{mlk}\psi_{mlk}[n] \tag{1.4}$$

where

$$\psi_{mlk}[n] = e^{j2\pi\frac{k}{T_U}(nT_B-\Delta-lT_S-mLT_S)}w_{ml}[n] \tag{1.5}$$

and

- m denotes the OFDM frame number
- l denotes the OFDM symbol number
- k denotes the carrier number
- c_{mlk} is the complex-valued modulation symbol for the carrier k of the data symbol number l in the frame number m. Here, we have $c_{mlk} = 0$ for inactive carriers
- K is the number of subcarriers, K_{2K} (2K mode) or K_{8K} (8K mode)
- L is the number of symbols per frame, which is equal to 68 in the DVB-T standard
- $w_{ml}[n] = w[n - l\frac{T_S}{T_B} - mL\frac{T_S}{T_B}]$ is a rectangular windowing function representing the duration of symbol.

Figure 1.9 shows the AAF of the 8K mode for 8 MHz channel of the DVB-T signal containing the complete set of deterministic components-guard interval, pilot, and TPS

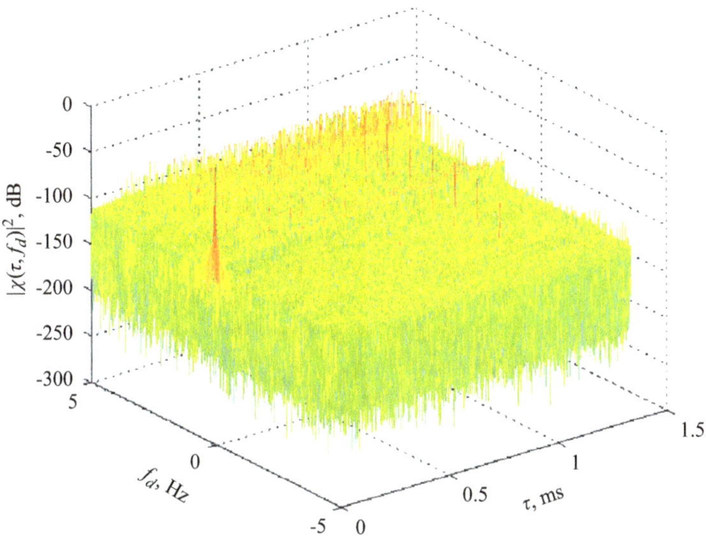

Figure 1.9 AAF of the 8K mode for 8 MHz channel of the DVB-T signal with characteristic shown in Table 1.2

carriers. The relevant parameters are listed in Table 1.2. As shown in Figure 1.9, there are some points in order.

- The guard interval insertion generates unwanted deterministic peaks in the ACF. As expected, the peak generated by guard interval occurs at $\frac{1}{4}T_U$, which is

Table 1.2 Simulated DVB-T signal parameters

Description	**Parameter**	**Value**
Total number of carriers	K_{8K}	8,192
Number of active carriers	$K_{8K}^{(C)}$	6,817
Number of data carriers	$K_{8K}^{(C)}$	6,048
Number of scattered pilot cells	N_S	524
Number of continual pilot carriers	N_C	177
Number of TPS carriers	N_{TPS}	68
Number of symbols per frame	L	68
Useful part of symbol	T_U	896 μs
Guard time	Λ	224 μs
Useful part of symbol	T_S	1,120 μs
Carrier spacing	$\frac{1}{T_U}$	1,116 Hz
Nominal channel bandwidth	B	8 MHz
Effective signal bandwidth	B_e	7.61 MHz

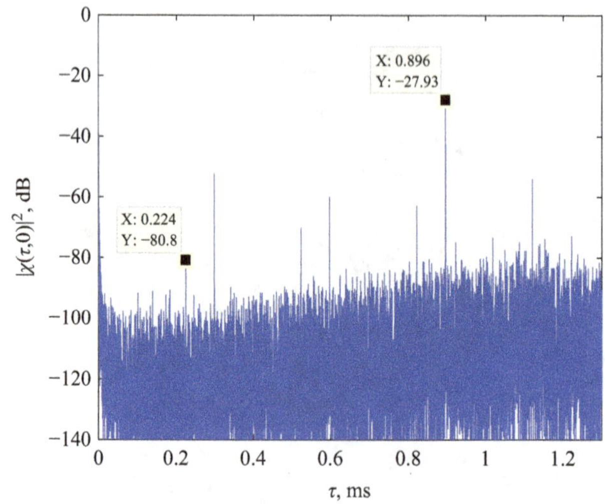

Figure 1.10 Unwanted deterministic peaks due to guard interval and cyclic prefix at zero Doppler frequency

equal to 224 μs. The cyclic prefix produces another ambiguity peak at $\tau = T_U$, where its amplitude is dependent on the length of the guard interval [25]. Thus, the longer the guard interval, the bigger the peak. These peaks at zero Doppler frequency are shown in Figure 1.10.

- Pilot signals give rise to both intrasymbol ambiguities (i.e., $\tau < T_S$), and intersymbol ambiguities (i.e., $\tau > T_S$) [25], where intersymbol ambiguities are outside of the feasible detection range for a PR system; thus, they are neglected in our study.
  - Ambiguities from continual pilots can potentially occur for $\tau = pT_U + qT_S$ and $f_d = \frac{\nu}{T_S}$ for integers p, q, and ν. Among them, the ambiguity peaks at $\tau = T_U$ and $\tau = T_S$ when $f_d = 0$ are significant and clearly seen in Figure 1.11, where other peaks in this figure cannot be assigned to the continual pilots.
  - The intra symbol ambiguities resulting from the scattered pilots at the (τ, f_d) combination are given by [25]

$$\tau = T_U\left(\frac{u}{12} + \frac{r}{3}\right) + 4bT_S, \quad f_d = \frac{1}{T_S}\left(\frac{u}{4} + h\right) \tag{1.6}$$

for $u = 0, 1, 2, 3$, and for integers r, b, and h. Significant intrasymbol ambiguities resulting from the scattered pilots are specified in Figures 1.12–1.19 for different values f_d of 0 Hz, 223.2 Hz, –223.2 Hz, 446.4 Hz, 669.6 Hz, –669.6 Hz, 892.8 Hz, and –892.8 Hz, respectively. All of these are obtained with (1.6) for $b = 0$. For another values of b, we can obtain intersymbol ambiguity peaks resulting from the scattered pilots, where their pattern of ambiguities will be

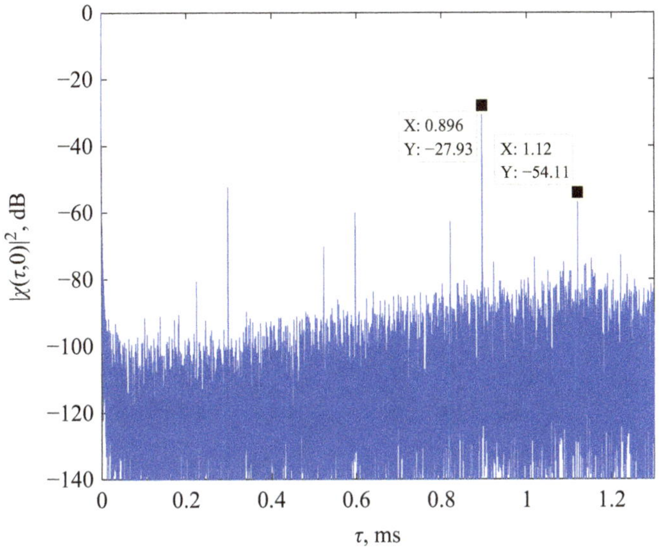

Figure 1.11 *Ambiguities from continual pilots of the DVB-T signal at zero Doppler frequency*

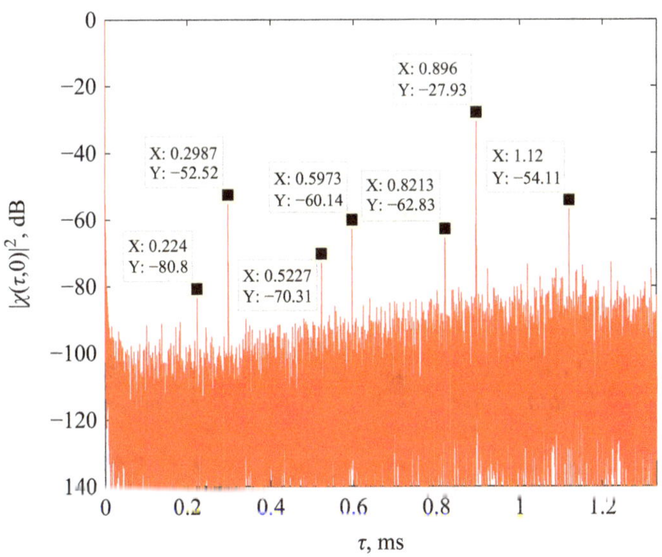

Figure 1.12 *DVB-T signal ambiguities due to the scattered pilots in delay domain when Doppler frequency is equal to* $f_d = 0$ *Hz*

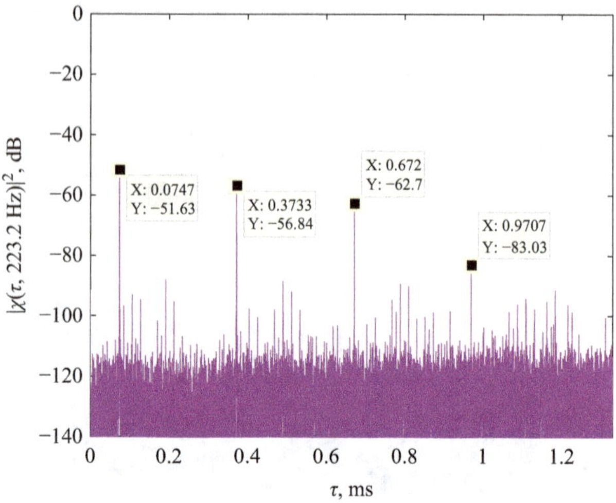

Figure 1.13 *DVB-T signal ambiguities due to the scattered pilots in delay domain when Doppler frequency is equal to $f_d = 223.2$ Hz*

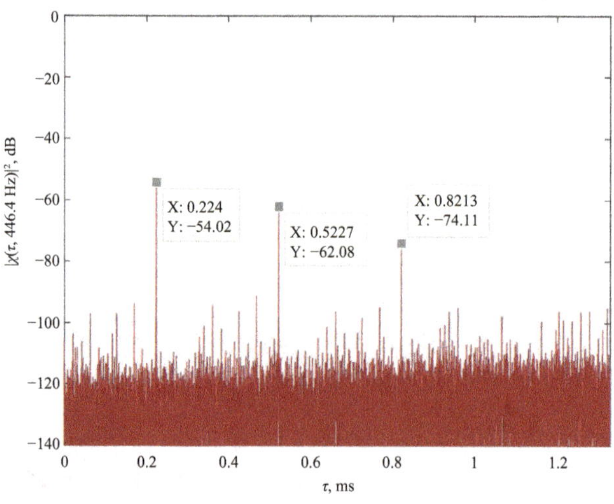

Figure 1.14 *DVB-T signal ambiguities due to the scattered pilots in delay domain when Doppler frequency is equal to $f_d = -223.2$ Hz*

repeated at delays of multiples of $4T_S$, which are outside of the feasible detection range for a PR system.

- Ambiguities due to frame (68 symbols) and super-frame (4 frames) structure can be expected. However, they are outside the feasible operating region of the PR and are ignored [25].

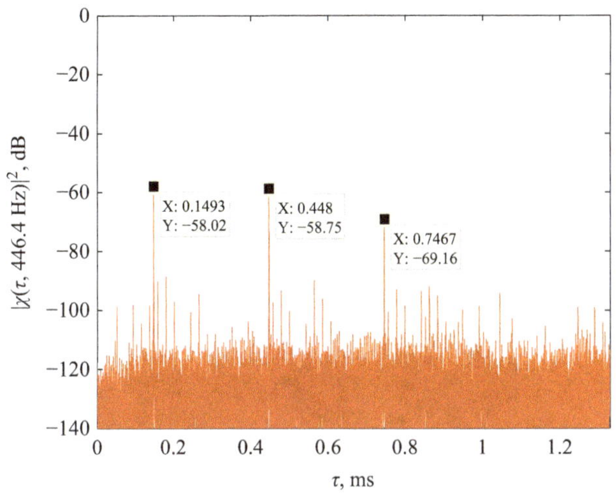

Figure 1.15 *DVB-T signal ambiguities due to the scattered pilots in delay domain when Doppler frequency is equal to* $f_d = 446.4$ *Hz*

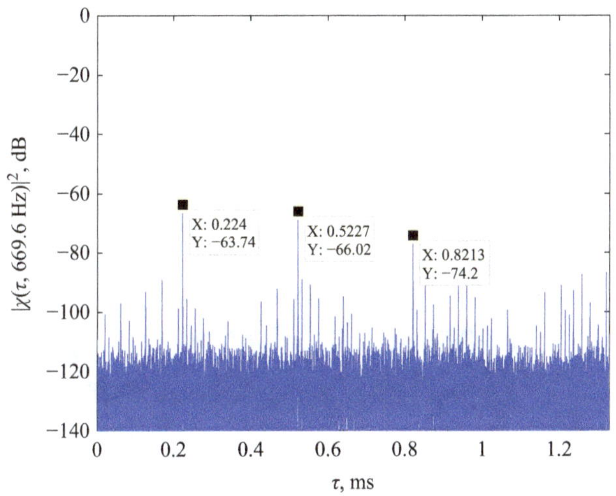

Figure 1.16 *DVB-T signal ambiguities due to the scattered pilots in delay domain when Doppler frequency is equal to* $f_d = 669.6$ *Hz*

1.1.3 ATV signal characteristic

ATV transmissions are typically transmitted within the frequency range of 500–600 MHz. There are three main modulation formats used in ATV: phase alternating line (PAL), National Television System Committee, and sequential couleurs avec memoire [1]. Among these, PAL is the most common modulation format, which uses 625 lines to make up the picture information, with a total duration of 64 μs, including a

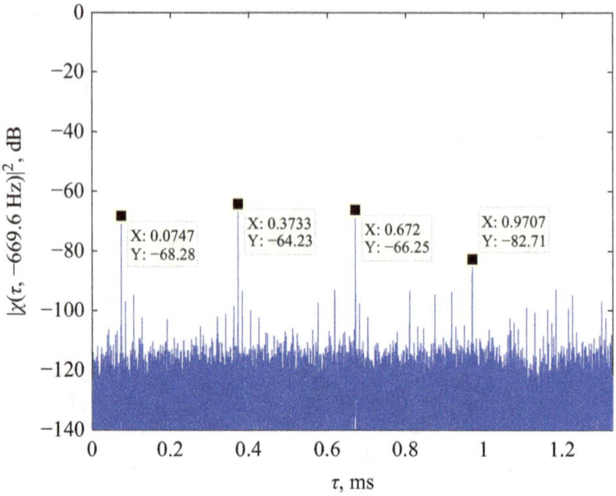

Figure 1.17 *DVB-T signal ambiguities due to the scattered pilots in delay domain when Doppler frequency is equal to* $f_d = -669.6$ *Hz*

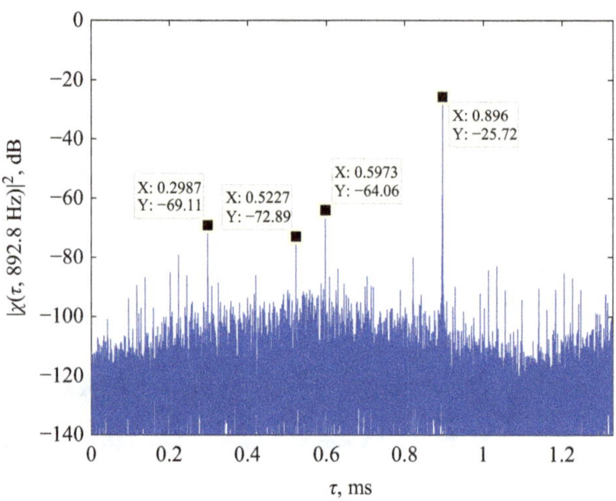

Figure 1.18 *DVB-T signal ambiguities due to the scattered pilots in delay domain when Doppler frequency is equal to* $f_d = 892.8$ *Hz*

16-μs sync pulse at the beginning of each line. The picture information is encoded as two interlaced scans with a frame rate of 50 Hz. The ATV signal uses two different modulation formats, vestigial-sideband amplitude modulation for vision signal and FM for sound signal, with a bandwidth of about 6 MHz. However, despite the signal occupying a bandwidth of 6 MHz, the vast majority of the signal's energy is concentrated within a narrow bandwidth of ±125 kHz around the carrier frequency [26, 27]. In this book, the focus is on this narrow frequency range and the study of its

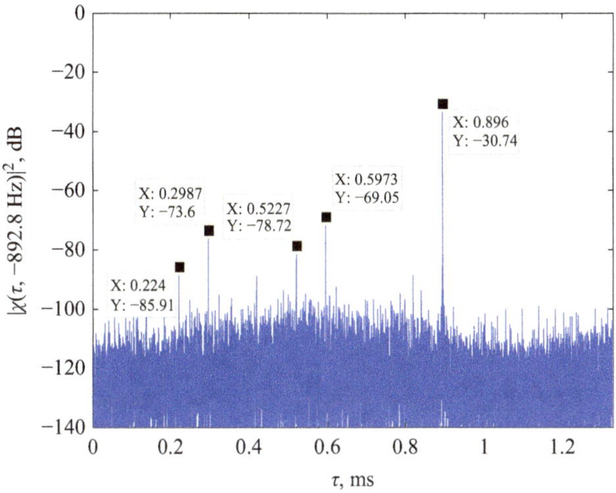

Figure 1.19 *DVB-T signal ambiguities due to the scattered pilots in delay domain when Doppler frequency is equal to $f_d = -892.8$ Hz*

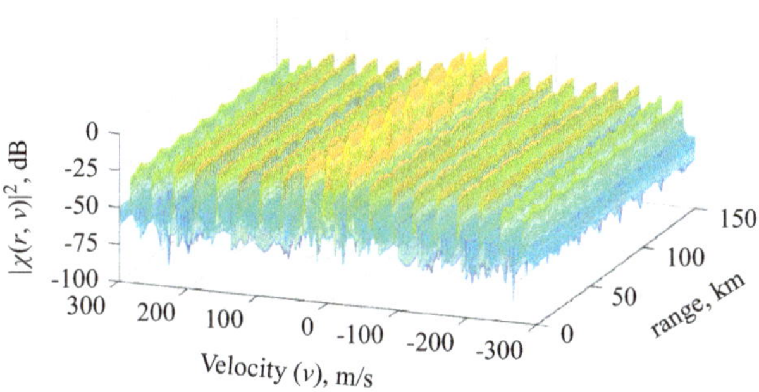

Figure 1.20 *AAF of ATV signal with integration time of 0.6 seconds*

characteristics. The authors present Figures 1.20 and 1.21 to illustrate the properties of the ATV signal. Figure 1.20 shows the AAF of the ATV signal, while Figure 1.21 presents the contour of AAF. Both of these figures are based on a specific integration time of 0.6 seconds. This information is crucial to understanding the technical aspects of ATV transmissions, including the bandwidth and modulation formats used in these transmissions. As can be seen, high ambiguities in range and velocity (or frequency) are clear. The first is due to the similarity between successive lines of picture information and because of the presence of sync pulse at the beginning of the lines, while the second is due to the frame scan rate of 50 Hz. To get a clear visual sense of the high ambiguity range sidelobe due to 64-μs line flyback time of the ATV signal, the range-domain cut of AAF is plotted in Figure 1.22. It is seen that the 64-μs line flyback time of the ATV signal gives rise to high range ambiguity at ranges corresponding to multiples of 19.2 km. This may imply that the relatively

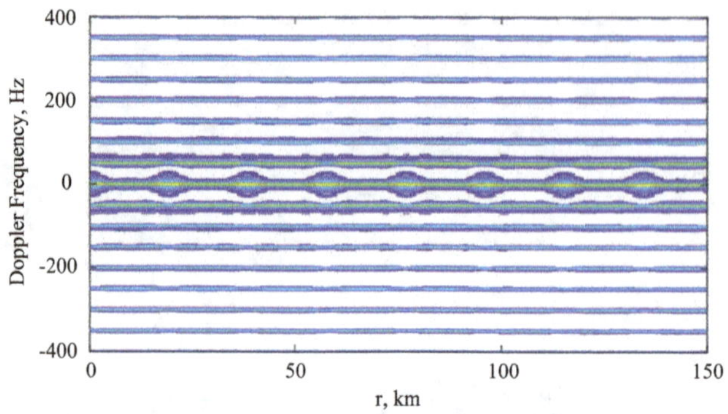

Figure 1.21 *Contour of AAF of ATV signal with integration time of 0.6 seconds*

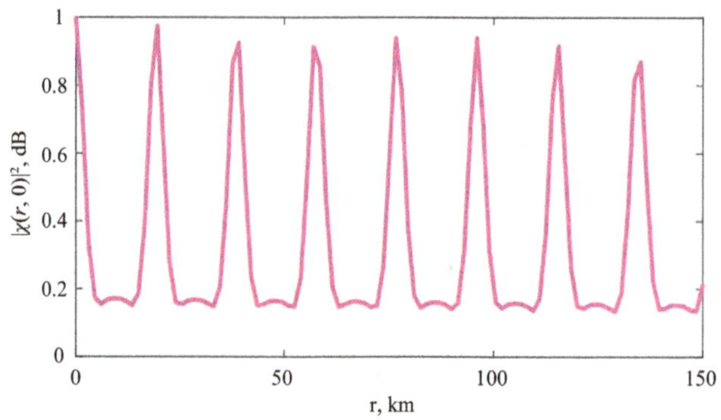

Figure 1.22 *Range-domain cut of AAF of ATV signal with integration time of 0.6 seconds*

short range can be expected from ATV-based PR, and the severity of this ambiguity makes application of the ATV-based PR much more likely to be limited. Besides, we can see the line frequency 50 Hz, resulting in ambiguities in frequency (or velocity) domain, as shown in Figure 1.23.

1.2 Geometry of passive radars

PRs may utilize one or more transmitter and one or more receivers as well. By considering this and the fact that there is a distance between each transmitter and receivers, we can form bistatic pair by using any one transmitter and any one receiver, where the distance between a transmitter and a receiver is called baseline. Given the knowledge of the transmitter positions, the resulting detection across bistatic pairs can be fused to localize targets in Cartesian space. Based on the numbers of the receivers and transmitters, PRs can be categorized as bistatic, multistatic, and multiple-input multiple-output (MIMO). In the bistatic configuration, there are one

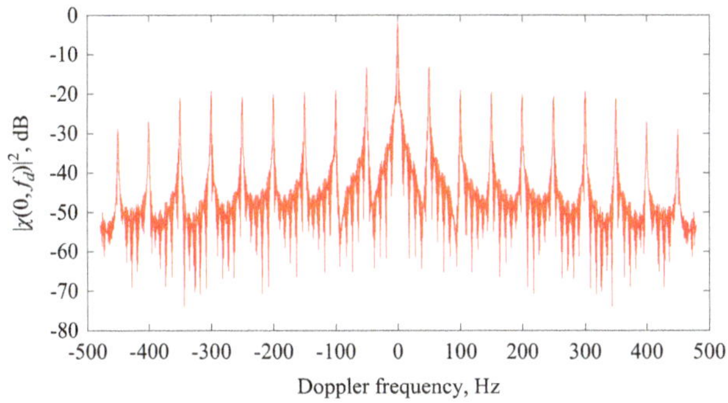

Figure 1.23 Frequency-domain cut of AAF of ATV signal with integration time of 0.6 seconds

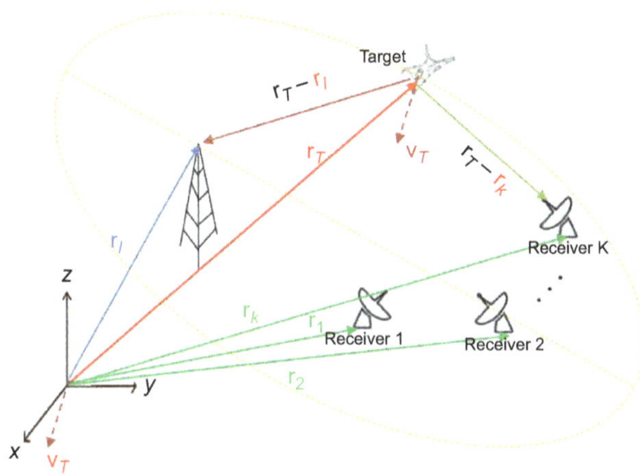

Figure 1.24 Geometry of PR

receiver and one transmitter, while a multistatic PR system comprises one transmitter and multiple receivers or multiple transmitters and one receiver. Finally, an MIMO system contains multiple receivers and transmitters. This is in comparison to the conventional active radar systems, which typically operate in a monostatic mode. Generally, the multistatic system performs better than a monostatic one. In other words, a multistatic system with spatially separated receivers makes it possible not only to collect more data samples over a given integration time but also to obtain spatial diversity of a target's radar cross-section (RCS) to achieve potential gains for target detection purposes.

Multistatic and MIMO PR systems can be treated as an interconnected set of bistatic PRs. This allows us to use basic bistatic radar theory to discuss all PR configurations. Consider a single-target detection scenario in a multistatic configuration, as shown in Figure 1.24. Let the position of the illuminator, receivers, and the target be $\mathbf{r}_I = [x_I, y_I, z_I]^T$, $\mathbf{r}_k = [x_k, y_k, z_k]^T$ for $k = 1, \ldots, K$, and $\mathbf{r} = [x, y, z]^T$, respectively. This single-target scenario further assumes the target moves with velocity vector $\mathbf{v} = [v_x, v_y, v_z]^T$. In the presence of a single target, the baseband equivalent received signal at the kth receiver can be described as

$$x_k(t) = \alpha_k s(t - \tau_k(t))e^{-j2\pi f_c \tau_k(t)} + w_k(t), \qquad 0 \leqslant t \leqslant T \tag{1.7}$$

where $s(t)$ and $w_k(t)$ are baseband equivalent transmitted signal and the additive thermal noise received at the kth receiver, respectively. Here, T is the integration time, α_k represents the unknown complex target reflectivity received at the kth receiver, $\tau_k(t)$ is the time-varying delay caused by velocity and acceleration of target seen at the kth receiver, and f_c is the carrier frequency. In what follows, it is assumed that the integration time starts at time 0 and ends at time T. According to the geometry of Figure 1.24, $\tau_k(t)$ can be computed as

$$\tau_k(t) = \frac{\| \mathbf{r} - \mathbf{r}_I \| + \| \mathbf{r} - \mathbf{r}_k \|}{c} \tag{1.8}$$

where c is the light speed. In the following, we assume that the transmitter and receivers are stationary, while target moves with velocity of $\mathbf{v}$ such that

$$\mathbf{r} = \mathbf{r}_0 + \mathbf{v}t \tag{1.9}$$

where $\mathbf{r}_0 = [x_0, y_0, z_0]^T$ is the location of the target at the beginning of the integration time. By substituting (1.9) in (1.8), we find

$$\tau_k(t) = \frac{\| \mathbf{r}_0 + \mathbf{v}t - \mathbf{r}_I \| + \| \mathbf{r}_0 + \mathbf{v}t - \mathbf{r}_k \|}{c} \tag{1.10}$$

By using the first-order Taylor approximation of $\tau_T(t)$ at $t = 0$, we obtain

$$\tau_k(t) = \tau_k(0) + \left.\frac{\partial \tau_k(t)}{\partial t}\right|_{t=0} \frac{(t-0)}{1!} \tag{1.11}$$

It can be shown that

$$\tau_k(0) = \frac{\| \mathbf{r}_0 - \mathbf{r}_I \| + \| \mathbf{r}_0 - \mathbf{r}_k \|}{c} \tag{1.12}$$

and

$$\left.\frac{\partial \tau_k(t)}{\partial t}\right|_{t=0} = \frac{1}{c}\left(\frac{\mathbf{v}^T(\mathbf{r}_0 - \mathbf{r}_I)}{\| \mathbf{r}_0 - \mathbf{r}_I \|} + \frac{\mathbf{v}^T(\mathbf{r}_0 - \mathbf{r}_k)}{\| \mathbf{r}_0 - \mathbf{r}_k \|}\right) \tag{1.13}$$

By substituting (1.12) and (1.13) into (1.11), we get

$$\tau_k(t) = \frac{R_0^{(k)}}{c} + \frac{v_b^{(k)}}{c} t \tag{1.14}$$

where

$$R_0^{(k)} = \| \mathbf{r}_0 - \mathbf{r}_I \| + \| \mathbf{r}_0 - \mathbf{r}_k \| \tag{1.15}$$

and

$$v_b^{(k)} = \frac{\mathbf{v}^T(\mathbf{r}_0 - \mathbf{r}_I)}{\| \mathbf{r}_0 - \mathbf{r}_I \|} + \frac{\mathbf{v}^T(\mathbf{r}_0 - \mathbf{r}_k)}{\| \mathbf{r}_0 - \mathbf{r}_k \|} \tag{1.16}$$

where $v_b^{(k)}$ is called the bistatic velocity of the considered target sensed in the kth receiver at the start of the integration time, respectively. Similarly, $R_0^{(k)}$ is termed the sum range sensed in the kth receiver at the beginning of the integration time. Finally, by substituting (1.14) into (1.7), we obtain

$$x_k(t) = \alpha_k e^{-j2\pi \frac{R_0^{(k)}}{\lambda}} s\left(t - \frac{R_0^{(k)}}{c} - \frac{v_b^{(k)}}{c}t\right) e^{-j2\pi (f_b^{(k)} t)} + w_k(t) \tag{1.17}$$

where $f_b^{(k)}$ is the bistatic Doppler frequency defined as $f_b^{(k)} = \frac{v_b^{(k)}}{\lambda}$, λ being the carrier frequency wavelength, i.e., $\lambda = \frac{c}{fc}$. If the received signal $x_k(t)$ is sampled with a sampling frequency f_s at the time instants $t_n = \frac{n}{fs} = nT_s$, $n = 0, \ldots, N-1$, the nth sample of $x_k(t)$ can be described as

$$x[n] = \beta_k s[n - \lceil \frac{R_0^{(1)} f_s}{c} + \frac{v_b^{(1)}}{c} n \rceil] e^{-j2\pi (f_b{}^{(1)} nT_s)} + w[n] \tag{1.18}$$

where $\beta_k = \alpha_k e^{-j2\pi \frac{R_0^{(k)}}{\lambda}}$ is the complex received signal amplitude. Here, the assumption that the reference signal is delayed by an integer number of samples is a reasonable approximation for sufficiently fast sampling. From (1.18), it is seen that very high-speed targets can lead to range migration (RM). More precisely, the ratio of distance traveled by a target during the integration time to the RR length, denoted as ΔR, can be defined as

$$\delta R_k = \frac{v_b^{(k)} T}{\Delta R} \tag{1.19}$$

The condition for limiting RM is given by $\delta R_k \leq 1$. This is equivalent to

$$T \leq \frac{\Delta R}{v_b^{(k)}} = T_{\max} \tag{1.20}$$

Thus, the maximum integration time, allowed without experiencing RM, is equal to $T_{\max} = \frac{\Delta R}{v_b^{(k)}}$. If the maximum velocity of interested targets is assumed equal to 300 m/s^2, the maximum integration times are 10 seconds and 133 milliseconds for an FM-based PR system with $\Delta R = 3$ km, and a DVB-T-based PR with $\Delta R = 40$ m, respectively. In Part 1, we will show that the target acceleration is more limiting, which is ignored here for simplicity. In the radar applications with $T < T_{\max}$, (1.18) reduces to

$$x_k[n] = \beta_k s[n - n_0^{(k)}] e^{-j2\pi f_b^{(k)} nT_s} + w_k[n] \tag{1.21}$$

where $n_0^{(k)} = \lceil \frac{R_0^{(k)} f_s}{c} \rceil$. This is the received signal model in the presence of a single target and the noise, which may be used to study the bistatic range and bistatic velocity parameters of interested targets.

Most of the PR target detection algorithms are based on the cross-correlation of the sequence $s[n]$ and that of $x_k[n]$ in the kth receiver. The sequence $s[n]$ is not known in PR. A conventional method for the unknown signal detection problem is to deploy a reference channel (RC) for collecting the unknown transmitted sequence to serve as a reference. Then, the target detection algorithm may be implemented based on the cross-correlation of the RC signal and that of $x_k[n]$ in the kth receiver. Let $y_k[n]$ denotes the RC signal at the kth receiver, given by

$$y_k[n] = \gamma_k s[n - n_L^{(k)}] + n_k[n] \tag{1.22}$$

where γ_k is an unknown scaling parameter accounting for the kth baseline direct path channel, and $n_k[n]$ represents the RC noise. Here, $n_L^{(k)}$ is the time sample associated with the kth baseline direct path, computed by $n_L^{(k)} = \lceil \frac{L_k f_s}{c} \rceil$ with L_k being the baseline distance between the transmitter and the kth receiver, given by

$$L_k = \| \mathbf{r}_I - \mathbf{r}_k \| \tag{1.23}$$

In the high signal-to-noise ratio (SNR) regime of the RC, we can rewrite the signal model of (1.21) based on the known sequence $y_k[n]$; i.e.,

$$x_k[n] = \frac{\beta_k}{\gamma_k} y_k[n - \lceil \frac{R_b^{(k)} f_s}{c} \rceil] e^{-j2\pi(f_b^{(k)} nT_s + \frac{1}{2}\frac{a_b^{(k)}}{\lambda}(nT_s)^2)} + w_k[n] \tag{1.24}$$

where $R_b^{(k)} = \frac{c(n_0^{(k)} - n_L^{(k)})}{f_s}$ with $R_b^{(k)}$ being relative bistatic range (RBR). It is seen that the RBR is defined as the range difference between directly received signal across baseline and that of the sum range, i.e., $R_b^{(k)} = R_0^{(k)} - L_k$. By using the signal model of (1.24), the cross-correlation processing behaves like the matched filter (MF). In the considered detection domain, which in this formulation is position velocity, we expect to have a large peak on the output of the MF at the true position velocity of the target. In contrast to the monostatic radar, where the locus of constant ranges forms a ball, contours of the constant $R_b^{(k)}$ form an ellipsoid in three-dimensional space. In two-dimensional space, constructed from a cross-section of the ellipsoid with a plane containing the transmitter and the receiver, we have an ellipse.

Figure 1.25 depicts iso-ranges for different bistatic range (i.e., $R_0^{(k)}$) values from 0 km to 70 km, every 10 km. Here, the zero bistatic range corresponds to a target located at the baseline, while the larger ellipse results in the larger bistatic range. The transmitter is located at $\mathbf{r}_I = [0, 0, 0]^T$, and the receiver is located at $\mathbf{r}_k = [70, 10, 0]^T$ km.

In a similar manner, we can define bistatic iso-velocity, given by $v_b^{(k)} = \zeta_k$. From (1.16), it is seen that the bistatic velocity depends on $\mathbf{r}_I$, $\mathbf{r}_1$, $\mathbf{r}$, and $\mathbf{v}$. Figures 1.26–1.28 show iso-Doppler contours when the transmitter is located at $\mathbf{r}_I = [0, 0, 0]^T$, the receiver is located at $\mathbf{r}_k = [70, 10, 0]^T$ km, and target velocity vector is equal to $\mathbf{v} = [1, 1, 0]^T$ m/s, $\mathbf{v} = [\frac{1}{\sqrt{2}}, \frac{1}{\sqrt{2}}, 0]^T$ m/s, and $\mathbf{v} = [0, 0, 0]^T$ m/s, respectively. In all cases, for simplicity, we set $\|\mathbf{v}\| = 1$. Here, there are some points in order. First, regardless of the target velocity vector $\mathbf{v}$, $v_b^{(k)}$ gets zero values on the baseline, where we get

$$\mathbf{v}^T \frac{(\mathbf{r}_0 - \mathbf{r}_I)}{\| \mathbf{r}_0 - \mathbf{r}_I \|} = -\mathbf{v}^T \frac{(\mathbf{r}_0 - \mathbf{r}_k)}{\| \mathbf{r}_0 - \mathbf{r}_k \|} \tag{1.25}$$

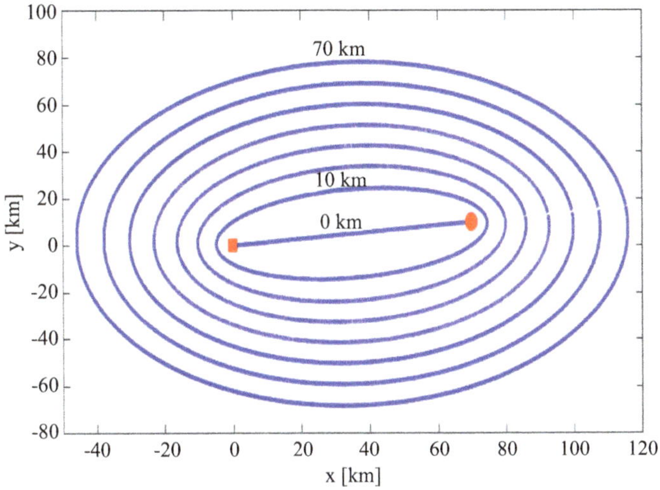

Figure 1.25 *Locus of constant bistatic range, i.e., iso-ranges for different bistatic range values of [0:10:70] km, when* $\mathbf{r}_l = [0, 0, 0]^T$ *and* $\mathbf{r}_k = [70, 10, 0]^T$ *km*

This means that the transmitter-to-target range is changing in an opposite way and is equal to that of the target-to-receiver. Second, it is easy to rewrite (1.16) as

$$v_b^{(k)} = \|\mathbf{v}\| \cos\left(\frac{\phi_k}{2}\right) \cos(\delta_k) \tag{1.26}$$

where ϕ_k is the angle subtended at the target by the transmitter and the kth receiver, termed as bistatic angle, and δ_k denotes angle between target velocity vector $\mathbf{v}$ and bistatic bisector associated to the kth transmitter–receiver. As a result, for the target position with $\delta_k = \frac{\pi}{2}$, we obtain zero bistatic velocity. This corresponds to the moving of the target along the ellipses in two-dimensional Cartesian scenarios, i.e., the target velocity vector is tangential to the bistatic ellipse in Cartesian coordinates. Thus, we get $\delta_k = 0$ when a target is moving in a direction orthogonal to the lines of constant bistatic range, resulting in maximum bistatic velocity. It can be shown that lines of maximum velocity in PRs take hyperbolic forms, while they are radii in monostatic radars.

Let $\tau_k^{(i)}$ denotes the delay measured in the kth receiver due to the ith target. The delay measurement resolution of a pulsed active radar with signal bandwidth B is equal to $\Delta\tau_k = \tau_k^{(2)} - \tau_k^{(1)} = \frac{1}{B}$. In general, the target RR is related to the delay measurement resolution and the radar and target geometry. In monostatic radar, the RR is easily related to the delay measurement resolution through $\Delta R_k = c\Delta\tau_k$ with c being the propagation velocity. This is concluded from the fact that constant monostatic ranges are concentric circles, as shown in Figure 1.29. In this figure, two targets are placed with a range difference of 5 km. The considered monostatic system is able to distinguish them as two separated targets when $B \geq 30$ kHz, i.e., $\Delta R_k \geq 5$ km. In multistatic

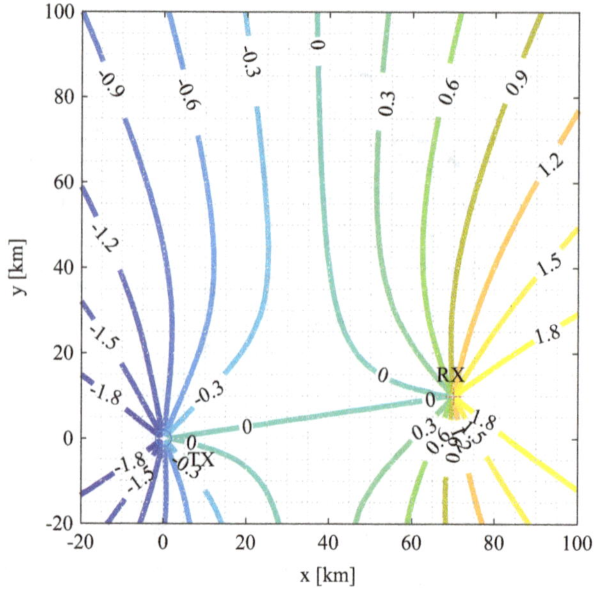

Figure 1.26 *Contours of constant bistatic velocities when* $\mathbf{r}_l = [0,0,0]^T$ *and* $\mathbf{r}_k = [70,10,0]^T$ *km, and* $\mathbf{v} = [1,0,0]^T$ *m/s*

geometry, the problem is not as simple as that of monostatic ones. In this case, we again can use $\Delta\tau_k = \frac{1}{B}$, but it is not straightforward to obtain the relationship between the bistatic delay resolution (i.e., $\Delta\tau_k$) and the Cartesian target RR in the general case. In Figure 1.30, the locus of the two constant bistatic ranges of 141.4 km and 116.7 km is plotted. It is seen that the constant bistatic range difference of 24.7 km leads to varying range differences in Cartesian coordinates. The separation of the bistatic ellipses at a line that is an extension of the baseline is minimum (i.e., it is about 12.4 km), whereas the separation at points that are colinear with respect to the bistatic bisector is maximum (i.e., it is about 22.78 km). In general, we can approximate the PR target RR in Cartesian coordinate as follows [1, 28]:

$$\Delta R_k^{(C)} = \frac{\Delta R_k}{2\cos\dfrac{\phi_k}{2}\cos(\theta_k)} \tag{1.27}$$

where θ_k is the angle between the bistatic bisector and a line joining the two targets, as shown in Figure 1.30. In the sequel, we call ΔR_k as bistatic target RR, but $\Delta R_k^{(C)}$ is termed as Cartesian bistatic target RR. There are some points in order. First, it is clearly seen that the Cartesian bistatic target RR depends on the signal bandwidth, which is not under the control of the radar designer, especially in FM-based PRs, bistatic angle ϕ_k (i.e., it depends on the relative of target position with respect to positions of transmitter and receiver) and angle θ_k, defined before. Second, we get $\Delta R_k^{(C)} \to \infty$ when $\phi_k = \frac{\pi}{2}$; thus, all resolution is lost. This case is a

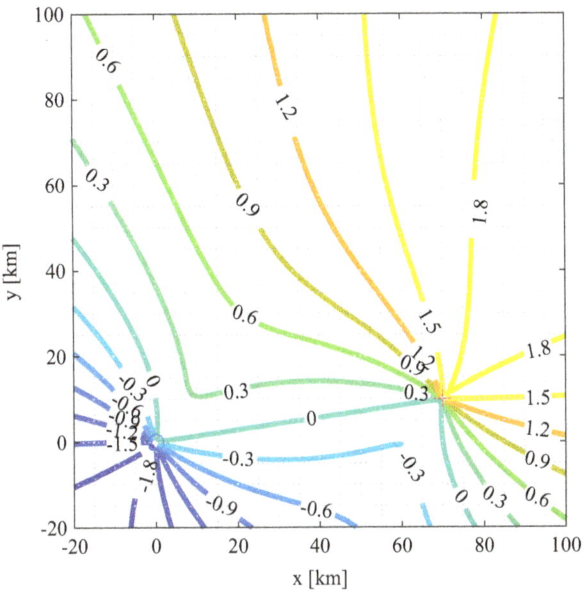

Figure 1.27 *Contours of constant bistatic velocities when* $\mathbf{r}_I = [0, 0, 0]^T$ *and* $\mathbf{r}_k = [70, 10, 0]^T$ *km, and* $\mathbf{v} = [\frac{1}{\sqrt{2}}, \frac{1}{\sqrt{2}}, 0]^T$ *m/s*

so-called forward scatter regime [28]. In a multistatic configuration, it is unlikely that a target is seen in the forward scatter regime for more than one transmitter–receiver pair; thus, target RR performance is determined by transmitter–receiver pairs, which are not in the forward scatter regime. Third, the Cartesian target RR of bistatic geometry is coarser than that of the equivalent monostatic geometry. It is known as a role of thumb that the Cartesian bistatic RR is almost twice the equivalent monostatic one [1, 28].

Similarly, passive bistatic radar velocity resolution can be computed as

$$\Delta v_k = \frac{\lambda}{T} \tag{1.28}$$

where T and λ are the integration time and wavelength, respectively. Thus, it is inversely proportional to the integration time. The longer the integration time, the better the velocity resolution. However, there is a limit to the extension of the integration time to prevent Doppler frequency migration and RM, discussed in Part 2. Besides, Δv_k depends on the wavelength λ computed as $\lambda = \frac{c}{f_c}$ with f_c being the carrier frequency. Again, the wavelength λ is not under the control of a radar designer like bandwidth B in the RR. In this case, the Cartesian bistatic velocity resolution can be computed from [1, 28]

$$\Delta v_k^{(C)} = \frac{\Delta v_k}{2\cos\dfrac{\phi_k}{2}} \tag{1.29}$$

This shows that the Cartesian bistatic velocity resolution is a function of wavelength, integration time, and target position with respect to transmitter and receiver, i.e., the

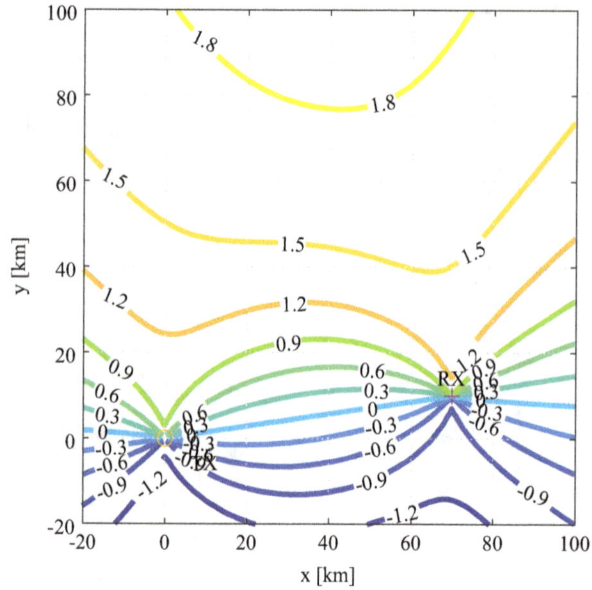

Figure 1.28 *Contours of constant bistatic velocities when* $\mathbf{r}_I = [0, 0, 0]^T$ *and* $\mathbf{r}_k = [70, 10, 0]^T$ *km, and* $\mathbf{v} = [0, 1, 0]^T$ *m/s*

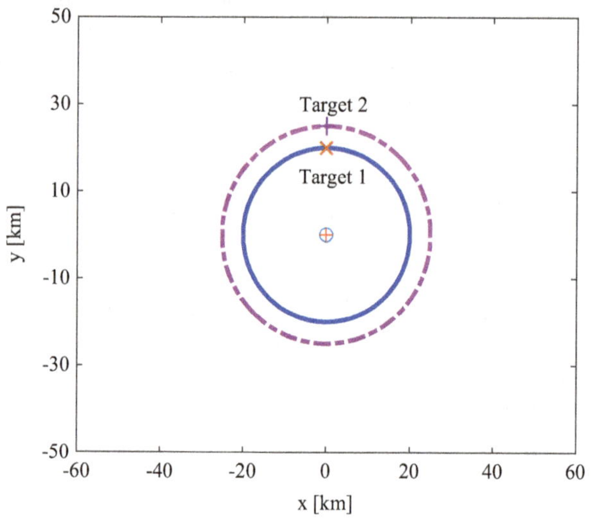

Figure 1.29 *Target RR in a monostatic geometry with two constant range lines of 20 km and 25 km*

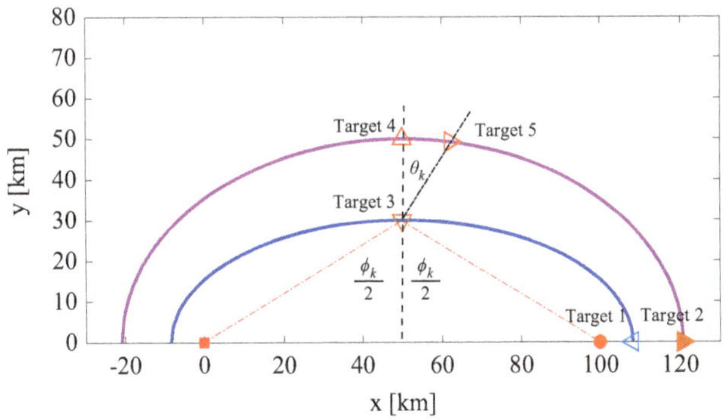

Figure 1.30 *Bistatic target RR concept*

bistatic angle ϕ_k. In Figures 1.31 and 1.32, the Cartesian bistatic velocity resolution is shown for FM-based and DVB-T PRs with carrier frequencies of 100 MHz and 600 MHz and integration times of 0.8 seconds and 28 milliseconds, respectively. Here, bistatic velocity resolutions for the FM- and DVB-T-based PRs are, respectively, equal to 3.75 m/s and 17.86 m/s, and it is assumed that the transmitter and receiver are located at $\mathbf{r}_I = [0, 0, 0]^T$ and $\mathbf{r}_k = [70, 10, 0]^T$ km, respectively.

1.3 Power budget and passive radar coverage

The opportunity transmitter of a PR illuminates PR receivers and the target as well. Under free space approximation, the direct signal power received by the kth RC antenna can be given by

$$P_k^{(D)} = \frac{P_I G_{IR}^{(k)} G_{RI}^{(k)} \lambda^2}{(4\pi)^2 \| \mathbf{r}_I - \mathbf{r}_k \|^2 L_{\text{loss-b}}} \tag{1.30}$$

where

- P_I = average transmitted power of illuminators of opportunity
- $G_{IR}^{(k)}$ = gain of transmit antenna toward the kth receiver
- $G_{RI}^{(k)}$ = gain of the kth receiver antenna toward the transmitter
- λ = signal wavelength
- $L_{\text{loss-b}}$ = losses (propagation and system) along the kth baseline

In a similar way, the received power from a hypothesized target at the kth surveillance channel antenna, depending on the target RCS σ_t and the transmitter–target–receiver geometry, is given by

$$P_k^{(T)} = \frac{P_I G_{IT}^{(k)} \sigma_t G_{TR}^{(k)} \lambda^2}{(4\pi)^3 \| \mathbf{r}_0 - \mathbf{r}_I \|^2 \| \mathbf{r}_0 - \mathbf{r}_k \|^2 L_{\text{loss}-T}} \tag{1.31}$$

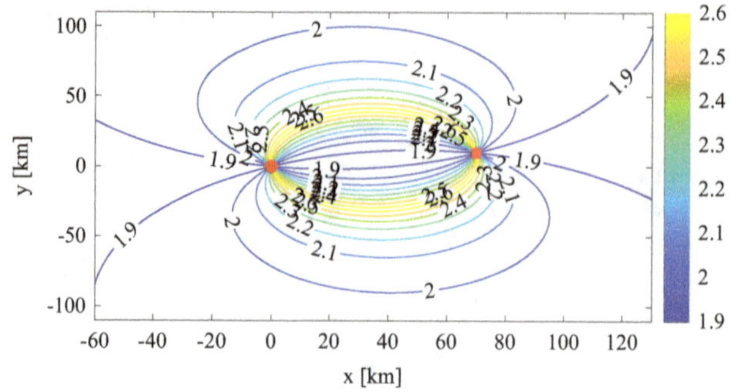

Figure 1.31 *Cartesian bistatic velocity resolution of the FM-based PR with bistatic velocity resolution of 3.75 m/s*

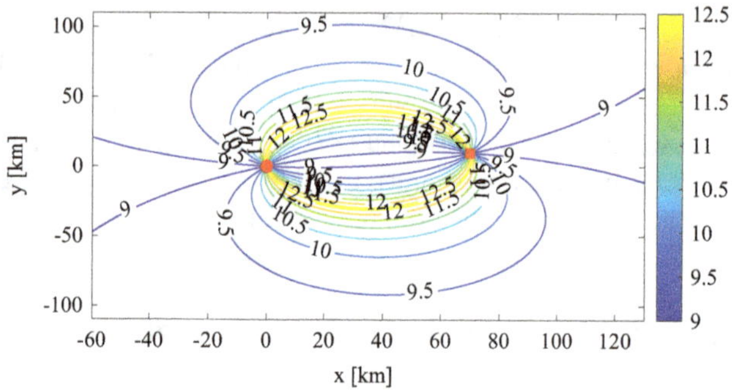

Figure 1.32 *Cartesian bistatic velocity resolution of the DVB-T PR with bistatic velocity resolution of 17.86 m/s*

where the new terms are as follows:

- $G_{IT}^{(k)}$ = gain of transmit antenna toward the hypothesized target
- $G_{TR}^{(k)}$ = gain of the kth receiver antenna toward the hypothesized target
- σ_t = bistatic RCS
- $L_{\text{loss}-\text{T}}$ = losses (propagation and system) along the transmitter–target–receiver and in the front end of the kth receiver path

To understand the performance potential of PR systems, we need to work with bistatic radar equation, which expresses the received signal-to-noise ratio (SNR) as a function of transmitted power P_I. To do so, we first define noise power, where its standard formulation is given by

$$P_n^{(k)} = k_b T_s^{(k)} B_k \tag{1.32}$$

where $T_s^{(k)}$ is the system noise temperature in kelvin (K), k_b is the Boltzmann's constant (1.38×10^{-23} Joules/K), and B_k is the noise bandwidth in Hertz. The system noise temperature includes the effect of antenna temperature T_a and receiver effective noise temperature $T_e^{(k)}$ [29], i.e.,

$$T_s^{(k)} = \frac{0.876T_a + 0.12T_0 + T_0(L_a - 1)}{L_a L_r^{(k)}} + \frac{T_0(L_r^{(k)} - 1)}{L_r^{(k)}} + T_e^{(k)} \tag{1.33}$$

where

$$T_e^{(k)} = T_0(F_k - 1) \tag{1.34}$$

In (1.33), T_0 is the standard or reference noise temperature (usually 290 K), L_a is the antenna radiation pattern loss, related to the antenna's voltage standing wave ratio, computed as [30]

$$L_a = \frac{(\mathrm{VSWR} + 1)^2}{4\mathrm{VSWR}} \tag{1.35}$$

$L_r^{(k)}$ is the receiving line loss. In (1.34), F_k is the noise factor of the kth surveillance or RC when it is converted to decibel (dB), called noise figure. Here, for simplicity, we assume that the noise powers of the reference and surveillance channels are equal.

Combining (1.31) and (1.32), we obtain the maximum available signal-to-noise power ratio, which reveals the expected radar performance for a particular scenario, given by

$$\mathrm{SNR}_k = \frac{\mathrm{EIRP}_k \sigma_t G_{TR}^{(k)} \lambda^2}{(4\pi)^3 \parallel \mathbf{r}_0 - \mathbf{r}_I \parallel^2 \parallel \mathbf{r}_0 - \mathbf{r}_k \parallel^2 L_{\mathrm{loss}-T} k_b T_s^{(k)} B_k} \tag{1.36}$$

where $\mathrm{EIRP}_k = P_I G_{IT}^{(k)}$ is termed as equivalent isotropic radiated power (EIRP), where it is used when two parameters P_I and $G_{IT}^{(k)}$ are not known separately. Most PR transmitters have omnidirectional patterns in azimuth, and they have biased patterns toward the earth's surface. Similarly, the maximum direct signal-to-noise ratio (DNR) received at the kth RC can be calculated as

$$\mathrm{DNR}_k = \frac{\mathrm{EIRP}_k G_{RI}^{(k)} \lambda^2}{(4\pi)^2 L_k^2 L_{\mathrm{loss}-\mathrm{b}} k_b T_s^{(k)} B_k} \tag{1.37}$$

It should be noted that received DNR in the surveillance channel is less than that of the DNR_k since the direct signal received from antenna sidelobe.

Note that SNR_k is termed as the received SNR before the signal is processed in the PR receiver. Besides, the parameters of the opportunity transmitters such as the transmitter antenna gain, the transmitter power, and the receiver bandwidth are not under the control of a PR system designer. Thus, PR generally obtains a short detection range, especially for low RCS targets. To increase the detection range, we can just focus on increasing the integration time and using large antennas in receivers. Thus, integration gain is an important factor when considering the performance limits of a PR system. The matched filtering type of correlation processing can be seen as coherent integration. Now, the SNR at the output of the signal processing blocks of the kth receiver, termed as $\mathrm{SNR}_k^{(SP)}$, can be related to SNR_k as

$$\mathrm{SNR}_k^{(SP)} = \mathrm{SNR}_k G_k^{(SP)} \tag{1.38}$$

where $G_k^{(SP)}$ is referred to as signal processing (integration) gain and is given by $G_k^{(SP)} = \mathrm{TB}_k$.

From (1.36), the minimum detectable RCS can be obtained as

$$\sigma_t^{(\min)} = \frac{(4\pi)^3 \parallel \mathbf{r}_0 - \mathbf{r}_I \parallel^2 \parallel \mathbf{r}_0 - \mathbf{r}_k \parallel^2 L_{\mathrm{loss}-T} P_n^{(k)} \mathrm{SNR}_k^{(\min)}}{\mathrm{EIRP}_k G_{TR}^{(k)} \lambda^2} \tag{1.39}$$

where $\mathrm{SNR}_k^{(\min)}$ denotes the minimum SNR required for target detection. Similarly, the bistatic maximum range product, known as bistatic radar maximum range equation, can be obtained as

$$\kappa = \sqrt{\frac{\mathrm{EIRP}_k \sigma_t G_{TR}^{(k)} \lambda^2}{(4\pi)^3 L_{\mathrm{loss}-T} P_n^{(k)} \mathrm{SNR}_k^{(\min)}}} \tag{1.40}$$

where $\kappa = \max\{\parallel \mathbf{r}_0 - \mathbf{r}_I \parallel \parallel \mathbf{r}_0 - \mathbf{r}_k \parallel\}$ and $\sqrt{\kappa}$ is sometimes called the equivalent monostatic range (EMR). For a realistic FM-based PR scenario with EIRP_k = 10 kw, $\mathbf{r}_I = [10, 10]^T$ km, $\mathbf{r}_k = [80, 10]^T$ km, $G_{TR}^{(k)}$ = 3 dBi, λ = 3 m, $L_{\mathrm{loss}-T}$ = 5 dB, $T_s^{(k)}$ = 2,401 K, B_k = 200 kHz, T = 1 second, and $\mathrm{SNR}_k^{(\mathrm{SP},\min)}$ = 14 dB (i.e., $\mathrm{SNR}_k^{(\min)}$ = –39 dB), detectable RCS in dBms as a function of the target position is plotted in Figure 1.33. This scenario with the same parameters but with $\mathbf{r}_k = [160, 10]^T$ km is repeated and its results are depicted in Figure 1.34. There are some points in order. First, it is seen that if the target is far from the PR receiver, but close to the transmitter, the received power is high enough. Figures with such characteristics are called *Cassini ovals* [28]. In a typical monostatic scenario, however, an inevitably small power is received when the target is far from the radar. Second, total detection coverage can be obtained from this figure for a specific RCS. By comparing the results of Figures 1.33 and 1.34, it can be concluded that the total detection coverage is decreased as L_k increases. If the maximum RCS is –5 dBms, the coverage area is approximately an elliptic when $\mathbf{r}_k = [80, 10]^T$ km, while it is two separate ellipses around transmitter and receiver for $\mathbf{r}_k = [160, 10]^T$ km. In Figure 1.35, EMR, say $\sqrt{\kappa}$, as a function of EIRP is depicted where the system parameters were the same as in Figures 1.33 and 1.34. The EMR would correspond to the detection range of a monostatic radar, where the baseline is equal to zero [28].

Let us define the dynamic range of the kth receiver, termed as DR_k, be the ratio of the direct signal to the noise power level, given by

$$\mathrm{DR}_k = \frac{\mathrm{EIRP}_k G_{RI}^{(k)} \lambda^2}{(4\pi)^2 L_k^2 L_{\mathrm{loss}-\mathrm{b}} P_n^{(k)}} \tag{1.41}$$

Figure 1.36 plots the required kth receiver dynamic range in an FM-based PR with parameters EIRP_k = 10 kw, $G_{TR}^{(k)}$ = 3 dBi, λ = 3 m, $L_{\mathrm{loss}-\mathrm{b}} = 3 + 10^{-5} L_k$, $T_s^{(k)}$ = 2,401 K, and B_k = 200 kHz. As can be seen, the required receiver dynamic range can be reduced by increasing the baseline distance, which is about 80 dB for the considered system setup. Knowing the system's dynamic range can assist us in placing the PR receivers in a suitable location [31].

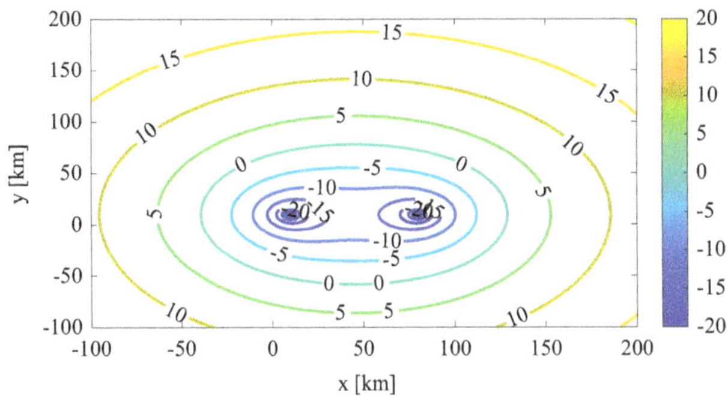

Figure 1.33 Detectable RCS of FM-based PR with parameters: $EIRP_k = 10$ kw, $\mathbf{r}_l = [10, 10]^T$ km, $\mathbf{r}_k = [80, 10]^T$ km, $G_{TR}^{(k)} = 3$ dBi, $\lambda = 3$ m, $L_{\text{loss}-T} = 5$ dB, $T_s^{(k)} = 2{,}401$ K, $B_k = 200$ kHz, $T = 1$ second, and $SNR_k^{(SP,min)} = 14$ dB (i.e., $\text{SNR}_k^{(\text{min})} = -39$ dB)

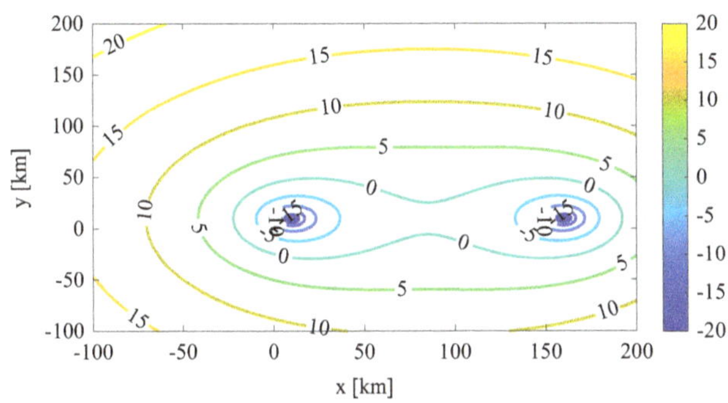

Figure 1.34 Detectable RCS of FM-based PR with parameters: $EIRP_k = 10$ kw, $\mathbf{r}_l = [10, 10]^T$ km, $\mathbf{r}_k = [160, 10]^T$ km, $G_{TR}^{(k)} = 3$ dBi, $\lambda = 3$ m, $L_{loss-T} = 5$ dB, $T_s^{(k)} = 2\,401$ K, $B_k = 200$ kHz, $T = 1$ second, and $SNR_k^{(SP,min)} = 14$ dB (i.e., $SNR_k^{(min)} = -39$ dB)

1.4 Summary

This chapter provides a tutorial style exploration of the fundamental principles of PRs, which include analyzing the properties of illuminators of opportunity, understanding the geometry of PRs, and evaluating the coverage of passive systems. These concepts are essential to gain a comprehensive understanding of PR systems.

The aim of this book is to focus on the issue of PR target detection using a detection theory framework. The latest research advancements in this field will be presented in the following chapters, which are divided into four parts. Part 1, Chapter 2 deals with conventional passive target detection algorithms. Part 2, which comprises seven chapters, covers the target detection problem in passive bistatic radar under a high SNR of the direct path signal in the RC. In Part 3, which consists of

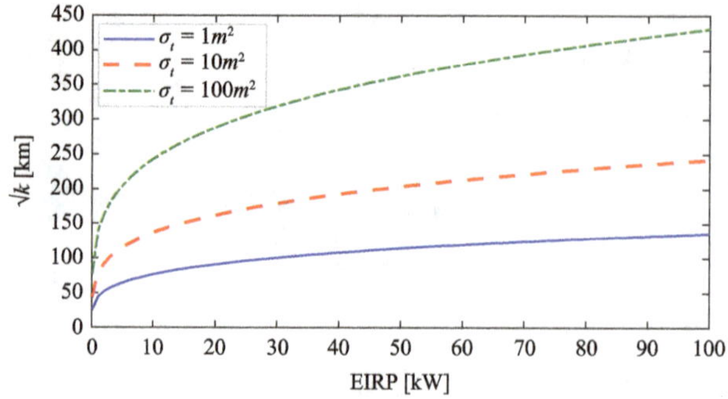

Figure 1.35 *EMR as a function of EIRP for FM-based PR with parameters: $G_{TR}^{(k)} = 3$ dBi, $\lambda = 3$ m, $L_{loss-T} = 5$ dB, $T_s^{(k)} = 2{,}401$ K, $B_k = 200$ kHz, $T = 1$ second, and $SNR_k^{(SP,min)} = 14$ dB (i.e., $SNR_k^{(min)} = -39$ dB)*

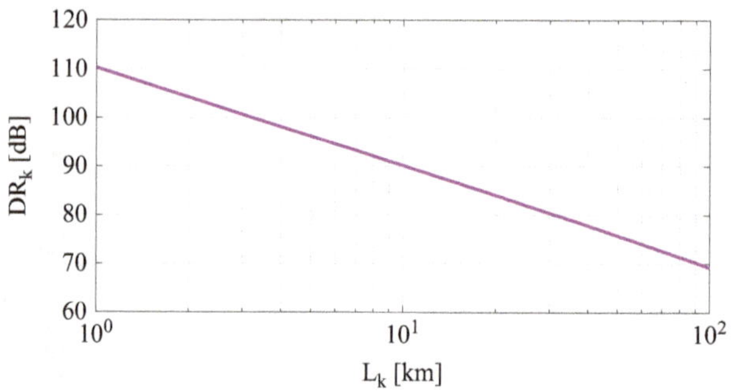

Figure 1.36 *Required* kth *receiver dynamic range as a function of baseline distance for FM-based PR with parameters: $EIRP_k = 10$ kw, $G_{TR}^{(k)} = 3$ dBi, $\lambda = 3$ m, $L_{loss-b} = 3 + 0.01 \times L_k$ with L_k in km, $T_s^{(k)} = 2{,}401$ K, and $B_k = 200$ kHz*

two chapters, the focus will be on multistatic PR target detection under noisy RCs. Part 4, with two chapters, addresses the target detection problem in multistatic and MIMO PRs.

References

[1] Griffiths H.D., Baker C.J. 'An introduction to passive radar' in Artech House; 2017.

[2] Howland P. 'Editorial: Passive radar systems'. *IEE Proceedings – Radar, Sonar and Navigation*. 2005, vol. 152, pp. 105–06.

[3] Farina A., Kuschel H. 'Guest editorial special issue on passive radar (Part I)'. *IEEE Aerospace and Electronic Systems Magazine*. 2012, vol. 27(10), pp. 5–5.

[4] Antoniou M. 'Guest editorial special issue on advanced passive radar techniques and applications'. *Sensors*. 2020.

[5] Griffiths H., Willis N. 'Klein Heidelberg – the first modern bistatic radar system'. *IEEE Transaction on Aerospace and Electronic Systems*; 2010. pp. 1571–88.

[6] Rittenbach O.E., Fishbein W. 'Semi-active correlation radar employing satellite-borne illumination'. *IRE Transactions on Military Electronics*. 1960, vol. 4(2/3), pp. 268–69.

[7] Willis N.J., Griffiths H.D. 'Advances in bistatic radar' in Vol. 2. SciTech Publishing; 2007.

[8] Griffiths H.D., Long N.R.W. 'Television-based bistatic radar'. *IEE Proceedings F Communications, Radar and Signal Processing*. 1986, vol. 133(7), pp. 649–57.

[9] Howland P.F. 'Target tracking using television-based bistatic radar'. *IEE Proceedings – Radar, Sonar and Navigation*. 1999, vol. 146, pp. 166–74.

[10] Sahr J.D., Lind F.D. 'The manastash ridge radar: a passive bistatic radar for upper atmospheric radio science'. *Radio Science*. 1997, vol. 32(6), pp. 2345–58.

[11] Baniak J., Baker G., Cunningham A.M., Martin L. 'Silent sentry passive surveillance'. *Aviation Week and Space Technology*. 1999.

[12] Howland P.E., Maksimiuk D., Reitsma G. 'FM radio based bistatic radar'. *IEE Proceedings – Radar, Sonar and Navigation*. 2005, vol. 152(3), pp. 107–15.

[13] Ferrier J., Klein M., Allam S. 'Frequency and waveform complementarities for passive radar applications'. *International Radar Symposium*. 2009, pp. 1–6.

[14] Fulcoli R., Sedehi M., Tilli,E, *et al.* 'AULOS: Italian passive coherent location (PCL) radar'. *Microwave Journal*. 2013, pp. 72–75.

[15] Lallo A.D., Farina A., Fulcoli R., *et al.* 'AULOS: Finmeccanica family of passive sensors'. *IEEE Aerospace and Electronic Systems Magazine*. 2016, vol. 31(11), pp. 24–29.

[16] Farina A., Lombardo P., Colone F. 'Passive radar: harvesting e.m. radiations for surveillance'. *4-hours Tutorial Presented at the IEEE Radar Conference 2010*; Washington, US, 2010.

[17] Zaimbashi A. 2013. 'Target detection in passive radars based on commercial FM radio signals'. [PhD thesis]. Iran, Shiraz University.

[18] Zaimbashi A., Derakhtian M., Sheikhi A. 'GLRT-based CFAR detection in passive bistatic radar'. *IEEE Transactions on Aerospace and Electronic Systems*. 2013, vol. 49(1), pp. 134–59.

[19] 'Transmission standard for FM sound broadcasting at VHF' in Rec. ITU-R BS; pp. 450–53.

[20] Baker C.J., Griffiths H.D., Papoutsis I. 'Passive coherent location radar systems. Part 2: Waveform properties'. *IEE Proceedings – Radar, Sonar and Navigation*. 2005, vol. 152(3), pp. 160–68.

[21] Edrich M., Schroeder A., Meyer F. 'Design and performance evaluation of a mature FM/DAB/DVB-T multi-illuminator passive radar system'. *IET Radar, Sonar and Navigation*. 2014, vol. 8(2), pp. 114–22.
[22] Griffiths H.D., Baker C.J. 'Passive coherent location radar systems. Part 1: performance prediction'. *IEE Proceedings – Radar, Sonar and Navigation*. 2005, vol. 152(3), pp. 153–59.
[23] O'Hagan D.W., Griffiths H.D., Ummenhofer S.M., Paine S.T. 'Elevation pattern analysis of common passive bistatic radar illuminators of opportunity'. *IEEE Transactions on Aerospace and Electronic Systems*. 2017, vol. 53(6), pp. 3008–19.
[24] *Digital video broadcasting (DVB); framing structure, channel coding and modulation for digital terrestrial television* [European telecommunications standard institute]. EN 300 744, VI. 1.2. 1997.
[25] Palmer J.E., Harms H.A., Searle S.J., Davis L. 'DVB-T passive radar signal processing'. *IEEE Transactions on Signal Processing*. 2015, vol. 61(8), pp. 2116–26.
[26] Wang H., Wang J., Zhong L. 'Mismatched filter for analogue TV-based passive bistatic radar'. *IET Radar, Sonar & Navigation*. 2011, vol. 5(5), pp. 573–81.
[27] Zaimbashi A. 'Target detection in analog terrestrial TV-based passive radar sensor: joint delay-Doppler estimation'. *IEEE Sensors Journal*. 2017, vol. 17(17), pp. 5569–80.
[28] Willis N.J. 'Bistatic radar' in 2nd ed. SciTech Publishing; 2005.
[29] Zaimbashi A. 'Introduction to radar system design' in lecture notes, Part 5: Radar receiver. Shahid Bahonar University of Kerman, Kerman, Iran; 2015.
[30] Balanis C.A. 'Antenna theory: Analysis and design' in John Wiley and Sons; 2016.
[31] Malanowski M., Kulpa K., Kulpa J., Samczynski P., Misiurewicz J. 'Analysis of detection range of FM-based passive radar'. *IET Radar, Sonar & Navigation*. 2014, vol. 8(2), pp. 153–59.

Chapter 2
Passive radar conventional target detection algorithms

Signal processing for radar systems involves various techniques and touches on multiple application areas. Radar technology has become a critical technology for civilian applications including air, maritime, and ground traffic control, as well as urban sensing and indoor monitoring. The development of radar technology has enabled accurate detection and tracking of targets and has become an integral part of modern society's infrastructure.

In contrast to active radars, passive radars face challenges in signal processing, particularly in target detection. This is primarily due to the fact that passive radars (PRs) rely on non-cooperative illuminators of opportunity, which generate waveforms not specifically designed for radar applications. Consequently, target echoes in PRs are often masked by interference signals. Considering this fact, in the following sections, we will first model the surveillance channel (SC) and reference channel (RC) of conventional passive radar systems to demonstrate the difficulties involved in target detection within these systems. Specifically, the SC is responsible for receiving the signal that has been reflected from the target, while the RC is responsible for receiving the signal transmitted by the illuminator of opportunity. The received signal in the SC is then correlated with the reference signal to detect the presence of the target. However, since the illuminator of opportunity is not designed for radar applications, the surveillance signal may be contaminated with interference signals, which can mask the target echo in the SC. To address this challenge, various signal processing techniques have been developed for PRs including interference removal, target detection, tracking, and multichannel processing. These techniques are designed to reduce the impact of interference signals and enhance the detectability of the target echo. In addition, new techniques such as compressive sensing, which allow for sparse signal processing, are being investigated for use in PR systems.

Overall, the development of signal processing techniques for PR systems is crucial for enhancing their performance and improving their target detection capabilities. As the demand for PR technology continues to grow, the use of advanced signal processing techniques will become increasingly important in enabling effective and reliable operation of these systems.

2.1 Surveillance and reference signal models

A passive bistatic radar (PBR) in its basic form involves two antennas – a surveillance antenna and a reference antenna. The surveillance antenna is used to monitor a particular sector, while the reference antenna is used to obtain a reference of the transmitted signal. The reference antenna is assumed to have a directional pattern aimed at the transmitter, while the surveillance antenna has a broad beam antenna aimed at the surveyed sector. This setup is depicted in Figure 2.1. When a transmitter emits a signal $s(t)$, we can expect to receive not only a direct signal but also reflections from various targets. These reflections are characterized by delays τ_m and Doppler shifts f_{dm}, and there may be a large number of ground scatterers that contribute to clutter and multipath effects [1–3]. To model this clutter, we can consider a set of discrete scatterers that represent nearby ground scatterers, distributed over a range from zero to a maximum clutter range of $R^{(c)}$. If we denote the received signal in the SC as $x_s(t)$, we can express it as

$$x_s(t) = \beta_0 s(t-\tau_d) + \sum_{i=1}^{N_c} \beta_i(t) s(t-\tau_i^{(c)}) + \sum_{m=0}^{K-1} a_m s(t-\tau_m) e^{j2\pi f_{dm} t} + n_s(t) \tag{2.1}$$

The thermal noise at the SC is denoted by $n_s(t)$, and $s(t)$ represents the transmitted signal, where $0 \leq t < T$. The delay corresponding to the direct path between the transmitter and the receiver is denoted by τ_d, which is equal to R_{tr}/c, where R_{tr} is the distance between the transmitter and the receiver, and c is the speed of light. The target's complex amplitudes, time delays, and Doppler frequency shifts are represented by a_m, τ_m, and for $m = 1, ..., M$, respectively. The direct-path signal complex amplitude received by the sidelobe/backlobe of the surveillance antenna is represented by β_0, while $\beta_i(t)$ and $\tau_i^{(c)}$ for $i = 1, \ldots, N_c$ represent the complex amplitudes and time delays of the clutter, respectively. The total number of targets is denoted by K, and N_c represents the number of clutter scatterers.

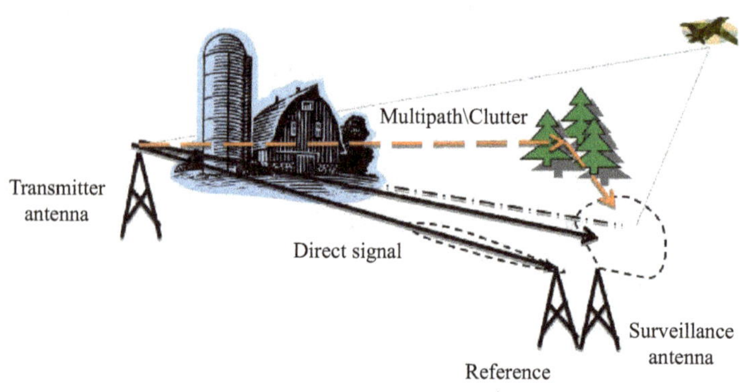

Figure 2.1 Simple configuration of a PBR system

In radar signal processing, it is common to assume that the clutter has a constant amplitude over the integration time. This assumption is valid when the echoes received from fixed scatterers such as the ground, rocks, and tree trunks are considered. However, when there are moving scatterers, such as leaves and branches of trees, the clutter's amplitude can vary over time. Therefore, it is more practical to model clutter as a random process with a specific power spectrum density (PSD). In the presented model, the assumption is made that the multipath/clutter with a specific PSD can be modeled by a few frequency components around the zero Doppler frequency as [1–3]

$$\beta_i(t) = \sum_{k=-Q}^{Q} \beta_{i,k} e^{j2\pi f_k^{(c)} t} \tag{2.2}$$

where $f_k^{(c)}$ are constant frequencies that are uniformity spaced with $f_k^{(c)} = (k-1-Q)\Delta f_c$ for $k = 1, ..., 2Q+1$ where Q and Δf_c are two parameters that should be set properly in a practical situation. Here, $\beta_{i,k}$s are considered as clutter's complex amplitudes at delays $\tau_i^{(c)}$ and frequencies $f_k^{(c)}$ for $i = 1, ..., N_c$.

Practically speaking, the reference signal used in a PBR is susceptible to interference from both multipath and target echoes, which can negatively impact interference removal and target detection performance [4, 5]. To address this issue, it is necessary to condition the reference signal using techniques such as beamforming and channel equalization prior to SC target detection. Previous works [4] and [5] have shown that the space–time constant modulus algorithm is effective at mitigating performance degradation caused by multipath in the reference signal. Until appropriate signal conditioning is performed, the signal received in the RC can be expressed as

$$x_r(t) = c_r s(t-\tau_d) + \sum_{i=1}^{N_c'} \beta_i'(t) s(t-\tau_i^{(c)'}) + n_r'(t), \qquad 0 \le t < T \tag{2.3}$$

where c_r is the complex amplitude of the direct signal received by the mainlobe of the reference antenna, $\beta_i'(t)$, $\tau_i^{(c)'}$ and N_c' have similar definitions as that of $\beta_i(t)$, $\tau_i^{(c)}$ and N_c are in the SC, and $n_r(t)$ is the receiver noise.

The signals $x_s(t)$ and $x_r(t)$ are sampled with a sampling frequency f_s at the time instants $t_n = n/f_s = nT_s$, $n = 0, ..., N-1$. If $x_s[n]$ and $x_r[n]$ are the nth time sample of the received signal at the SC and RC, respectively, then we can write

$$\begin{aligned} x_s[n] = \beta_0 s[n-n_d] + \textstyle\sum_{i=1}^{N_c} \sum_{k=-Q}^{Q} \beta_{i,k} s[n-n_i^{(c_0)}] e^{j2\pi f_k^{(c)} T_s n} \\ + \textstyle\sum_{m=0}^{k-1} a_m s[n-n_m^{(t)}] e^{jn\Omega_m^{(t)}} + n_s[n] \end{aligned} \tag{2.4}$$

and

$$x_r[n] = c_r s[n-n_d] + \sum_{i=1}^{N_c'} \beta_i'[n] s[n-n_i^{(c_0)'}] + n_r'[n] \tag{2.5}$$

with $n = 0, ..., N-1$, where $\Omega_m^{(t)} = 2\pi f_{dm} T_s$ is the normalized Doppler frequency of target, and we have $\tau_d = n_d T_s$, $\tau_m = n_m^{(t)} T_s$, $\tau_i^{(c)} = n_i^{(c_0)} T_s$, and $\tau_i^{(c)'} = n_i^{(c_0)'} T_s$. Here, $n_i^{(c_0)}$s for $i = 1, ..., N_c$ are the corresponding delays of clutter/multipath scatterers resulting from nearby ground scatterers distributed from the bistatic range of zero to $R^{(c)}$, i.e., $N_c = \frac{R_c f_s}{c}$. After accomplishing signal conditioning in the RC, we may obtain

$$x_r[n] = c_r s[n - n_d] + n_r[n] \tag{2.6}$$

The term $n_r[n]$ represents the total amount of receiver noise and interference residuals. Specifically, in the RC, it is assumed that after adequate signal conditioning, the interference residuals have a power level lower than that of the receiver noise, even though they still exhibit noisy behavior.

To detect PR targets, it appears essential to have a separate RC to provide a reference of transmitted signal for detecting the target signal in the SC. However, there are other ways to address the PR target detection problem without using an RC. Thus, PR target detection techniques can be categorized into two primary groups based on whether they use an RC or not. The first group includes two-channel PR target detection techniques, whereas the second group involves single-channel approaches that rely only on the SC.

1. The first primary approach, abbreviated as Ca.1, uses an RC to collect the transmitted signal, which is contaminated with thermal noise and multipath signals. To remove the effects of multipath signals, an equalization algorithm called signal conditioning is applied in the RC. If perfect signal conditioning is achieved, the only limiting interference in the RC is thermal noise. There are two general viewpoints on dealing with the noisy RC. The first approach assumes that a high direct-path signal-to-noise ratio (DNR) is accessible in the RC, leading to an ideal reference (IR) channel with a high-DNR condition. In this case, IR-based techniques (Ca.1-IR) are used for detection. The second approach takes into account the effect of thermal noise in the RC, which leads to a noisy reference (NR) channel. The target detection algorithms are then designed to properly handle this noise. This approach is referred to as NR-based techniques (Ca.1-NR). In summary, the first category of PR systems assumes that a signal conditioning method is utilized on the RC to remove multipath effects, and then either Ca.1-IR or Ca.1-NR is used for the target detection process, depending on whether the noise in the RC is assumed to be negligible or significant.
2. The second category refers to situations where obtaining an acceptable RC quality may not be possible due to several reasons. One of the reasons may be the blockage of the line-of-sight paths between transmitters and receivers, which can result in low DNRs. Another reason may be highly directive transmit antenna patterns or a severe clutter multipath environment. Under these challenging circumstances, it may not be feasible to use a dedicated RC. In such cases, the best solution is to employ multireceiving channels instead of relying on a dedicated RC. This approach, known as multistatic PR detection without the RC, is referred to as Ca.2. The use of multireceiving channels enables the collection of more data samples over a given observation time. It also allows for the exploitation of the spatial diversity of the target radar cross-section and the interchannel correlation among different receivers. This approach can lead to a

better understanding of the radar environment and more accurate target detection, even in challenging conditions where obtaining an acceptable RC quality is not possible.

The first category, especially Ca.1-IR approaches, has been widely explored in References 1–23. Among the category Ca.1-IR, there are various methods to remove interference signals and detect targets of interest. These methods can be divided into two general cases. The first one is based on the concept of detection theory [1–15], while the second one is based on conventional signal processing [6, 16–21]. In the first case, the passive target detection problem is modeled as a binary (or M-ary) hypothesis testing problem and then solved by resorting to the detection theory framework [1–5]. In contrast, in the second case, some conventional radar signal processing algorithms have been exploited, which are not optimal for PR target detection purposes. In the conventional passive radar target detection (C-PRTD), the interference removal and target detection steps are separately performed. This is in contrast to the detection theory-based methods where they are done under same criteria, i.e., maximizing the target detection probability for a given false alarm probability. For example, in the second case, adaptive methods such as least mean squares (LMS) or recursive least squares (RLS) filters are recommended to be used in order to remove interference signals, and two-dimensional cell averaging (2D-CA) constant false alarm rate (CFAR) detector is exploited for target detection purpose [20, 21]. Some other conventional methods utilize subspace-based methods, known as extensive cancellation algorithm (ECA) and sequential cancellation algorithm (SCA), in conjunction with the 2D-CA detector for this task [6, 20, 21].

Although the Ca.1-IR approaches have some particular advantages, their performances would be significantly degraded especially when the DNR of the RC is low. To remedy this situation, some other researchers claimed that improved detection performance can be possible if we resort to Ca.1-NR approaches, where the noise in the RC was properly taken into account [24–26]. Recently, it has been shown that it is required to properly set detection thresholds of the Ca.1-NR approaches to achieve acceptable detection performances in practical situations [27, 28]. In Part 3, we will provide a detailed analysis of the research conducted in References 27, 28.

In References 29–34, various target detection methods for Ca.2 with multiple SCs have been explored, which can be categorized based on factors such as interference signal presence, Bayesian or deterministic modeling of transmitted signal, calibrated or uncalibrated receivers, noise level knowledge, and single-input multiple-output (SIMO) or multiple-input multiple-output (MIMO) configurations. Recently, a unified framework for multistatic PR target detection was proposed in Reference 35, where target detection methods were developed under the assumption that the received signal was a noisy version of the target echo. However, in practical situations, direct-path signals can be stronger than the target echo and must be considered in PR target detection problems. In Reference 36, a closed-form Rao-based target detector was proposed that considered uncalibrated receivers and direct-path signal interference. These works will be examined in Part 4 of the book.

In this chapter, we analyze and assess the C-PRTD methods through simulation results. These methods involve separate interference removal and target detection steps, and their interrelationships are discussed in detail. Additionally, the subsequent parts will scrutinize the detection theory-based methods.

2.2 Conventional interference removal approaches

Although there are many benefits to using PRs, most broadcast illuminators that use bistatic geometries and continuous wave transmissions can lead to increased complexity and require additional signal processing stages. This is because interference signals need to be removed, including both direct and multipath signals. It is crucial to choose an appropriate interference removal algorithm to ensure the interference signals are removed as effectively as possible. While analog interference suppression methods like antenna null steering and physical shielding are somewhat effective, they do not provide sufficient results due to the high dynamic ranges of these interference signals. Therefore, it is necessary to use digital methods like subspace-based and adaptive filtering to remove these interference signals [37].

In this section, a comprehensive analysis of classical adaptive filtering methods and subspace-based techniques is presented through extensive simulation results. These results investigate the amounts of interference cancellation, the null width and depth of clutter removal algorithms in the range-Doppler region of interference signals, false alarm regulation, and detection performance evaluation in multi-target scenarios. In particular, we will focus on the last two parameters discussed for the first time in this section. Alghough there has been significant research on PR interference removal techniques in recent years, none have investigated the impact of interference removal techniques on target detection. The subsequent subsections will delve into this crucial aspect in more detail.

2.2.1 Subspace-based interference removal approaches

In this section, we investigate the ECA and SCA methods among subspace-based techniques. These methods assume that the clutter and multipath can be modeled as a group of $(N_c + 1)$ scatterers, which are produced by nearby ground scatterers within a maximum range of $R^{(c)}$ from the bistatic range of zero. To model the PSD of the multipath echoes, $(2P + 1)$ discrete frequency components can be considered around the zero Doppler, as shown in (2.2). Now, we can define a $(N_c + 1) \times (2P + 1)$ -dimensional clutter subspace matrix $\mathbf{V}$, where its columns span the clutter subspace. Specifically, the mth column of $\mathbf{V}$ can be defined as

$$[\mathbf{v}]_m = \mathbf{P}^{nm}\mathbf{x}_r \odot \mathbf{e}_m \tag{2.7}$$

where

$$\mathbf{x}_r = [x_r[0], \ldots, x_r[N-1]]^T \tag{2.8}$$

The vector $\mathbf{e}_m$ is a Doppler steering vector that corresponds to the mth column of matrix $\mathbf{V}$. The nth element of $\mathbf{e}_m$ is defined as $[\mathbf{e}_m]_n = e^{j2\pi f_c^{(m)} nT_s}$, where $n = 0, \ldots, N-1$, and $m = 1, \ldots, M$ with $M = (N_c + 1)(2P + 1)$. The matrix $\mathbf{P}$ is a permutation matrix consisting of zeros and ones, and it delays the input signal vector $\mathbf{x}_r$ by a single time sample. The product $\mathbf{P}^{n_m}\mathbf{x}_r$ represents a delayed version of the reference signal vector $\mathbf{x}_r$ with n_m time delay samples. Similarly, the vector $[\mathbf{v}]_m$ is a sampled delayed and Doppler replica of the reference signal vector, where the delay and Doppler shift are, respectively, given by n_m and $f_c{}^m$.

The ECA filter is a signal processing technique that is commonly used in surveillance applications to enhance the quality of the received signal. When this filter is applied to the surveillance signal vector $\mathbf{x}_s$, we obtain

$$\mathbf{x}_E = \mathbf{x}_s - \mathbf{V}\hat{\beta} \tag{2.9}$$

where

$$\mathbf{x}_s = [x_s[0], \ldots, x_s[N-1]]^T \tag{2.10}$$

To estimate the clutter direction vector, denoted by $\hat{\boldsymbol{\beta}}$, we can resort to the well-known least squares (LS) estimation method to obtain

$$\hat{\beta} = (\mathbf{V}^H\mathbf{V})^{-1}\mathbf{V}^H\mathbf{x}_s \tag{2.11}$$

Plunging (2.11) into (2.9) yields

$$\mathbf{x}_E = \Pi_{\mathbf{V}}^{\perp}\mathbf{x}_s \tag{2.12}$$

where matrix $\Pi_{\mathbf{V}}^{\perp}$ is the orthogonal projection matrix onto the columns of $\mathbf{V}$ and is defined as $\Pi_{\mathbf{V}}^{\perp} = \mathbf{I} - \mathbf{V}(\mathbf{V}^H\mathbf{V})^{-1}\mathbf{V}^H$. If we think of the subspace spanned by the columns of $\mathbf{V}$ as a "clutter/multipath (disturbance) subspace" and the orthogonal subspace as the "signal subspace," then $\mathbf{x}_E$ represents the projection of the received vector $\mathbf{x}_s$ onto the signal subspace. It is worth noting that the ECA method is based on LS and can accommodate Doppler shifts of the reference signal to model the effects of slow-moving clutter echoes. As demonstrated in the simulation section of this chapter, the ECA method for mitigating interference has a satisfactory performance, and its computational complexity is $\mathcal{O}(NM^2)$.

To simplify the computational complexity and memory requirements of the ECA, one can utilize a sequential version called the SCA. The introduction of this algorithm begins with the utilization of the following theorem.

Theorem 1. *Suppose we have two matrices* $\mathbf{X}$ *and* $\mathbf{Y}$ *with dimensions* $N \times M_1$ *and* $N \times M_2$*, respectively. We define the partitioned matrix* $\mathbf{Z}$ *as the concatenation of* $\mathbf{X}$ *and* $\mathbf{Y}$*, such that* $\mathbf{Z} = [\mathbf{X}, \mathbf{Y}]$*. We can represent the orthogonal projection matrix onto the columns of* $\mathbf{Z}$ *as follows*:

$$\Pi_{\mathbf{Z}}^{\perp} = \Pi_{\Pi_{\mathbf{X}}^{\perp}\mathbf{Y}}^{\perp}\Pi_{\mathbf{X}}^{\perp} \tag{2.13}$$

Through the utilization of Theorem 1, we can condense the SCA method for computing $\mathbf{x}_{\mathrm{SCA}} = \Pi^{\perp}\mathbf{V}\mathbf{x}_s$ into a sequential process, as follows:

1. Set $k = 1$: Initialize the value of k as 1. Consider $\mathbf{Y}^{(1)} = [\mathbf{V}, \mathbf{x}_s]$: Concatenate the matrix $\mathbf{V}$ and the signal vector $\mathbf{x}_s$ to form the matrix $\mathbf{Y}^{(1)}$, where $\mathbf{V} = [\mathbf{v}_1, \mathbf{v}_2, ..., \mathbf{v}_M]$.
2. Partition obtained matrix $\mathbf{Y}^{(k)}$ as $\mathbf{Y}^{(k)} = [\mathbf{y}, \mathbf{Y}]$: Split the matrix $\mathbf{Y}^{(k)}$ obtained in Step 1 into two parts: a row vector $\mathbf{y}$ of dimension $N \times 1$ and a submatrix $\mathbf{Y}$ of dimension $N \times (M - k + 1)$. Compute $\mathbf{Y}^{(k+1)} = \Pi_{\mathbf{y}}^{\perp}\mathbf{Y}$: Project the submatrix $\mathbf{Y}$ orthogonally onto the complement of the span of $\mathbf{y}$ and store the result in $\mathbf{Y}^{(k+1)}$.
3. Set $k + 1 \leftarrow k$: Increment k by 1. For $k \leq M$ return back to Step 2: Repeat Steps 2 and 3 until k becomes equal to M.
4. Compute a cleaned surveillance signal vector: After completing Step 3, the final value of $\mathbf{Y}^{(k+1)}$ is $\mathbf{Y}^{(M+1)}$, which is a submatrix (vector) of the original matrix $\mathbf{Y}^{(1)}$. Compute a cleaned surveillance signal vector as $\mathbf{x}_{\mathrm{SCA}} = \mathbf{Y}^{(M+1)}$ by taking this submatrix (vector) as the output.

The computational complexity of the SCA and ECA is the same, but the SCA can be executed in parallel to decrease the workload for each CPU. Additionally, the SCA does not require the construction of large matrices, resulting in reduced memory requirements. However, since the ECA and SCA are mathematically equivalent and have the same performance, only the ECA will be considered in the following discussion.

2.2.2 Adaptive filter-based interference removal approaches

Many traditional adaptive filtering methods, such as the LMS algorithm, normalized least mean squares, and RLS algorithms, can be modified to eliminate PR interference, rather than using the subspace-based method. These adaptive techniques are more resistant to deterioration in time-varying clutter environments because they continuously adapt [37]. In general, statistical or deterministic frameworks can be used to formulate the filter design problem for estimating a desired signal from an interference-contaminated signal [38, 39].

The LMS-based methods approach interference removal from a statistical viewpoint, while the least-square-based methods use a deterministic approach. A PR system's interference removal problem is shown in Figure 2.2, where signal conditioning is performed on the RC to obtain a clean version of the transmitted signal, and beamforming is applied in the SC to reduce interference signals like multipath and direct-path signals. The cleaned RC and SC signals are denoted as $x_r[n]$ and $x[n]$, respectively. To obtain an output signal $x_r^F[n]$ with desirable features, an M-tap transversal filter is used as an adaptive filter that reshapes the input signal $x_r[n]$. To eliminate multipath signals in PR applications, it is necessary to generate delayed versions of $x_r[n]$ in the output of the adaptive filter. The desired signal of the adaptive filter in this scenario is the direct-path and multipath signals of the SC, resulting in the error signal $x_c[n]$, which is obtained by taking the difference between the output of the filter and that of the SC, i.e., $x_c[n] = x[n] - x_r^F[n]$. To design the filter, the filter coefficients must be chosen such that the desired signal and filter output match as closely as possible. This is typically achieved by optimizing a performance

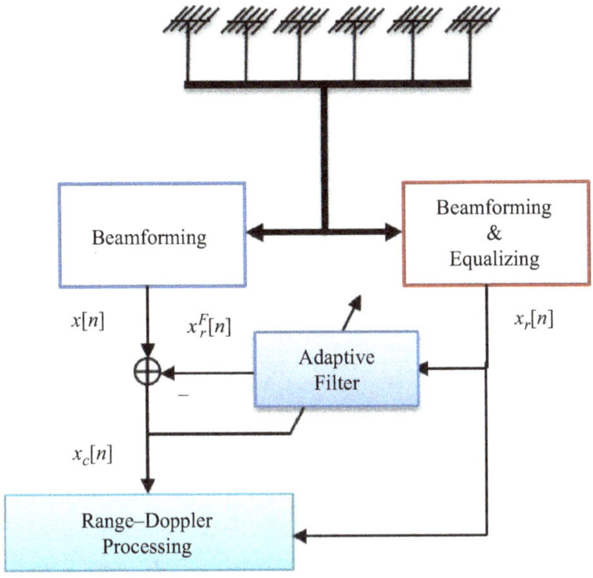

Figure 2.2 General structure of adaptive filter in PBR configuration

function that quantifies the error signal. The most commonly used performance function in the statistical and deterministic frameworks is the mean square value of the error signal and a weighted sum of the squared error signal, respectively.

2.2.2.1 Least mean squares algorithms

The mean squared error performance function to design the Wiener filter is given by

$$J[n] = E\{|x_c[n]|^2\} \tag{2.14}$$

where $E\{.\}$ denotes the statistical expectation. By considering M-tap transversal filter, the tap-weight vector and filter input can be, respectively, defined as

$$\mathbf{w} = [w_0, w_1, ..., w_{(M-1)}]^T \tag{2.15}$$

and

$$\underline{\mathbf{x}}_r[n] = [x_r[n], x_r[n-1], ..., x_r[n-M+1]]^T \tag{2.16}$$

The filter output is

$$x_r^F[n] = \mathbf{w}^H \underline{\mathbf{x}}_r[n] \tag{2.17}$$

The error signal and the performance function can be given by

$$x_c[n] = x[n] - x_r^F[n] = x[n] - \mathbf{w}^H \underline{\mathbf{x}}_r[n] \tag{2.18}$$

and

$$J = \sigma_x^2 - \mathbf{w}^H\mathbf{p} - \mathbf{p}^H\mathbf{w} + \mathbf{w}^H\mathbf{R}\mathbf{w} \tag{2.19}$$

where $\sigma_x^2 = E\{|x[n]|^2\}$, $\mathbf{p} = E\{\underline{\mathbf{x}}_r[n]x^*[n]\}$, and $\mathbf{R} = E\{\underline{\mathbf{x}}_r[n]\underline{\mathbf{x}}_r[n]^H\}$. This function is quadratic in $\mathbf{w}$ and has a unique global minimum, which can be found at the optimal solution given by

$$\mathbf{w}_0 = \mathbf{R}^{-1}\mathbf{p} \tag{2.20}$$

Suppose that the inverse of $\mathbf{R}$ exists, which is the solution to the Wiener–Hopf equation, as mentioned in Reference 38. The LMS method, introduced by Widrow and Hoff in 1960, is a useful approach for implementing the Wiener filter without directly solving the Wiener–Hopf equation. It is a step-by-step algorithm that enables the adjustment of filter coefficients $\mathbf{w}$ Reference 38.

$$\mathbf{w}[n+1] = \mathbf{w}[n] + \mu\underline{\mathbf{x}}_r[n]x_c^*[n] \tag{2.21}$$

where μ is the step-size parameter and can be selected to guarantee the convergence and accuracy of the LMS algorithm. To achieve convergence of the LMS algorithm, it is necessary to have both the convergence of $E|\mathbf{w}[n]|$ toward $\mathbf{w}_0$ and the convergence of the variance of elements of $\mathbf{w}[n]$ to some limited values. This can be accomplished by selecting an appropriate value for μ such that Reference 38

$$0 \leq \mu \leq \frac{2}{\| \underline{\mathbf{x}}_r[n] \|^2} \tag{2.22}$$

Typically, the performance of the LMS algorithm's convergence is mainly influenced by the eigenvalue spread of the covariance matrix $\mathbf{R}$. This matrix's eigenvalues relate to the input signal's power spectral density. The frequency modulation (FM)-based PBR exhibits a high eigenvalue spread in $\mathbf{R}$, resulting in slow convergence, even though the LMS algorithm's computational complexity is low. The LMS algorithm's summary is presented in Algorithm 1. Two parameters, μ and M, must be appropriately selected to use the LMS algorithm to eliminate the interference signal of the SC. The first parameter determines the convergence rate and the LMS algorithm's accuracy, while the second parameter is based on the clutter's maximum range bin.

2.2.2.2 Recursive least squares algorithms

This section introduces a deterministic-based adaptive algorithm that eliminates interference signals from the SC of a PR. The algorithm employs the method of LS. In contrast to the LMS algorithm aimed to minimize the mean square of the estimation error, in the method of LS, at any time instant n (i.e., $n > 0$) the adaptive filter coefficients are optimized so that the performance function

$$\zeta[n] = \sum_{k=1}^{n} \gamma[n,k]|x_c[n,k]|^2 \tag{2.23}$$

Algorithm 2.1 LMS Algorithm

Input: Sequences $\underline{x}_r[n]$, $x[n]$ and $\mathbf{w}[n]$,
Output: $x_r^F[n]$ and and $\mathbf{w}[n+1]$,
1: **Filtering:** $x_r^F[n] = \mathbf{w}^{\mathrm{H}}\underline{\mathbf{x}}_r[n]$
2: **Error Estimation:** $x_c[n] = x[n] - x_r^F[n]$,
3: **Tap-weight vector adaptation:** $\mathbf{w}[n+1] = \mathbf{w}[n] + \mu\underline{\mathbf{x}}_r[n]x_c^*[n]$

is minimized [38]. Thus, it is called LS. In (2.23), $\gamma[n, k]$s are weighting function of the performance function at time instant n, and $x_c[n, k]$ is defined as

$$x_c[n, k] = x[k] - \mathbf{w}^H[n]\underline{\mathbf{x}}_r[k] \tag{2.24}$$

It is seen that $x_c[n, 1]$ is the error signal at the beginning of the algorithm, and $x_c[n, k]$ for $k = 1, ..., n$ is the error signal computed based on the tap-weight vector $\mathbf{w}[n]$. Thus, this method exploits all the observations from the time filter starts until the present time.

Before proceeding, it is useful to examine the solution of the LS method to determine the connection between the ECA and the above LS-based method. To do so, let us define the input signal matrix up to time instant n, given by

$$\mathbf{X}[n] = [\underline{\mathbf{x}}_r[1], \underline{\mathbf{x}}_r[2], ..., \underline{\mathbf{x}}_r[n]] \tag{2.25}$$

where $\underline{\mathbf{x}}_r[k] = [\mathbf{0}_l^T, x_r[k]]^T$ with $l = n - k - 1 \geq 0$ where 0 means that we do not need any zero vector at the beginning of vector $\underline{\mathbf{x}}_r[k]$. Similarly, let us define $\mathbf{x}[n] = [x[1], x[2], ..., x[n]]^T$ as desired signal vector of the filter up to instant n, and error signal vector up to time n as $\mathbf{x}_c[n] = [x_c[1], x_c[2], ..., x_c[n]]^T$. By substituting $\beta[n, k] = 1$, the performance function can be rewritten as

$$\zeta[n] = \mathbf{x}_c[n]^H\mathbf{x}_c[n] \tag{2.26}$$

where

$$\mathbf{x}_c[n] = \mathbf{x}[n] - \mathbf{X}^T[n]\mathbf{w}^*[n] \tag{2.27}$$

Plugging (2.27) into (2.26) yields

$$\begin{aligned}\zeta[n] = {} & |\mathbf{x}[n]|^2 - \mathbf{x}^H[n]\mathbf{X}^T[n]\mathbf{w}^*[n] - \mathbf{w}^T[n]\mathbf{X}^*[n]\mathbf{x}[n] \\ & + \mathbf{w}^T[n](\mathbf{X}^*[n]\mathbf{X}[n]^T)\mathbf{w}^*[n]\end{aligned} \tag{2.28}$$

The optimum solution of this problem can be obtained as

$$\hat{\mathbf{w}}[n] = (\mathbf{X}[n]\mathbf{X}^H[n])^{-1}\mathbf{X}[n]\mathbf{x}^*[n] \tag{2.29}$$

By substituting (2.29) into (2.27), we obtain

$$\mathbf{x}_c[n] = \Pi^{\perp}_{\mathbf{X}[n]}\mathbf{x}[n] \tag{2.30}$$

where

$$\Pi^{\perp}_{\mathbf{X}[n]} = \mathbf{I} - \mathbf{X}^T[n](\mathbf{X}^*[n]\mathbf{X}^T[n])^{-1}\mathbf{X}^*[n] \tag{2.31}$$

To obtain the cleaned signal $\mathbf{x}_c[n]$, we project it orthogonally onto the subspace spanned by the rows of matrix $\mathbf{X}[n]$. It is worth noting that the LS method can be related to the ECA method. Assuming that the clutter has zero Doppler frequencies and a maximum range of M samples, the clutter matrix $\mathbf{V}$ of the ECA can be expressed as follows:

$$\mathbf{V} = [\mathbf{P}^M \mathbf{x}_r, \mathbf{P}^{M-1} \mathbf{x}_r, \ldots, \mathbf{P}^1 \mathbf{x}_r, \mathbf{x}_r] \tag{2.32}$$

In the LS method and in the time instant N, we have

$$\mathbf{X}[N] = [\underline{\mathbf{x}}_r[1], \underline{\mathbf{x}}_r[2], \ldots, \underline{\mathbf{x}}_r[N]] \tag{2.33}$$

By comparing (2.32) and (2.33), we find that

$$\mathbf{V} = \mathbf{X}^T[N] = [\underline{\mathbf{x}}_r[1], \underline{\mathbf{x}}_r[2], \ldots, \underline{\mathbf{x}}_r[N]]^T \tag{2.34}$$

This indicates that both the ECA and LS techniques coincide at time N. To achieve this, a matrix $\mathbf{X}[n]$ with dimensions of $N \times N$ must be built, and the inverse of the covariance matrix, which is $M \times M$ in size, must be computed. These processes contribute to the high computational complexity and memory requirements of the ECA and LS methods. Consequently, a recursive version of either the LS or ECA method will be introduced in the following. In RLS, the weighting function beta $[n,k]$ is selected as

$$\gamma[n, k] = \lambda_R^{n-k} \tag{2.35}$$

where λ_R is called forgetting factor with $\lambda_R > 0$, resulting in more emphasizing on the recent samples and tends to forget the past ones [38]. As such, the performance function can be rewritten as

$$\zeta[n] = \mathbf{x}_c^H[n]\Lambda[n]\mathbf{x}_c[n] \tag{2.36}$$

where

$$\Lambda[n] = \mathrm{diag}\{\lambda_R^{n-1}, \ldots, \lambda_R^1, \lambda_R^0\} \tag{2.37}$$

Thus, the optimum solution of (2.29) can be rewritten as

$$\hat{\mathbf{w}}[n] = (\mathbf{X}[n]\Lambda[n]\mathbf{X}^H[n])^{-1}\mathbf{X}[n]\Lambda[n]\mathbf{x}^*[n] \tag{2.38}$$

Algorithm 2 provides a recursive implementation of (2.38), which is known as the RLS algorithm in the following discussion. The convergence speed of the RLS algorithm is affected by the spread of eigenvalues in the input signal correlation matrix $\mathbf{R}$. In comparison to the LMS algorithm, the RLS algorithm can achieve faster convergence. However, the computational complexity of the RLS algorithm is $\mathcal{O}(M^2N)$, while that of the LMS algorithm is $\mathcal{O}(MN)$.

Algorithm 2.2 RLS Algorithm

Input: Sequences $x_r[n]$, $x[n]$, $\mathbf{w}[0] = \mathbf{0}$ and $\mathbf{G}[n] = \delta^{-1}\mathbf{I}_M$ with $\delta = 0.0001\sigma_{\mathbf{x}_r}$.
Output: $x_r^F[n]$ and and $\mathbf{w}[n]$,

1: **Compute gain vector:** $\mathbf{K}[n] = \frac{\lambda_R^{-1}\mathbf{G}[n-1]\mathbf{x}_r[n]}{1+\lambda_R^{-1}\mathbf{x}_r^{\mathrm{H}}[n]\mathbf{G}[n-1]\mathbf{x}_r[n]}$ with $0.99 \leq \lambda_R \leq 0.999999$
2: **Filtering:** $x_r^F[n] = \mathbf{w}^{\mathrm{H}}[n-1]\mathbf{x}_r[n]$

Error Estimation: $x_c[n-1,n] = x[n] - x_r^F[n]$,
Tap-weight vector adaptation: $\mathbf{w}[n] = \mathbf{w}[n-1] + \mathbf{K}[n]x_c^*[n-1,n]$
Update: $\mathbf{G}[n] = \lambda_R^{-1}(\mathbf{G}[n-1] - \mathbf{K}[n]\mathbf{x}_r^T[n]\mathbf{G}[n-1])$

2.3 Conventional PR target detection approaches

The traditional approach to PR involves assuming that RC provides an approximation of the transmitted signal, which is then used as a matched filter to calculate the cross-ambiguity function (CAF) between the received surveillance signal and the reference signal. This produces a range–Doppler map, which is processed using a 2D CFAR processor to determine the presence or absence of targets. The most commonly used CFAR detection method is the 2D-CA-CFAR, which uses the envelope of the CAF as input. Each sample in the CAF's range and Doppler dimensions is referred to as a cell. In the 2D-CA-CFAR detector, a decision about the presence or absence of a target is made in the cell under test (CUT), and the interference level is estimated using the surrounding cells, known as reference cells. A few nearby cells on each side of the CUT, referred to as guard cells, are excluded from the interference level estimation to prevent possible power spillover from the CUT. It is assumed that G_x range cells and G_y Doppler cells in the surrounding area of the CUT are used as guard cells, where the total range cells (Doppler cells), including reference cells and guard cells, are L_x (L_y). To detect targets, the magnitude of the CAF in the CUT is compared to a threshold formed by multiplying the interference level estimate with a constant η, which is determined by the desired false alarm probability. Figure 2.3 shows the reference cells and guard cells of the 2D-CA-CFAR detector.

The process of PR signal processing can be seen in Figure 2.4 and typically involves several steps. These include beamforming, canceling out direct and reflected

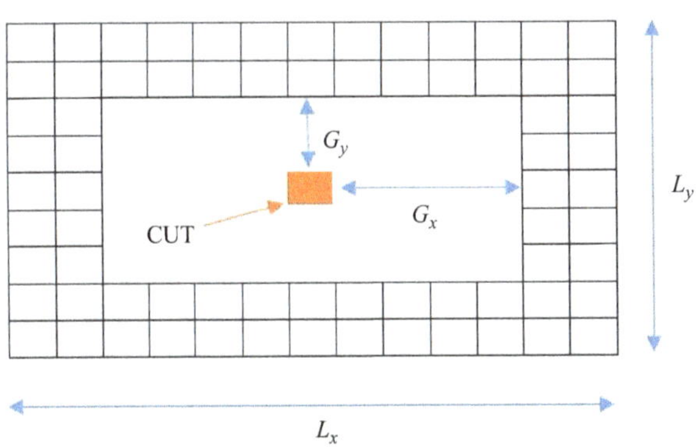

Figure 2.3 Reference cells and guard cells in 2D-CA-CFAR detection, where the columns represent the range dimension and Doppler ones are represented by rows

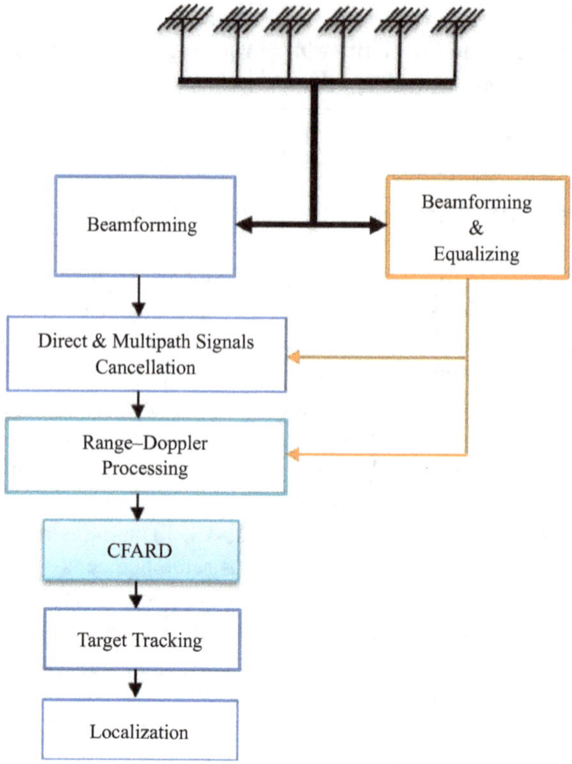

Figure 2.4 *General signal processing steps in PR*

signals, processing the range and Doppler information, performing constant false alarm rate detection (CFARD), tracking targets in the SC as well as the signal conditioning in the RC through techniques such as beamforming and equalization.

2.4 Performance results

In this section, we demonstrate how traditional signal processing techniques in the PR perform in terms of eliminating direct and multipath signals. We start by analyzing the efficiency of ECA, LMS, and RLS methods in removing these interference signals. We evaluate their performance in a situation where there are six targets, as detailed in Table 2.1. Here, the input signal-to-noise ratio (SNR_i) is defined as

$$\mathrm{SNR}_i = \frac{|\alpha|^2}{N_0 B} \tag{2.39}$$

where α is the complex amplitude received from a desired target, N_0 is the noise power per unit bandwidth, and B is the receiver bandwidth. As such, the input noise power is given by $\sigma^2 = N_0 B$. Note that, in our simulation, SNR_i is reported for the receiver bandwidth of 200 kHz. Besides, we assumed that a direct-path signal with

Table 2.1 Target echo parameters for the first scenario (S1)

	T_1	T_2	T_3	T_4	T_5	T_6
R_B, km	5	20	45	70	100	135
f_d, Hz	2.93	−8.8	17.6	−29.3	58.67	88
SNR_i, dB	−36	−35	−33	−31	−30	−36

an input DNR of 45 dB is received in the SC (i.e., $\mathrm{DNR}_s = 45$ dB), while it is 65 dB in the RC (i.e., $\mathrm{DNR}_r = 65$ dB). Note that DNR_s and DNR_r can be defined by replacing the target amplitude of (2.39) with direct-path amplitudes of the SC and RC, respectively. The coherent processing interval or integration time is set equal to one second, say $T = 1$ s. Likewise, we can define the output SNR as

$$\mathrm{SNR}_o = \frac{|\alpha|^2 T}{N_o} \tag{2.40}$$

It is easy to show that the output signal-to-noise ratio (SNR_o) is equal to the SNR_i multiplied by the detector gain (G_p), which is also known as the matched filter gain. The detector gain is given by $G_p = BT$, where B is the bandwidth and T is the observation time. For $B = 200$ kHz and $T = 1$, the detector gain is $G_p = 53$ dB.

To evaluate how well direct-path and multipath removal algorithms work, we establish an interference cancellation (IC) parameter, defined as:

$$\mathrm{IC} = \frac{P_{c,\mathrm{input}}}{P_{c,\mathrm{output}}} \tag{2.41}$$

where the power of the interference signals at the input and output of the adaptive filter is denoted by $P_{c,\,\mathrm{input}}$ and $P_{c,\mathrm{output}}$, respectively. In order to evaluate the effectiveness of the LMS, RLS, and ECA algorithms, a study was conducted in which a six-target scenario was examined under low-SNR conditions, as well as two multipath scenarios. The first multipath scenario assumed that the multipath signals resulted from stationary scatterers, while the second scenario considered moving multipath scatterers.

2.4.1 *Multipath removal capability: stationary multipath scatterers*

Here, we consider a scenario including a surveillance direct-path signal with $\mathrm{DNR}_s = 65$ dB, ten fixed scatterers with $5\,\mathrm{dB} \leq \mathrm{CNR}_i \leq 35\,\mathrm{dB}$ distributed in the range between 0 km and 55 km, and six target echoes whose characteristics are described in Table 2.1. Note that CNR_i is defined by replacing the target amplitude of (2.39) with multipath signal amplitude. In Figure 2.5, the output of the square law of the CAF as the function of bistatic range–Doppler is depicted without applying any interference cancellation algorithms. It is observed that a strong peak corresponding to the direct-path signal located at the bistatic range–Doppler coordinate of (0, 0) is present, resulting in masking the considered targets and the multipath echoes.

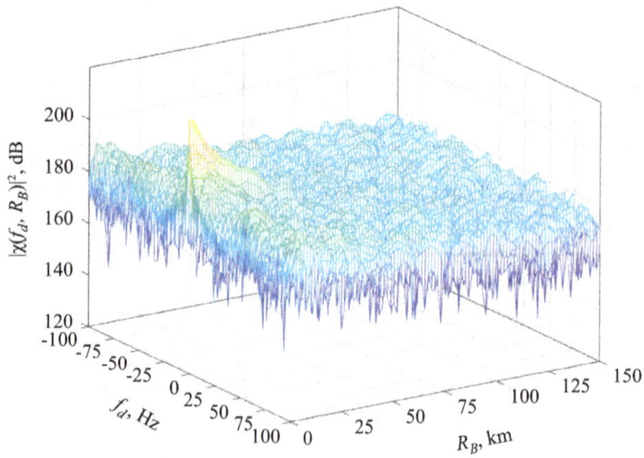

Figure 2.5 Square-law-detected output of the CAF output as a function of bistatic range–Doppler without applying any interference removal algorithms

The choice of removal cancellation algorithm parameters (i.e., μ_x for the LMS algorithm and λ_R for the RLS one) is a key point in algorithm implementation and represents a compromise between convergence rate and removal performance. To do so, first, we use the IC criteria and choose cancellation algorithm parameters to obtain maximum IC values. Due to the low convergence rate of the LMS algorithm, we use a training time, say T_t, to initialize its weighting taps. Indeed, the weighting taps of the LMS are not initialized with zero values, but they are initialized with values obtained at the end of the training time. The results of this simulation are shown in Figures 2.6 and 2.7 for the LMS and RLS algorithm, respectively. From

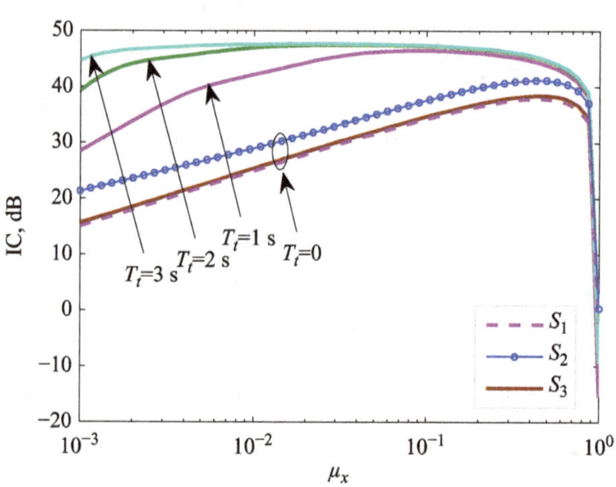

Figure 2.6 IC as a function of μ_x for the LMS algorithm for different values of T_t and different types of the FM signal

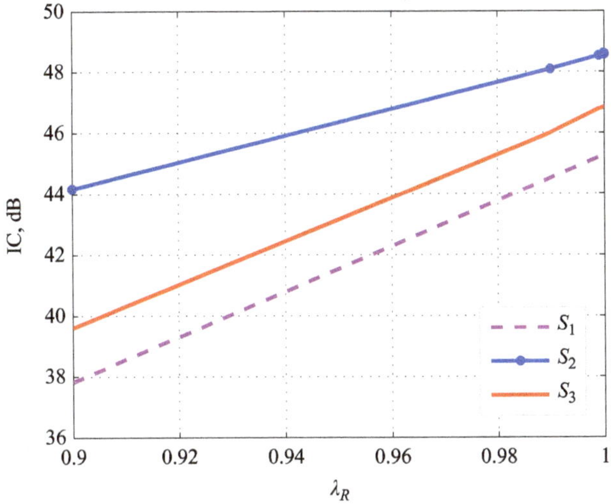

Figure 2.7 IC as a function of μ_x for the RLS algorithm for different types of the FM signal

Figure 2.6, it is seen that the values of μ_x should be selected equal to 0.4, 0.1, and 0.04 to achieve maximum IC for $T_t = 0$, $T_t = 1$, and $T_t > 1$, respectively. For the RLS algorithm, as shown in Figure 2.7, the IC increases as the λ_R increases. In the sequel, we use $\lambda_R = 0.999999$ for the RLS algorithm. It is important to examine the effect of the time-varying feature of the FM signals on the selection of parameters μ_x and λ_R. This is done in Figures 2.6 and 2.7 for different FM signals S_1, S_2, and S_3. It is seen that the obtained values of IC depend on the type of FM signals, but the optimum values of μ_x and λ_R are not changed.

Figure 2.8 shows the output of the square law of the CAF as the function of bistatic range–Doppler evaluated when the ECA is applied. It is seen that the ECA algorithm performs properly, where a deep null appears at the zero Doppler corresponding to the frequencies of the considered stationary multipath echoes. Besides, it is observed that all six targets are easily recognized with IC = 47.81 dB. This simulation for the LMS algorithm is repeated. The obtained values of IC for $T_t = 0$, $T_t = 1$ s, $T_t = 2$ s, and $T_t = 3$ s are, respectively, 38.4 dB, 46.46 dB, 47.48 dB, and 47.63 dB. This shows that the LMS and ECA algorithms can obtain the same values of IC by increasing the value of T_t in the LMS algorithm. To more precisely compare the performance of the ECA and LMS algorithms, we depict the square law of the CAF as the function of the bistatic range–Doppler of the LMS algorithm in Figures 2.9–2.12 for different values of T_t and their corresponding optimum value of μ_x. There are some points in order. First, it is clear that using training time is required to achieve acceptable multipath echo cancellation in the LMS algorithm. Second, it is observed that the range–Doppler coordinates with ranges less than that corresponding to the maximum range of multipath echoes are affected by the LMS algorithm, resulting in detection performance degradation, as discussed in the following. In this case for $T_t \geq 1$ s, it is seen that only four targets with ranges larger than

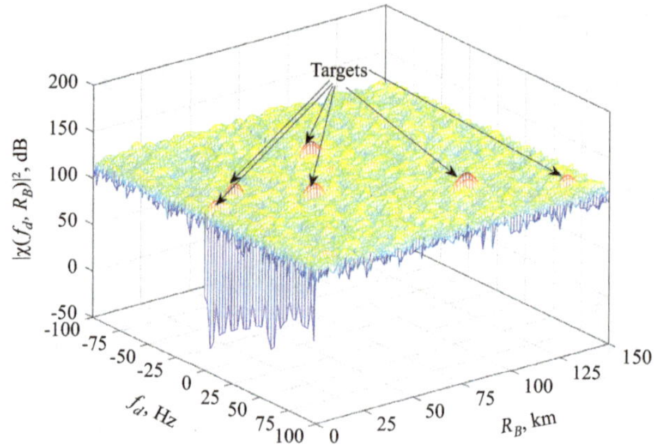

Figure 2.8 Square-law-detected output of the CAF output as a function of bistatic range–Doppler with applying the ECA cancellation algorithm in the presence of stationary multipath signals

that of the maximum multipath range are detected. To examine the effect of the training time on the performance of the LMS algorithm, the absolute value of the weight taps corresponding to that of the multipath echoes as a function of time is shown in Figure 2.13. It is seen that estimated multipath coefficients approach their actual values for $T > 2$ s. This clearly shows the low convergence of the LMS algorithm when it is used for FM-based PR interferenc removal purposes.

For the RLS algorithm, the results are depicted in Figures 2.14 and 2.15 for different values of $\lambda_R = 0.999999$ and $\lambda_R = 0.9995$, respectively. In this case, IC obtains values equal to 46.83 dB and 46.82 dB for $\lambda_R = 0.999999$ and $\lambda_R = 0.9995$,

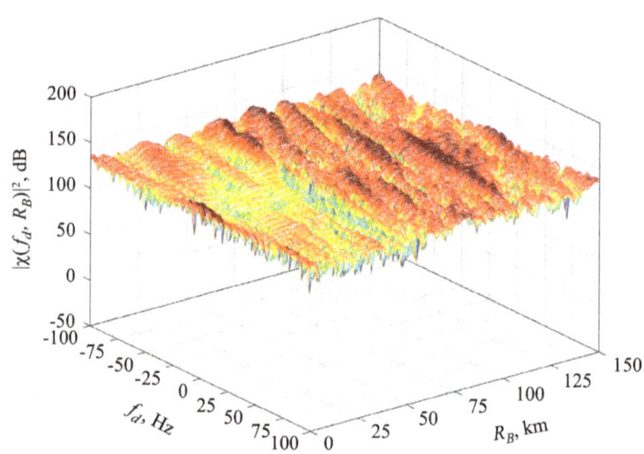

Figure 2.9 Square-law-detected output of the CAF output as a function of bistatic range–Doppler with applying the LMS cancellation algorithm with $\mu_x = 0.5$ and $T_t = 0$ s

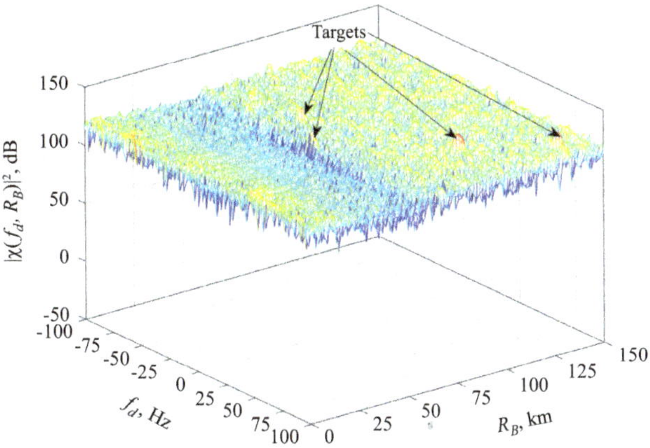

Figure 2.10 Square-law-detected output of the CAF output as a function of bistatic range–Doppler with applying the LMS cancellation algorithm with $\mu_x = 0.1$ and $T_t = 1$ s

respectively. This shows that the RLS algorithm attains lower values of IC, about 1 dB, as compared with the ECA algorithm. This is a reasonable result since the RLS algorithm can be considered as a recursive implementation of the ECA algorithm when we set $\lambda_R = 1$. By comparing the results for different values of λ_R, we observe that the values near 1 result in better IC values and better detection performance. For example, it is seen that all targets are detected when the λ_R is set equal to 0.999999, while only five targets are declared for 0.9995. Similar to Figure 2.13, the absolute value of the weight taps corresponding to that of the multipath echoes as a function of time is shown in Figures 2.16 and 2.17 for

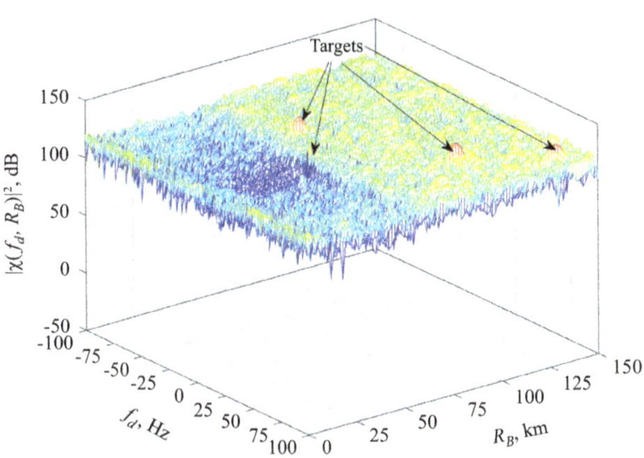

Figure 2.11 Square-law-detected output of the CAF output as a function of bistatic range–Doppler with applying the LMS cancellation algorithm with $\mu_x = 0.04$ and $T_t = 2$ s

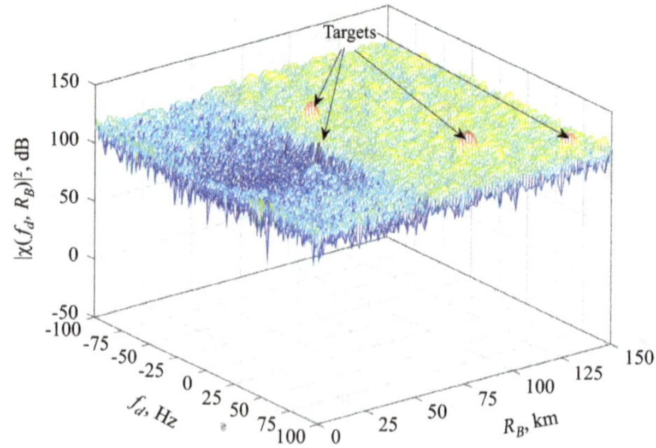

Figure 2.12 *Square-law-detected output of the CAF output as a function of bistatic range–Doppler with applying the LMS cancellation algorithm with $\mu_x = 0.04$ and $T_t = 3$ s*

$\lambda_R = 0.999999$ and $\lambda_R = 0.9995$ of the RLS algorithm, respectively. An acceptable convergence can be seen for $\lambda_R = 0.999999$ value, but it is not for $\lambda_R = 0.9995$, leading to detection performance degradation as shown in Figure 2.17. The zero Doppler cut of the square law of the CAF output as a function of the bistatic range is depicted in Figure 2.18 when the ECA and the RLS cancellation algorithm with $\lambda_R = 0.999999$ are used for interference removal stage. As is apparent a deep null appears at the zero Doppler when the ECA is exploited as compared to the RLS algorithm. Besides, the ranges corresponding to the direct-signal and multipath echoes can be recognized from the peaks seen in Figure 2.18, i.e., one peak at

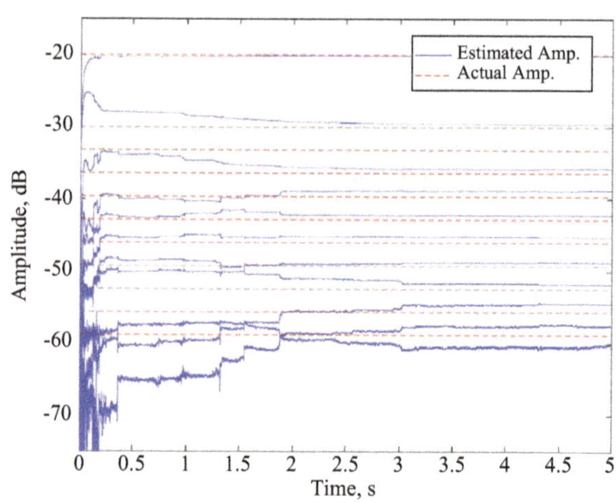

Figure 2.13 *Absolute values of LMS filter's coefficients corresponding to multipath echoes as a function of time*

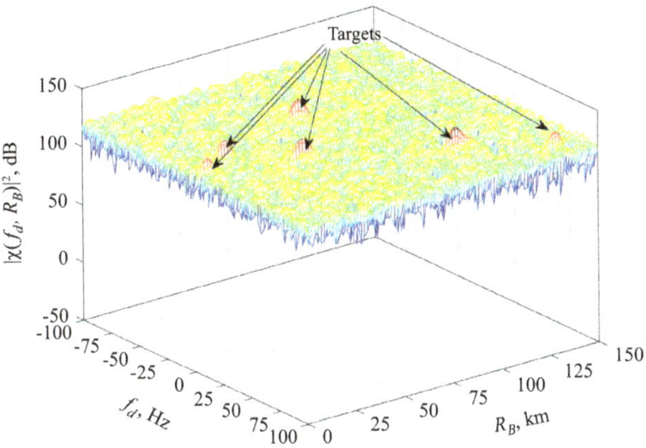

Figure 2.14 *Square-law-detected output of the CAF output as a function of bistatic range–Doppler with applying the RLS cancellation algorithm with* $\lambda_R = 0.999999$

zero bistatic range corresponds to direct-path signal, and the other nine peaks correspond to multipath echoes.

2.4.2 Multipath removal capability: moving multipath scatterers

If we receive echoes from fixed scatterers, assuming constant amplitude of clutter over integration time is a valid assumption even in the presence of multipath or clutter. However, when moving scatterers are present, the clutter's amplitude varies over time. To account for this variation, we assume that each clutter amplitude has an exponential PSD over integration time. The exponential PSD is defined as [1–3]

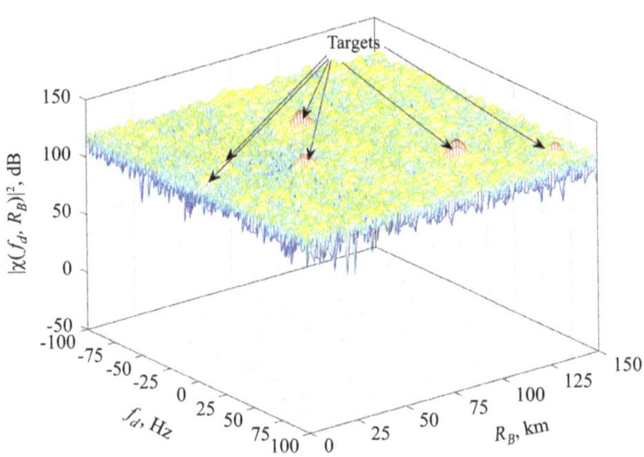

Figure 2.15 *Square-law-detected output of the CAF output as a function of bistatic range–Doppler with applying the RLS cancellation algorithm with* $\lambda_R = 0.9995$

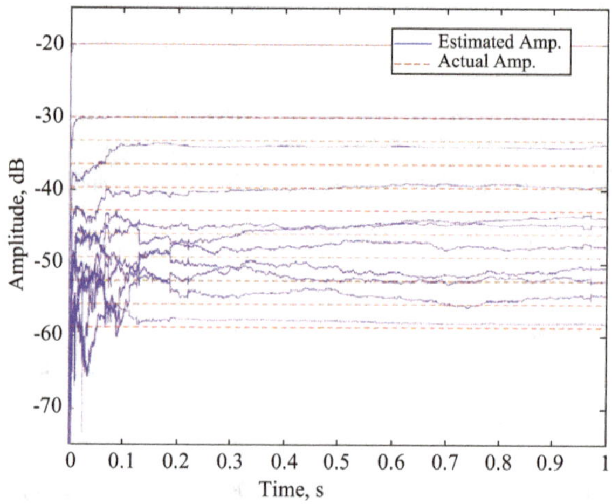

Figure 2.16 Absolute values of RLS filter's coefficients corresponding to multipath echoes as a function of time for $\lambda_R = 0.999999$

$$S(v) = \mathrm{CNR}_i\left[\frac{r}{r+1}\delta(v) + \frac{1}{r+1}\frac{\rho\lambda}{4}\exp(-\rho|v|)\right] \tag{2.42}$$

where v is the clutter velocity in m/s, r is the ratio of DC power to AC power in the spectrum which is dependent on the radar frequency and wind speed, λ is the radar wavelength, and ρ is the shape parameter which is a function of wind speed. In (2.42), $\delta(v)$ is the Dirac delta function that characterizes the steady component of the clutter, while the second term corresponds to the varying components in the

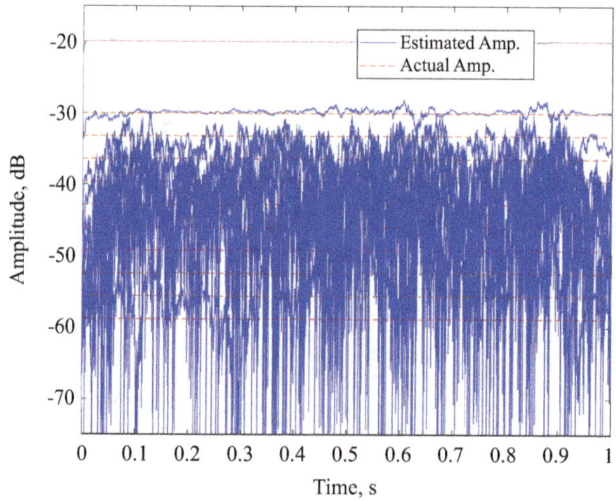

Figure 2.17 Absolute values of RLS filter's coefficients corresponding to multipath echoes as a function of time for $\lambda_R = 0.9995$

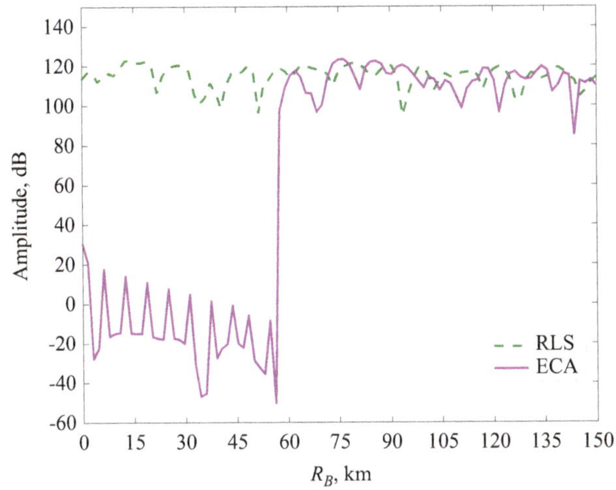

Figure 2.18 *Zero Doppler cut of the square-law-detected output of the CAF output as a function of bistatic bistatic range when the RLS cancellation algorithm with* $\lambda_R = 0.999999$ *and ECA are used for interference removal*

clutter power spectral density. For our simulation, we adopt an exponential power spectral density with parameters of $\rho = 7$ and $r = 90$ for all clutter amplitudes. The other simulation parameters are similar to those used in the previous subsection.

The LMS algorithm's results indicate that it does not perform well when used for FM-based PRs. Therefore, the ECA and RLS algorithms will be the focus of further examination. Without the clutter removal algorithm, Figure 2.19 shows

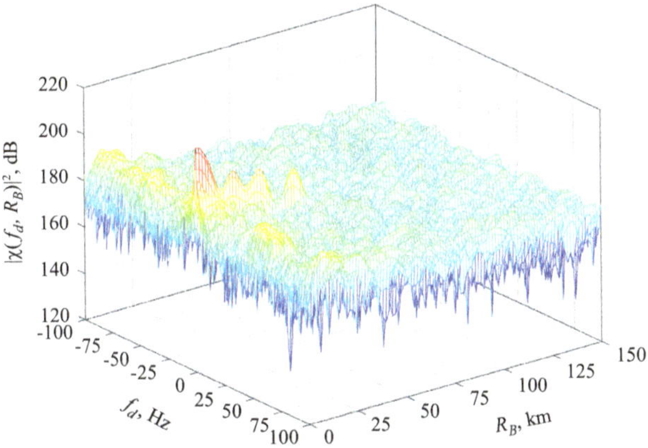

Figure 2.19 *Square-law-detected output of the CAF output as a function of bistatic range–Doppler in the presence of moving multipath signals but in the absence of multipath signal removal algorithm*

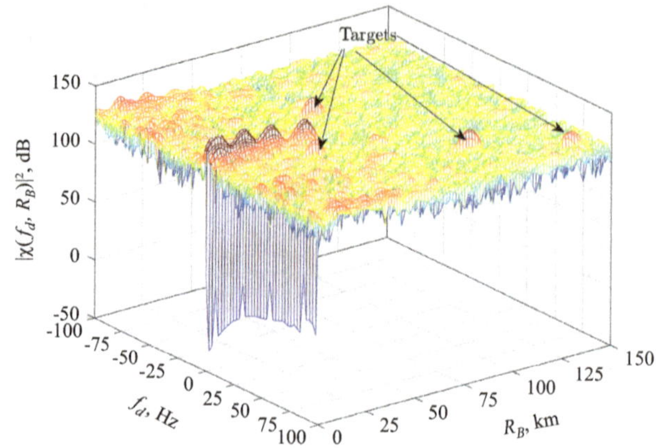

Figure 2.20 *Square-law-detected output of the CAF output as a function of bistatic range–Doppler with applying the ECA cancellation algorithm in the presence of moving multipath signals, and when* $\Delta f_c = 0$

that targets cannot be detected, and the peak of the direct-path and multipath signals is visible. The clutter can be modeled using a specific PSD with a few frequency components around the zero Doppler at frequencies $f_k^{(c)} = (k - 1 - P)\Delta f_c$ for $k = 1, \ldots, 2P + 1$, as described in (2.2). This modeling assumes that clutter removal can be performed by selecting appropriate values of Δf_c and P using the ECA algorithm. Figures 2.20–2.22 depict the ECA clutter removal performance for different values of Δf_c and P, namely $\Delta f_c = 0$, $(\Delta f_c = 0.25, P = 1)$, and

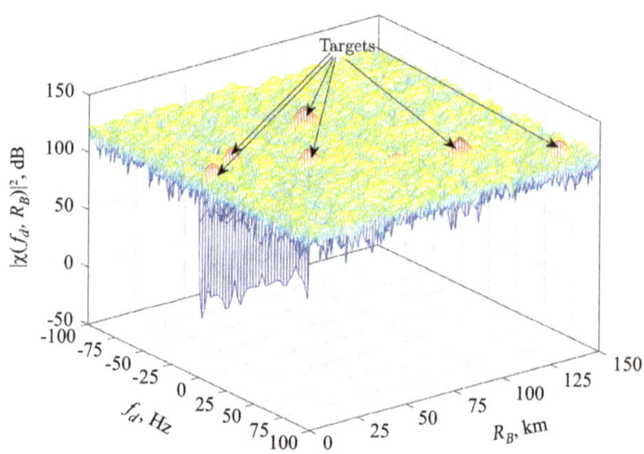

Figure 2.21 *Square-law-detected output of the CAF output as a function of bistatic range–Doppler with applying the ECA cancellation algorithm in the presence of moving multipath signals, and when* $(\Delta f_c = 0.25, P = 1)$

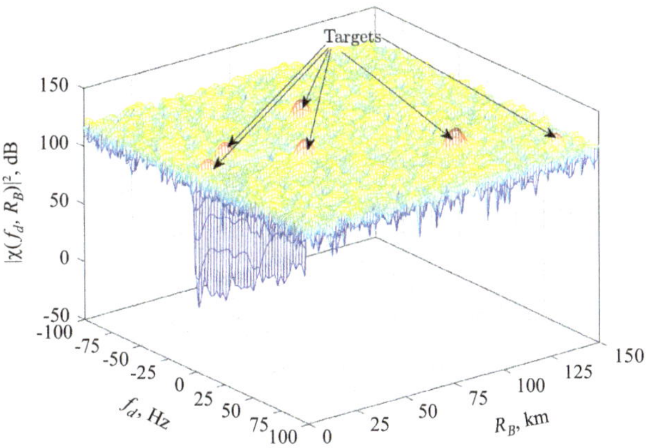

Figure 2.22 *Square-law-detected output of the CAF output as a function of bistatic range–Doppler with applying the ECA cancellation algorithm in the presence of moving multipath signals, and when* $(\Delta f_c = 0.25, P = 2)$

$(\Delta f_c = 0.25, P = 2)$. In Figure 2.20, it is seen that a deep null is inserted in zero Doppler, but the clutter echoes are not removed completely. The results show only four targets are detected, and we obtain IC = 46.18 dB. It is seen from Figure 2.21 with $(\Delta f_c = 0.25, P = 1)$ that the clutter removal capability of the ECA is increased to 46.48 dB, and all six targets can be declared. In the case of $(\Delta f_c = 0.25, P = 2)$, we obtain IC = 46.5 dB and six interested targets are detected, as reported in Figure 2.22. It should be noted that the target T_1 can be removed by increasing the value of P. This implies that the clutter region, defined by parameters Δf_c and P, should be properly determined.

For the RLS algorithm, the results are depicted in Figures 2.23 and 2.24 for different values of $\lambda_R = 0.999999$ and $\lambda_R = 0.9995$, respectively. There are some points in order. First, it is seen that the lower value of λ_R results in a bigger value of IC, where we obtain IC of 45.50 dB and 45.75 dB for λ_R of 0.999999 and 0.9995, respectively. Note that a reverse behavior can be seen when the results of this subsection are compared with that of the previous subsection. Second, if the value of λ_R is set equal to 0.999999, then the clutter components with nonzero frequencies are not removed properly. In this case, as shown in Figure 2.24, the lower value of λ_R equal to 0.9995 results in better multipath signal cancellation and target detection as well. For $\lambda_R = 0.999999$, the number of detected targets is three, while it is six for $\lambda_R = 0.9995$. Besides, for targets with ranges smaller than that of the maximum range of clutter, the target detection performance is degraded by decreasing the value of λ_R. In the case of $\lambda_R = 0.9995$, for example, targets T_1 and T_2 with ranges of 5 km and 20 km are also weakened by the RLS algorithm. Third, ECA's null depth is significantly larger than that of the RLS algorithm.

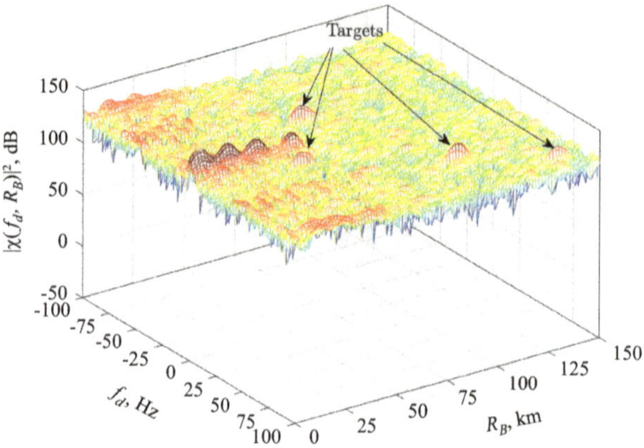

Figure 2.23 *Square-law-detected output of the CAF output as a function of bistatic range–Doppler with applying the RLS cancellation algorithm with $\lambda_R = 0.999999$ in the presence of moving multipath signals*

2.4.3 Detection performance evaluation

First of all, it is important to note that the waveforms utilized in PR systems are not intended for radar applications. To gain a better understanding of this, the range–Doppler coordinate of $(R_{\text{CUT}}, f_{\text{CUT}})$ is considered as the CUT and it is assumed that there is a target at the range–Doppler of $(R_{\text{CUT}} + \delta R_B, f_{\text{CUT}})$ with $\text{SNR}_i = -30$ dB. The purpose of this is to assess false alarm regulation at the CUT in the presence of an interference target located at $(R_{\text{CUT}} + \delta R_B, f_{\text{CUT}})$. This helps us to comprehend

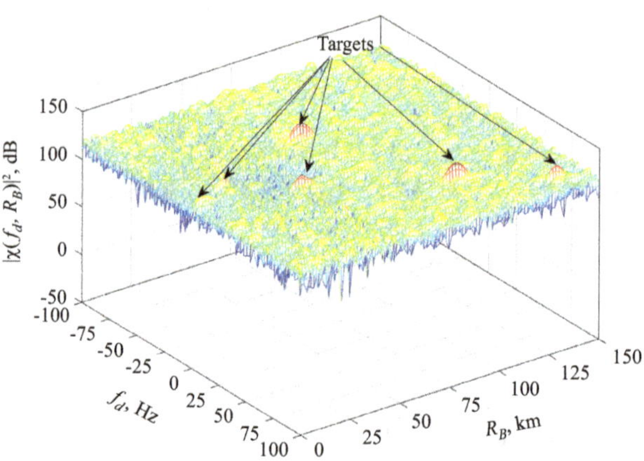

Figure 2.24 *Square-law-detected output of the CAF output as a function of bistatic range–Doppler with applying the RLS cancellation algorithm with $\lambda_R = 0.9995$ in the presence of moving multipath signals*

the impact of using waveforms not designed for radar applications. To accomplish this, we employ 2D-CA-CFAR for target detection after utilizing ECA for clutter removal. In the following, this detection approach is referred to as the ECA-CA detection algorithm. Figure 2.25 shows how the empirical false alarm probability varies with changes in δR_B when using the ECA-CA method with a detection threshold set to achieve a false alarm probability of 10^{-2}. It is seen that the presence of target near to the CUT significantly changes the false alarm regulation, resulting to either excessive false alarm rate or detection performance degradation. This implies that exploiting detection methods like ECA-CA or RLS-CA are not effective approaches for target detection in PRs since the waveforms used for target detection in PR systems are not designed for radar applications. In the following discussion, targets that interfere with the detection of a CUT are referred to as interfering targets. In Reference 6, a clean-based detection method was proposed to eliminate interfering targets similar to clutter echoes, allowing for control of false alarm probability and detection of weak targets near strong ones. However, for example, this method has a limitation to detect and remove a target with $\text{SNR}_i = -30$ dB as an interfering target. Thus, the PR target detection in the presence of strong targets remains unsolved. In the upcoming section, a new detection method based on detection theory framework will be introduced. The proposed detection algorithm will be shown to be a more sophisticated version of the simplified and heuristic approach presented in Reference 6.

In order to assess the detection capabilities of the ECA-CA and RLS-CA methods, a new multi-target scenario named S_2 is introduced, which is characterized in Table 2.2. In this scenario, target T_2 is an interfering target for detecting target T_3 and vice versa. Aditionally, the bistatic ranges of T_1 and T_2 are smaller than the maximum clutter range

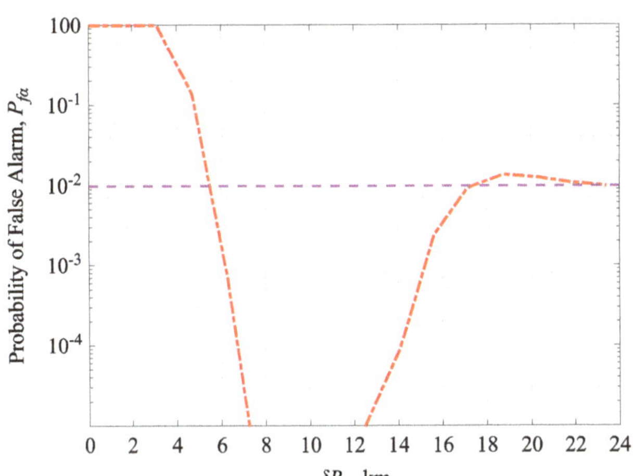

Figure 2.25 Empirical false alarm probability as a function of δR_B when detection threshold of the ECA-CA method is set for desired false alarm probability of 10^{-2}

Table 2.2 Target echo parameters for the second scenario (S2)

Targets	T_1	T_2	T_3	T_4
R_B, km	12.50	40.62	53.12	129.68
f_d, Hz	7.04	−35.2	−35.2	82.13
SNR_i, dB	−31	−28	−33	−30

of 50 km. This allows us to evaluate the capability of interfrence removal of the RLS algorithm on the detection of these targets. Figure 2.26 depicts the detection probability as a function of SNR_o in this scenario. According to the results in Figure 2.26, the RLS-CA method shows a noticeable decline in detection performance for targets T_1, T_2, and T_3, while the ECA-CA method performs significantly better than the RLS-CA method. This disparity is primarily due to the characteristics of the RLS algorithm, which affect range-Doppler cells with ranges that are less than the maximum clutter range during the removal of multipath signals. Additionally, the impact of interfering targets T_2 and T_3 on each other's detection performance is evident from Figure 2.26. The detection performance of T_3 is more negatively affected than that of T_2 since the latter has a smaller SNR. However, both ECA-CA and RLS-CA methods exhibit similar detection performance for detecting the target T_4 as it is located far from the clutter region and other targets. In scenarios with multiple targets, both ECA-CA and RLS-CA methods experience significant detection performance degradation due to power spillover from the test cell and its surrounding cells, which are used to estimate the interference level.

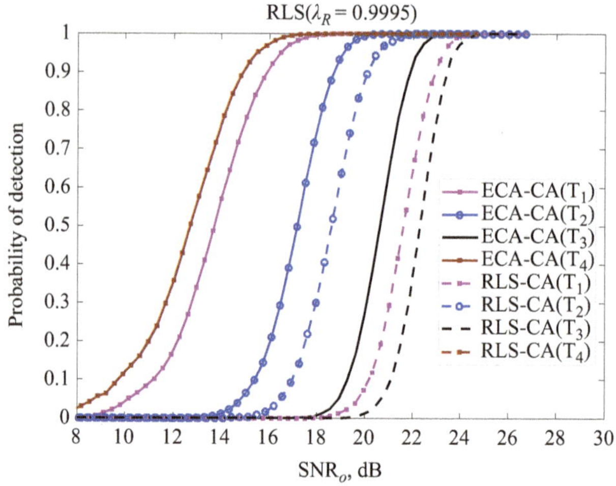

Figure 2.26 Probability of detection as a function of SNR_o for two detection approaches ECA-CA and RLS-CA with $\lambda_R = 0.9995$ and multiple target scenario with characteristics listed in Table 2.2, and when the desired false alarm probability is set equal to 10^{-3}

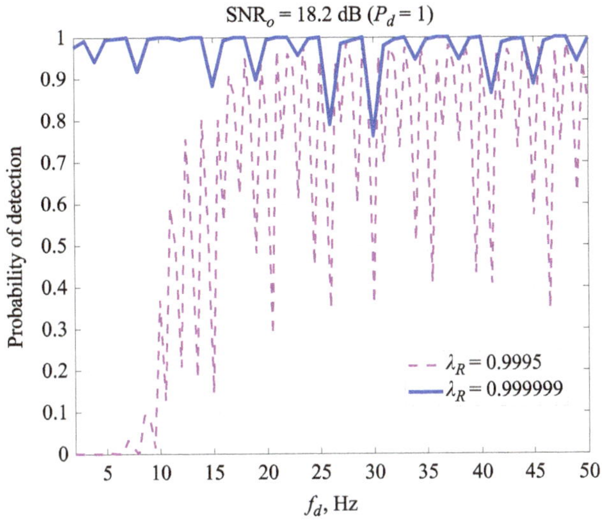

Figure 2.27 *Detection probability as a function of Doppler frequency for different values of λ_R of the RLS-CA detection approach with SNR$_o$ = 18.20 dB*

To get more insight about target performance degradation of the RLS-based method on range–Doppler coordinates with ranges smaller than that of the maximum range of clutter, we depict the curve of detection probability as a function of Doppler frequency for two values of λ_R. The results of this simulation are shown in Figure 2.27. It is seen that this attenuation is more when we adjust the RLS algorithm with $\lambda_R = 0.9995$ as compared to that of $\lambda_R = 0.99999$, where the first results in more values of IC, as discussed before. Thus, more multipath cancellation leads to more target detection performance degradation in the low-Doppler frequency region.

2.5 Summary

In this chapter, we first model the received signal in the surveillance and reference of a bistatic passive radar. Then, we classify passive radar target detection techniques into two main categories, based on whether a reference channel is employed or not. In this chapter, we concentrate on the conventional signal processing approaches in passive radars and provide extensive simulation results to show that these methods are not optimal for the passive radar detection problem. Therefore, it is necessary to devise a target detection algorithm that considers the nature of opportunity signals.

References

[1] Zaimbashi A. 2013. Target detection in passive radars based on commercial FM radio signals. [PhD thesis]. Iran, Shiraz University.

[2] Zaimbashi A., Derakhtian M., Sheikhi A. 'GLRT-based CFAR detection in passive bistatic radar'. *IEEE Transactions on Aerospace and Electronic Systems*. 2013, vol. 49(1), pp. 134–59.

[3] Zaimbashi A., Derakhtian M., Sheikhi A. 'Invariant target detection in multiband FM-based passive bistatic radar'. *IEEE Transactions on Aerospace and Electronic Systems.* 2014, vol. 50(1), pp. 720–36.

[4] Colone F., Cardinali R., Lombardo P, *et al.* 'Space-time constant modulus algorithm for multipath removal on the reference signal exploited by passive bistatic radar'. *IET Radar, Sonar and Navigation*. 2009, vol. 3(3), pp. 253–64.

[5] Zaimbashi A. 'Target detection in FM-based passive radars in the presence of interference signals in reference channel'. *Journal of "Radar."* 2016, vol. 10(2), pp. 1–15.

[6] Colone F., O'Hagan D.W., Lombardo P., Baker C.J. 'A multistage processing algorithm for disturbance removal and target detection in passive Bistatic radar'. *IEEE Transactions on Aerospace and Electronic Systems*. 2009, vol. 45(3), pp. 698–722.

[7] Zaimbashi A. 'Multiband FM-based passive bistatic radar: target range resolution improvement'. *IET Radar, Sonar and Navigation*. 2016, vol. 10(1), pp. 174–85.

[8] Zaimbashi A. 'Broadband target detection algorithm in FM-based passive bistatic radar systems'. *IET Radar, Sonar and Navigation*. 2016, vol. 10(8), pp. 1485–99.

[9] Zaimbashi A. 'Forward M-ary hypothesis testing approach for target detection in FM and DVB-T-based passive bistatic radars'. *IEEE Transactions on Signal Processing*. 2017, vol. 65(10), pp. 2659–71.

[10] Zaimbashi A. 'Target detection in analog terrestrial TV-based passive radar sensor: joint delay-Doppler estimation'. *IEEE Sensors Journal*. 2017, vol. 17(17), pp. 5569–80.

[11] Zaimbashi A. 'Target detection in FM-based passive radars in the presence of interference signals in reference channel'. *Journal of "Radar"*. 2015, vol. 3(3), pp. 1–15.

[12] Solatzadeh Z., Zaimbashi A. 'Accelerating target detection in passive radar sensors: delay-Doppler-acceleration estimation'. *IEEE Sensors Journal*. 2018, vol. 18(13), pp. 5445–54.

[13] Zaimbashi A. 'Integration gain limitations in passive coherent location radars for manoeuvring target detection'. *Electronics Letters*. 2016, vol. 52(21), pp. 1801–04.

[14] Zaimbashi A. 'Multiband FM-based PBR system in presence of model mismatch'. *Electronics Letters*. 2016, vol. 52(18), pp. 1563–65.

[15] Zaimbashi A., Derakhtian M., Sheikhi A. 'Evaluation of detection performance of passive bistatic radar detectors based on commercial FM radio signals'. *Journal of "Radar."* 2014, vol. 1(2), pp. 23–34.

[16] Griffiths H.D., Long N.R.W. 'Television based Bistatic radar'. *IEE Proceedings F Communications, Radar and Signal Processing*. 1986, vol. 133(7), pp. 649–57.

[17] Howland P.E. 'A passive metric radar using a transmitter of opportunity'. *International 1994 Conference on Radar*; Paris, 1994. pp. 251–56.

[18] Howland P.E., Maksimiuk D., Reitsma G. 'FM radio based bistatic radar'. *IEE Proceedings on Radar, Sonar and Navigation*. 2005, vol. 152(3), pp. 107–15.

[19] Kulpa K.S., Czekała Z. 'Masking effect and its removal in PCL radar'. *IEE Proceedings on Radar, Sonar and Navigation*. 2005, vol. 152(3), pp. 174–78.

[20] Colone F., Cardinali R., Lombardo P. 'Cancellation of clutter and multipath in passive radar using a sequential approach'. *IEEE 2006 Radar Conference*; Verona, NY, 2006. pp. 393–99.

[21] Cardinali R., Colone F., Ferretti C., Lombardo P. 'Comparison of clutter and multipath cancellation techniques for passive radar'. *IEEE 2007 Radar Conference*; Boston, MA, 2007. pp. 469–74.

[22] Griffiths H.D., Baker C.J. 'Passive coherent location radar systems. Part 1: Performance prediction'. *IEE Proceedings – Radar, Sonar and Navigation*. 2005, vol. 152(3), pp. 153–59.

[23] Baker C.J., Griffiths H.D., Papoutsis I. 'Passive coherent location radar systems. Part 2: Waveform properties'. *IEE Proceedings on Radar, Sonar and Navigation*. 2005, vol. 152(3), pp. 160–168.

[24] Hack D.E., Patton L.K., Himed B., Saville M.A. 'A detection in passive MIMO radar networks'. *IEEE Transactions on Signal Processing*. 2014, vol. 62(11), pp. 2999–3012.

[25] Cui G., Liu J., Li H., Himed B. 'Signal detection with noisy reference for passive sensing'. *Signal Processing*. 2015, vol. 108, pp. 389–99.

[26] Gogineni S., Setlur P., Rangaswamy M., Nadakuditi R.R. 'Passive radar detection with noisy reference channel using principal subspace similarity'. *IEEE Transactions on Aerospace and Electronic Systems*. 2018, vol. 54(1), pp. 18–36.

[27] Zaimbashi A. 'Multistatic passive radar sensing algorithms with calibrated receivers'. *IEEE Sensors Journal*. 2020, vol. 20(14), pp. 7878–85.

[28] Javidan M.H., Zaimbashi A., Liu J. 'Target detection in passive radar under noisy reference channel: a new threshold-setting strategy'. *IEEE Transactions on Aerospace and Electronic Systems*. 2020, vol. 56(6), pp. 4711–22.

[29] Liu J., Li H., Himed B. 'Two target detection algorithms for passive multistatic radar'. *IEEE Transactions on Signal Processing*. 2014, vol. 62(22), pp. 5930–39.

[30] Hack D.E., Patton L.K., Himed B., Saville M.A. 'Centralized passive MIMO radar detection without direct-path reference signals'. *IEEE Transactions on Signal Processing*. 2014, vol. 62(11), pp. 3013–23.

[31] Zhang X., Li H., Himed B. 'Multistatic detection for passive radar with direct-path interference'. *IEEE Transactions on Aerospace and Electronic Systems*. 2017, vol. 53(2), pp. 915–25.

[32] Zhang X., Li H., Himed B. 'A direct-path interference resistant passive detector'. *IEEE Signal Processing Letters*. 2017, vol. 24(6), pp. 818–22.

[33] Zhang X., Li H., Himed B. 'Multistatic detection for passive radar with direct-path interference'. *IEEE Transactions on Aerospace and Electronic Systems*. 2017, vol. 53(2), pp. 915–25.

[34] Zhang X., Li H., Himed B. 'Multistatic passive detection with parametric modeling of the IO waveform'. *Signal Processing*. 2017, vol. 141, pp. 187–98.

[35] Zaimbashi A. 'A unified framework for multistatic passive radar target detection under uncalibrated receivers'. *IEEE Transactions on Signal Processing*. 2021, vol. 69(1), pp. 695–708.

[36] Zaimbashi A., Greco M.S. 'Multistatic passive radar target detection under uncalibrated receivers with direct-path interference'. *IEEE Transactions on Aerospace and Electronic Systems*. 2022, vol. 58(6), pp. 5443–55.

[37] Garry J.L., Baker C.J., Smith G.E. 'Evaluation of direct signal suppression for passive radar'. *IEEE Transactions on Geoscience and Remote Sensing*. 2017, vol. 55(7), pp. 3786–99.

[38] Farhang-Boroujeny B. 'Adaptive filters: Theory and applications' in John Wiley and Sons; 2013 Apr. 19.

[39] Haykin S.O. 'Adaptive filter theory' in 4th ed. Englewood Cliffs, NJ, USA: Prentice-Hall; 2001.

Part II

Target detection in passive bistatic radar under high-SNR reference channel

Passive bistatic radar (PBR) systems commonly employ a reference channel (RC) to capture transmitted signal and a surveillance channel (SC) to receive echoes from the interested targets. In the first category of passive radar target detection approaches, it is assumed that the RC provides a high direct-path signal-to-noise ratio (SNR). This is often referred to as the ideal reference (IR) channel or high-SNR RC. This assumption allows for efficient target detection in the SC, even in the presence of the receiver noise, the direct-path signal, multipath/clutter echoes, and interfering targets. All target detection algorithms based on this assumption fall into the first category, known as Ca.1-IR approaches.

In this part, we provide a comprehensive review of detection-theory-based target detection approaches in Ca.1-IR. Specifically, this part is structured in the following way: Chapter 3 explores the problem of detecting multiple targets in a single-channel FM-based PBR. Chapters 4 and 5 investigate this problem in the context of multiband FM-based PBR to enhance target detection quality and target range resolution, respectively. In Chapter 5, a novel approach for multiband target detection is proposed, which simultaneously improves target detection quality and target range resolution. The multi-target detection problem in passive radar is then modeled as an M-ary hypothesis testing problem and presented in Chapters 6 and 7 for target detection in FM-, DVB-T-, and analog TV-based passive radars. Finally, Chapter 8 focuses on multi-accelerating-target detection problem of passive radars.

Chapter 3

Multitarget detection problem in single-band FM-based passive radar

3.1 Introduction

The design of transmitted waveforms in active radars is carefully tailored to meet the operational modes and requirements, whereas passive radars (PRs) rely on non-cooperative illuminators of opportunity, which may not be specifically designed for radar applications. Consequently, PRs may struggle with waveforms that are not optimized for their needs, which can result in target echoes being obscured by echoes from other strong targets in multitarget scenarios. This chapter focuses on the detection of multiple targets in a single-band FM-based PR. Specifically, the chapter only covers recent research developments in the Ca.1-IR category and examines the problem of detecting multiple targets in the presence of interfering signals, such as direct-path signals, multipath echoes, interfering targets, and thermal noise.

Most research on Ca.1-IR has typically focused on detecting a single target or a limited set of interference signals using intuitive methods. These conventional detection processes in PRs rely on thresholding the delay-Doppler cross-correlation function (CCF) between the cleaned surveillance channel signal and reference channel signal. However, these methods lack optimality. In contrast, in this chapter, the authors use detection theory concepts to address the more complex problem of detecting a target in the presence of direct-path signal, multipath echoes, interfering targets, and thermal noise. The problem of detecting targets is formulated as a composite hypothesis-testing problem, and the generalized likelihood ratio test (GLRT) criterion is used to solve it. The detection probability and false alarm probability are obtained through closed-form expressions, which reveal how the low peak-sidelobe of the transmitted FM signals can impact the detection performance of other targets of interest. In addition, a comprehensive evaluation of the performance of the detector is provided, demonstrating its superior performance in both detection accuracy and false alarm control. The detector is shown to effectively cancel out direct signals, multipath echoes, and interfering targets, and is compared with classical approaches in the Ca.1-IR. The materials of this chapter are based on the findings published in Reference 1.

The arrangement of this chapter is as follows: The problem of detecting targets using FM-based single-band PR is introduced in section 3.2, while the analytical performance analysis is provided in section 3.3. Section 3.4 evaluates the effectiveness of the FM-based single-band PR target detection algorithm. Finally, a summary is presented in section 3.5.

3.2 High-SNR RC-based problem formulation

The surveillance channel (SC) received signal of (2.4) versus noisy reference channel (RC) signal of (2.6) can be rewritten as

$$\begin{aligned} x[n] &= \tfrac{\beta_0}{c_r} y[n] + \textstyle\sum_{i=1}^{N_c} \sum_{k=-Q}^{Q} \tfrac{\beta_{i,k}}{c_r} y[n-n_i^{(co)}]e^{j2\pi f_k^{(c)} T_s n} \\ &+ \textstyle\sum_{m=0}^{K-1} \tfrac{a_m}{c_r} y[n-n_m^{(t)}]e^{jn\Omega_m^{(t)}} + \tfrac{\beta_0}{c_r} n_r[n] \\ &+ \textstyle\sum_{i=1}^{N_c} \sum_{k=-Q}^{k=Q} \tfrac{\beta_{i,k}}{c_r} n_r[n-n_i^{(co)}]e^{j2\pi f_k^{(c)} T_s n} \\ &+ \textstyle\sum_{m=0}^{K-1} \tfrac{a_m}{c_r} n_r[n-n_m^{(t)}]e^{jn\Omega_m^{(m)}} + n_s[n] \end{aligned} \tag{3.1}$$

The RC antenna is assumed to be directional and pointed toward the transmitter, while the SC antenna has a broad beam pointed at the surveyed sector. As a result, the direct-path signal is received by the mainlobe of the RC antenna, but it is received by the sidelobes/backlobes of the broad-beam SC antenna, leading to a much smaller value of $\frac{\beta_0}{c_r} \ll 1$. It is also assumed that the direct-path signal-to-noise ratio (SNR) is high, indicating a high-SNR RC (ideal RC) condition, allowing us to ignore the $\frac{\beta_0}{c_r} n_r[n]$ term compared with $n_s[n]$. Furthermore, due to receiving the reflections of the direct signal from targets or clutter scatterers, we can also conclude that $\frac{\beta_{i,k}}{c_r} \ll 1$ and $\frac{a_m}{c_r} \ll 1$. In addition, if we make the assumption that the noise floor is the same in both the RCs and SCs, we can assume that the sum of the last four terms in (3.1) is approximately equal to $n_s[n]$ Reference 2. Therefore, we can represent (3.1) as follows:

$$\begin{aligned} x[n] &= c_1 y[n] + \textstyle\sum_{i=1}^{N_c} \sum_{k=-Q}^{Q} c_{i,k} y[n-n_i^{(co)}]e^{j2\pi f_k^{(c)} T_s n} \\ &+ \textstyle\sum_{m=0}^{K-1} \alpha_m^{(t)} y[n-n_m^{(t)}]e^{jn\Omega_m^{(t)}} + n_s[n] \end{aligned} \tag{3.2}$$

where $c_1 \triangleq \frac{\beta_0}{c_r}$, $c_{i,k} \triangleq \frac{\beta_{i,k}}{c_r}$, and $\alpha_m^{(t)} \triangleq \frac{a_m}{c_r}$. By defining matrices $\mathbf{\Lambda}^{(c)}$ and $\mathbf{D}^{(c)}$ such that

$$[\mathbf{\Lambda}^{(c)}]_{ij} = 2\pi f_i^{(c)} T_s \qquad i = 1, \ldots, 2Q+1, \quad j = 1, \ldots, N_c \tag{3.3}$$

$$[\mathbf{D}^{(c)}]_{ij} = n_j^{(co)} \qquad i = 1, \ldots, 2Q+1, \quad j = 1, \ldots, N_c \tag{3.4}$$

and by defining vectors $\mathbf{\Omega}^{(c)} \triangleq [0, \text{vec}(\mathbf{\Lambda}^{(c)})^T]^T$ and $\mathbf{n}^{(c)} \triangleq [0, \text{vec}(\mathbf{D}^{(c)})^T]^T$, where vec(.) stacks the column of its argument in a column vector, we can express (3.2) as

$$x[n] = \sum_{k=1}^{P} c_k y[n-n_k^{(c)}]e^{jn\Omega_k^{(c)}} + \sum_{m=0}^{K-1} \alpha_m^{(t)} y[n-n_m^{(t)}]e^{jn\Omega_m^{(t)}} + n_s[n] \tag{3.5}$$

where $P = (2Q+1)N_c + 1$, and $\Omega_k^{(c)}$ and $n_k^{(c)}$ are the kth elements of vector $\mathbf{\Omega}^{(c)}$ and $\mathbf{n}^{(c)}$, respectively. Note that, in (3.4) c_k's for $k = 2, \ldots, P$ are the clutter complex amplitudes corresponding to delay-Doppler coordinates $(n_k^{(c)}, \Omega_k^{(c)})$, and c_1 is the complex amplitude of the direct signal received in the SC.

Consider testing for the presence of a target within a delay-normalized Doppler frequency cell (n_0, Ω_0). This composite detection problem may be formulated as a binary hypothesis test, given by

$$\begin{cases} \mathcal{H}_0: x[n] = \sum_{k=1}^{P} c_k y[n - n_k^{(c)}] e^{jn\Omega_k^{(c)}} + \sum_{m=1}^{K-1} \alpha_m^{(t)} y[n - n_m^{(t)}] e^{jn\Omega_m^{(t)}} \\ \qquad + n_s[n] \\ \mathcal{H}_1: x[n] = \alpha_0 y[n - n_0] e^{jn\Omega_0} + \sum_{k=1}^{P} c_k y[n - n_k^{(c)}] e^{jn\Omega_k^{(c)}} \\ \qquad + \sum_{m=1}^{K-1} \alpha_m^{(t)} y[n - n_m^{(t)}] e^{jn\Omega_m^{(t)}} + n_s[n] \end{cases} \tag{3.6}$$

with $n = 0, \ldots, N-1$. To solve the detection problem, the following assumptions are made:

1. The scenario involves $(K-1)$ interfering targets whose exact number K is unknown. The signal amplitudes of these targets, denoted as $\alpha_m^{(t)}$, are considered complex-valued, deterministic, and unknown. Additionally, the delays and normalized Doppler frequencies of these interfering targets, represented as $n_m^{(t)}$ and $\Omega_m^{(t)}$, respectively, are also unknown. Meanwhile, the complex amplitude of the hypothesized target, denoted as $\alpha_0^{(t)}$ and labeled as α_0, is also considered unknown.
2. The amplitudes of clutter/multipath signals, denoted by c_i for $i = 2, \ldots, P$, as well as the amplitude of the direct signal, denoted by c_1, are assumed to be complex-valued, deterministic, and unknown. The delays of clutter/multipath signals, denoted by $n_k^{(c)}$, and their normalized Doppler frequencies, denoted by $\Omega_k^{(c)}$ for $k = 1, \ldots, P$, are also assumed to be unknown. However, it is reasonable to assume that the multipath and ground clutter scatterers are generated by nearby ground scatterers within a certain bistatic range, denoted by $R^{(c)}$, and that their band-limited spectra are centered around zero frequency. The values of $R^{(c)}$ and the bandwidth of the clutter spectrum depend on various factors such as the radar location, radar frequency, and wind speed. Thus, it is practical to consider a limited range of delays and Doppler shifts, referred to as the clutter/multipath region. Since there may be clutter components in the clutter region with any unknown combination of delays and Doppler shifts, it is necessary to consider all possible delay-Doppler coordinates in this region for designing the detector.
3. It is assumed that the samples of $n_s[n]$ are white noise with a complex Gaussian distribution and the variance σ^2 is unknown.

Now, we need to specify the probability density function (pdf) of the complex observation $\mathbf{x} \triangleq [x[0], x[1], \ldots, x[N-1]]^T$ under both hypotheses, given by Reference 2:

$$f(\mathbf{x}; \mathscr{H}_0) = \frac{1}{(\pi\sigma^2)^N} e^{\left(-\frac{1}{\sigma^2}\Sigma_{n=0}^{N-1}\left|x[n] - x_c[n] - x_l[n]\right|^2\right)} \tag{3.7}$$

$$f(\mathbf{x}; \mathscr{H}_1) = \frac{1}{(\pi\sigma^2)^N} e^{\left(-\frac{1}{\sigma^2}\Sigma_{n=0}^{N-1}\left|x[n] - x_c[n] - x_l[n] - \alpha_0 y[n-n_0]e^{jn\Omega_0}\right|^2\right)} \tag{3.8}$$

where

$$x_l[n] = \sum_{m=1}^{K-1} \alpha_m^{(t)} y[n - n_m^{(t)}] e^{jn\Omega_m^{(t)}} \tag{3.9}$$

$$x_c[n] = \sum_{k=1}^{P} c_k y[n - n_k^{(c)}] e^{jn\Omega_k^{(c)}} \tag{3.10}$$

Based on the Neyman–Pearson criterion, the best solution to the problem of hypothesis testing in (3.6) is the likelihood ratio test. However, this test cannot be utilized in practical situations because it requires knowledge of various parameters, such as $\mathbf{c} = [c_1, \ldots, c_P]^T$, $\alpha^{(t)} = [\alpha_1^{(t)}, \ldots, \alpha_{K-1}^{(t)}]^T$, α_0, $\{n_m^{(t)}\}_{m=1}^{K-1}$, $\{\Omega_m^{(t)}\}_{m=1}^{K-1}$, and σ^2, which are typically unknown. To overcome this limitation, one can use the GLRT, as described in References 3 and 4. The GLRT is similar to the likelihood ratio test, but it replaces the unknown parameters with their maximum likelihood estimates. In the following sections, we derive the GLRT-based detector for three different cases: (1) when σ^2 is known and there are no interfering targets ($K = 1$), (2) when σ^2 is unknown and there are no interfering targets ($K = 1$), and (3) when there are $(K - 1)$ interfering targets and σ^2 is unknown, using a two-step GLRT-based detector.

3.2.1 GLRT-based detector: known σ^2 and when $K = 1$

In this section, the GLRT for a non-interfering target scenario with a known σ^2 is presented. To obtain the GLRT, the unknown parameters in each hypothesis are substituted with their maximum likelihood estimations (MLEs) specific to that hypothesis. This process can be expressed as

$$L_{\mathrm{GLR}}(\mathbf{x}) = \frac{\max_{\alpha_0,\mathbf{c}} f(\mathbf{x}; \alpha_0, \mathbf{c}, \mathscr{H}_1)}{\max_{\mathbf{c}} f(\mathbf{x}; \mathbf{c}, \mathscr{H}_0)} \underset{\mathscr{H}_0}{\overset{\mathscr{H}_1}{\gtrless}} \eta \tag{3.11}$$

To ensure the probability of false alarm is met, the threshold η is chosen. To obtain the GLRT, we first need to calculate the MLEs of the unknown parameters for each hypothesis. The MLEs of the unknown parameters $\mathbf{c} = [c_1, \ldots, c_P]^T$ and α_0 under $\mathscr{H}_1$ can be obtained by taking the derivative of the exponential argument in (3.8) with respect to the unknown parameters α_0 and c_q for $q = 1, \ldots, P$, and setting the derivative to zero, resulting in

$$\hat{\alpha}_0 \sum_{n=0}^{N-1} |y[n-n_0]|^2 + \sum_{k=1}^{P} \hat{c}_{k1} \sum_{n=0}^{N-1} y[n-n_k^{(c)}]y^*[n-n_0]e^{jn(\Omega_k^{(c)}-\Omega_0)} \\ = \sum_{n=0}^{N-1} x[n]y^*[n-n_0]e^{-jn\Omega_0} \tag{3.12}$$

and

$$\sum_{n=0}^{N-1} x[n]y^*[n-n_q^{(c)}]e^{-jn\Omega_q^{(c)}} = \hat{\alpha}_0 \sum_{n=0}^{N-1} y[n-n_0]y^*[n-n_q^{(c)}]e^{jn(\Omega_0-\Omega_q^{(c)})} \\ + \sum_{k=1}^{P} \hat{c}_{k1} \sum_{n=0}^{N-1} y[n-n_k^{(c)}]y^*[n-n_q^{(c)}]e^{jn(\Omega_k^{(c)}-\Omega_q^{(c)})} \tag{3.13}$$

where $q = 1, \ldots, P$, and hats denote estimators and the second subscripts denote the hypothesis to which it belongs. Similarly, under $\mathscr{H}_0$, maximizing the related probability density function (pdf) by differentiating the exponential argument in (3.7) with respect to c_q for $q = 1, \ldots, P$, and setting the derivative to zeros, which yields

$$\sum_{k=1}^{P} \hat{c}_{k0} \sum_{n=0}^{N-1} y[n-n_k^{(c)}]y^*[n-n_q^{(c)}]e^{jn(\Omega_k^{(c)}-\Omega_q^{(c)})} \\ = \sum_{n=0}^{N-1} x[n]y^*[n-n_q^{(c)}]e^{-jn\Omega_q^{(c)}} \quad q = 1, \ldots, P \tag{3.14}$$

we can rewrite (3.12) and (3.14) in matrix form respectively, given by

$$\begin{bmatrix} \mathbf{R}_c & \mathbf{r}_{sc} \\ \mathbf{r}_{sc}^H & r_{ss} \end{bmatrix} \begin{bmatrix} \hat{\mathbf{c}}_1 \\ \hat{\alpha}_0 \end{bmatrix} = \begin{bmatrix} \mathbf{r}_{xc} \\ r_{xs} \end{bmatrix} \tag{3.15}$$

$$\mathbf{R}_c \hat{\mathbf{c}}_0 = \mathbf{r}_{xc}. \tag{3.16}$$

In (3.15), $\hat{\mathbf{c}}_1$ and $\hat{\alpha}_0$ are the MLEs of $\mathbf{c}$ and α_0 under $\mathscr{H}_1$ hypothesis. Similarly in (3.16), $\hat{\mathbf{c}}_0$ is the MLE of $\mathbf{c}$ under $\mathscr{H}_0$ hypothesis. In this case, $\mathbf{R}_c$ can be thought as a $P \times P$ related clutter/multipath correlation matrix whose the (q, k) entry $[\mathbf{R}_c]_{qk}$ is defined as

$$[\mathbf{R}_c]_{qk} = \sum_{n=0}^{N-1} y[n-n_k^{(c)}]y^*[n-n_q^{(c)}]e^{-jn(\Omega_q^{(c)}-\Omega_k^{(c)})} \qquad q, k = 1, \ldots, P \tag{3.17}$$

From (3.17), it is easy to see that the matrix $\mathbf{R}_c$ is Hermitian, i.e., $\mathbf{R}_c^H = \mathbf{R}_c$. $\mathbf{r}_{sc}$ is a $P \times 1$ vector whose the qth element is

$$[\mathbf{r}_{sc}]_q = \sum_{n=0}^{N-1} y[n-n_0]y^*[n-n_q^{(c)}]e^{-jn(\Omega_q^{(c)}-\Omega_0)} \qquad q = 1, \ldots, P \tag{3.18}$$

Note that $[\mathbf{r}_{sc}]_q$ is the related cross-correlation between the qth possible clutter/multipath scatter and the desired target signal at delay-Doppler (n_0, Ω_0). $\mathbf{r}_{xc}$ is a $P \times 1$ vector whose the qth element is

$$[\mathbf{r}_{xc}]_q = \sum_{n=0}^{N-1} x[n]y^*[n-n_q^{(c)}]e^{-jn\Omega_q^{(c)}} \qquad q = 1, \ldots, P \tag{3.19}$$

The $[\mathbf{r}_{xc}]_q$ represents the related cross-correlation between the received signal in the SC and the qth possible clutter/multipath scatter. r_{ss} is the related auto-correlation of the desired target signal, i.e.,

$$r_{ss} = \sum_{n=0}^{N-1} |y[n-n_0]|^2 \tag{3.20}$$

and r_{xs} represents the related correlation between the received signal in the SC and the desired target signal at delay-Doppler (n_0, Ω_0), and is

$$r_{xs} = \sum_{n=0}^{N-1} x[n]y^*[n-n_0^{(l)}]e^{-jn\Omega_0^{(l)}} \tag{3.21}$$

In order to derive the MLEs of the unknown parameters in (3.15) and (3.16), it is necessary for the matrix $\mathbf{R}_c$ to have an inverse. It can be proven that $\mathbf{R}_c$ is a positive semidefinite matrix, i.e., $\mathbf{R}_C \succeq 0$ [1], making it necessary to apply a proper diagonal loading before inverting the matrix. Once this is done, the MLEs for the unknown parameters in each hypothesis can be determined as

$$\begin{bmatrix} \hat{\mathbf{c}}_1 \\ \hat{\alpha}_0 \end{bmatrix} = \begin{bmatrix} \mathbf{R}_c & \mathbf{r}_{sc} \\ \mathbf{r}_{sc}^H & r_{ss} \end{bmatrix}^{-1} \begin{bmatrix} \mathbf{r}_{xc} \\ r_{xs} \end{bmatrix} \tag{3.22}$$

$$\hat{\mathbf{c}}_0 = \mathbf{R}_c^{-1}\mathbf{r}_{xc}. \tag{3.23}$$

The partitioned matrix inversion Lemma states that if a matrix is partitioned and all of the individual matrices within the partition have inverses that exist for the appropriate dimensions, then the inverse of the entire partitioned matrix also exists. For example, we have [5]

$$\begin{bmatrix} \mathbf{A} & \mathbf{B} \\ \mathbf{C} & \mathbf{D} \end{bmatrix}^{-1} = \begin{bmatrix} \mathbf{A}^{-1} + \mathbf{A}^{-1}\mathbf{BKCA}^{-1} & -\mathbf{A}^{-1}\mathbf{BK} \\ -\mathbf{KCA}^{-1} & \mathbf{K} \end{bmatrix}$$

where

$$\mathbf{K} = (\mathbf{D} - \mathbf{CA}^{-1}\mathbf{B})^{-1}$$

Making use of the partitioned matrix inversion Lemma in (3.22), the MLEs of unknown parameters under hypothesis $\mathscr{H}_1$ are given by

$$\begin{bmatrix} \hat{\mathbf{c}}_1 \\ \hat{\alpha}_0 \end{bmatrix} = \begin{bmatrix} \mathbf{R}_c^{-1} + g\mathbf{R}_c^{-1}\mathbf{r}_{sc}\mathbf{r}_{sc}^H\mathbf{R}_c^{-1} & -g\mathbf{R}_c^{-1}\mathbf{r}_{sc} \\ -g\mathbf{r}_{sc}^H\mathbf{R}_c^{-1} & g \end{bmatrix} \begin{bmatrix} \mathbf{r}_{xc} \\ r_{xs} \end{bmatrix} \tag{3.24}$$

where

$$g \triangleq \frac{1}{r_{ss} - \mathbf{r}_{sc}^H\mathbf{R}_c^{-1}\mathbf{r}_{sc}}. \tag{3.25}$$

using (3.23), (3.24) can be rewritten as

$$\begin{bmatrix} \hat{\mathbf{c}}_1 \\ \hat{\alpha}_0 \end{bmatrix} = \begin{bmatrix} \hat{\mathbf{c}}_0 + g\mathbf{R}_c^{-1}\mathbf{r}_{sc}\mathbf{r}_{sc}^H\hat{\mathbf{c}}_0 - gr_{xs}\mathbf{R}_c^{-1}\mathbf{r}_{sc} \\ -g\mathbf{r}_{sc}^H\hat{\mathbf{c}}_0 + gr_{xs} \end{bmatrix} \tag{3.26}$$

After replacing the MLEs of α_0 and $\mathbf{c}$ in the pdfs (3.7) and (3.8) based on each hypothesis, we can create the likelihood ratio. Then, by taking the logarithm of this ratio and performing some simplification, we obtain the test statistic

$$\ln L_{GLR} = \frac{\mathbf{r}_{xc}^H(\hat{\mathbf{c}}_1 - \hat{\mathbf{c}}_0) + \hat{\mathbf{c}}_1^H(\mathbf{r}_{xc} - \mathbf{R}_c\hat{\mathbf{c}}_1) + r_{ss}|\hat{\alpha}_0|}{\sigma^2} \underset{\mathcal{H}_0}{\overset{\mathcal{H}_1}{\gtrless}} \eta \tag{3.27}$$

using (3.26), the test statistic becomes

$$\frac{1}{\sigma^2}\frac{|r_{xs} - \mathbf{r}_{sc}^H\hat{\mathbf{c}}_0|^2}{(r_{ss} - \mathbf{r}_{sc}^H\mathbf{R}_c^{-1}\mathbf{r}_{sc})} \underset{\mathcal{H}_0}{\overset{\mathcal{H}_1}{\gtrless}} \eta \tag{3.28}$$

To achieve a more intuitive form of this test statistic, we can recast the test as

$$\frac{1}{\sigma^2}\frac{|\sum_{n=0}^{N-1}(x[n] - \sum_{k=1}^{P}[\hat{\mathbf{c}}_0]_k y[n - n_k^{(c)}]e^{jn\Omega_k^{(c)}})y^*[n - n_0]e^{-jn\Omega_0}|^2}{(r_{ss} - \mathbf{r}_{sc}^H\mathbf{R}_c^{-1}\mathbf{r}_{sc})} \underset{\mathcal{H}_0}{\overset{\mathcal{H}_1}{\gtrless}} \eta \tag{3.29}$$

The numerator of the test indicates the 2D-CCF between the reference signal and the cleaned surveillance signal at a particular point (n_0, Ω_0). It is worth noting that in References 6 and 7, this numerator is used as a detection test, despite being derived from a distinct framework. To gain a new perspective on this test statistic, it is easy to show that $\mathbf{r}_{sc} = \mathbf{V}^H\mathbf{s}_0$, $\mathbf{r}_{xc} = \mathbf{V}^H\mathbf{x}$, $r_{xs} = \mathbf{s}_0^H\mathbf{x}$, $r_{ss} = \mathbf{s}_0^H\mathbf{s}_0$, and $\hat{\mathbf{c}}_0 = (\mathbf{V}^H\mathbf{V})^{-1}\mathbf{V}^H\mathbf{x}$, where $\mathbf{V} = [\mathbf{v}_1, \dots, \mathbf{v}_q, \dots, \mathbf{v}_P]$ is an $N \times P$ matrix with $\mathbf{v}_q = (\mathbf{a}_{\Omega_q^{(c)}} \odot \mathbf{y}_{n_q^{(c)}})$, and $\mathbf{s}_0 = (\mathbf{a}_{\Omega_0} \odot \mathbf{y}_{n_0})$, where $\mathbf{y}_l = \mathbf{P}^l\mathbf{y}$ and $[\mathbf{a}_\Omega]_k = e^{jk\Omega}$ for $k = 0, \dots, N-1$. In this case, $\mathbf{P}$ is an $N \times N$ permutation matrix defined as $[\mathbf{P}]_{ij} = 1$ if $i = j+1$ and 0 otherwise for $i = 0, \dots, N-1$ and $j = 0, \dots, N-1$. Hence, (3.29) can be expressed as

$$\frac{1}{\sigma^2}\frac{|\mathbf{s}_0^H\mathbf{P}_\mathbf{V}^\perp\mathbf{x}|^2}{\mathbf{s}_0^H\mathbf{P}_\mathbf{V}^\perp\mathbf{s}_0} \underset{\mathcal{H}_0}{\overset{\mathcal{H}_1}{\gtrless}} \eta \tag{3.30}$$

where $\mathbf{P}_V = \mathbf{V}(\mathbf{V}^H\mathbf{V})^{-1}\mathbf{V}^H$ is the projection matrix that projects a vector onto the columns of $\mathbf{V}$, and $\mathbf{P}_\mathbf{V}^\perp = \mathbf{I}_N - \mathbf{V}(\mathbf{V}^H\mathbf{V})^{-1}\mathbf{V}^H$ is the orthogonal projection matrix that projects a vector onto the space orthogonal to that spanned by the columns of $\mathbf{V}$. Using the Hermitian and idempotent property of matrix $\mathbf{P}_\mathbf{V}^\perp$, i.e., $\mathbf{P}_\mathbf{V}^\perp = \mathbf{P}_\mathbf{V}^{\perp H}$ and $\mathbf{P}_\mathbf{V}^\perp = \mathbf{P}_\mathbf{V}^{\perp 2}$, we can rewrite (3.30) with lower complexities as

$$\frac{1}{\sigma^2}\frac{|\mathbf{s}_0^H\mathbf{P}_\mathbf{V}^\perp\mathbf{x}|^2}{\|\mathbf{P}_\mathbf{V}^\perp\mathbf{s}_0\|^2} \underset{\mathcal{H}_0}{\overset{\mathcal{H}_1}{\gtrless}} \eta \tag{3.31}$$

If we think of the subspace spanned by the columns of $\mathbf{V}$ as a "clutter/multipath subspace" and the orthogonal subspace as the "signal subspace," then the numerator represents the square of correlation between the projection of the received vector $\mathbf{x}$ onto the signal subspace and the desired target signal in the given delay n_0 and normalized Doppler Ω_0.

3.2.2 GLRT-based detector: unknown σ^2 and when $K = 1$

The goal of this subsection is to obtain the GLRT when dealing with unknown noise and clutter/multipath signals variance (σ^2). To achieve this, we substitute the MLEs of $\mathbf{c}$, α_0, and σ^2 for each hypothesis into the pdfs presented in (3.7) and (3.8). We then use these pdfs to construct the likelihood ratio. The MLEs of $\mathbf{c}$ and α_0 are computed in a similar way as in (3.23) and (3.26) for each hypothesis. Next, by substituting these MLEs into the pdfs of (3.7) and (3.8) and setting the derivative of the resulting pdfs with respect to σ^2 equal to zero, we can determine the MLEs of σ^2, given by

$$\begin{aligned} \hat{\sigma}_1^2 &= \frac{1}{N}(\|\mathbf{x}\|^2 - 2Re\{\hat{\mathbf{c}}_1^H\mathbf{r}_{xs}\} - r_{ss}|\alpha_0|^2 + \hat{\mathbf{c}}_1^H\mathbf{R}_c\hat{\mathbf{c}}_0) \\ &= \frac{1}{N}(\|\mathbf{x}\|^2 - \hat{\mathbf{c}}_0^H\mathbf{R}_c\hat{\mathbf{c}}_0 - g|r_{xs} - \mathbf{r}_{sc}^H\hat{\mathbf{c}}_0|^2) \end{aligned} \tag{3.32}$$

under $\mathscr{H}_1$ hypothesis, and

$$\hat{\sigma}_0^2 = \tfrac{1}{N}(\|\mathbf{x}\|^2 - \hat{\mathbf{c}}_0^H\mathbf{R}_c\hat{\mathbf{c}}_0) \tag{3.33}$$

under $\mathscr{H}_0$ hypothesis. Substituting these into the pdfs under the two hypotheses and constructing the likelihood ratio gives the GLRT statistic:

$$\begin{aligned} L_{\mathrm{GLR}_1}(\mathbf{x}) &= \frac{\max_{\alpha_0,\mathbf{c},\sigma^2} f(\mathbf{x};\alpha_0,\mathbf{c},\sigma^2,\mathscr{H}_1)}{\max_{\mathbf{c},\sigma^2} f(\mathbf{x};\mathbf{c},\sigma^2,\mathscr{H}_0)} = \left(\frac{\hat{\sigma}_0^2}{\hat{\sigma}_1^2}\right)^N \\ &= \left(\frac{\|\mathbf{x}\|^2 - \hat{\mathbf{c}}_0^H\mathbf{R}_c\hat{\mathbf{c}}_0}{\|\mathbf{x}\|^2 - \hat{\mathbf{c}}_0^H\mathbf{R}_c\hat{\mathbf{c}}_0 - g|r_{xs} - \mathbf{r}_{sc}^H\hat{\mathbf{c}}_0|^2}\right)^N \\ &= \left(\frac{1}{1 - \dfrac{g|r_{xs} - \mathbf{r}_{sc}^H\hat{\mathbf{c}}_0|^2}{\|\mathbf{x}\|^2 - \hat{\mathbf{c}}_0^H\mathbf{R}_c\hat{\mathbf{c}}_0}}\right)^N \end{aligned} \tag{3.34}$$

Similar to the GLRT derivation in the previous subsection, we can rewrite the GLRT statistic as

$$L_{\mathrm{GLR}_1}(\mathbf{x}) = \left(\frac{1}{1 - \dfrac{|\mathbf{s}_0^H\mathbf{P}_{\mathbf{V}}^{\perp}\mathbf{x}|^2}{(\mathbf{s}_0^H\mathbf{P}_{\mathbf{V}}^{\perp}\mathbf{s}_0)(\mathbf{x}^H\mathbf{P}_{\mathbf{V}}^{\perp}\mathbf{x})}}\right)^N \tag{3.35}$$

Using Cauchy–Schwartz inequality [5] and the idempotent property of the matrix $\mathbf{P}_{\mathbf{V}}^{\perp}$, i.e., $\mathbf{P}_{\mathbf{V}}^{\perp} = \mathbf{P}_{\mathbf{V}}^{\perp 2}$, we have

$$|\mathbf{s}_0^H\mathbf{P}_{\mathbf{V}}^{\perp}\mathbf{x}|^2 = |\mathbf{s}_0^H\mathbf{P}_{\mathbf{V}}^{\perp 2}\mathbf{x}|^2 < (\mathbf{s}_0^H\mathbf{P}_{\mathbf{V}}^{\perp}\mathbf{s}_0)(\mathbf{x}^H\mathbf{P}_{\mathbf{V}}^{\perp}\mathbf{x}) \tag{3.36}$$

It follows that

$$0 < \frac{|\mathbf{s}_0^H \mathbf{P}_{\mathbf{V}}^{\perp} \mathbf{x}|^2}{(\mathbf{s}_0^H \mathbf{P}_{\mathbf{V}}^{\perp} \mathbf{s}_0)(\mathbf{x}^H \mathbf{P}_{\mathbf{V}}^{\perp} \mathbf{x})} \leq 1 \tag{3.37}$$

Since $(\frac{1}{1-z})^N$ for $z < 1$ is an increasing function of z, the GLRT in (3.35) is statistically equivalent to

$$L_{\mathrm{GLR}_1}(\mathbf{x}) = \frac{|\mathbf{s}_0^H \mathbf{P}_{\mathbf{V}}^{\perp} \mathbf{x}|^2}{(\mathbf{s}_0^H \mathbf{P}_{\mathbf{V}}^{\perp} \mathbf{s}_0)(\mathbf{x}^H \mathbf{P}_{\mathbf{V}}^{\perp} \mathbf{x})} = \frac{|\mathbf{s}_0^H \mathbf{P}_{\mathbf{V}}^{\perp} \mathbf{x}|^2}{\|\mathbf{P}_{\mathbf{V}}^{\perp} \mathbf{s}_0\|^2 \|\mathbf{P}_{\mathbf{V}}^{\perp} \mathbf{x}\|^2} \underset{\mathcal{H}_0}{\overset{\mathcal{H}_1}{\gtrless}} \eta_1 \tag{3.38}$$

In the following sections, we will use the term GLR_1 to refer to this detector. The subscript of GLR_1 indicates the target number that was taken into account while developing the GLR test. A threshold value, η_1, is chosen for GLR_1 such that the probability of false detection is equal to a predefined value.

In practical scenarios, the goal is to find a detector that maximizes the probability of detecting a target while dealing with interference signals such as noise, clutter/multipath, and interfering targets, while keeping the probability of false alarms under control. The current approach, referred to as GLR_1, only accounts for the impact of noise and clutter/multipath on the detection process but does not consider interfering targets. Interfering targets may cause an increase in false alarms or mask other targets, and this problem is particularly noticeable in passive radar systems due to the unpredictable nature of transmitted waveforms. To address this issue, the detection problem should take into account the presence of interfering targets ($K > 1$). Therefore, in the following subsection, we derive the GLR detector that considers the presence of noise, clutter/multipath, and interfering targets. Although the GLR_1 detector is not practical for implementation, it serves as a benchmark for evaluating the performance of the GLR detector derived in the subsequent subsection.

3.2.3 Two-stage GLRT-based detector: general problem

In this section, the goal is to obtain a target detection algorithm based on the GLRT criterion in the presence of noise, clutter/multipath, and interfering targets with unknown parameters. As such, the GLRT cannot be straightforwardly obtained since there are lots of unknowns including $\mathbf{c} = [c_1, ..., c_P]^T$, $\alpha^{(t)} = [\alpha_1^{(t)}, ..., \alpha_{K-1}^{(t)}]^T$, α_0, σ^2, and interfering target's delay-Doppler coordinate pairs (n_m, Ω_m) for $m = 1, ..., K-1$. To address this, we obtain a two-stage GLRT-based detector rather than a one-stage detector. In this regard, it first assumes that all (n_m, Ω_m)'s are known and drive the GLRT detector over the remaining unknowns $\mathbf{c}$, $\alpha^{(t)}$, α_0, and σ^2. Then, the estimated (n_m, Ω_m) pairs from the second stage are substituted in place of the true ones into the GLRT test. To do this, a multistage algorithm is proposed in the second stage of detector design. In what follows, the first and second stages are described in detail.

3.2.4 First stage of 2S-GLRT-based detector

The first stage assumes that the delay-Doppler pairs (n_m, Ω_m) for $m = 1, ..., K-1$ of the interfering targets are already known. Using this assumption, the GLRT is obtained by replacing the unknown parameters, including $\mathbf{c} = [c_1, ..., c_P]^T$, $\alpha^{(t)} = [\alpha_1^{(t)}, ..., \alpha_{K-1}^{(t)}]^T$ (complex amplitudes of interfering targets), α_0 and σ^2, with their MLEs under the given hypothesis. For the current problem, the GLRT is expressed as

$$\frac{\max_{\alpha_0,\alpha^{(t)},\mathbf{c}} f(\mathbf{x};\alpha_0,\alpha^{(t)},\mathbf{c},\mathcal{H}_1)}{\max_{\alpha^{(t)},\mathbf{c}} f(\mathbf{x};\alpha^{(t)},\mathbf{c},\mathcal{H}_0)} \underset{\mathcal{H}_0}{\overset{\mathcal{H}_1}{\gtrless}} \eta_K \tag{3.39}$$

where the threshold η_K is selected to satisfy the probability of false alarm requirement. Similar to the previous subsection, the MLEs of unknown parameters $\alpha^{(t)}$, α_0, and $\mathbf{c}$ can be obtained as

$$\begin{bmatrix} \mathbf{R}_c & \mathbf{R}_{tc} & \mathbf{r}_{sc} \\ \mathbf{R}_{tc}^H & \mathbf{R}_{tt} & \mathbf{r}_{st} \\ \mathbf{r}_{sc}^H & \mathbf{r}_{st}^H & r_{ss} \end{bmatrix} \begin{bmatrix} \hat{\mathbf{c}}_1 \\ \hat{\alpha}_1^{(t)} \\ \hat{\alpha}_0 \end{bmatrix} = \begin{bmatrix} \mathbf{r}_{xc} \\ \mathbf{r}_{xt} \\ r_{xs} \end{bmatrix} \tag{3.40}$$

In (3.40), $\hat{\mathbf{c}}_1$ and $\hat{\alpha}_1^{(t)}$ are the MLEs of $\mathbf{c}$ and $\alpha^{(t)}$ under $\mathcal{H}_1$ hypothesis, respectively. Similarly, under $\mathcal{H}_0$ hypothesis, we have

$$\begin{bmatrix} \mathbf{R}_c & \mathbf{R}_{tc} \\ \mathbf{R}_{tc}^H & \mathbf{R}_{tt} \end{bmatrix} \begin{bmatrix} \hat{\mathbf{c}}_0 \\ \hat{\alpha}_0^{(t)} \end{bmatrix} = \begin{bmatrix} \mathbf{r}_{xc} \\ \mathbf{r}_{xt} \end{bmatrix} \tag{3.41}$$

where $\hat{\mathbf{c}}_0$ and $\hat{\alpha}_0^{(t)}$ are the MLEs of $\mathbf{c}$ and $\alpha^{(t)}$ under $\mathcal{H}_0$ hypothesis, respectively. Similarly, $\mathbf{R}_c$ is the $P \times P$ related clutter/multipath correlation as defined in (3.17), $\mathbf{r}_{sc}$ is the $P \times 1$ vector whose qth element $[\mathbf{r}_{sc}]_q$ is defined in (3.18), $\mathbf{r}_{xc}$ is the $P \times 1$ vector whose the qth element $[\mathbf{r}_{xc}]_q$ is defined in (3.19). Here, r_{ss} and r_{xs} are defined in (3.20) and (3.21), respectively. Also, $\mathbf{r}_{st}$ is a $(K-1) \times 1$ vector whose the mth element $[\mathbf{r}_{st}]_m$ is the related correlation between the mth interfering target at delay-Doppler of $(n_m^{(t)}, \Omega_m^{(t)})$ and that of the desired target signal at (n_0, Ω_0), and is given by

$$[\mathbf{r}_{st}]_m = \sum_{n=0}^{N-1} y[n-n_0]y^*[n-n_m^{(t)}]e^{-jn(\Omega_m^{(t)}-\Omega_0)} \tag{3.42}$$

$\mathbf{R}_{tc}$ is a $P \times (K-1)$ matrix whose the (q, m) the entry, $[\mathbf{R}_{tc}]_{qm}$, is defined as

$$[\mathbf{R}_{tc}]_{qm} = \sum_{n=0}^{N-1} y[n-n_m^{(t)}]y^*[n-n_q^{(c)}]e^{-jn(\Omega_q^{(c)}-\Omega_m^{(t)})} \tag{3.43}$$

where $q = 1, ..., P$ and $m = 1, ..., K-1$. From (3.43), it is easy to see that $[\mathbf{R}_{tc}]_{qm}^* = [\mathbf{R}_{tc}]_{mq}$. Also, $\mathbf{R}_{tt}$ is a $(K-1) \times (K-1)$ matrix whose the (k, m) entry $[\mathbf{R}_{tt}]_{km}$ is defined as

$$[\mathbf{R}_{tt}]_{km} = \sum_{n=0}^{N-1} y[n-n_m^{(t)}]y^*[n-n_k^{(t)}]e^{-jn(\Omega_k^{(t)}-\Omega_m^{(t)})} \tag{3.44}$$

where $k = 1, ..., K-1$. In order to cast the results in a more compact form, it is worthwhile to introduce a partitioned correlation matrix of the interference signals containing clutter/multipath and interfering targets as

$$\mathbf{R}_i \triangleq \begin{bmatrix} \mathbf{R}_c & \mathbf{R}_{tc} \\ \mathbf{R}_{tc}^H & \mathbf{R}_{tt} \end{bmatrix} \tag{3.45}$$

a vector containing the correlation of the desired target signal and the interference signals as

$$\mathbf{r}_{si} \triangleq \begin{bmatrix} \mathbf{r}_{sc}^T & \mathbf{r}_{st}^T \end{bmatrix}^T \tag{3.46}$$

a vector containing the correlation of the SC signal and the interference signals as

$$\mathbf{r}_{xi} \triangleq \begin{bmatrix} \mathbf{r}_{xc}^T & \mathbf{r}_{xt}^T \end{bmatrix}^T \tag{3.47}$$

and vectors containing the MLEs of the $\mathbf{c}$ and $\boldsymbol{\alpha}^{(t)}$ under $\mathscr{H}_k$ hypothesis for $k = 1, 2$ as

$$\hat{\xi}_k \triangleq \begin{bmatrix} \hat{\mathbf{c}}_k^T & \hat{\alpha}_k^{(t)^T} \end{bmatrix}^T \quad k = 0, 1 \tag{3.48}$$

By these definitions, we can rewrite (3.40) and (3.41), respectively, as

$$\begin{bmatrix} \mathbf{R}_i & \mathbf{r}_{si} \\ \mathbf{r}_{si}^H & r_{ss} \end{bmatrix} \begin{bmatrix} \hat{\xi}_1 \\ \hat{\alpha}_0 \end{bmatrix} = \begin{bmatrix} \mathbf{r}_{xi} \\ r_{xs} \end{bmatrix} \tag{3.49}$$

and

$$\mathbf{R}_i \hat{\xi}_0 = \mathbf{r}_{xi} \tag{3.50}$$

The matrix $\mathbf{R}_i$ is positive semidefinite. Therefore, it is necessary to apply appropriate diagonal loading to the matrix before performing its inversion. Once this is done, the MLEs of the unknown parameters in each hypothesis can be calculated.

$$\begin{bmatrix} \hat{\xi}_1 \\ \hat{\alpha}_0 \end{bmatrix} = \begin{bmatrix} \mathbf{R}_i & \mathbf{r}_{si} \\ \mathbf{r}_{si}^H & r_{ss} \end{bmatrix}^{-1} \begin{bmatrix} \mathbf{r}_{xi} \\ r_{xs} \end{bmatrix} \tag{3.51}$$

$$\hat{\xi}_0 = \mathbf{R}_i^{-1} \mathbf{r}_{xi} \tag{3.52}$$

Making use of the partitioned matrix inversion Lemma, we can rewrite (3.51) as

$$\begin{bmatrix} \hat{\xi}_1 \\ \hat{\alpha}_0 \end{bmatrix} = \begin{bmatrix} \hat{\xi}_0 + g_i \mathbf{R}_i^{-1} \mathbf{r}_{si} \mathbf{r}_{si}^H \hat{\xi}_0 - g_i r_{xs} \mathbf{R}_i^{-1} \mathbf{r}_{si} \\ -g_i \mathbf{r}_{si}^H \hat{\xi}_0 + g_i r_{xs} \end{bmatrix} \tag{3.53}$$

where

$$g_i \triangleq \frac{1}{r_{ss} - \mathbf{r}_{si}^H \mathbf{R}_i^{-1} \mathbf{r}_{si}} \tag{3.54}$$

Substituting the MLEs of α_0, $\hat{\xi}$ into pdfs (3.7) and (3.8) and maximizing them with respect to σ^2, the MLEs of σ^2 can be expressed as

$$\hat{\sigma}_1^2 = \frac{1}{N}(\|\mathbf{x}\|^2 - \hat{\xi}_0^H \mathbf{R}_i \hat{\xi}_0 - g_i |r_{xs} - \mathbf{r}_{si}^H \hat{\xi}_0|^2) \tag{3.55}$$

under $\mathscr{H}_1$ hypothesis, and

$$\hat{\sigma}_0^2 = \frac{1}{N}(\|\mathbf{x}\|^2 - \hat{\xi}_0^H \mathbf{R}_i \hat{\xi}_0) \tag{3.56}$$

under $\mathcal{H}_0$ hypothesis. Substituting these into the pdfs under two hypotheses and constructing the likelihood ratio, the test (3.39) becomes

$$\left(\frac{1}{1 - \dfrac{g_i |r_{xs} - \mathbf{r}_{si}^H \hat{\xi}_0|^2}{\|\mathbf{x}\|^2 - \hat{\xi}_0^H \mathbf{R}_i \hat{\xi}_0}} \right)^N \underset{\mathcal{H}_0}{\overset{\mathcal{H}_1}{\gtrless}} \eta_K \tag{3.57}$$

Similar to the GLRT derivation in the previous subsection, we can express (3.57) as follows:

$$\left(\frac{1}{1 - \dfrac{|\mathbf{s}_0^H \mathbf{P}_{\mathbf{U}}^{\perp} \mathbf{x}|^2}{(\mathbf{s}_0^H \mathbf{P}_{\mathbf{U}}^{\perp} \mathbf{s}_0)(\mathbf{x}^H \mathbf{P}_{\mathbf{U}}^{\perp} \mathbf{x})}} \right)^N \underset{\mathcal{H}_0}{\overset{\mathcal{H}_1}{\gtrless}} \eta_K \tag{3.58}$$

where matrix $\mathbf{U}$ is an $N \times P$ matrix defined as $\mathbf{U} = [\mathbf{V}, \mathbf{T}]$. Note that, in this case, we have $P = (2Q+1)N_c + K$. More precisely, the matrix $\mathbf{U}$ is partitioned into two sub-matrices; the clutter/multipath matrix $\mathbf{V} = [\mathbf{v}_1, ..., \mathbf{v}_q, ..., \mathbf{v}_P]$, which defined in the previous section, and an interfering targets matrix $\mathbf{T} = [\mathbf{t}_1, ..., \mathbf{t}_m, ..., \mathbf{t}_{(K-1)}]$, in which $\mathbf{t}_m = (\mathbf{a}_{\Omega_m^{(t)}} \odot \mathbf{y}_{n_m^{(t)}})$, where $\mathbf{y}_{n_m^{(t)}} = \mathbf{P}^{n_m^{(t)}} \mathbf{y}$ and $[\mathbf{a}_{\Omega_m^{(t)}}]_k = e^{jk\Omega_m^{(t)}}$ for $k = 0, ..., N-1$. Generally speaking, every column of the matrix $\mathbf{U}$ is built by a proper time-delayed and Doppler-shifted version of the reference signal $\mathbf{y}$ according to range-Doppler coordinates of the clutter region and the interfering targets. $\mathbf{P}_U = \mathbf{U}(\mathbf{U}^H\mathbf{U})^{-1}\mathbf{U}^H$ is a projection matrix that projects a vector onto the columns of $\mathbf{U}$, and $\mathbf{P}_{\mathbf{U}}^{\perp} = \mathbf{I}_N - \mathbf{U}(\mathbf{U}^H\mathbf{U})^{-1}\mathbf{U}^H$ is the orthogonal projection matrix that projects a vector onto the space orthogonal to that spanned by the columns of $\mathbf{U}$. Similarly, since $(\frac{1}{1-z})^N$ for $z < 1$ is an increasing function of z, the GLRT in (3.58) is statistically equivalent to

$$L_{\mathrm{GLR}_K}(\mathbf{x}) = \frac{|\mathbf{s}_0^H \mathbf{P}_{\mathbf{U}}^{\perp} \mathbf{x}|^2}{(\mathbf{s}_0^H \mathbf{P}_{\mathbf{U}}^{\perp} \mathbf{s}_0)(\mathbf{x}^H \mathbf{P}_{\mathbf{U}}^{\perp} \mathbf{x})} = \frac{|\mathbf{s}_0^H \mathbf{P}_{\mathbf{U}}^{\perp} \mathbf{x}|^2}{\|\mathbf{P}_{\mathbf{U}}^{\perp} \mathbf{s}_0\|^2 \|\mathbf{P}_{\mathbf{U}}^{\perp} \mathbf{x}\|^2} \underset{\mathcal{H}_0}{\overset{\mathcal{H}_1}{\gtrless}} \eta_K \tag{3.59}$$

This detector is known as the GLR$_K$ detector, where the subscript K corresponds to the number of targets in the GLRT derivation. The columns of the matrix $\mathbf{U}$ can be considered as the basis for the "interference subspace," which includes both clutter and interfering target signals, while the orthogonal subspace is referred to as the "signal subspace." The numerator of the GLRT represents the correlation squared between the projection of the received vector $\mathbf{x}$ onto the signal subspace and the desired target signal for a specific delay n_0 and normalized Doppler Ω_0.

3.2.5 Second stage of 2S-GLRT-based detector

The first step in designing a detector using the GLRT approach assumes that the receiver has prior knowledge of the correct delay-Doppler pairs $(n_m^{(t)}, \Omega_m^{(t)})$ for all interfering targets, where $m = 1, ..., K-1$. However, in practical situations, this information is unknown and needs to be estimated. To achieve this, a multistage algorithm is proposed that uses an approximate and simplified version of the GLRT (GLR_K), known as ASGLR. The algorithm is executed consecutively to estimate the unknown parameters by computing ASGLR over the desired delay-Doppler plane. To reduce the computational complexity of realizing the numerator of the test (3.59) for each range (delay) of interest, the Fast Fourier Transform (FFT) is utilized to obtain

$$\mathbf{l}_{\mathrm{GLR}_K}(\mathbf{x}, n) = \frac{\mathbf{f}(n) \odot \mathbf{f}^*(n)}{\|\mathbf{P}_{\mathrm{U}}^{\perp}\mathbf{s}_n\|^2 \|\mathbf{P}_{\mathrm{U}}^{\perp}\mathbf{x}\|^2} \qquad n = 1, ..., N_r \tag{3.60}$$

where N_r is a time delay index corresponding to the maximum bistatic range of interest obtained from $R_{\max} = cN_mT_s$, and $\mathbf{f}(n) = \mathrm{FFT}(\mathbf{x}_c \odot \mathbf{s}_n, N_F)$ where $\mathbf{s}_n = \mathbf{P}^n\mathbf{y}^*$ and $\mathbf{x}_c \triangleq \mathbf{P}_{\mathrm{U}}^{\perp}\mathbf{x}$ is a vector containing the N_F-point FFT of the vector $\mathbf{x}_c \odot \mathbf{s}_n$ for any value of n. In our study, we consider $N_F = 2N$.

The sampling frequency of the signal is much higher than the Doppler frequency of the target of interest. This enables the use of decimation to reduce the processing load needed to calculate the N_F-point FFT, while still maintaining high signal processing gain. The decimation process can be implemented efficiently using a cascaded integrator-comb (CIC) filter [8, 9]. The decimated signal is then filtered to remove signal components outside of the desired frequency band. The N_F-point FFT can be substituted with an N_d-point FFT, where N_d is determined by dividing the original FFT size N by the decimation factor R, as described in Reference 7. In order to simplify (3.60) and decrease its computational complexity, we can use an approximation by replacing $|\mathbf{P}_{\mathrm{U}}^{\perp}\mathbf{s}_n|^2$ with $|\mathbf{s}_n|^2$. This approximation is especially valid for delay-Doppler coordinates that are distant from those of interfering targets and clutter. Hence, we can simplify (3.60) to obtain

$$\mathbf{l}_{\mathrm{GLR}_K}(\mathbf{x}_c, n) = \frac{\mathbf{f}(n) \odot \mathbf{f}^*(n)}{\|\mathbf{s}_n\|^2 \|\mathbf{x}_c\|^2} \qquad n = 1, ..., N_r \tag{3.61}$$

This test statistic is termed as the ASGLR, as mentioned before. Now, we can calculate the ASGLR according to range-Doppler plane of interest, resulting in an $N_f \times N_r$ matrix, denoted by $\mathbf{L}(\mathbf{x}_c)$ whose the nth column is $\mathbf{l}_{\mathrm{GLR}_H}(\mathbf{x}_c, n)$. Here, N_f takes smaller values than N_d for a desired range-Doppler plane. The main steps of the interfering targets positioning algorithm can be summarized as follows:

1. Set $m \leftarrow 0$, $\mathbf{U}^{(0)} = \mathbf{V}$ and $\mathbf{x}_c^{(0)} = \mathbf{x}$.
2. To remove the interference signals, we perform using

$$\mathbf{x}_c^{(m+1)} = \mathbf{x}_c^{(m)} - \mathbf{U}^{(m)}\hat{\xi}_0^{(m)} \tag{3.62}$$

where $\hat{\xi}_0^{(m)} = (\mathbf{U}^{(m)^H}\mathbf{U}^{(m)})^{-1}\mathbf{U}^{(m)^H}\mathbf{x}_c^{(m)}$. Therefore, $\mathbf{x}_c^{(m+1)} = \mathbf{P}_{\mathbf{U}}^{\perp(m)}\mathbf{x}_c^{(m)}$ in which $\mathbf{P}_{\mathbf{U}}^{\perp(m)} = \mathbf{I} - \mathbf{U}^{(m)}(\mathbf{U}^{(m)^H}\mathbf{U}^{(m)})^{-1}\mathbf{U}^{(m)^H}$. Let us define matrix $\mathbf{R}_i^{(m)}$ such that

$$\mathbf{R}_i^{(m)} \triangleq \mathbf{U}^{(m)^H}\mathbf{U}^{(m)} = \begin{bmatrix} \mathbf{U}^{(m-1)^H}\mathbf{U}^{(m-1)} & \mathbf{U}^{(m-1)^H}\mathbf{T}^{(m)} \\ \mathbf{T}^{(m)^H}\mathbf{U}^{(m-1)} & \mathbf{T}^{(m)^H}\mathbf{T}^{(m)} \end{bmatrix} = \begin{bmatrix} \mathbf{R}_i^{(m-1)} & \mathbf{R}_{ti}^{(m)} \\ \mathbf{R}_{ti}^{(m)^H} & \mathbf{R}_{tt}^{(m)} \end{bmatrix} \tag{3.63}$$

The partitioned matrix inversion Lemma can be used to obtain the inverse of $\mathbf{R}_i^{(m)^{-1}}$, given by

$$\mathbf{R}_i^{(m)^{-1}} = \begin{bmatrix} \mathbf{R}_i^{(m-1)^{-1}} + \mathbf{R}_i^{(m-1)^{-1}}\mathbf{R}_{ti}^{(m)}\mathbf{K}\mathbf{R}_{ti}^{(m)^H}\mathbf{R}_i^{(m-1)^{-1}} & -\mathbf{R}_i^{(m-1)^{-1}}\mathbf{R}_{ti}^{(m)}\mathbf{K} \\ -\mathbf{K}\mathbf{R}_{ti}^{(m)^H}\mathbf{R}_i^{(m-1)^{-1}} & \mathbf{K} \end{bmatrix} \tag{3.64}$$

where

$$\mathbf{K} = \left(\mathbf{R}_{tt}^{(m)} - \mathbf{R}_{ti}^{(m)^H}\mathbf{R}_i^{(m-1)^{-1}}\mathbf{R}_{ti}^{(m)}\right)^{-1} \tag{3.65}$$

This implies that $(\mathbf{U}^{(m)^H}\mathbf{U}^{(m)})^{-1}$ in $\mathbf{P}_{\mathbf{U}}^{\perp(m)}$ can be recursively updated to further reduce the computational complexity of matrix inversion.

3. Compute the matrix $\mathbf{L}(\mathbf{x}_c^{(m+1)})$ corresponding to the delay-Doppler plane of interest.
4. To determine if there are any interfering targets, we need to compare the highest value of $\mathbf{L}(\mathbf{x}_c^{(m+1)})$ across the time delay and Doppler shift indices with a preset threshold η_a. If this maximum value is greater than η_a, it indicates that the range-Doppler coordinates associated with this maximum value represent the estimated bistatic range and Doppler shift of an interfering target. In such a scenario, we can continue executing the remaining steps to locate other interfering targets. The algorithm stops running when the maximum value obtained is lower than the threshold η_a. It is crucial to choose η_a according to the desired false alarm probability of the system.
5. The procedure involves updating the value of variable m to $m+1$, and using the delay-Doppler coordinates $(\hat{n}_m^{(t)}, \frac{\hat{k}f_s}{N_F})$ of the previously detected maximum value as an estimate for the position of the mth interfering target. However, this estimation alone is not accurate enough for super interfering target cancellation. To address this issue, a small area around the range-Doppler coordinates of the previously detected maximum is considered. Specifically, the Doppler and range extents are selected to cover seven frequency elements equal to $(\hat{k} \pm q)\frac{f_s}{6N_F}$ for $q = 0, 1, 2, 3$, spread over three delay indices $\hat{n}_m^{(t)} - 1$, $\hat{n}_m^{(t)}$, and $\hat{n}_m^{(t)} + 1$. This yields a mask of dimensions 7×3 or a total number of range-Doppler bins equal to $N_b = 21$. It has been shown through simulations that this extension is sufficient for strong target cancellation.

6. To cancel out the interference caused by the mth target, the interfering matrix needs to be updated. This can be done by adding a new column to the matrix. The new column is a replica of the reference signal that has been time-delayed and Doppler shifted based on the extended Doppler-range coordinates. This process can be written mathematically as $\mathbf{U}^{(m)} = [\mathbf{U}^{(m-1)}, \mathbf{T}^{(m)}]$, where $\mathbf{T}^{(m)}$ is a matrix whose ith column is denoted by $\mathbf{t}_i^{(m)}$ and is a proper time-delayed and Doppler-shifted replica of the reference signal based on the extended Doppler-range coordinates in step 4.
7. Go back to step 2.

Step 4 of the algorithm outlines the stopping criterion. The proposed Imperative Target Positioning (ITP) algorithm can identify targets that are above the noise level of the SC, and the number and positions of imperative targets in the range-Doppler plane of interest can be determined. The GLR_K can then be utilized by selecting an imperative target as the testing target and the remaining $(K-1)$ targets as interfering targets. This serves as the confirmation step for the detection method. While this detection algorithm may require multiple iterations to detect all possible targets and confirm their presence in the presence of other targets, it ensures optimal performance in terms of false alarm probability regulation and superior detection performance. Chapter 7 introduces a new detection algorithm that eliminates the confirmation step, which is based on the M-ary hypothesis-testing formulation of the single-band PR target detection problem.

3.3 Analytical performance analysis

In this section, closed-form expressions for both false alarm probability and detection probability are obtained. In Appendix 3A, it is shown that the detection statistic:

$$L(\mathbf{x}) = \frac{N-(P+1)}{\frac{1}{L_{\text{GLR}_K}(\mathbf{x})} - 1} \underset{\mathcal{H}_0}{\overset{\mathcal{H}_1}{\gtrless}} \eta \tag{3.66}$$

follows the distribution

$$L(\mathbf{x}) \sim \begin{cases} F_{1,N-(P+1)} & \text{under } \mathcal{H}_0 \\ F'_{1,N-(P+1)}(\delta) & \text{under } \mathcal{H}_1 \end{cases} \tag{3.67}$$

where $\delta = \frac{|\alpha_0|^2}{\sigma^2}\|\mathbf{P}_{\mathbf{U}}^{\perp}\mathbf{s}_0\|^2$. In (3.66), $P = (2Q+1)N_c + K$ is the column dimension of the interference matrix $\mathbf{U}$. In (3.67), $F_{1,N-(P+1)}$ denotes a central complex F distribution with 1 numerator complex degree of freedom and $N-(P+1)$ denominator complex degree of freedom, and $F'_{1,N-(P+1)}(\delta)$ denotes a noncentral complex F distribution with 1 numerator complex degree of freedom and $N-(P+1)$ denominator complex degree of freedom and the noncentrality parameter δ.

The false alarm and detection probabilities of the proposed detector can be respectively expressed as

$$P_{fa} = Q_{F_{1,N-(P+1)}}(\eta), \tag{3.68}$$

and

$$P_d = Q_{F'_{1,N-(P+1)}(\delta)}(\eta) \tag{3.69}$$

where $Q_{F_{1,N-(P+1)}}$ and $Q_{F'_{1,N-(P+1)}(\delta)}$ are the right-tail probability of central and noncentral complex F distribution, respectively. As can be seen from (3.68), the threshold setting is feasible with no prior knowledge of any unknown parameters. As a result, the GLRT ensures the constant false alarm rate (CFAR) property.

For the problem at hand, N gets a large value to give a desired integration gain. As a consequence, $F_{1,N-(P+1)}$ and $F'_{1,N-(P+1)}(\delta)$ tend to the complex χ_1^2 and $\chi_1^{2'}(\delta)$, respectively. Note that if a random variable x has a complex noncentral chi-squared distribution with k complex degree of freedom and noncentrality parameter δ then its pdf is $f(x;k,\delta) = e^{-(\delta+x)}(\frac{x}{\delta})^{\frac{k-1}{2}} I_{k-1}(2\sqrt{x\delta})u(x)$, where $u(x) = 1$ if $x > 0$ and zero otherwise and $I_\alpha(x)$ is the αth-order modified Bessel function of first kind [10]. Therefore, the P_{fa} and P_d can be evaluated as

$$P_{fa} = Q_{\chi_1^2}(\eta) = e^{-\eta} \tag{3.70}$$

and

$$P_d = Q_{\chi_1^{2'}(\delta)}(\eta) \tag{3.71}$$

respectively. For a predetermined P_{fa} value, we should choose $\eta = -\ln(P_{fa})$. Hence, for test statistic $L_{\mathrm{GLR}_K}(\mathbf{x})$ in (3.59), the threshold must be set equal to $\eta_K = \frac{\eta}{N-(P+1)+\eta}$. Moreover, it is worthwhile to note that $N \gg P$, so an incorrect estimate of the number of interfering targets cannot affect the GLR_K threshold. Therefore, we can set the threshold η_a in the ITP algorithm with an assumed maximum number of targets. The detection probability of (3.69) can also be written as

$$P_d = e^{-(\delta+\eta)} \sum_{k=0}^{\infty} \left(\frac{\delta^k}{k!} \sum_{n=0}^{k} \frac{\eta^n}{n!} \right) \tag{3.72}$$

Using (3.72), it is easy to see that P_d is an increasing function of $\delta = \frac{|\alpha_0|^2}{\sigma^2} \|\mathbf{P}_{\mathrm{U}}^{\perp}\mathbf{s}_0\|^2$. If $\frac{|\alpha_0|^2}{\sigma^2}$ is regarded as the input SNR, then the SNR gain provided by the GLR detector can be considered as $\|\mathbf{P}_{\mathrm{U}}^{\perp}\mathbf{s}_0\|^2$ or $(\mathbf{s}_0^H \mathbf{P}_{\mathrm{U}}^{\perp}\mathbf{s}_0)$. It is worth to focus on this term to gain a deeper insight into the detector performance of the proposed GLRT-based detector as a function of range and Doppler.

Now the orthogonal projection matrix $\mathbf{P}_{\mathrm{U}}^{\perp}$ is an idempotent matrix. Therefore its eigenvalues are either zero or one [11]. This helps us to obtain $0 \le \mathbf{s}_0^H \mathbf{P}_{\mathrm{U}}^{\perp}\mathbf{s}_0 \le \|\mathbf{s}_0\|^2$. Besides, $\mathbf{P}_{\mathrm{U}}^{\perp}\mathbf{U} = \mathbf{0}$, i.e., $\mathbf{P}_{\mathrm{U}}^{\perp}\mathbf{u}_i = 0$ for $i = 1, ..., P$. Thus, $\|\mathbf{P}_{\mathrm{U}}^{\perp}\mathbf{s}_0\|^2$ takes its minimum if $\mathbf{s}_0 \in \{\mathbf{u}_1, ..., \mathbf{u}_P\}$. This means that the proposed detector can fully remove the interference signal with corresponding delay-Doppler coordinates as that of the interference signals (clutter and interfering targets). In a same way, we can conclude that $\|\mathbf{P}_{\mathrm{U}}^{\perp}\mathbf{s}_0\|^2$ takes the value of $\|\mathbf{s}_0\|^2$

when the delay-Doppler coordinate of $\mathbf{s}_0$ is far enough from those of the interference signals. Similarly, $\frac{\|\mathbf{P}_{\mathrm{U}}^{\perp}\mathbf{s}_0\|^2}{\|\mathbf{s}_0\|^2}$ takes values close to zero and one respectively when the delay-Doppler coordinates of $\mathbf{s}_0$ are near or far enough from those of the interference signals. This mainly depends on the range-Doppler resolution of the opportunity signal embedded in $\mathbf{s}_0$. Based on this fact, in the proposed subspace-based detection algorithm, the detection loss (DL) associated with an opportunity signal can be defined as

$$\mathrm{DL} = \frac{\|\mathbf{P}_{\mathrm{U}}^{\perp}\mathbf{s}_0\|^2}{\|\mathbf{s}_0\|^2} \tag{3.73}$$

where $0 \leq \mathrm{DL} \leq 1$. In essence, if the delay-Doppler coordinates of a target are significantly different from those of interfering signals, there will be no loss of detection. However, when the coordinates of the target are similar to those of the interference, DL is high. This implies that the proposed detector effectively cancels out the interference signals to achieve the desired false alarm probability.

3.4 Simulation results

In this section, we present comprehensive simulation findings to assess the effectiveness of the proposed detectors in managing false alarms, canceling interference, and evaluating detection permanence. It is important to mention that the detectors proposed in this chapter fall under the Ca.1-IR category, hence we will use the terms IR-GLR and GLR detectors interchangeably. To evaluate performance, a signal based on FM with a 200 kHz channel bandwidth is used. To sample this signal correctly, a Nyquist frequency of 200 kHz is required. For evaluation purposes, an integration time of 1 second is assumed. Figures 3.1 and 3.2 display the spectrum of the radio FM signal and its complex envelope, respectively.

Three different multitarget scenarios were considered to evaluate the performance capability of the proposed detector in terms of false alarm regulation, interference removal, and detection. In the first scenario, six targets were included, each having far enough range-Doppler coordinates from each other. This scenario is similar to the one used in Reference 12, and the characteristics of the targets are provided in Table 3.1. In the second scenario, a different multitarget scenario was created, where targets 1, 2, and 3 were assumed to have the same Doppler bin with varying ranges, while targets 4, 5, and 6 had the same range with almost different Doppler frequencies. Additionally, target 7 was considered to have a high range with a low received SNR. The characteristics of the targets in this scenario are provided in Table 3.2.

In Tables 3.1–3.3, R_B and f_d respectively denote the relative bistatic range and bistatic Doppler frequency, and the input signal-to-noise ratio, termed as SNR_i, is given by

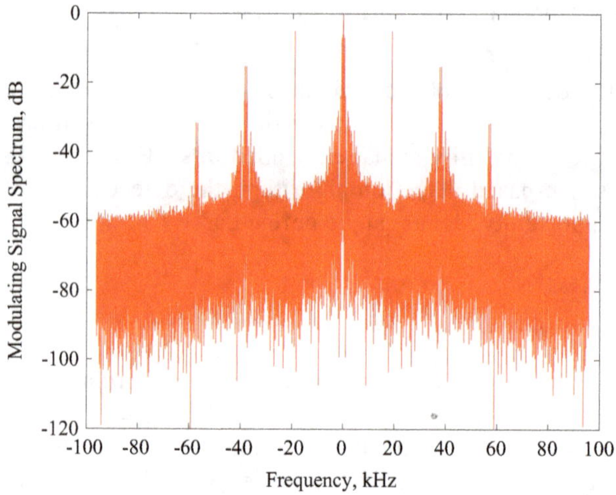

Figure 3.1 Modulating signal spectrum of a stereo FM radio signal

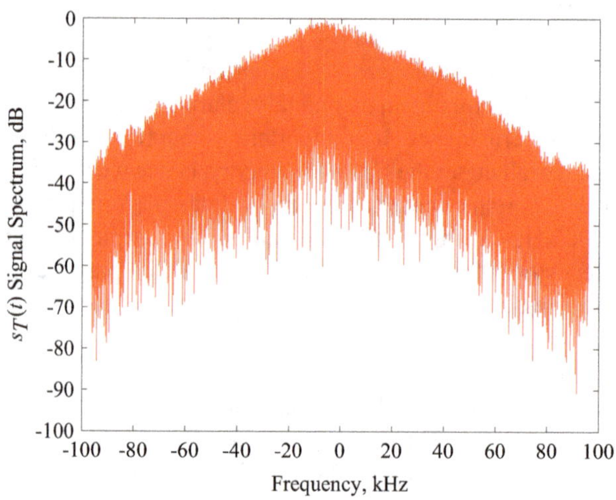

Figure 3.2 FM radio signal spectrum

Table 3.1 First scenario (S#1): target characteristics

Target	T_1	T_2	T_3	T_4	T_5	T_6
R_B (km)	20.25	60.00	60.00	99.75	110.25	129.75
f_d^y (Hz)	44.00	–73.33	14.67	88.00	–44.00	–88.00
SNR_i (dB)	4.1	–3.8	–20.8	–21.1	–21.6	–22.1

Table 3.2 Second scenario (S#2): target characteristics

Target	T_1	T_2	T_3	T_4	T_5	T_6	T_7
R_B (km)	7.03	17.97	40.04	59.57	59.57	59.57	150.00
f_d (Hz)	44.00	44.00	44.00	–14.67	–19.07	–26.40	73.33
SNR_i (dB)	12	–14	–32	10	–5	–31	–35

Table 3.3 Third scenario (S#3): target characteristics

Target	T_1	T_2	T_3	T_4	T_5	T_6
R_B (km)	9.37	21.87	129.69	59.57	59.57	59.57
f_d (Hz)	44.00	44.00	73.33	–14.67	–20.53	–26.40
SNR_i (dB)	–30	–31	–35	–32	–33	–34

$$\text{SNR}_i = \frac{|\alpha_0|^2}{N_0 B} \tag{3.74}$$

The noise power per unit bandwidth is denoted by N_0, and the receiver bandwidth is denoted by B. Therefore, the input noise power can be calculated as $\sigma^2 = N_0 B$. It is important to note that the reported SNR_i in our simulation is for a receiver bandwidth of 200 kHz. Additionally, we assume that there is a direct input signal-to-noise ratio (DNR) of approximately 63 dB, and there are ten clutter spikes received with varying clutter-to-noise ratios (CNR_i) between 5 dB and 35 dB over the bistatic ranges of 0–55 km. Although assuming clutter with a constant amplitude over the integration time is accurate for echoes from stationary scatterers, the presence of moving scatterers causes clutter to vary over time. To account for this, our simulation assumes that each clutter amplitude has an exponential power spectrum density over the integration time, described as [13]:

$$S(v) = \text{CNR}\left(\frac{r}{r+1}\delta(v) + \frac{1}{r+1}\frac{\rho\lambda}{4}\exp(-\rho|v|)\right) \tag{3.75}$$

The clutter velocity is denoted as v and measured in meters per second. The ratio of DC power to AC power in the spectrum is represented by the parameter r and depends on the radar frequency and wind speed. The radar wavelength is indicated by λ, while the shape parameter ρ is a function of wind speed. The expression (3.75) describes the clutter power spectral density (PSD), where the Dirac delta function $\delta(v)$ characterizes the steady component and the second term represents the varying components of clutter PSD. In our simulation, we use an exponential PSD with the specific values of $\rho = 7$ and $r = 90$ for all clutter amplitudes.

3.4.1 False alarm regulation evaluation

Detecting targets in a PR system can be challenging due to the presence of noise, clutter, and interfering targets. It is important to maintain a CFAR while achieving maximum detection probability, which requires effectively canceling out clutter/multipath and interfering targets. To demonstrate the importance of this issue, three simulations were conducted and the detection threshold of the proposed detector was analytically set to 10^{-3}. To reduce computational complexity, the false alarm probability was fixed at 10^{-3}.

The first examination involved a simulation where clutter was placed at a non-integer multiple of the sampling frequency f_s in the bistatic range from 0 km to 55 km, and the Doppler frequency was zero with $\rho = 0$. Figure 3.3 was used to analyze how different sampling frequencies affect the false alarm regulation of the proposed detector. The false alarm probability was plotted against the relative bistatic range for different sampling frequencies when the Doppler frequency was 8 Hz. The results indicated that a sampling frequency of around four times the Nyquist frequency f_N was needed to regulate false alarms and achieve the desired probability of 10^{-3}. This was due to the sampling frequency's impact on modeling the clutter subspace and the proposed detector's ability to effectively remove clutter signals. Additionally, an effective receiver bandwidth needed to be chosen to ensure that the noise remained temporally white.

The clutter subspace, which can be represented by a few frequency components around zero Doppler, is characterized by frequencies $f_k^{(c)} = (k - 1 - Q)\Delta f_c$, where $k = 1, ..., 2Q + 1$. In Figure 3.4, the false alarm probability is shown as a function of bistatic Doppler frequency for a cell with a range of 2 km and a sampling frequency of $4f_N$. The aim is to determine the appropriate frequency bins for the

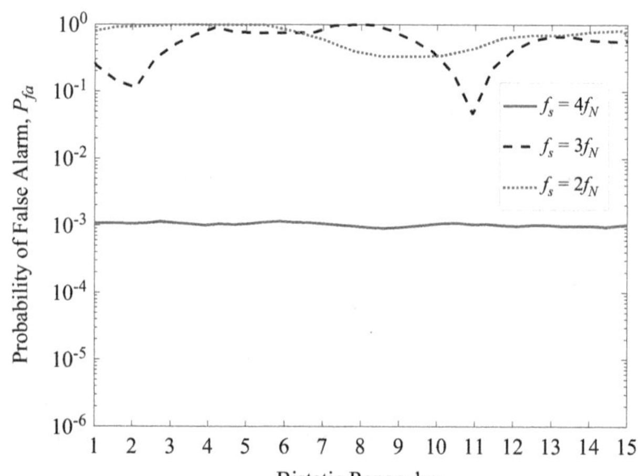

Figure 3.3 False alarm regulation of the GLR_1 detector as a function of relative bistatic range when the Doppler frequency of the cell under test is equal to 8 Hz, and taking the sampling frequency f_s as a parameter

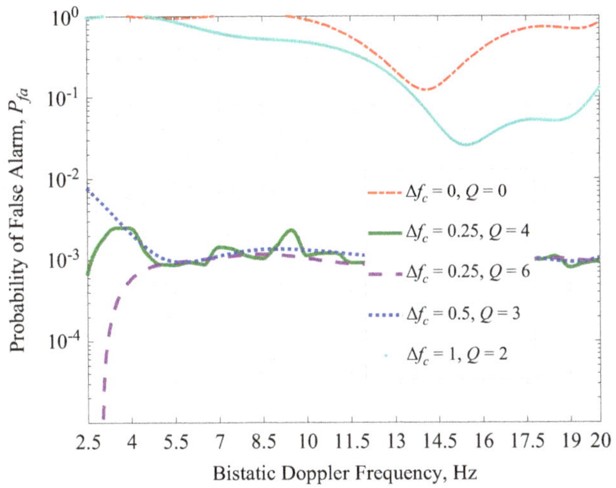

Figure 3.4 False alarm regulation of the proposed GLR_1 *detector as a function of relative bistatic Doppler frequency when the bistatic range of the cell under test is 2 km, and taking the* Δf_c *and* Q *as parameters*

clutter subspace of the proposed detector to effectively remove the clutter signals and achieve a desired false alarm probability. The parameters Δf_c and Q are chosen to specify the clutter subspace, and the clutter is modeled using (3.75) with $\rho = 7$ and $r = 90$. The results in Figure 3.4 demonstrate that selecting a clutter subspace with a zero frequency bin ($\Delta f_c = 0$ and $Q = 0$) is insufficient to remove the received clutter signal, leading to an empirical false alarm probability of one whithin certain Doppler frequency bins. The proposed detector can achieve the desired false alarm probability of 10^{-3} by selecting $\Delta f_c = 0.25$ Hz and $Q = 6$ as the clutter subspace parameters, resulting in the removal of the received clutter signal.

The comparison between the false alarm regulation of GLR_1 and GLR_2 is shown in Figure 3.5, assuming the existence of an interfering target, with a range difference of ΔR from the cell under test, and their Doppler frequencies are equal. In this simulation, the SNR_i of the interfering target is set to –30 dB. The proposed GLR_1 does not take into account the presence of any interfering targets during its development, leading to a significant increase in the false alarm probability. On the other hand, the proposed GLR_2 is able to detect and remove one interfering target, making it possible to regulate the false alarm probability. These simulation results not only show the difference between target detection in passive radars and that of active radars but also indicate that applying conventional passive radar target detection approaches from Chapter 2 may lead to an excessive false alarm probability and significant detection losses. This inspired the development of the proposed GLR_K detector, which takes into account the existence of $(K-1)$ interfering targets before detecting the presence or absence of a new target.

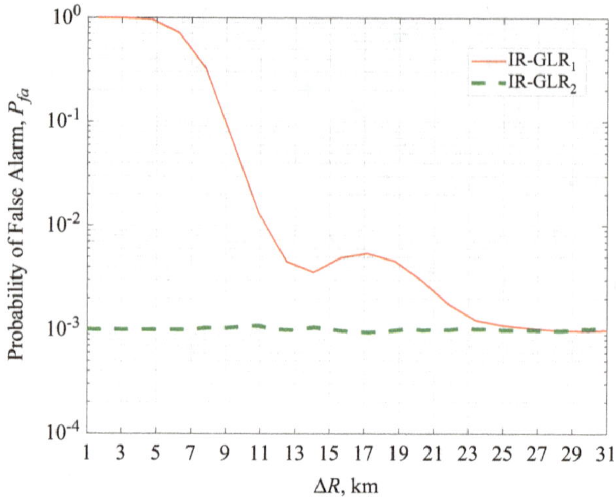

Figure 3.5 False alarm regulation comparison of the proposed GLR$_1$ and GLR$_2$ detectors in range dimension (i.e., ΔR) when there is an interfering target with $\text{SNR}_i = -30$ *dB and the desired false alarm probability is* $P_{fa} = 10^{-3}$

3.4.2 Detection performance evaluation

The aim of designing the detector is to devise a detector that maximizes the probability of detection while maintaining a fixed-size constraint. The CFAR test is an example of a fixed-size detector that has an empirical false alarm probability equal to a pre-determined false alarm probability in the presence of system uncertainty. In the previous section, it was demonstrated that the GLR$_K$ proposed exhibits CFAR behavior against clutter. In this section, we first evaluate the effectiveness of the proposed ITP algorithm in scenarios involving multiple targets, and then demonstrate that the performance of the GLR$_K$ is similar to that of the benchmark detector (GLR$_1$).

The first multitarget scenario from Table 3.1 was processed using the ITP algorithm, which terminated after six iterations. The results are illustrated in Figure 3.6, where the estimated and actual positions of the targets are represented by symbols □ and +, respectively. The results indicate that all targets were accurately located in the same number of iterations as the number of targets. In Table 3.2 second multitarget scenario, the ITP algorithm stopped after seven iterations and the results are depicted in Figure 3.7. One could be enticed to substitute the interference matrix $\mathbf{U}^{(m)}$ with $\mathbf{T}^{(m)}$ instead of $[\mathbf{U}^{(m-1)}, \mathbf{T}^{(m)}]$ to simplify the ITP algorithm's computation complexity, as suggested in References 14 and 15. The results of Figure 3.8, however, indicates that numerous false targets can emerge as a consequence, leading to the termination of the ITP algorithm after 65 iterations rather than the expected 7 iterations. Although the GLR$_K$ detector provides a confirmation stage, increasing

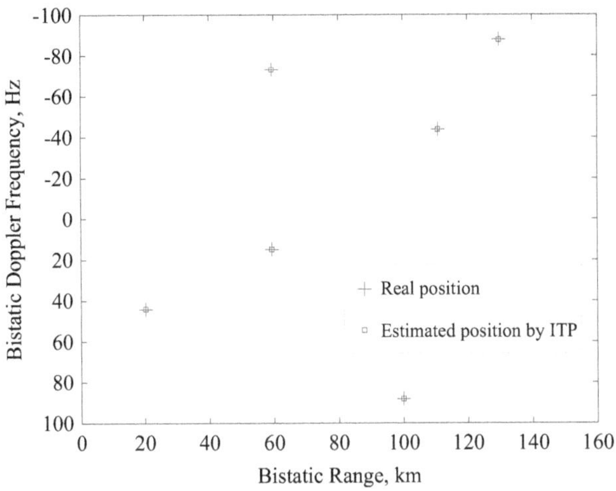

Figure 3.6 Result of applying the proposed ITP algorithm when the first multitarget scenario is considered

the number of false targets would significantly increase the multitarget detector's overall computational complexity.

To examine the detection performance of the proposed detector in the presence of a multitarget scenario, the SNR_i of the kth testing target, say T_k, is changed in the presence of other targets, namely interfering targets, with characteristics of Tables 3.1–3.3. Let us define the output signal-to-noise ratio (SNR_o), given as

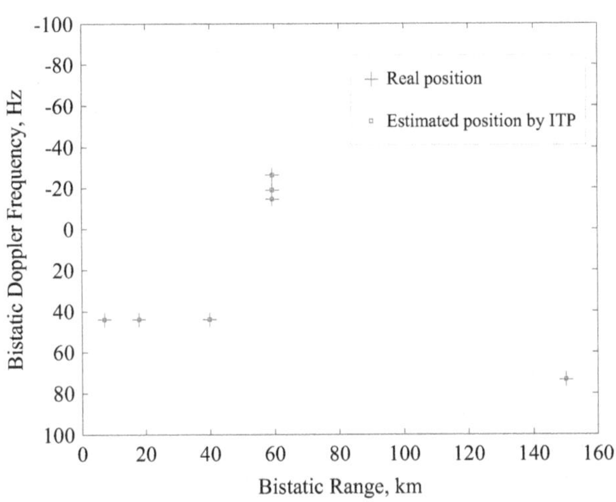

Figure 3.7 Result of applying the proposed ITP algorithm when the second multitarget scenario is considered

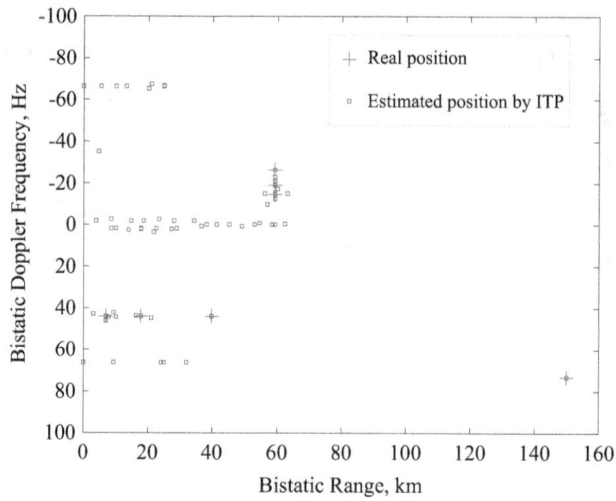

Figure 3.8 Result of applying the proposed ITP algorithm when the second multitarget scenario is considered. Here, the interference matrix $\mathbf{U}^{(m)}$ *is replaced with* $\mathbf{T}^{(m)}$*.*

$$\mathrm{SNR}_o = \frac{|\alpha_0|^2 T}{N_0} \tag{3.76}$$

The relationship between the SNR_o and SNR_i can be expressed as $\mathrm{SNR}_o = \mathrm{SNR}_i G_p$, where G_p is the maximum integration/processing gain and is equal to the product of the bandwidth B and the integration time T, i.e., $G_p = BT$. In the case of the FM signal with a bandwidth of 200 kHz and an integration time of 1 second, the processing gain is approximately 53 dB.

Figure 3.9 compares the theoretical performance of (3.72) with the Monte Carlo (MC) simulation in a scenario with a single target. The results indicate that the theoretical performance aligns with that obtained from MC simulations. Figure 3.10 shows the detection probability as a function of SNR_o for $P_{fa} = 10^{-6}$, where the aim is to detect the kth target (i.e., T_k) of the first scenario using the GLR_6 detector. The results of GLR_1 detector are included as a benchmark for comparison. This figure illustrates that the proposed GLR_K has the capability to detect the test target T_k and remove any interference from other interfering targets. This highlights the superiority of the proposed GLR_6 in scenarios with multiple targets. More precisely, the GLR_6 detector shows a detection loss of approximately 0.24 dB when detecting targets 4 and 5 in comparison to the GLR_1 benchmark detector. The simulation was repeated for the second multitarget scenario that includes six interfering targets, where the results are shown in Figure 3.11 for the proposed GLR_7 detector. It can be observed that the detection performance varies among different testing targets. The GLR_7 detector follows the performance of GLR_1 benchmark detector with degradation of approximately 0.01, 0.07, 0.10, 0.12, 0.22, 0.52, and 0.56 dB when $P_d = 0.9$. To provide further clarification, the DL of (3.73) is examined in Figure 3.12. This

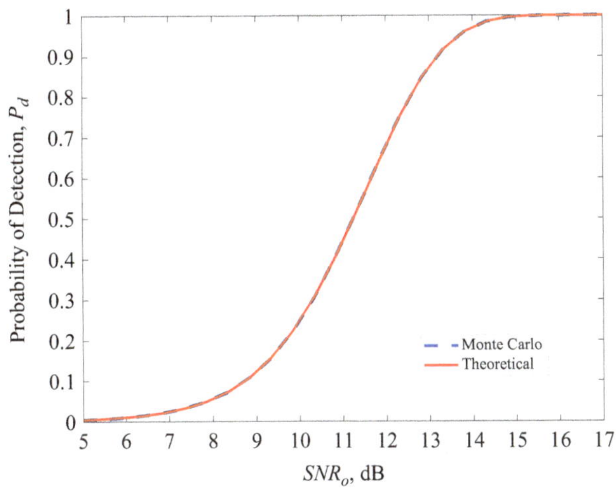

Figure 3.9 Detection probability comparison of (3.72) with that obtained by the Monte Carlo simulation when $P_{fa} = 10^{-6}$

figure shows DL as a function of relative bistatic range for a cell under test with zero Doppler. To study the clutter removal capability of the proposed detector, we assume a zero Doppler for the cell under test. From Figure 3.12, it is seen that the proposed detector places a deep and wide null in the range-dimension of the clutter region. This means that our proposed detector has a superior clutter removal capability to regulate the false alarm probability. Figure 3.13 indicates DL as a function

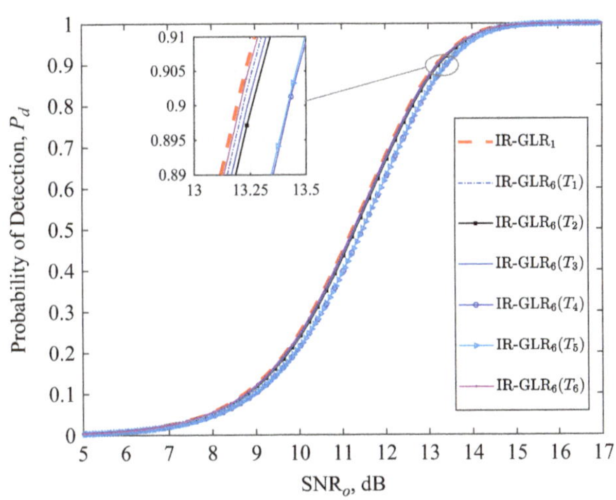

Figure 3.10 Detection probability as a function of SNR_o *for* $P_{fa} = 10^{-6}$*. The aim here is to evaluate the detection performance of the* GLR_6 *detector when detecting* kth *target (i.e.,* T_k*) in the first multitarget scenario.*

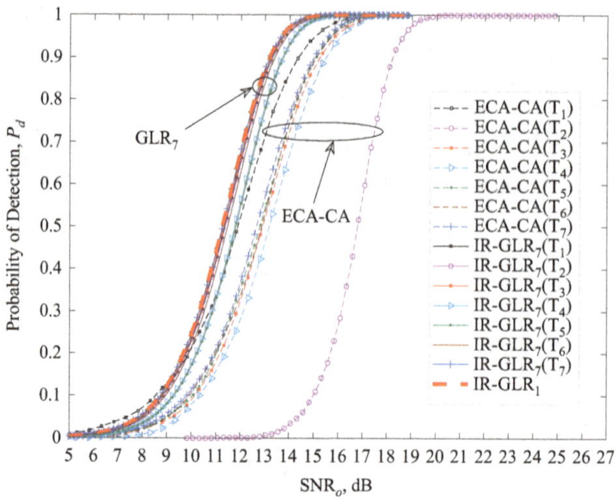

Figure 3.11 *Detection probability as a function of* SNR$_o$ *for* $P_{fa} = 10^{-6}$. *The aim here is to evaluate the detection performance of the GLR$_7$ and ECA-CA detectors when detecting kth target (i.e., T_k) in the second multitarget scenario.*

of bistatic Doppler frequency of the cell under test with the relative bistatic range of 500 m. Again, a deep null can be seen in the Doppler dimension of the clutter region. Figures 3.14 and 3.15 also provide some insight into the DL around a target which is placed at the range of 20 km and the Doppler frequency of $f_d = 44$ Hz. The simulation was repeated for a new scenario in which there are clutter and a single

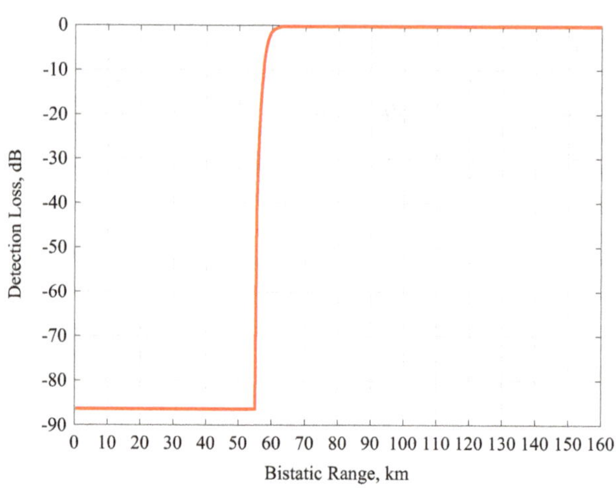

Figure 3.12 *Detection loss as a function of relative bistatic range for a cell under test with zero Doppler*

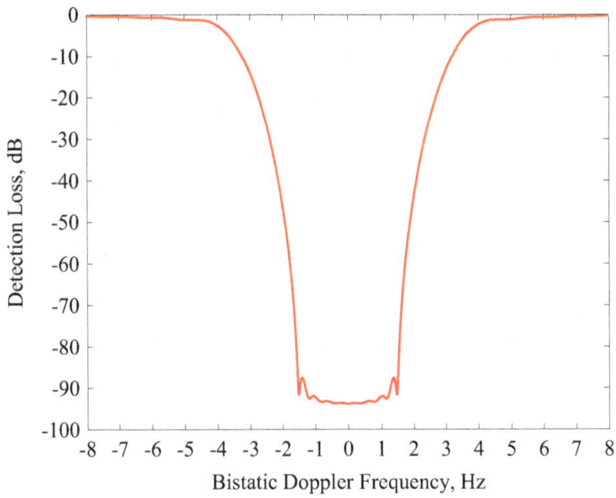

Figure 3.13 *Detection loss as a function of bistatic Doppler frequency for a cell under test with the relative bistatic range of 500 m*

interfering target, and the results are illustrated in Figure 3.16. The results of this are illustrated in Figure 3.16. It is observed that the proposed detector is able to place two deep nulls in the delay-Doppler coordinates corresponding to the clutter region and that of the interfering target. In general, these losses are specified by the range-Doppler characteristics of the exploited FM signals, and it is particularly significant when dealing with multitarget scenarios.

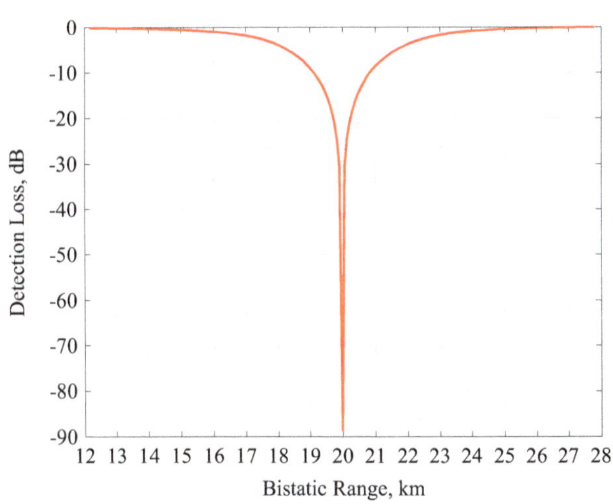

Figure 3.14 *Detection loss as a function of relative bistatic range around an interfering target with the range of 20 km*

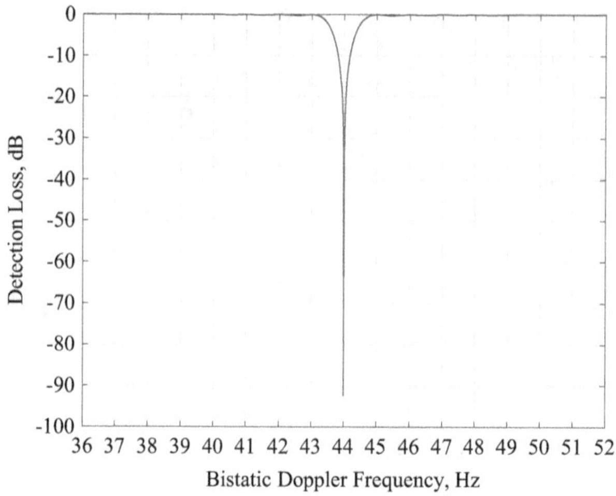

Figure 3.15 Detection loss as a function of bistatic Doppler frequency around an interfering target with the Doppler frequency of $f_d = 44$ Hz

In Chapter 2, it was discussed that the conventional detectionapproaches in PBRs involve setting a threshold on the 2D-CCF between the cleaned surveillance signal and a reference signal. To achieve a predetermined false alarm probability, a conventional CFAR must be employed. For example, a multistage processing algorithm was presented in Reference 12 that combines the extensive cancellation algorithm (ECA) for interference removal with the 2D-CA-CFAR detector for detection. This algorithm is referred to as ECA-CA in the sequel. To evaluate how well the GLR_K detector performs in comparison to the ECA-CA detector, we consider the second and third multitarget scenarios. In the ECA-CA detector, appropriate guard range-Doppler cells are used to prevent self-masking and capture effect, as described in Reference 16. Figures 3.11 and 3.17 depict a comparison of the detection capabilities of the GLR_K and ECA-CA detectors for the second and third multitarget scenarios, respectively. These figures clearly illustrate the superior performance of the proposed detector over the ECA-CA detector. In comparison to the GLR_7 detector, the ECA-CA detector experiences an additional DL of approximately 1 dB, 5 dB, 1.9 dB, 1.7 dB, 1.2 dB, 1.6 dB, and 1.8 dB to detect testing targets T_k for $k = 1, \ldots, 7$ of the second multitarget scenario, respectively, where they are 6 dB, 7 dB, 1.1 dB, 2.5 dB, 2.7 dB, and 2.2 dB for the third multitarget scenario. The third scenario corresponds to a situation where the ECA-CA algorithm is unable to detect and eliminate interfering targets, leading to significant detection losses. The high DLs in this scenario may be attributed to the 2D-CA-CFAR, which is susceptible to the influences of self-masking and capture effects.

It would be beneficial to assess the computational complexity of our proposed detection method in comparison to the ECA-CA approach when dealing with K targets. The comparison assumes that K is a multiple of 3. It can be shown that most of the computational complexity of the proposed detection process depends on $NN_b^2K^2 + 2NN_bK^2 + 4NN_cK$ in terms of the number of

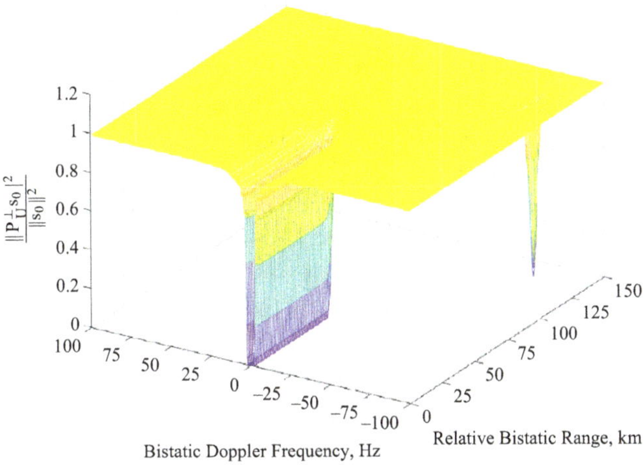

Figure 3.16 Detection loss as a function of bistatic range and Doppler frequency in the presence of clutter and interfering target with the range-Doppler coordinate of (120 km, –80 Hz).

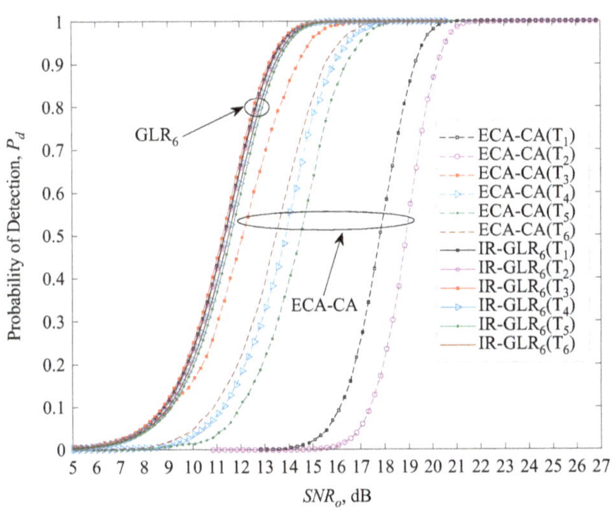

Figure 3.17 Detection probability as a function of SNR_o *for* $P_{fa} = 10^{-6}$. *The aim here is to evaluate the detection performance of the* GLR_6 *and ECA-CA detectors when detecting kth target (i.e.,* T_k*) in the third multitarget scenario.*

complex multiplications (CMs). In the ECA-CA detection process, it is assumed that only $\frac{2K}{3}$ targets can be detected and removed by its multistage algorithm under $\frac{K}{3}$ stages. As such, two targets can be declared in each stage, and other $\frac{K}{3}$

targets are need to be detected by using the 2D-CA-CFAR detector. In this case, most of the computational complexity of the ECA-CA detection process depends on $\frac{4}{9}NN_b^2K^2 + \frac{4}{9}NN_bK^2 + \frac{4}{3}NN_cK$. Hence, it can be concluded that although the proposed detector has a slightly higher computational complexity compared to the ECA-CA detector, it is still capable of achieving a significant enhancement in target detection performance of approximately 7 dB. This improvement is evident even when the detector operates with high-quality FM signals, as illustrated in Figure 3.17. The increased complexity is justified especially in situations where FM signals exhibit limited range resolution and low peak-to-sidelobe levels.

3.5 Summary

A systematic framework is presented in this chapter, which covers signal modeling, detection method, and statistical analysis, for bistatic PR target detection. In this chapter, a two-stage GLR detector has been developed to detect targets in the presence of interference, which includes noise, direct signal, clutter/multipath, and interfering targets. To detect targets in a multitarget scenario, a multistage algorithm has been proposed, which detects and eliminates the strongest target in order of decreasing signal strength. The false alarm probability and detection probability are calculated through closed-form expressions, which support the CFAR behavior and the detection performance loss of the proposed detector in the presence of interfering targets. Simulation results indicate that the proposed method is effective in controlling false alarm probability and achieving superior detection performance in the presence of noise, clutter, and interfering target signals. The detection quality of the multistage algorithm is shown to depend on the content of broadcasted FM signals, as demonstrated by the concept of DL. Next chapters will focus on the multiband detection problem to enhance target detection quality and target range resolution.

From our discussions, it is clear that detecting targets through PR involves intricate computational signal processing to achieve comparable performance to traditional active radar. However, we are fortunate that recent advancements in both digital signal processor hardware and algorithms have been made to help deal with these added complexities.

References

[1] Zaimbashi A., Derakhtian M., Sheikhi A. 'GLRT-based CFAR detection in passive bistatic radar'. *IEEE Transactions on Aerospace and Electronic Systems*. 2013, vol. 49(1), pp. 134–59.

[2] Zaimbashi A. 2013. 'Target detection in passive radars based on commercial FM radio signals'. [PhD thesis]. Shiraz University, Iran.

[3] Van Trees H.L. 'Detection, estimation, and modulation theory'. 1968, New York: Wiley.

[4] Kay S.M. 'Fundamentals of statistical signal processing: detection theory'. Upper Saddle River, NJ: Prentice-Hall. 1998, vol. 2.

[5] Horn R.A., Johnson C.R. 'Matrix analysis' in Cambridge University Press; 1985.

[6] Kulpa K.S., Czekała Z. 'Masking effect and its removal in PCL radar'. *IEE Proceedings – Radar, Sonar and Navigation*. 2005, vol. 152(3), p. 174.

[7] Cherniakov M. 'Bistatic radar emerging technology' in Wiley and Sons: Chichester, UK; 2007.

[8] Harris F.J. 'On the use of windows for harmonic analysis with the discrete Fourier transform'. *Proc. IEEE*. 1978, vol. 66(1), p. 5183.

[9] Hogenauer E.B. 'An economical class of digital filters for decimation and interpolation'. *IEEE Transactions on Acoustics, Speech, and Signal Processing*. 1981, vol. 29(2), pp. 155–162.

[10] Kelly E.J., Forsythe X.M. 'Adaptive detection and parameter estimation for Multidimensional signal models'. *Lincoln Lab., Mass. Inst Technol., Lexington, MA, Tech. Rep.* 1989, p. 848.

[11] Golub G.H., Van Loan C.F. 'Matrix computations', 2nd ed. John Hopkins University Press; 1996.

[12] Colone F., O'Hagan D.W., Lombardo P., Baker C.J. 'A multistage processing algorithm for disturbance removal and target detection in passive bistatic radar'. *IEEE Transactions on Aero Space and Electronic Systems*. 2009, vol. 45(2), pp. 698–722.

[13] Billingsley J.B. 'Low-Angle land clutter measurements and empirical models' in William Andrew Publishing; 2002.

[14] Bolvardi H., Derakhtian M., Sheikhi A. 'Dynamic clutter suppression and Multitarget detection in a DVB-T-based passive radar'. *IEEE Trans. on Aerospace and Electronic Systems*. 2017, vol. 53(4), pp. 1812–1825.

[15] Bolvardi H., Derakhtian M., Sheikhi A. 'Reduced complexity generalised likelihood ratio detector for digital video broadcasting terrestrial-based passive radar'. *IET Radar, Sonar & Navigation*. 2015, vol. 9(8), pp. 1021–29. Available from https://onlinelibrary.wiley.com/toc/17518792/9/8

[16] Zaimbashi A., Norouzi Y. 'Automatic dual censoring cell-averaging CFAR detector in non-homogenous environments'. *EURASIP Journal on Signal Processing*. 2008, vol. 88(11), pp. 2611–2621.

[17] Muirhead R.J. 'Aspect of multivariant statistical theory' in John Wiley Sons; 2005.

[18] Alvin C., Rencher G., Bruce S. 'Linear models in statistics' in John Wiley Sons; 2008.

Appendix 3A
Derivation of detection performance

This appendix provides the derivation for the distribution of the proposed GLR_K detector. To do so, we first recast the detection statistic (3.59) as

$$L(\mathbf{x}) = \frac{N-(P+1)}{\dfrac{1}{L_{\text{GLR}_K}(\mathbf{x})}-1} \underset{\mathscr{H}_0}{\overset{\mathscr{H}_1}{\gtrless}} \eta \tag{3A.1}$$

where $P = (2Q+1)N_c + K$ is the column dimension of the interference matrix $\mathbf{U}$. It should be noted that (3A.1) and (3.59) are related through strictly monotonic transformations. Therefore, while these two test statistics are not identical, they are equivalent in terms of statistical testing. In other words, they will produce the same results in terms of performance measures. The use of (3A.1) is beneficial in demonstrating the independence of the numerator and denominator of the statistical test. Moreover, it simplifies the process of obtaining the probability density function of the test statistic under both hypotheses. Substituting (3.59) into (3A.1), after some manipulations, yields

$$L(\mathbf{x}) = \frac{N-(P+1)}{1}\frac{\mathbf{x}^H(\mathbf{P}_{\mathbf{U}}^{\perp}-\mathbf{P}_{\mathbf{K}}^{\perp})\mathbf{x}}{\mathbf{x}^H\mathbf{P}_{\mathbf{K}}^{\perp}\mathbf{x}} \tag{3A.2}$$

where $\mathbf{P}_{\mathbf{K}}^{\perp} = \mathbf{I}_N - \mathbf{K}(\mathbf{K}^H\mathbf{K})^{-1}\mathbf{K}^H$ with $\mathbf{K} = [\mathbf{U}, \mathbf{s}_0]$. It is easy to show that

$$\mathbf{P}_{\mathbf{U}}^{\perp} - \mathbf{P}_{\mathbf{K}}^{\perp} = \mathbf{P}_K - \mathbf{P}_U = -\frac{\mathbf{P}_{\mathbf{U}}^{\perp}\mathbf{s}_0\mathbf{s}_0^H\mathbf{P}_{\mathbf{U}}^{\perp}}{\mathbf{s}_0^H\mathbf{P}_U\mathbf{s}_0} \tag{3A.3}$$

Equation (3A.2) can also be written as

$$L(\mathbf{x}) = \frac{N-(P+1)}{1}\frac{L_n(\mathbf{x})}{L_d(\mathbf{x})} = \frac{N-(P+1)}{1}\frac{(\frac{\mathbf{x}}{\sigma})^H(\mathbf{P}_K-\mathbf{P}_U)(\frac{\mathbf{x}}{\sigma})^H}{(\frac{\mathbf{x}}{\sigma})^H\mathbf{P}_{\mathbf{K}}^{\perp}(\frac{\mathbf{x}}{\sigma})} \tag{3A.4}$$

The test comprises of two quadratic forms in the numerator and denominator, as evident from (3A.4). In order to determine the probability density function of the numerator and denominator of the test, we refer to Theorem 1.

Theorem 1. *Let* $\mathbf{z} \sim \mathscr{CN}(\mu, \mathbf{I})$ *be a complex* $N \times 1$ *vector, and let* $\mathbf{A}$ *be a Hermitian matrix of dimension* $N \times N$. *Then* $\mathbf{z}^H\mathbf{A}\mathbf{z}$ *has a noncentral complex chi-squared random variable with* k *complex degree of freedom and noncentrality parameters* δ *if and only if* $\mathbf{A}$ *is an idempotent matrix* ($\mathbf{A}^2 = \mathbf{A}$), *in which case the complex degree of freedom and the noncentrality parameters are respectively,* $k = \text{rank}(\mathbf{A}) = \text{tr}(\mathbf{A})$ *and* $\delta = \mu^H\mathbf{A}\mu$ *[10] and [17].*

Let k_n and k_d be the complex degree of freedom of the numerator and denominator of the test statistic (3A.4), respectively, that are given by

$$k_n = \operatorname{tr}\left(\frac{\mathbf{P}_{\mathbf{U}}^{\perp}\mathbf{s}_0\mathbf{s}_0^H\mathbf{P}_{\mathbf{U}}^{\perp}}{\mathbf{s}_0^H\mathbf{P}_{\mathbf{U}}^{\perp}\mathbf{s}_0}\right) = \frac{1}{\mathbf{s}_0^H\mathbf{P}_{\mathbf{U}}^{\perp}\mathbf{s}_0}\operatorname{tr}(\mathbf{P}_{\mathbf{U}}^{\perp}\mathbf{s}_0\mathbf{s}_0^H) = \frac{1}{\mathbf{s}_0^H\mathbf{P}_{\mathbf{U}}^{\perp}\mathbf{s}_0}\operatorname{tr}(\mathbf{s}_0^H\mathbf{P}_{\mathbf{U}}^{\perp}\mathbf{s}_0) = 1 \tag{3A.5}$$

and

$$\begin{aligned} k_d &= \operatorname{tr}(\mathbf{P}_K^{\perp}) = \operatorname{tr}(\mathbf{I}_N) - \operatorname{tr}((\mathbf{K}^H\mathbf{K})^{-1}\mathbf{K}^H\mathbf{K}) \\ &= \operatorname{tr}(\mathbf{I}_N) - \operatorname{tr}(\mathbf{I}_{P+1}) = N - (P+1) \end{aligned} \tag{3A.6}$$

It is easy to show that the matrices $(\mathbf{P}_K - \mathbf{P}_U)$ and $\mathbf{P}_{\mathbf{K}}^{\perp}$ are idempotent. The numerator $L_n(\mathbf{x})$ can be written under $\mathscr{H}_0$ as follows:

$$L_n(\mathbf{x}) = \left(\frac{\mathbf{x}}{\sigma}\right)^H (\mathbf{P}_K - \mathbf{P}_U)\left(\frac{\mathbf{x}}{\sigma}\right)^H = \left(\frac{\mathbf{U}\mathbf{c}_0 + \mathbf{n}_s}{\sigma}\right)^H (\mathbf{P}_{\mathbf{U}}^{\perp} - \mathbf{P}_{\mathbf{K}}^{\perp})\left(\frac{\mathbf{U}\mathbf{c}_0 + \mathbf{n}_s}{\sigma}\right) \tag{3A.7}$$

since $\mathbf{P}_{\mathbf{U}}^{\perp}\mathbf{U} = 0$ and $\mathbf{P}_{\mathbf{K}}^{\perp}\mathbf{U} = 0$, (3A.7) reduces to

$$L_n(\mathbf{x}) = \left(\frac{\mathbf{n}_s}{\sigma}\right)(\mathbf{P}_{\mathbf{U}}^{\perp} - \mathbf{P}_{\mathbf{K}}^{\perp})\left(\frac{\mathbf{n}_s}{\sigma}\right) \tag{3A.8}$$

where $(\frac{\mathbf{n}_s}{\sigma}) \sim \mathscr{CN}(0, \mathbf{I})$. It follows from Theorem 1 that $L_n(\mathbf{x}) \sim \chi_1^2$ under $\mathscr{H}_0$. It is easy to show that, under $\mathscr{H}_1$, we have $(\frac{\mathbf{x}}{\sigma}) \sim \mathscr{CN}(\frac{\mathbf{U}\xi + \mathbf{s}_0\alpha_0}{\sigma}, \mathbf{I})$ and, in turn, we get $L_n(\mathbf{x}) \sim \chi_1^2(\delta)$. Consequently,

$$L_n(\mathbf{x}) \sim \begin{cases} \chi_1^2 & \text{under } \mathscr{H}_0 \\ \chi_1^2(\delta) & \text{under } \mathscr{H}_1 \end{cases} \tag{3A.9}$$

From Theorem 1, the noncentrality parameters δ can be obtained as

$$\delta = \left(\frac{\alpha_0^*\mathbf{s}_0^H + \xi^H\mathbf{U}^H}{\sigma}\right)(\mathbf{P}_{\mathbf{U}}^{\perp} - \mathbf{P}_{\mathbf{K}}^{\perp})\left(\frac{\mathbf{s}_0\alpha_0 + \mathbf{U}\xi}{\sigma}\right) = \frac{|\alpha_0|^2}{\sigma^2}(\mathbf{s}_0^H\mathbf{P}_{\mathbf{U}}^{\perp}\mathbf{s}_0) \tag{3A.10}$$

since $\mathbf{P}_{\mathbf{U}}^{\perp}\mathbf{U} = 0$, $\mathbf{P}_{\mathbf{K}}^{\perp}\mathbf{U} = 0$, and $\mathbf{P}_{\mathbf{K}}^{\perp}\mathbf{s}_0 = 0$, we have

$$\delta = \frac{|\alpha_0|^2}{\sigma^2}(\mathbf{s}_0^H\mathbf{P}_{\mathbf{U}}^{\perp}\mathbf{s}_0) \tag{3A.11}$$

since $\mathbf{P}_{\mathbf{K}}^{\perp}\mathbf{U} = 0$ and $\mathbf{P}_{\mathbf{K}}^{\perp}\mathbf{K} = 0$, the denominator of the test under $\mathscr{H}_i$ can be written as

$$L_d(\mathbf{x}) = \left(\frac{\mathbf{n}_s}{\sigma}\right)^H \mathbf{P}_{\mathbf{K}}^{\perp}\left(\frac{\mathbf{n}_s}{\sigma}\right)^H \tag{3A.12}$$

From Theorem 1 and (3A.6), it follows that

$$L_d(\mathbf{x}) \sim \begin{cases} \chi_{N-(P+1)}^2 & \text{under } \mathscr{H}_0 \\ \chi_{N-(P+1)}^2 & \text{under } \mathscr{H}_1 \end{cases} \tag{3A.13}$$

The independence of the numerator and denominator of the test can be demonstrated by utilizing Theorem 2.

Theorem 2. *Let* $\mathbf{x} \sim \mathscr{CN}(\mu, \mathbf{C})$ *be a complex* $N \times 1$ *vector, and let* $\mathbf{A}$ *and* $\mathbf{B}$ *be two given* $N \times N$ *Hermitian matrices. If* $\mathbf{ACB} = \mathbf{0}$, *then the quadratic forms* $\mathbf{x}^H\mathbf{Ax}$ *and* $\mathbf{x}^H\mathbf{Bx}$ *are independent [18].*

Making use of Theorem 2, we have from (3A.4) that

$$\mathbf{P}_{\mathbf{K}}^{\perp}(\mathbf{P}_K - \mathbf{P}_U) = -\mathbf{P}_{\mathbf{K}}^{\perp}\mathbf{P}_U = \left(\mathbf{P}_{\mathbf{U}}^{\perp} + \frac{\mathbf{P}_{\mathbf{U}}^{\perp}\mathbf{s}_0\mathbf{s}_0^H\mathbf{P}_{\mathbf{U}}^{\perp}}{\mathbf{s}_0^H\mathbf{P}_{\mathbf{U}}^{\perp}\mathbf{s}_0}\right)\mathbf{P}_U = 0. \tag{3A.14}$$

This means that the numerator and denominator of the test are independent to obtain

$$L(\mathbf{x}) \sim \begin{cases} F_{1,N-(P+1)} & \text{under } \mathscr{H}_0 \\ F'_{1,N-(P+1)}(\delta) & \text{under } \mathscr{H}_1 \end{cases} \tag{3A.15}$$

where $\delta = \frac{|\alpha_0|^2}{\sigma^2}\|\mathbf{P}_{\mathbf{U}}^{\perp}\mathbf{s}_0\|^2 = \frac{|\alpha_0|^2}{\sigma^2}(\mathbf{s}_0^H\mathbf{P}_{\mathbf{U}}^{\perp}\mathbf{s}_0)$. Here, $F_{1,N-(P+1)}$ denotes a central complex F distribution with 1 numerator complex degree of freedom and $N-(P+1)$ denominator complex degree of freedom, and $F'_{1,N-(P+1)}(\delta)$ denotes an noncentral complex F distribution with 1 numerator complex degree of freedom and $N-(P+1)$ denominator complex degree of freedom and the noncentrality parameter δ.

Chapter 4

Multitarget detection in multiband FM-based passive bistatic radar: target detection quality improvement

4.1 Introduction

Passive bistatic radars (PBRs) that rely on single-band frequency modulation (FM) suffer from inconsistent detection performance due to the varying content of the transmitted signal. To overcome this issue, researchers have demonstrated that the joint utilization of multiple channels broadcasted from a single FM transmitter can improve target detection quality in FM-based PBR (FBPBR) systems References 1–6. In these works, two experimental FBPBR prototypes have been developed to show the effectiveness of multi-frequency detection to improve target detection performance or target range resolution (RR). The authors of these works have relied on conventional target detection algorithms, and the results they have presented are derived only from empirical observations without any mathematical validation. However, in this chapter, a systematic framework covering signal modeling, detection method, and statistical analysis is presented.

In this chapter, we aim to improve the quality of target detection by utilizing multiple broadcasted channels from a single transmitter. The problem of detecting a target in the presence of interference signals is formulated as a composite hypothesis test. To address this, we derive a multiband uniformly most powerful invariant (UMPI) test, which is an optimal invariant detector. False alarm and detection probabilities are calculated for the Swerling 0 (SW0) and Swerling 1 (SW1) target models. The proposed multiband detector is shown to be robust against the time-varying program content of FM radio channels. Additionally, a multiband PBR system offers combined diversity gain and frequency diversity gain, thanks to the joint exploitation of multiple broadcasted channels and independent returns from one target over different frequency channels. This chapter is based on previous works published in Reference 7.

The sections of this chapter are structured as follows: Section 4.2 presents the formulation of the detection problem as a composite hypothesis-testing problem. Section 4.3 discusses the derivation of a two-step UMPI test for the multiband passive radar target detection problem. Section 4.4 provides an analytical performance assessment for both false alarm and detection probabilities. Section 4.5 examines the performance of the proposed UMPI detector. Last, section 4.6 summarizes this chapter.

4.2 Signal modeling

We can consider a PBR system that takes advantage of M channels broadcasted from a single transmitter. In order to represent the signal received from the mth channel, where $m = 1, \ldots, M$, we will use the signal model that has been proposed for the case of a single channel as discussed in Chapter 3. This model will be adapted to account for multiple channels. Specifically, we will denote the nth time sample of the baseband equivalent signal received by the surveillance channel due to the mth broadcasted channel as $r_s^{(m)}[n]$, where

$$\begin{aligned} r_s^{(m)}[n] &= c_d^{(0,m)} r_r^{(m)}[n] + \sum_{i=1}^{N_c^{(m)}} c_n^{(i,m)} r_r^{(m)}[n - n_{c0}^{(i,m)}] \\ &+ \sum_{k=0}^{K_m-1} \alpha_t^{(k,m)} r_r^{(m)}[n - n_t^{(k,m)}] e^{j2\pi(\frac{v_t^{(k)}}{\lambda_m})nT_s} + n_s^{(m)}[n] \end{aligned} \tag{4.1}$$

with $n = 0, \ldots, N_m - 1$ and $m = 1, \ldots, M$, where

- The set of wavelengths transmitted by radio channels is represented by $\{\lambda_m\}_{m=1}^{M}$, where M denotes the number of channels utilized in the detection algorithm. The sampling frequency, f_s, is equal to $\frac{1}{T_s}$ and satisfies the Nyquist rate. The length of the data vector, N_m, corresponds to the integration time of the mth received surveillance channel, denoted by T_m, which is equal to $N_m T_s$.
- The nth time sample of the reference channel resulting from the mth broadcasted channel is denoted by $r_r^{(m)}[n]$. Prior to this, proper signal conditioning is carried out to eliminate the influence of multipath and target echoes in the reference channel. The parameters $\alpha_t^{(k,m)}$, $n_t^{(k,m)}$, and $v_t^{(k)}$ indicate the complex amplitudes, bistatic delays, and velocities of the kth target in the mth frequency channel, respectively. The value of k ranges from 1 to K_m, where K_m represents the maximum number of targets to be detected in the mth frequency channel.
- The complex amplitude of the direct-path signal picked up by the sidelobe/backlobe antenna in the surveillance channel from the mth broadcasted channel is denoted as $c_d^{(0,m)}$.
- The variables $c_n^{(i,m)}$ and $n_{c0}^{(i,m)}$ denote the time-varying amplitudes and bistatic delays, respectively, of the clutter scatterers resulting from the mth broadcasted channel. These clutter scatterers are caused by nearby ground scatterers, and $N_c^{(m)}$ represents the equivalent number of clutter echoes, where $i = 1, \ldots, N_c^{(m)}$.
- $n_s^{(m)}[n]$ represents the noise sample obtained from the mth surveillance channel at the nth instance.

When echoes from stationary objects such as ground and buildings are received, it is accurate to consider clutter components with constant amplitudes over the integration time. However, the presence of slow-moving objects like grass and forest can lead to fluctuations in clutter amplitudes over time. To address this, the time-varying amplitudes of clutter are modeled as a slowly changing function of time, denoted by $c_n^{(i,m)} = \sum_{q=-Q_i}^{Q_i} c^{(i,q,m)} e^{j2\pi f_c^{(i,q,m)} nT_s}$. In this model, $c^{(i,q,m)}$ represents constant clutter amplitudes at the delay–Doppler coordinates of $(n_{c_0}^{(i,m)}, f_c^{(i,q,m)})$, where $f_c^{(i,q,m)} = \frac{v_c^{(i,q)}}{\lambda_m}$ and $v_c^{(i,q)}$ denote the Doppler velocities of clutter scatterers, which are determined by the velocity spread in the clutter spectrum. To illustrate, surface clutter typically has a small range of velocities due to the movement of the clutter scatterers, which results in a narrow Doppler velocity range for stationary radar. For terrain, the velocity range varies from nearly zero for rocky surfaces to about 0.33 m/s for trees affected by wind [8]. Alternatively, assuming an upper limit for the velocity range of surface clutter, we can choose $v_c^{(i,q)}$ uniformly distributed across the clutter velocity range [9]. For example, we can select $v_c^{(i,q)} = q\Delta v_c$ for $q = -Q_i, ..., Q_i$, with Q_i and Δv_c being two parameters that need to be appropriately determined based on the velocity range of the clutter spectrum [8, 9]. For ease of notation, we henceforth consider the maximum velocity range for all delays of clutter, i.e., $Q = \max_i\{Q_i\}$ for $i = 1, ..., N_c^{(m)}$ and $m = 1, ..., M$. Additionally, it is generally understood that the clutter echoes result from nearby ground scatterers distributed from zero to a maximum bistatic range of $R_c^{(m)}$ [9–11] and that the spectrum of a clutter scatterer can be approximated by a set of velocities denoted by $\{v_c^{(i,q)}\}_{q=-Q}^{Q}$. Taking these factors into consideration, we can define a limited delay–velocity extent, known as the clutter region, for receiving possible clutter echoes. In this scenario, we can assume that there are P_m components with unknown amplitudes $\alpha_c^{(p,m)}$ located at specific delay–velocity coordinates $(n_c^{(p,m)}, v_c^{(p)})$ within the clutter region, including the direct signal and clutter echoes. P_m is determined by $P_m = (2Q+1)N_c^{(m)} + 1$, where $N_c^{(m)}$ represents the number of ground scatterers up to the bistatic clutter cut-off range of $R_c^{(m)}$, and $(2Q+1)$ represents the maximum number of possible Doppler velocity components of a clutter scatterer.

In the presence of the system thermal noise, direct signal, clutter/multipath echoes, and $K_m - 1$ interfering targets, the passive radar target detection problem at the hypothesized delay–velocity coordinate of (v_0, n_0) can be represented as the binary composite hypothesis-testing problem, given by

$$\begin{cases} \mathscr{H}_0\colon r_s^{(m)}[n] = \sum_{p=1}^{P_m-1} \alpha_c^{(p,m)} r_r^{(m)}[n - n_c^{(p,m)}] e^{jn\Omega_c^{(p,m)}} \\ \qquad + \sum_{k=1}^{K_m-1} \alpha_t^{(k,m)} r_r^{(m)}[n - n_t^{(k,m)}] e^{jn\Omega_t^{(k,m)}} + n_s^{(m)}[n], \\ \mathscr{H}_1\colon r_s^{(m)}[n] = \alpha_0^{(m)} r_r^{(m)}[n - n_0] e^{jn\Omega_t^{(0,m)}} \\ \qquad + \sum_{p=1}^{P_m-1} \alpha_c^{(p,m)} r_r^{(m)}[n - n_c^{(p,m)}] e^{jn\Omega_c^{(p,m)}} \\ \qquad + \sum_{k=1}^{K_m-1} \alpha_t^{(k,m)} r_r^{(m)}[n - n_t^{(k,m)}] e^{jn\Omega_t^{(k,m)}} \\ \qquad + n_s^{(m)}[n] \end{cases} \tag{4.2}$$

for $n = 0, \ldots, N_m - 1$ and $m = 1, \ldots, M$. Here, $\Omega_c^{(p,m)} = 2\pi(\frac{v_c^{(p)}}{\lambda_m})T_s$ and $\Omega_t^{(k,m)} = 2\pi(\frac{v_t^{(k)}}{\lambda_m})T_s$ are normalized Doppler frequencies of clutter and target echoes, respectively. To solve this multiband detection problem, the following assumptions are assumed to hold:

1. The interfering target signal amplitudes $\alpha_t^{(k,m)}$, $k = 1, \ldots, K_m - 1$ and $m = 1, \ldots, M$ are assumed to be complex-valued, deterministic, and unknown. Furthermore, the number of interfering targets, denoted as $K_m - 1$, their respective delays denoted as $n_t^{(k,m)}$, and their bistatic velocities denoted as $v_t^{(k)}$ are also assumed to be unknown.
2. In practical situations, the actual velocity–delay coordinates of the ground scatterers are not known and can be situated anywhere within the clutter region. Therefore, we assume that we are dealing with the worst-case scenario of receiving clutter echoes, and thus, we take into account all of the delay–velocity coordinates of the clutter region. However, it is worth noting that the clutter amplitudes $\alpha_c^{(p,m)}$ for any delay–velocity coordinate within the clutter region are considered unknown complex parameters. Additionally, three parameters of the clutter region, namely Q, Δv_c, and $R_c^{(m)}$, should be set appropriately in practical situations based on the maximum Doppler velocity spread of the clutter spectrum and the radar's location.
3. It is assumed that the random variables $n_s^{(m)}[n]$ for $n = 1, ..., N_m$ are independent and identically distributed samples of complex Gaussian white noise with zero mean and an unknown variance of σ^2. Additionally, it is assumed that the thermal noise of the surveillance channels within the system is also independent.

The received signal (4.1) has the following compact form:

$$\mathbf{r} = \mathbf{S}\mathbf{a}_0 + \mathbf{C}\mathbf{a}_c + \mathbf{T}\mathbf{a}_t + \mathbf{n}_s \tag{4.3}$$

where $\mathbf{r} = [\mathbf{r}_s^{(1)^T}, ..., \mathbf{r}_s^{(M)^T}]^T$ is the N_e-dimensional column vector containing all observed data from the M broadcasted channels, in which $N_e = \sum_{m=1}^{M} N_m$ and $\mathbf{r}_s^{(m)} = [r_s^{(m)}[0], ..., r_s^{(m)}[N_m - 1]]^T$ for $m = 1, ..., M$; $\mathbf{a}_0 = [\alpha_0^{(1)}, ..., \alpha_0^{(M)}]^T$ is the M-dimensional column vector containing all complex amplitudes of the desired target in the M broadcasted channels; $\mathbf{a}_t = [\mathbf{a}_t^{(1)^T}, ..., \mathbf{a}_t^{(M)^T}]^T$ is the K_e-dimensional column vector containing all complex amplitudes of the interfering targets from the M broadcasted channels, in which $K_e = \sum_{m=1}^{M}(K_m - 1)$ and $\mathbf{a}_t^{(m)} = [\alpha_t^{(1,m)}, ..., \alpha_t^{(K_m-1,m)}]^T$ for $m = 1, ..., M$; $\mathbf{a}_c = [\mathbf{a}_c^{(1)^T}, ..., \mathbf{a}_c^{(M)^T}]^T$ is the P_e-dimensional column vector containing all complex amplitudes of direct signal and clutter scatterers received from the M broadcasted channels with $P_e = \sum_{m=1}^{M} P_m$ and $\mathbf{a}_c^{(m)} = [\alpha_c^{(1,m)}, ..., \alpha_c^{(P_m,m)}]^T$ for $m = 1, ..., M$; and $\mathbf{n}_s = [\mathbf{n}_s^{(1)^T}, ..., \mathbf{n}_s^{(M)^T}]^T$ is the N_e-dimensional column vector containing all noise vectors from the M broadcasted channels in which $\mathbf{n}_s^{(m)} = [n_s^{(m)}[0], ..., n_s^{(m)}[N_m - 1]]^T$ for $m = 1, ..., M$. In (4.3), the matrices **S**, **C**, and **T** are, respectively, $N_e \times M$, $N_e \times P_e$, and $N_e \times K_e$ block diagonal matrices, represented by

$$\mathbf{S} \triangleq \text{Diag}\left(\mathbf{s}^{(1)}, ..., \mathbf{s}^{(M)}\right) \tag{4.4}$$

and

$$\mathbf{C} \triangleq \text{Diag}\left(\mathbf{C}^{(1)}, ..., \mathbf{C}^{(M)}\right) \tag{4.5}$$

and

$$\mathbf{T} \triangleq \mathrm{Diag}\left(\mathbf{T}^{(1)}, ..., \mathbf{T}^{(M)}\right) \tag{4.6}$$

where for $m = 1, ..., M$, we have $\mathbf{s}^{(m)} = (\mathbf{r}_r^{(n_0,m)} \odot \mathbf{e}^{(v_0,m)})$ in which $\mathbf{r}_r^{(n_0,m)} = \mathbf{P}^{n_0}\mathbf{r}_r^{(m)}$ and $[\mathbf{e}^{(v_0,m)}]_n = e^{j2\pi(\frac{v_0}{\lambda^{(m)}})nT_s}$ for $n = 0, ..., N_m - 1$; $\mathbf{C}^{(m)} = [\mathbf{c}_1^{(m)}, \ldots, \mathbf{c}_p^{(m)}, \ldots, \mathbf{c}_{P_m}^{(m)}]$ is an $N_m \times P_m$ matrix with $\mathbf{c}_p^{(m)} = (\mathbf{r}_r^{(n_c^{(p,m)},m)} \odot \mathbf{e}^{(v_c^{(p,m)},m)})$ in which $n_c^{(p,m)}$ and $v_c^{(p,m)}$ are determined based on the delay–velocity coordinates of the clutter region for $p = 1, ..., P_m$; $\mathbf{T}^{(m)} = [\mathbf{t}_1^{(m)}, ..., \mathbf{t}_k^{(m)}, ..., \mathbf{t}_{K_m-1}^{(m)}]$ is an $N_m \times (K_m - 1)$ matrix with $\mathbf{t}_k^{(m)} = (\mathbf{r}_r^{(n_t^{(k,m)},m)} \odot \mathbf{e}^{(v_t^{(k,m)},m)})$ for $n = 0, ..., N_m - 1$. In this case, $\mathbf{P}$ is an $N_m \times N_m$ permutation matrix defined as $[\mathbf{P}]_{ij} = 1$ if $i = j + 1$ and 0 otherwise for $i = 0, ..., N_m - 1$ and $j = 0, ..., N_m - 1$.

4.3 Design of UMPI-based detector

The composite binary hypothesis-testing problem (4.2) can be compactly formulated as

$$\begin{cases} \mathcal{H}_0\colon \mathbf{r} = \mathbf{C}\mathbf{a}_c + \mathbf{T}\mathbf{a}_t + \mathbf{n}_s \\ \mathcal{H}_1\colon \mathbf{r} = \mathbf{S}\mathbf{a}_0 + \mathbf{C}\mathbf{a}_c + \mathbf{T}\mathbf{a}_t + \mathbf{n}_s \end{cases} \tag{4.7}$$

The goal is to develop a target detection algorithm that utilizes the UMPI test in the presence of noise, clutter/multipath, and interfering targets with unknown parameters. However, obtaining the UMPI test directly is not possible due to the large number of unknowns such as $\mathbf{T}$, $\mathbf{a}_0$, $\mathbf{a}_c$, $\mathbf{a}_t$, and σ^2. To overcome this challenge, a two-stage UMPI-based detector is proposed instead of a one-stage detector. Initially, the detector assumes that the matrix $\mathbf{T}$ is known and applies the UMPI detector using the classical approach on the remaining unknowns $\mathbf{a}_0$, $\mathbf{a}_c$, $\mathbf{a}_t$, and σ^2. Subsequently, the estimated matrix from the second stage is used in place of the true matrix $\mathbf{T}$ in the UMPI test. To accomplish this, a multistage algorithm is introduced in the second stage of the detector design. The first and second stages are explained in detail below.

4.3.1 First stage of UMPI-based detector

In the first stage of the detector design, it is assumed that the matrix $\mathbf{T}$ is known, and the UMPI detector is derived according to the classical approach over the remaining unknowns $\mathbf{a}_0$, $\mathbf{a}_c$, $\mathbf{a}_t$, and σ^2. The classical approach to derive a UMPI test is as follows [12, 13]:

1. Express the problem invariance in terms of a transformation group,
2. Determine the maximal invariant (MI) statistic,
3. Calculate the probability density functions (pdfs) of the MI for each hypothesis,
4. Create the likelihood ratio of the MI,
5. Establish that the detector statistic has a monotone likelihood ratio (MLR).

Even though this method may appear challenging to implement in numerous scenarios, we employ it to devise a passive radar UMPI detector. However, prior to proceeding, we apply some transformations to the observation data $\mathbf{r}$ to simplify the detection problem. Initially, we examine the $NN_e \times (N_e - P_e)$ matrix, which is orthogonal and given by

$$\mathbf{U}_C \triangleq \text{Diag}\left(\mathbf{U}_C^{(2,1)}, \ldots, \mathbf{U}_C^{(2,M)}\right) \tag{4.8}$$

where the diagonal element of $\mathbf{U}_C$ can be obtained using the singular eigenvalue decomposition (SVD) of $\mathbf{C}^{(m)}$ such that $\mathbf{U}_C^{(2,m)^H}\mathbf{C}^{(m)} = \mathbf{0}$ for $m = 1, \ldots, M$ [14]. Here, $\mathbf{U}_C^{(2,m)}$ is $N_m \times (N_m - P_m)$ orthogonal matrix that spans the orthogonal complement of the columns space of matrix $\mathbf{C}^{(m)}$. Left-multiplying the observation vector $\mathbf{r}$ by $\mathbf{U}_C^H$ results in

$$\mathbf{x}_C \triangleq \mathbf{U}_C^H\mathbf{r} = \mathbf{S}_C\mathbf{a}_0 + \mathbf{W}\mathbf{a}_t + \mathbf{n}_C \tag{4.9}$$

where $\mathbf{x}_C$ is the $(N_e - P_e)$-dimensional column vector, block diagonal matrix $\mathbf{S}_C$ is defined as $\mathbf{S}_C \triangleq \mathbf{U}_C^H\mathbf{S}$ with diagonal element defined by $\mathbf{s}_C{}^{(m)} = \mathbf{U}_C^{(2,m)^H}\mathbf{s}^{(m)}$, block diagonal matrix $\mathbf{W}$ is defined as $\mathbf{W} \triangleq \mathbf{U}_C^H\mathbf{T}$ with an $(N_m - P_m) \times (K_m - 1)$ diagonal element $\mathbf{W}^{(m)} = \mathbf{U}_C^{(2,m)^H}\mathbf{T}^{(m)}$, and $\mathbf{n}_C \triangleq \mathbf{U}_C^H\mathbf{n}$ with subvector $\mathbf{n}_C^{(m)} = \mathbf{U}_C^{(2,m)^H}\mathbf{n}_s^{(m)}$ for $m = 1, \ldots, M$. Since $\mathbf{U}_C^{(2,m)^H}\mathbf{U}_C^{(2,m)} = \mathbf{I}_{(N_m - P_m)}$ and $\mathbf{n}_s^{(m)}$ is zero mean white Gaussian noise with the covariance matrix of $\sigma^2\mathbf{I}_{N_m}$, the transformed noise $\mathbf{n}_C^{(m)}$ is also zero mean, white, and Gaussian noise with the covariance matrix of $\sigma^2\mathbf{I}_{(N_m - P_m)}$.

Again, let us define another orthogonal $(N_e - P_e) \times K_e$ matrix, given by

$$\mathbf{U}_w \triangleq \text{Diag}\left(\mathbf{U}_w^{(2,1)}, \ldots, \mathbf{U}_w^{(2,M)}\right) \tag{4.10}$$

The diagonal element of $\mathbf{U}_w$ can be obtained using the SVD of $\mathbf{W}^{(m)}$ such that $\mathbf{U}_w^{(2,m)^H}\mathbf{W}^{(m)} = \mathbf{0}$ for $m = 1, \ldots, M$. Here, $\mathbf{U}_w^{(2,m)}$ is $(N_m - P_m) \times (N_m - P_m - K_m + 1)$ orthogonal matrix that spans the orthogonal complement of the columns space of matrix $\mathbf{W}^{(m)}$. Left-multiplying the observation data $\mathbf{x}_C$ by $\mathbf{U}_w^H$ leads to

$$\mathbf{x} \triangleq \mathbf{U}_w^H\mathbf{x}_C = \mathbf{S}_w\mathbf{a}_0 + \mathbf{n}_w \tag{4.11}$$

where $\mathbf{x} = [\mathbf{x}^{(1)^T}, \ldots, \mathbf{x}^{(M)^T}]^T$ is the $(N_e - P_e - K_e)$-dimensional column vector, block diagonal matrix $\mathbf{S}_w$ is defined as $\mathbf{S}_w \triangleq \mathbf{U}_w^H\mathbf{S}_c$ with diagonal element $\mathbf{s}_w^{(m)} = \mathbf{U}_w^{(2,m)^H}\mathbf{s}_c^{(m)}$ and $\mathbf{n}_w \triangleq \mathbf{U}_w^H\mathbf{n}_c$ with subvector $\mathbf{n}_w^{(m)} = \mathbf{U}_w^{(2,m)^H}\mathbf{n}_c^{(m)}$ for $m = 1, \ldots, M$. Since $\mathbf{U}_w^{(2,m)^H}\mathbf{U}_w^{(2,m)} = \mathbf{I}_{(N_m - P_m - K_m + 1)}$ and $\mathbf{n}_c^{(m)}$ is zero mean white Gaussian noise with the covariance matrix of $\sigma^2\mathbf{I}_{(N_m - P_m)}$, the transformed noise $\mathbf{n}_w^{(m)}$ is also zero mean, white, and Gaussian noise with the covariance matrix of $\sigma^2\mathbf{I}_{(N_m - P_m - K_m + 1)}$. Now, the detection problem (4.7) can be simplified as

$$\begin{cases} \mathcal{H}_0: \mathbf{x} = \mathbf{n}_w \\ \mathcal{H}_1: \mathbf{x} = \mathbf{S}_w\mathbf{a}_0 + \mathbf{n}_w \end{cases} \tag{4.12}$$

To express the detection problem (4.12) in terms of transformation groups, we will show that this detection problem is invariant under the composition of the transformation groups G_Q and G_d, given by

$$G_Q = \{g_Q\colon g_Q(\mathbf{x}) = \mathbf{Q}^l_{\mathbf{s}_w}\mathbf{x}\} \tag{4.13}$$

where l is an integer value, and $\mathbf{Q}_{\mathbf{s}_w}$ is a block diagonal matrix defined as

$$\mathbf{Q}_{\mathbf{s}_w} \triangleq \mathrm{Diag}\left(\mathbf{Q}_{\mathbf{s}_w^{(1)}}, \ldots, \mathbf{Q}_{\mathbf{s}_w^{(M)}}\right) \tag{4.14}$$

Here, $\mathbf{Q}_{\mathbf{s}_w^{(m)}}$ matrices are Householder matrices with Householder vectors $\mathbf{s}_w^{(m)}$ defined as $\mathbf{Q}_{\mathbf{s}_w^{(m)}} = \mathbf{I}_{(N_m-P_m-K_m+1)} - 2\frac{\mathbf{s}_w^{(m)}\mathbf{s}_w^{(m)H}}{\|\mathbf{s}_w^{(m)}\|^2}$ for $m = 1, \ldots, M$, and

$$G_d = \{g_d\colon g_d(\mathbf{x}) = d\mathbf{x},\ d \text{ is a complex scalar}\} \tag{4.15}$$

The transformations, G_Q and G_d, meet the requirements for being groups as they are both closed sets, associative, and contain both the identity and inverse elements. It can be proven that when the transformation groups in (4.13) and (4.15) are combined, the distribution of the observation and parameter spaces remains unaltered for each hypothesis. This result can be proven due to the following:

- Under $\mathscr{H}_1$, we have $\mathbf{x}^{(m)} \sim \mathscr{CN}(\mathbf{s}_w^{(m)}\alpha_0^{(m)}, \sigma^2\mathbf{I}_{(N_m-P_m-K_m+1)})$. Thus, for the transformed data, we have

$$g_{Q^{(m)}}(\mathbf{x}^{(m)}) = \mathbf{Q}_{\mathbf{s}_w^{(m)}}\mathbf{x}^{(m)} \sim \mathscr{CN}(\mathbf{s}_w^{(m)}(-1)^l\alpha_0^{(m)}, \sigma^2\mathbf{I}_{(N_m-P_m-K_m+1)}) \tag{4.16}$$

  for $m = 1, \ldots, M$. Since $\alpha_0^{(m)}$, $m = 1, \ldots, M$, are unknown complex values, the distribution family of the transformed data does not change. Under the null hypothesis, the proof is similar, except that $\alpha_0^{(m)} = 0$. Thus, the detection problem is invariant under $g_{Q^{(m)}}(\mathbf{x}^{(m)})$ transformation for $m = 1, \ldots, M$.
- Under $\mathscr{H}_1$, from $\mathbf{x}^{(m)} \sim \mathscr{CN}(\mathbf{s}_w^{(m)}\alpha_0^{(m)}, \sigma^2\mathbf{I}_{(N_m-P_m-K_m+1)})$, we get $g_d(\mathbf{x}^{(m)}) = d\mathbf{x}^{(m)} \sim \mathscr{CN}(\mathbf{s}_w^{(m)}d\alpha_0^{(m)}, |d|^2\sigma^2\mathbf{I}_{(N_m-P_m-K_m+1)})$ for $m = 1, \ldots, M$. Since $\alpha_0^{(m)}$, $m = 1, \ldots, M$, and σ^2 are unknown complex and real values, respectively, the distribution family of the transformed data is not changed. Under the null hypothesis, the proof is similar, except that $\alpha_0^{(m)} = 0$. Therefore, the detection problem is invariant under $g_d(\mathbf{x}^{(m)})$ transformation for $m = 1, \ldots, M$.

The initial phase of deriving the classical UMPI is now finished, which demonstrates that the problem under consideration remains unchanged when subjected to transformation groups belonging to G_Q and G_d. The next stage involves identifying an MI statistic, which can be described as follows. Formally, the statistic $M(\mathbf{x})$ is MI under the group G if and only if

(Invariancy): $M(\mathbf{x}) = M(g(\mathbf{x}))$, for all $\mathbf{x}$, and $g \in G$

(Maximality): $M(\mathbf{x}_1) = M(\mathbf{x}_2)$ implies $\mathbf{x}_2 = g(\mathbf{x}_1)$ for some $g \in G$.

To find MI statistic in the case of the two transformation groups G_Q and G_d, we can resort to the following theorem:

Theorem 1. *Let G be a group of transformation, and let G_Q and G_d be two subgroups generating G. Suppose that there exists a maximal invariant $\mathbf{y} = M_Q(\mathbf{x})$ with respect to G_Q and for any $g_d \in G_d$ and any $\mathbf{x}_1, \mathbf{x}_2 \in \mathscr{C}_x^N$ we have*

$$M_Q(\mathbf{x}_1) = M_Q(\mathbf{x}_2) \Rightarrow M_Q(g_d(\mathbf{x}_1)) = M_Q(g_d(\mathbf{x}_2)) \tag{4.17}$$

and there exists an MI statistic $\mathbf{z} = M_H(\mathbf{y})$ under the group H of transformations defined by $g_H(\mathbf{y}) \triangleq M_Q(g_d(\mathbf{x}))$, then $\mathbf{z} = M_H(M_Q(\mathbf{x}))$ is MI with respect to G.
Proof. (see, e.g., Reference 12, pp. 217, Chapter 6, Theorem 6.2.2).

According to this theorem, it is possible to find the MI in two steps by only calculating the MI for one of the subgroups in each step. Additionally, this theorem allows us to confirm if the MI determination process is feasible. Initially, it can be demonstrated that an MI for G_Q can be obtained as

$$\mathbf{y} = M_Q(\mathbf{x}) = \left[\sum_{m=1}^{M} \frac{|\mathbf{s}_w^{(m)H}\mathbf{x}^{(m)}|^2}{\|\mathbf{s}_w^{(m)}\|^2}, \|\mathbf{x}\|^2\right] \tag{4.18}$$

This statistic is invariant since, for all $g_Q \in G_Q$, we have

$$M_Q(g_Q(\mathbf{x})) = \left[\sum_{m=1}^{M} \frac{\left|\left(\mathbf{Q}^l_{\mathbf{s}_w^{(m)}}\mathbf{s}_w^{(m)}\right)^H \mathbf{x}^{(m)}\right|^2}{\|\mathbf{s}_w^{(m)}\|^2}, \|\mathbf{Q}^l_{\mathbf{S}_w}\mathbf{x}\|^2\right] \tag{4.19}$$

where $\mathbf{Q}_{\mathbf{S}_w}$ is a block diagonal matrix defined as $\mathbf{Q}_{\mathbf{S}_w} \triangleq \text{Diag}\left(\mathbf{Q}_{\mathbf{s}_w^{(1)}}, \dots, \mathbf{Q}_{\mathbf{s}_w^{(M)}}\right)$. Here, $\mathbf{Q}_{\mathbf{s}_w^{(m)}} = \mathbf{I}_{(N_m - P_m - K_m + 1)} - 2\frac{\mathbf{s}_w^{(m)}\mathbf{s}_w^{(m)H}}{\|\mathbf{s}_w^{(m)}\|^2}$ for $m = 1, \dots, M$. Since $\mathbf{Q}^l_{\mathbf{s}_w^{(m)}}\mathbf{s}_w^{(m)} = (-1)^l\mathbf{s}_w^{(m)}$, $\mathbf{Q}^H_{\mathbf{s}_w^{(m)}} = \mathbf{Q}_{\mathbf{s}_w^{(m)}}$, and $\mathbf{Q}_{\mathbf{s}_w^{(m)}}\mathbf{Q}_{\mathbf{s}_w^{(m)}} = \mathbf{I}$ for $m = 1, \dots, M$ and any integer value of l, it can be concluded that $M_Q(g_Q(\mathbf{x})) = M_Q(\mathbf{x})$. This shows that the statistic $\mathbf{y}$ is invariant. To see the maximality of $\mathbf{y}$, suppose that for given $\mathbf{x}_1$ and $\mathbf{x}_2$, we have $M_Q(\mathbf{x}_1) = M_Q(\mathbf{x}_2)$. Now, it follows from (4.18) that

$$\begin{cases} \sum_{m=1}^{M} \frac{\left|\mathbf{s}_w^{(m)H}\mathbf{x}_1^{(m)}\right|^2}{\|\mathbf{s}_w^{(m)}\|^2} = \sum_{m=1}^{M} \frac{\left|\mathbf{s}_w^{(m)H}\mathbf{x}_2^{(m)}\right|^2}{\|\mathbf{s}_w^{(m)}\|^2} \\ \|\mathbf{x}_1\|^2 = \|\mathbf{x}_2\|^2 \end{cases} \tag{4.20}$$

where $\mathbf{x}_i = [\mathbf{x}_i^{(1)^T}, \dots, \mathbf{x}_i^{(M)^T}]^T$ for $i = 1, 2$. Let us define matrix $\mathbf{S}_q$ as $\mathbf{S}_q \triangleq \text{Diag}\left(\frac{\mathbf{s}_w^{(1)}}{\|\mathbf{s}_w^{(1)}\|}, \dots, \frac{\mathbf{s}_w^{(M)}}{\|\mathbf{s}_w^{(M)}\|}\right)$; this may help us to rewrite (4.20) as

$$\begin{cases} \|\mathbf{S}_q^H\mathbf{x}_2\|^2 = \|\mathbf{S}_q^H\mathbf{x}_1\|^2 \\ \|\mathbf{x}_1\|^2 = \|\mathbf{x}_2\|^2 \end{cases} \tag{4.21}$$

To proceed, we need to find some $g_Q \in G_Q$ such that $\mathbf{x}_2 = g_Q(\mathbf{x}_2) = \mathbf{Q}^l_{\mathbf{S}_w}\mathbf{x}_1$. It is important to note that $\mathbf{Q}_{\mathbf{S}_w}$ is a block diagonal matrix with diagonal elements related to the Householder matrices having Householder vectors $\mathbf{s}_w^{(m)}$ for $m = 1, \dots, M$. Thus, we must determine the value of l based on the given vectors $\mathbf{x}_1$ and $\mathbf{x}_2$ in order to prove maximality. We can easily observe that $\mathbf{Q}_{\mathbf{S}_w}^H = \mathbf{Q}_{\mathbf{S}_w}$, $\mathbf{Q}_{\mathbf{S}_w}^{2k+1}\mathbf{S}_q = -\mathbf{S}_q$, and $\mathbf{Q}_{\mathbf{S}_w}^{2k} = \mathbf{I}$. Therefore, for $\mathbf{x}_2 \neq \mathbf{x}_1$, l must be an odd integer such that $\mathbf{x}_2 = \mathbf{Q}_{\mathbf{S}_w}^{2k+1}\mathbf{x}_1$. On the other hand, for $\mathbf{x}_2 = \mathbf{x}_1$, l must be an even integer such that $\mathbf{x}_2 = \mathbf{Q}_{\mathbf{S}_w}^{2k}\mathbf{x}_1 = \mathbf{x}_1$. So far, we have demonstrated that the transformation $g_Q \in G_Q$ connects the two vectors $\mathbf{x}_1$ and $\mathbf{x}_2$. Thus, we can conclude that the statistic $\mathbf{y}$ is MI.

To drive an MI for second transformation group, it is necessary to confirm condition (4.17) by checking that $M_Q(g_d(\mathbf{x}_1)) = M_Q(d\mathbf{x}_1) = d\mathbf{Q}_{\mathbf{S}_w}^l\mathbf{x}_1 = dM_Q(\mathbf{x}_1)$. The assumption that $M_Q(\mathbf{x}_1) = M_Q(\mathbf{x}_2)$ leads to $M_Q(g_d(\mathbf{x}_1)) = M_Q(g_d(\mathbf{x}_2))$, thus fulfilling the theorem condition of (4.17). Subsequently, a group G_H that acts on $\mathbf{y}$ must be determined before finding an MI under that group. Theorem 1 states that the group can be represented as $g_H(\mathbf{y}) = M_Q(g_d(\mathbf{x})) = |d|^2\,\mathbf{y}$. The MI for group G_H is given by $z = M_H(\mathbf{y}) = M_H(y_1, y_2) = \frac{y_1}{y_2}$. This statistic is invariant under G_H, since $M_H(g_H(\mathbf{y})) = M_H(\mathbf{y})$. Moreover, if $M_H(\mathbf{y}^{(1)}) = M_H(\mathbf{y}^{(2)})$ for some given $\mathbf{y}^{(1)}$ and $\mathbf{y}^{(2)}$, we get $\frac{y_1^{(1)}}{y_2^{(1)}} = \frac{y_1^{(2)}}{y_2^{(2)}}$, implying $\frac{y_1^{(2)}}{y_1^{(1)}} = \frac{y_2^{(2)}}{y_2^{(1)}} = c$, where c is a positive scalar value. Therefore, $[y_1^{(2)}, y_2^{(2)}] = c[y_1^{(1)}, y_2^{(1)}]$, or $\mathbf{y}^{(2)} = g_H(\mathbf{y}^{(1)})$ for some $g_H \in G_H$. Thus, $M_H(\mathbf{y})$ is the MI under group G_H. Consequently, the MI under the composite group G can be expressed as

$$z = M_H(M_Q(\mathbf{x})) = \frac{\sum_{m=1}^{M} \frac{\left|\mathbf{s}_w^{(m)^H}\mathbf{x}^{(m)}\right|^2}{\|\mathbf{s}_w^{(m)}\|^2}}{\|\mathbf{x}\|^2} \tag{4.22}$$

In terms of original observation, the MI statistic can be written as

$$z = \frac{\sum_{m=1}^{M} \frac{\left|\mathbf{s}^{(m)^H}\mathbf{U}_C^{(2,m)}\mathbf{U}_W^{(2,m)}\mathbf{U}_W^{(2,m)^H}\mathbf{U}_C^{(2,m)^H}\mathbf{r}^{(m)}\right|^2}{\|\mathbf{U}_W^{(2,m)^H}\mathbf{U}_C^{(2,m)^H}\mathbf{s}^{(m)}\|^2}}{\sum_{m=1}^{M} \mathbf{r}^{(m)^H}\mathbf{U}_C^{(2,m)}\mathbf{U}_W^{(2,m)}\mathbf{U}_W^{(2,m)^H}\mathbf{U}_C^{(2,m)^H}\mathbf{r}^{(m)}} \tag{4.23}$$

After some matrix manipulations, we can rewrite the MI statistic in (4.23) as

$$z = \frac{\sum_{m=1}^{M} \frac{\left|\mathbf{s}^{(m)^H}\Pi_{\mathbf{U}^{(m)}}^{\perp}\mathbf{r}^{(m)}\right|^2}{\mathbf{s}^{(m)^H}\Pi_{\mathbf{U}^{(m)}}^{\perp}\mathbf{s}^{(m)}}}{\sum_{m=1}^{M} \mathbf{r}^{(m)^H}\Pi_{\mathbf{U}^{(m)}}^{\perp}\mathbf{r}^{(m)}} \tag{4.24}$$

where $\mathbf{U}^{(m)} = [\mathbf{C}^{(m)}, \mathbf{T}^{(m)}]$, and $\Pi_{\mathbf{U}^{(m)}}^{\perp}$ is the orthogonal projection onto the subspace span by columns of $\mathbf{U}^{(m)}$.

Appendix 4A presents the distribution of a 1–1 function of z denoted by t. This function is defined as $t = \frac{N_e - P_e - K_t}{M}\frac{z}{1-z}$, where $K_t = K_e + M$. According to the results presented, t follows a central complex F distribution under the null hypothesis $\mathcal{H}_0$, while it follows a noncentral complex F distribution under the alternative hypothesis $\mathcal{H}_1$. It is worth noting that previous studies [12] have shown that the noncentral complex F distribution satisfies the MLR property of t. This implies that instead of performing a threshold test on the likelihood ratio, it is possible to conduct the test by setting a threshold on the statistic t.

Therefore, the UMPI test can be stated as rejecting the null hypothesis $\mathcal{H}_0$ if $t > \eta_t$, or equivalently,

$$L_{\mathrm{UMPI}}(\mathbf{r}) = \frac{\sum_{m=1}^{M} \frac{\left|\mathbf{s}^{(m)H} \Pi_{\mathbf{U}^{(m)}}^{\perp} \mathbf{r}^{(m)}\right|^2}{\mathbf{s}^{(m)H} \Pi_{\mathbf{U}^{(m)}}^{\perp} \mathbf{s}^{(m)}}}{\sum_{m=1}^{M} \mathbf{r}^{(m)H} \Pi_{\mathbf{U}^{(m)}}^{\perp} \mathbf{r}^{(m)}} > \eta_{\mathrm{UMPI}} \tag{4.25}$$

where $\eta_{\mathrm{UMPI}} = \frac{\eta_t}{(N_e - P_e - K_t) + \eta_t}$. The threshold η_t is selected such that the desired P_{fa} requirement is satisfied in the test $t > \eta_t$.

4.3.2 Second stage of UMPI-based detector

In the first stage of UMPI-based detector design, it is assumed that the interfering target matrix (ITM) $\mathbf{T}^{(m)}$ for $m = 1, \ldots, M$ is available. This implies that the target delay–velocity pairs $(n_t^{(k)}, v_t^{(k)})$ for $k = 1, \ldots, K_m$ with $m = 1, \ldots, M$ are known for the receiver. However, in practical scenarios, this information is unknown and requires estimation. To estimate the target delay–velocity pairs of ITM, a multistage algorithm called multiband imperative target positioning (MITP) is proposed. MITP is an approximate version of the derived multiband UMPI algorithm in the first stage. The algorithm identifies imperative targets sequentially from the strongest level to the weakest level, thus preventing the strong targets from masking other targets of lower levels due to high side peaks of the FM signals.

The proposed MITP algorithm consists of multiple stages. In the first stage, clutter echoes in the surveillance signal are removed using the UMPI test. Then, each target is sequentially detected even in the presence of interfering targets. To eliminate the signal of an interfering target, the vector of the surveillance signal is projected orthogonally to the subspace of the interfering targets in each stage. The strongest target delay–velocity pairs are determined using the UMPI detector. This is achieved by applying $\Pi_{\mathbf{U}^{(m)}}^{\perp} = \Pi_{\Pi_{\mathbf{C}^{(m)}}^{\perp} \mathbf{T}^{(m)}}^{\perp} \Pi_{\mathbf{C}^{(m)}}^{\perp}$. By projecting the received vector in each channel $\mathbf{r}^{(m)}$ onto the space orthogonal to the subspace spanned by the columns of matrix $\mathbf{C}^{(m)}$, the clutter echoes can be removed from the surveillance signal, i.e.,

$$\mathbf{y}_0^{(m)} = \Pi_{\mathbf{C}^{(m)}}^{\perp} \mathbf{r}^{(m)} = \mathbf{r}^{(m)} - \mathbf{C}^{(m)} \mathbf{R}_0 \mathbf{C}^{(m)H} \mathbf{r}^{(m)} \tag{4.26}$$

where $\mathbf{R}_0 = (\mathbf{C}^{(m)H} \mathbf{C}^{(m)})^{-1}$. Here, the matrix $\mathbf{R}_0$ can be precomputed according to clutter region in the mth channel for real-time applications. Using (4.25), the UMPI test for a desired delay–velocity plane can be calculated as

$$\mathbf{L}^{(0)}(n, v) = \frac{\sum_{m=1}^{M} \frac{\left|\mathbf{t}_{(n,v)}^{(m)\,H} \mathbf{y}_0^{(m)}\right|^2}{\mathbf{t}_{(n,v)}^{(m)\,H} \Pi_{\mathbf{U}^{(m)}}^{\perp} \mathbf{t}_{(n,v)}^{(m)}}}{\sum_{m=1}^{M} \|\mathbf{y}_0^{(m)}\|^2} \tag{4.27}$$

for $n = 1, \ldots, N_r$ and $|v| \leq v_{\max}$, where n and v are the range and the Doppler bins, and N_r can be considered as the time delay index corresponding to the maximum bistatic range of interest computed from $R_{\max} = cN_rT_s$, $v_{\max}$ is also the maximum speed of interested targets, and $\mathbf{t}_{(n,v)}^{(m)} = (\mathbf{r}_r^{(n,m)} \odot \mathbf{e}^{(v,m)})$ with $[\mathbf{e}^{(v,m)}]_n = e^{j2\pi(\frac{v}{\lambda^{(m)}})nT_s}$ for $n = 0, \ldots, N_m - 1$. By comparing the maximum value of $\mathbf{L}^{(0)}(n, v)$ with the predetermined threshold η_a, if this maximum exceeds the threshold η_a, it means that the range–velocity coordinate corresponding to this maximum is the estimation of $(n_t^{(1)}, v_t^{(1)})$. Note that the threshold η_a should be selected according to the desired false alarm probability of the system (see section 4.6 for more details). At this stage, the first column of the ITM can, therefore, be constructed as $\mathbf{t}_1^{(m)} = (\mathbf{r}_r^{(n_t^{(1)},m)} \odot \mathbf{e}^{(v_t^{(1)},m)})$ for $m = 1, \ldots, M$. After constructing $\mathbf{t}_1^{(m)}$, similar to (4.26), we can proceed to obtain

$$\mathbf{y}_1^{(m)} = \Pi_{\mathbf{t}_1^{(m)\perp}} \mathbf{y}_0^{(m)} \tag{4.28}$$

where

$$\mathbf{t}_1^{(m)\perp} = \Pi_{C^{(m)}}^{\perp} \mathbf{t}_1^{(m)} = (\mathbf{I} - \mathbf{C}^{(m)}\mathbf{R}_0\mathbf{C}^{(m)^H})\mathbf{t}_1^{(m)} \tag{4.29}$$

In a similar manner, we can use the following test statistic to find the delay–velocity pairs of the kth interfering target, given by

$$\mathbf{L}^{(k)}(n, v) = \frac{\sum_{m=1}^{M} \dfrac{\left| \mathbf{t}_{(n,v)}^{(m)\,H} \mathbf{y}_{k-1}^{(m)} \right|^2}{\mathbf{t}_{(n,v)}^{(m)\,H} \Pi_{\mathbf{U}^{(m)}}^{\perp} \mathbf{t}_{(n,v)}^{(m)}}}{\sum_{m=1}^{M} \|\mathbf{y}_{k-1}^{(m)}\|^2} \tag{4.30}$$

where $\mathbf{y}_{k-1}^{(m)}$, $m = 1, \ldots, M$, are obtained after finding $\mathbf{t}_{k-1}^{(m)}$ of the ITM. Similarly, after finding $\mathbf{t}_k^{(m)}$, we can find

$$\mathbf{y}_k^{(m)} = \Pi_{\mathbf{t}_k^{(m)\perp}} \mathbf{y}_{k-1}^{(m)} \tag{4.31}$$

where $\mathbf{t}_k^{(m)\perp}$ can be obtained, after some manipulations, as follows:

$$\mathbf{t}_k^{(m)\perp} = \mathbf{t}_k^{(m)} - \mathbf{C}^{(m)}\mathbf{R}_0\mathbf{C}^{(m)^H}\mathbf{t}_k^{(m)} - \sum_{s=1}^{k-1} \mathbf{t}_s^{(m)\perp}\mathbf{R}_s\mathbf{t}_s^{(m)\perp^H}\mathbf{t}_k^{(m)} \tag{4.32}$$

where $\mathbf{R}_s = (\mathbf{t}_s^{(m)\perp^H}\mathbf{t}_s^{(m)\perp})^{-1}$ for $s = 1, \ldots, k - 1$.

Furthermore, it is easy to show that the value $\|\Pi_{\mathbf{U}^{(m)}}^{\perp}\mathbf{t}_{(n,v)}^{(m)}\|^2$ takes $\mathbf{t}_{(n,v)}^{(m)}$ or equivalently $\|\mathbf{r}_r^{(n,m)}\|^2$ for the delay–velocity coordinates far from that of the clutter and interfering targets. Therefore, to reduce the computational complexity of the calculation (4.30) for delay n, it is reasonable to approximate $\|\Pi_{\mathbf{U}}^{\perp(m)}\mathbf{t}_{(n,v)}^{(m)}\|^2$ with $\|\mathbf{r}_r^{(n,m)}\|^2$. Hence, an approximated version of (4.30) is given by

$$\mathbf{L}_a^{(k)}(n, v) = \frac{\sum_{m=1}^{M} \dfrac{\left| \mathbf{t}_{(n,v)}^{(m)\,H} \mathbf{y}_{k-1}^{(m)} \right|^2}{\|\mathbf{r}_r^{(n,m)}\|^2}}{\sum_{m=1}^{M} \|\mathbf{y}_{k-1}^{(m)}\|^2} \tag{4.33}$$

To make the calculation less computationally intensive for every combination of delay and velocity, it is recommended to use the fast Fourier transform (FFT) method to determine the numerator of test (4.33) for each desired range (delay). This can be achieved by applying the FFT method to each range (delay) separately in the following manner:

$$\mathbf{L}_a^{(k)}(n, l_v) = \frac{\sum_{m=1}^{M} \frac{\mathbf{f}^{(m)}(n) \odot \mathbf{f}^{(m)*}(n)}{\|\mathbf{r}_r^{(n,m)}\|^2}}{\sum_{m=1}^{M} \|\mathbf{y}_{k-1}^{(m)}\|^2}, \qquad n = 1, \ldots, N_r \tag{4.34}$$

where $\mathbf{f}^{(m)}(n) = \mathcal{F}(\mathbf{y}_{k-1}^{(m)} \odot \mathbf{r}_r^{(n,m)}, N_F^{(m)})$, in which operator $\mathcal{F}$ performs an $N_F^{(m)}$-point FFT of $\mathbf{y}_0^{(m)} \odot \mathbf{r}_r^{(n,m)}$ such that the zero-frequency component is moved to the center of $\mathbf{f}^{(m)}(n)$. In order to utilize the FFT implementation of (4.33) effectively, it is necessary to employ a varying number of FFT points across different frequency channels. This is because different broadcast channels will produce different Doppler shifts of $f_d^{(m)} = \frac{v}{\lambda_m}$ from the same target with a velocity of v. This occurs because each channel uses distinct carrier frequencies with wavelengths of λ_m for $m = 1, \ldots, M$. Therefore, the estimated Doppler frequencies can be obtained by performing $N_F^{(m)}$-point FFTs, with $\hat{f}_d^{(m)} = l_v^{(m)} \frac{f_s}{N_F^{(m)}}$, where f_s represents the baseband sampling frequency of the system, and $l_v^{(m)}$ corresponds to the frequency bins that match the actual Doppler frequencies $f_d^{(m)}$ for $m = 1, \ldots, M$. Consequently, if the number of FFT points, $N_F^{(m)}$, is constant across all channels, then $l_v^{(m)}$ will assume different values for a specific target velocity, or in other words, the $\mathbf{f}^{(m)}(n)$ peak will appear in different frequency bins for $m = 1, \ldots, M$. This is in conflict with the combination gain of the joint exploitation of the multiband detection of the proposed algorithm. To resolve this issue, it is necessary to have equal frequency bins for a given velocity across various frequency channels. This requires that $\frac{N_F^{(1)}}{\lambda_1} = \cdots = \frac{N_F^{(m)}}{\lambda_m} = \cdots = \frac{N_F^{(M)}}{\lambda_M}$. To simplify this, we can select a reference frequency in the FM-band and use $N_F^{(m)} = \frac{\lambda_m}{\lambda_r} N_F^{(r)}$, where $N_F^{(r)}$ is the number of FFT points assigned to the reference frequency. Setting $N_F^{(r)}$ equal to $N_r = T_r f_s$, where N_r is the data vector length corresponding to the reference frequency integration time denoted by T_r, allows us to compute the required integration time in the different channels based on $N_m = \frac{\lambda_m}{\lambda_r} N_r$ for $m = 1, \ldots, M$. To ensure that the number of FFT points is an integer value, we can round N_m to the nearest whole number and use this value as $N_F^{(m)}$. To minimize the impact of this approximation, we can multiply the rounded value of N_m by 2 to obtain $N_F^{(m)}$. Additionally, since the velocity of the targets we are interested in has a maximum value of $v_{\max}$, the range of possible values for l_v is limited to $|l_v| \le l_{v_{\max}}$. This allows us to use a decimation technique to reduce the computational load required for calculating the $N_F^{(m)}$-point FFT in (4.34) [9].

To summarize, the main steps of the MITP algorithm can be outlined as follows:

1. Set $k \leftarrow 0$, $\mathbf{U}_0^{(m)\perp} = \mathbf{C}^{(m)}$, $\mathbf{R}_0 = (\mathbf{C}^{(m)H}\mathbf{C}^{(m)})^{-1}$, and $\mathbf{y}_{-1}^{(m)} = \mathbf{r}^{(m)}$ for $m = 1, \ldots, M$.
2. Calculate $\mathbf{y}_k^{(m)} = \mathbf{y}_{k-1}^{(m)} - \mathbf{U}_k^{(m)\perp} \mathbf{R}_k \mathbf{U}_k^{(m)\perp H} \mathbf{y}_{k-1}^{(m)}$ for $m = 1, \ldots, M$.

3. Calculate the UMPI test statistic according to the range–Doppler plane of interest

$$\mathbf{L}_a^{(k)}(n, l_v) = \frac{\sum_{m=1}^{M} \frac{\mathbf{f}^{(m)}(n) \odot \mathbf{f}^{(m)*}(n)}{\|\mathbf{r}_r^{(n,m)}\|^2}}{\sum_{m=1}^{M} \|\mathbf{y}_k^{(m)}\|^2} \qquad n = 1, \ldots, N_r \tag{4.35}$$

where $\mathbf{f}^{(m)}(n) = \mathcal{F}(\mathbf{y}_k^{(m)} \odot \mathbf{r}_r^{(n,m)}, N_F^{(m)})$. Here, n and l_v are the time delay index and Doppler bin index, respectively.

4. Compare the maximum value of matrix $\mathbf{L}_a^{(k)}(n, l_v)$ over the range–Doppler plane of interest with the predetermined detection threshold η_a. If it does not exceed the detection threshold, there is not any target and the algorithm stops; otherwise, the coordinate $(\hat{n}_t^{(k)}, \hat{l}_v f_s \frac{\lambda_r}{N_F^{(r)}})$ associated with the maximum value of $\mathbf{L}_a^{(k)}(n, l_v)$ shows an estimation of the kth interfering target coordinates and should be preserved. As such, we need to continue in order to find other interfering targets; thus we set $k \leftarrow (k+1)$. The threshold η_a is required to be set according to the desired false alarm probability of the system (see section 4.4 for more details). Since the algorithm may erroneously find the target coordinate, a two-dimensional (2D) mask should be established around the estimated coordinate of the detected target. The signal associated with the delay–Doppler coordinates of this mask should then be removed. This enables the proposed algorithm to fully remove the effect of interfering targets. To that end, for each detected target, the number of $n'l'$ vectors is established, and the associated signal of each vector is canceled out in each iteration, where $n' \times l'$ is the dimension of the 2-D mask in the range and Doppler domain. The simulations, for the FM signal, show that selecting $n' = 3$ and $l' = 7$ suffices to effectively remove the detected target signal. This is corresponding to velocities of $(\hat{l}_v \pm q)\frac{f_s}{6}\frac{\lambda_r}{N_F^{(r)}}$ for $q = 0, 1, 2, 3$ and the delay index of $\hat{n}_t^{(k)} - 1$, $\hat{n}_t^{(k)}$, and $\hat{n}_t^{(k)} + 1$.

5. To remove the detected interfering target, we need to compute and store the matrices $\mathbf{U}_k^{(m)\perp}$ and $\mathbf{R}_k$; that are

$$\mathbf{U}_k^{(m)\perp} = \mathbf{U}_k^{(m)} - \sum_{s=0}^{k-1} \mathbf{U}_s^{(m)\perp} \mathbf{R}_s \mathbf{U}_s^{(m)\perp^H} \mathbf{U}_k^{(m)} \tag{4.36}$$

and

$$\mathbf{R}_k = (\mathbf{U}_k^{(m)\perp^H} \mathbf{U}_k^{(m)\perp})^{-1} \tag{4.37}$$

where the ith column of $\mathbf{U}_k^{(m)}$, $[\mathbf{U}_k^{(m)}]_i = \mathbf{t}_i^{(m)}$, is made of a proper time-delayed and Doppler-shifted replica of the reference signal in the mth channel based on the extended range–velocity coordinates at Step 5.

6. Go back to Step 2.

Step 4 outlines the stopping criterion for the algorithm. The proposed MITP algorithm is capable of locating targets above the noise level in the surveillance channel.

By terminating the algorithm, the matrix $\mathbf{T}^{(m)}$ for $m = 1, ..., M$ is obtained, along with the number of essential targets. The confirmation step of the detection method can then be implemented, where one imperative target is selected as the testing target and the other $(K-1)$ imperative targets are interfering targets. Despite the need to iterate several times to detect all potential targets and confirm the detection of a target in the presence of others, this detection algorithm offers outstanding detection performance while maintaining false alarm probability regulation. In Chapter 7, a new detection algorithm is introduced that eliminates the confirmation step. This approach is grounded in the M-ary hypothesis-testing formulation of the single-band passive radar target detection problem and is expandable to multiband scenarios.

4.4 Analytical performance analysis

This section involves deriving formulas for both false alarm probability and detection probability in a closed form. Additionally, the analytical investigation of the detection capabilities of the UMPI detector under the SW1 target model is discussed.

In Appendix 4A, it is shown that the statistic $t = \frac{N_e - P_e - K_t}{M} \frac{z}{1-z}$ is distributed as a central complex F distribution with M and $(N_e - P_e - K_t)$ complex degree of freedom under the $\mathcal{H}_0$ hypothesis to obtain

$$P_{fa} = Q_{F_{M,(N_e - P_e - K_t)}}(\eta_t) \tag{4.38}$$

The probability denoted as $Q_{F_{M,(N_e - P_e - K_t)}}$ represents the right-tail probability of the central complex F distribution. The threshold can be set without any prior knowledge of the unknown parameters, as shown in (4.38). This implies that the UMPI detector has the constant false alarm rate (CFAR) property against the number of unknowns. When N_e is large, $F_{M,(N_e - P_e - K_t)}$ approaches the complex $\frac{\chi_M^2}{M}$ [15], resulting in

$$P_{fa} = e^{-M\eta_t} \sum_{n=0}^{M-1} \frac{(M\eta_t)^n}{n!} \tag{4.39}$$

To achieve a desired value of P_{fa}, we must calculate η_t using (4.39). We can then determine the detection threshold for the UMPI test in (4.25) as $\eta_{\text{UMPI}} = \frac{\eta_t}{(N_e - P_e - K_t) + \eta_t}$. It is important to note that because N_e is significantly larger than $P_e + K_t$, an inaccurate estimate of the number of interfering targets will not affect the threshold-setting process for the UMPI test. As a result, we can establish the threshold η_a in the MITP algorithm (as described in section 4.3.2) by assuming a maximum number of mandatory targets.

Appendix 4A demonstrates that, if we assume the alternative hypothesis $\mathcal{H}_1$, the statistic t follows a noncentral complex F distribution. This distribution has complex degrees of freedom M and $(N_e - P_e - K_t)$, and a noncentrality parameter δ, given by

$$\delta \triangleq \sum_{m=1}^{M} \frac{|\alpha_0{}^{(m)}|^2}{\sigma^2} \| \Pi_{\mathrm{U}}^{(m)\perp} \mathbf{s}^{(m)} \|^2 \tag{4.40}$$

Using this distribution, we can determine the detection probability associated with the test statistic t, i.e.,

$$P_{d,0} = Q_{F'_{M,(N_e-P_e-K_t)}(\delta)}(\eta_t) \tag{4.41}$$

where $Q_{F'_{M,(N_e-P_e-K_t)}(\delta)}$ is the right-tail probability of noncentral complex F distribution. Similarly, due to the large value of N_e in our detection problem, we can approximate $F_{M,(N_e-P_e-K_t)}(\boldsymbol{\delta})$ with $\frac{\chi^2_M(\delta)}{M}$. Hence, the detection probability of the UMPI test can be simplified to obtain

$$P_{d,0} = e^{-(\delta+M\eta_t)} \sum_{k=0}^{\infty} \left(\frac{\delta^k}{k!} \sum_{n=0}^{M+k-1} \frac{(M\eta_t)^n}{n!} \right) \tag{4.42}$$

The function $P_{d,0}$ increases as δ increases, where δ is computed as the sum of $\frac{|\alpha_0{}^{(m)}|^2}{\sigma^2}$ multiplied by $\| \Pi_{\mathrm{U}}^{(m)\perp} \mathbf{s}^{(m)} \|^2$, for all m from 1 to M. The ratio $\frac{|\alpha_0{}^{(m)}|^2}{\sigma^2}$ represents the input target signal-to-noise ratio (SNR) of the mth broadcasting channel, and the term $\| \Pi_{\mathrm{U}}^{(m)\perp} \mathbf{s}^{(m)} \|^2$ can be considered as the SNR gain provided by the UMPI detector due to the mth broadcasting channel. Thus, the multiband detector has benefited from a weighted sum of the SNR gain of different channels, as represented by the parameter δ.

The targets being considered up to now have been referred to as Swerling 0 (SW0) target models, under the assumption that they do not experience any fluctuations. However, in reality, the amplitude of targets can fluctuate significantly, which can have a significant impact on detection performance. To account for this, the SW1 target model is also being considered, which assumes that target amplitude is completely correlated over the integration time but independent over different integration times. It is also important to define the statistical variation of target amplitude over different frequency bands in a multiband target detection scenario. This is done by considering two cases: SW1-C and SW1-I. Under SW1-C, the amplitude of a given target is assumed to have a completely correlated complex Gaussian pdf over different frequency bands, while under SW1-I, it is assumed to have independent complex Gaussian pdfs for the same target over different frequency bands. Using these models, we will determine the detection probability of the UMPI detector.

4.4.1 SW1-I target model

The SW1-I target model enables a multiband PBR system to achieve a maximum frequency diversity gain across multiple frequency bands. To approach this target model, the frequency separation between the FM channels used in the system must be sufficient. In the case where the signal amplitudes of a target are assumed to be independent complex Gaussian variables with zero mean and a variance of $\sigma_s^{(m)2}$ over the M frequency bands, the detection probability can be calculated as (for more information, refer to Appendix 4B)

$$P_{d,1}^{I} = e^{-M\eta_t} \sum_{k=0}^{\infty} \sum_{n=0}^{M+k-1} \frac{(M\eta_t)^n}{n!} \sum_{m=1}^{M} \frac{1}{\beta_m (1 + \frac{1}{\beta_m})^{k+1} \prod_{s=1, s\neq m}^{M} \left(1 - \frac{\beta_s}{\beta_m}\right)} \tag{4.43}$$

where $\beta_k = \frac{\sigma_s^{(k)2}}{\sigma^2} \| \Pi_{\mathrm{U}}^{(k)\perp} \mathbf{s}^{(k)} \|^2$.

4.4.2 SW1-C target model

The SW1-C target model is considered to be conservative in terms of performance because of the high fluctuation loss that occurs over a single frequency band, and there is no frequency diversity gain over different frequency bands. This assumption is only valid when FM radio channels are closely spaced in frequency. In this scenario, it is assumed that the amplitude of the target follows a complex Gaussian variable with zero mean and variance σ_s^2 across multiple frequency bands, denoted by ${\alpha_0}^{(m)} = \alpha_0$. As a result, the probability of detection can be determined using the details provided in Appendix 4B, i.e.,

$$P_{d,1}^C = e^{-M\eta_t} \sum_{k=0}^{\infty} \frac{1}{\beta(1+\frac{1}{\beta})^{k+1}} \sum_{n=0}^{M+k-1} \frac{(M\eta_t)^n}{n!} \tag{4.44}$$

where $\beta = \frac{{\sigma_s}^2}{\sigma^2} \sum_{m=1}^{M} \|\Pi_{\mathbf{U}}^{(m)^{\perp}} \mathbf{s}^{(m)}\|^2$.

The expression $0 \leq \|\Pi_{\mathbf{U}}^{(m)^{\perp}} \mathbf{s}^{(m)}\|^2 \leq \|\mathbf{s}^{(m)}\|^2$ holds for $m = 1, \ldots, M$. This has significant implications. First, when $\mathbf{s}^{(m)}$ belongs to the set $\{\mathbf{u}_1^{(m)}, \ldots, \mathbf{u}_{P_m+K_m-1}^{(m)}\}$, $\|\Pi_{\mathbf{U}}^{(m)^{\perp}} \mathbf{s}^{(m)}\|^2$ is zero. This implies that our proposed detector efficiently cancels out interference signals, such as clutter/multipath and interfering targets. Second, $\|\Pi_{\mathbf{U}}^{(m)^{\perp}} \mathbf{s}^{(m)}\|^2$ equals $\|\mathbf{s}^{(m)}\|^2$ in the delay–velocity coordinates, which are far from those of interference signals. This indicates that the detection loss (DL) experienced by the interested target is negligible in the mth broadcasting channel, provided that its delay–velocity is significantly different from that of interference signals. Third, $\|\Pi_{\mathbf{U}}^{(m)^{\perp}} \mathbf{s}^{(m)}\|^2$ varies between zero and one in the delay–velocity coordinates close to those of interference signals, depending on the delay–velocity resolution of the received signal in the mth channel. This means that a target placed near an interfering target experiences some (DL). These concepts can be utilized to evaluate the quality of signal (QoS) in multiband detection based on the (DL), which is defined as

$$\mathrm{DL}[n,v] = \frac{1}{M} \sum_{m=1}^{M} \frac{\|\Pi_{\mathbf{s}_{(0,0)}^{(m)}}^{\perp} \mathbf{s}_{(n,v)}^{(m)}\|^2}{\|\mathbf{s}_{(n,v)}^{(m)}\|^2} \tag{4.45}$$

or equivalently

$$\mathrm{DL}[n,v] = 1 - \frac{1}{M} \sum_{m=1}^{M} \frac{|\mathbf{r}_r^{(m)H} \mathbf{s}_{(n,v)}^{(m)}|^2}{\|\mathbf{r}_r^{(m)}\|^2 \|\mathbf{r}_r^{(n,m)}\|^2} \tag{4.46}$$

where n and v are the range index and velocity, respectively. In (4.45) and (4.46), we have $\mathbf{s}_{(n,v)}^{(m)} = (\mathbf{r}_r^{(n,m)} \odot \mathbf{e}^{(v,m)})$ in which $\mathbf{r}_r^{(n,m)} = \mathbf{P}^n \mathbf{r}_r^{(m)}$ and $[\mathbf{e}^{(v,m)}]_n = e^{j2\pi(\frac{v}{\lambda(m)})nT_s}$ for $n = 0, \ldots, N_m - 1$. It is also straightforward to conclude that $0 \leq \mathrm{DL}[n,v] \leq 1$.

In PBR systems, the attainable Doppler resolution and RR are influenced by the integration time and bandwidth of opportunity signals, respectively. Due to the changes in the effective bandwidth of the FBPBR, signal quality in the range fluctuates over time while signal quality in velocity is predetermined by integration time.

In what follows, we focus on evaluating the signal quality in the range. To do so, we set $v = 0$ in (4.46) to obtain

$$\mathrm{DL}[n] = 1 - \frac{1}{M}\sum_{m=1}^{M} \frac{|\mathbf{r}_r^{(m)H}\mathbf{r}_r^{(n,m)}|^2}{\|\mathbf{r}_r^{(m)}\|^2\|\mathbf{r}_r^{(n,m)}\|^2} \tag{4.47}$$

The computation of DL[n] based on reference signals is a simple normalized cross-correlation. A signal with no (DL) in its range neighborhood, i.e., DL[n] = 1 for $n \neq 0$, is considered a high-quality signal. This indicates that DL[n] provides information about the RR, peak-to-sidelobe-level ratio (PSLR), and integrated sidelobe level ratio (ISLR) of the multiband signal. In section 4.5, the quality of several signals is evaluated, and their impact on multiband detection cases is investigated.

4.5 Performance results

This section presents simulation results to confirm the effectiveness of the proposed detection process. Three simulated signals, namely S_1, S_2, and S_3, with carrier frequencies of 90.6, 99.5, and 103.5 MHz, are used as waveforms of opportunity. These signals are generated based on the international standard outlined in [16] to ensure their resemblance to real signals. The integration times assigned to S_1, S_2, and S_3 in our simulation are 1.1038, 1.0050, and 0.9662 seconds, respectively. This choice of integration times is associated with the FFT-based implementation of the MITP algorithm, which is explained in section 4.3.

To assess how the FM radio signal's quality changes over time, Figure 4.1 illustrates the variations of PSLR, ISLR, and RR as a function of detection batch

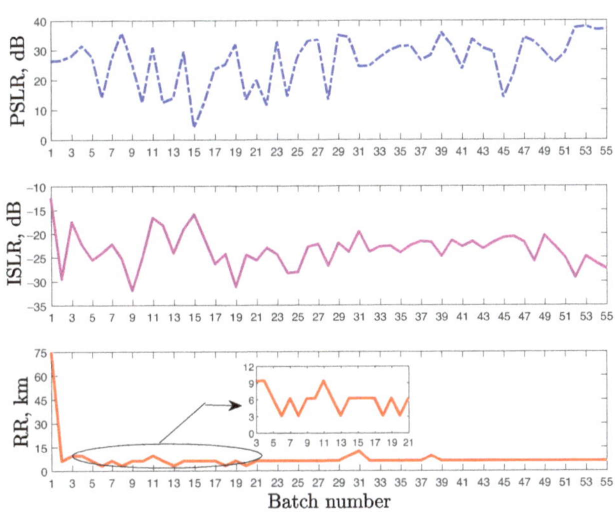

Figure 4.1 PSLR, ISLR, and RR as a function of time (or batch number) for an FM radio signal, containing silence, music, and speech sections

index. For each batch, we have set the integration time to 1 second. It is evident that the PSLR, ISLR, and RR change over time, indicating that the effectiveness of an FBPBR is time-dependent. To demonstrate this in terms of detection performance, we have chosen three signals (S_1, S_2, and S_3) that have their primary self-ambiguity function cutoff at zero velocity, as illustrated in Figure 4.2. In order to examine how the performance of an FBPBR system is affected by signal S_1 with limited RR and signal S_3 with poor PSLR and ISLR, we analyze a multitarget situation. The characteristics of the targets in this particular scenario are presented in Table 4.1. In order to make a comparison, the input SNR_i is defined as

$$\mathrm{SNR}_i = \frac{\sigma_s^2}{N_0 B} \tag{4.48}$$

where the noise power per unit bandwidth is denoted by N_0, and the receiver bandwidth is denoted by B. Thus, the input noise power can be calculated using the formula $\sigma^2 = N_0 B$. It is important to note that for the SW0 target model, the input

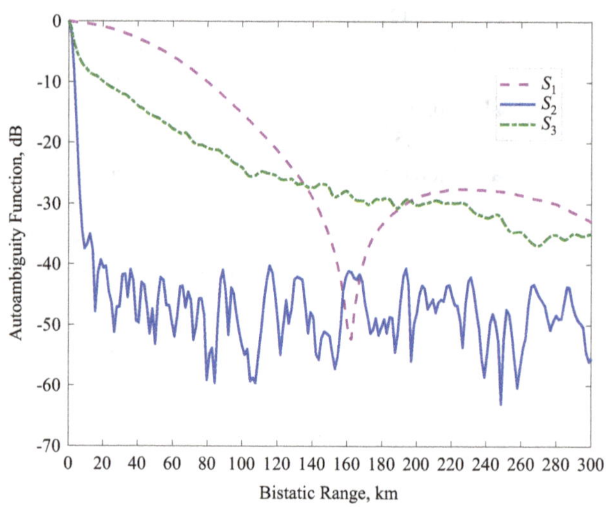

Figure 4.2 The main cut-off autoambiguity function at zero velocity for different signals S_1, S_2, *and* S_3 *corresponding to a silence, music, and speech part of an FM radio signal, respectively*

Table 4.1 Characteristic of the targets in the simulated multitarget scenario

Target	T_1	T_2	T_3	T_4	T_5	T_6	T_7
Bistatic range (km)	7.03	13	59.57	59.57	93	135.8	150
Bistatic velocity (m/s)	147.1	147.1	−55.3	−70.2	241.7	241.7	−273
SNR_i (dB)	3.4	−14	−11.3	−23.2	−5.1	−31.7	−34

noise power is given by $\sigma_s^2 = |\alpha_0^{(m)}|^2$. In this chapter, the reported values of SNR_i are based on a nominal assigned channel bandwidth of 200 kHz for the FM radio, which requires a Nyquist sampling frequency of 200 kHz. Consequently, the signal lengths for S_1, S_2, and S_3 are equal to $N_1 = 220{,}760$, $N_2 = 201{,}000$, and $N_3 = 193{,}240$, respectively.

We made the assumption that the direct signal and the multipath/clutter echoes over different channels were independent. In our simulation, we varied the input direct/path signal-to-noise ratio (DNR_i) of direct signals across different channels between 58 and 63 dB, and we considered ten clutter spikes with different input clutter-to-noise ratios (CNR_i) ranging from 5 to 35 dB, distributed in the bistatic ranges of 0–55 km. The clutter amplitudes followed an exponential power spectral density with a shape parameter of 7 and a dc-to-ac power ratio of 90, similar to Chapter 3. To evaluate the effectiveness of the clutter model, we set the detector threshold analytically to achieve $P_{fa} = 10^{-3}$. To apply the proposed UMPI test, we appropriately set the parameters Δv_c and Q to suppress the clutter. Our simulation results show that setting the clutter region parameters to $\Delta v_c = 0.75$ m/s and $Q = 7$ perfectly suppresses clutter and achieves the predetermined false alarm probability. We also provide results for $\Delta v_c > 0.75$ m/s ($\Delta v_c < 0.75$ m/s) and $Q > 7$ ($Q = 7$), all of them lead to $P_{fa} > 10^{-3}$. According to our simulation results, there are no benefits of multiband PBRs over their single-band counterparts in terms of clutter removal.

Figures 4.3 and 4.4 illustrate the probability of detection in relation to SNR_i for varying M values under the SW0 and SW1-I target models. The probability

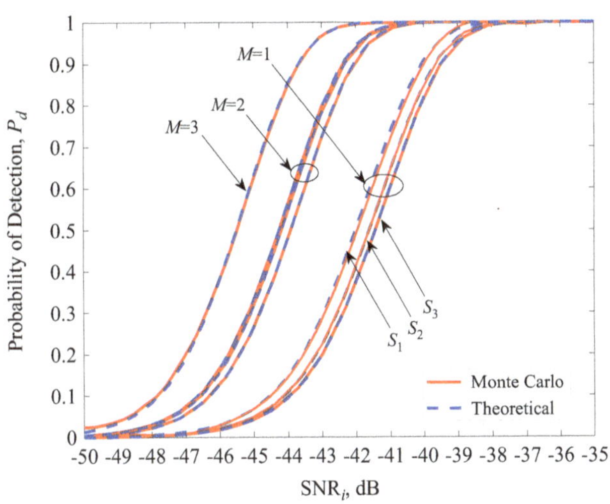

Figure 4.3 *Detection probability as a function of* SNR_i *for the SW0 target model when* $P_{fa} = 10^{-6}$*, and the number of channels (M) is considered as parameter. The lines denote Monte Carlo simulation results, whereas the dashed lines denote the corresponding analytical results.*

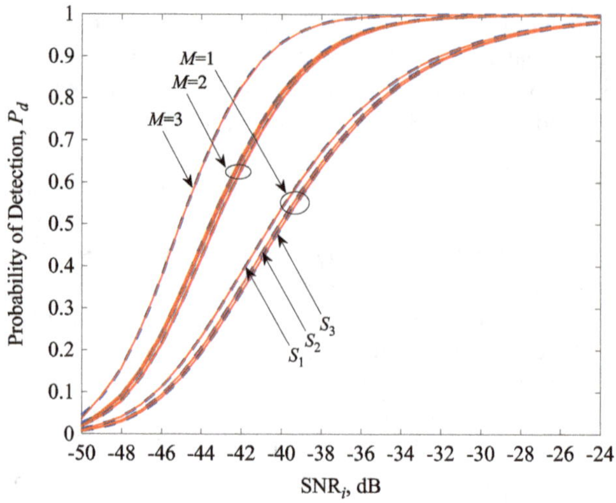

Figure 4.4 *Detection probability as a function of* SNR$_i$ *for the SW1-I target model when* $P_{fa} = 10^{-6}$*, and the number of channels (*M*) is considered as parameter. The lines denote Monte Carlo simulation results, whereas the dashed lines denote the corresponding analytical results.*

of false alarm (P_{fa}) has been set to 10^{-6} hereafter. The solid lines in these figures represent the theoretical results, while the dashed lines depict the results obtained from Monte Carlo (MC) simulations. These figures demonstrate that the MC simulation results align well with the theoretical results. As expected, an increase in the number of channels or prolonging the integration time enhances detection performance.

Figure 4.5 compares the detection performances of the UMPI detector in the presence of targets with the SW0, SW1-I, and SW1-C models. The detection probability is plotted against the SNR$_i$ for $M = 1, 2, 3$. The figure highlights two key diversity gains: combined diversity gain and frequency diversity gain. The former is achieved through the simultaneous use of multiple broadcasted channels in the detector derivation, while the latter is obtained from independent returns of a target received over different frequency channels. The SW1-I target model is observed to provide both diversity gains, whereas the SW1-C and SW0 target models are found to offer only the combined diversity gain. To be more specific, at $P_d = 0.8$ and for $M = 3$, the SW1-I target model demonstrates a combined diversity gain of roughly 3.6 dB and a frequency diversity gain of approximately 3.4 dB, while the SW1-C and SW0 target model scenarios only show a combined diversity gain of about 3.6 dB for $M = 3$. However, in practical applications, it may not be feasible to obtain completely independent target returns across different FM channels. This implies that achieving a frequency diversity gain ranging from 0 dB to 3.4 dB is

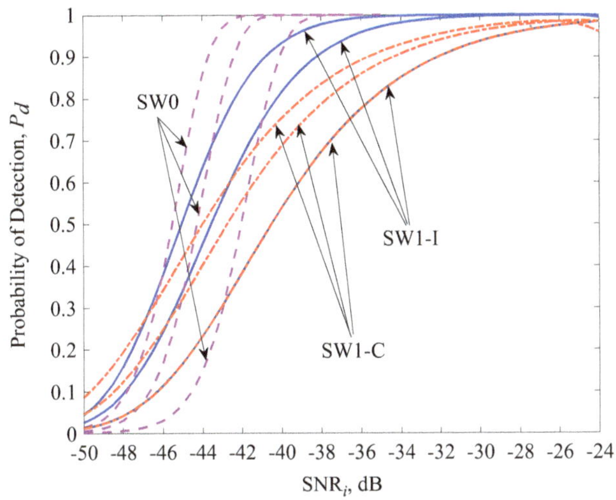

Figure 4.5 *Detection probability comparison as a function of* SNR$_i$ *under different target models (i.e., SW0, SW1-I, and SW1-C) for* $P_{fa} = 10^{-6}$ *and* $M = 1, 2, 3$.

possible. Additionally, detecting a fluctuating target requires a higher SNR compared with a target with a constant echo return, which is referred to as fluctuation loss. For example, the fluctuation loss is about 5.2 dB when $P_d = 0.8$ and $M = 1$. Fortunately, multiband detection can help mitigate the aforementioned detection performance fluctuation by acquiring independent returns of a target, as shown in Figure 4.5.

To see the time-varying performance of a single-channel UMPI target detection method, the detection performances of different targets of Scenario 1 are compared in Figures 4.6–4.8 for the FM signals S_1, S_2 and S_3, respectively. This is accomplished by treating the kth target as the testing target and the remaining imperative targets as interfering targets to create the matrix $\mathbf{T}$. In our case all interfering targets are successfully detected by the MITP algorithm to construct the matrix $\mathbf{T}$. Then, these figures are obtained by changing the input SNR$_i$ of the testing target T_k in the presence of other targets listed in Table 4.1. The results of Figure 4.6 indicate that targets T_5 and T_6 experience some detection degradation as compared with targets T_3 and T_4. Here, the targets T_1 (T_2) can not be detected by the proposed detector in the presence of T_2 (T_1) as interfering target, while targets T_3 and T_4 follows the detection performance of a single-target scenario (STS). All of these results can be justified by the poor range characteristic of the FM signal S_1.

To clarify, Figure 4.9 plots the DL as a measure of the signal's quality of service (QoS) against the bistatic range. The results show that when signal S_1 is used

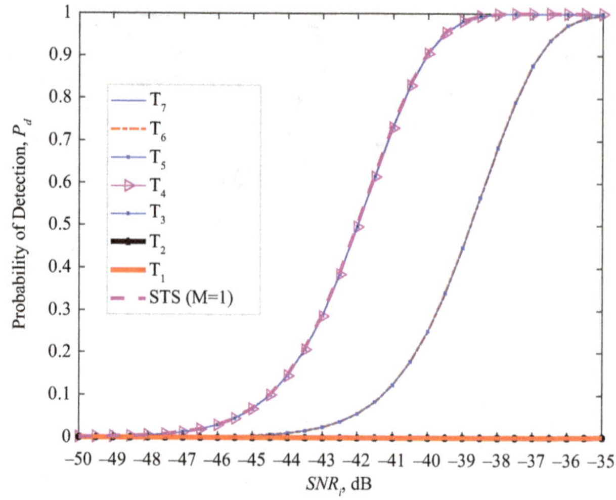

Figure 4.6 *Detection probability as a function of* SNR$_i$ *for the SW0 target model when* $P_{fa} = 10^{-6}$, *and the signal* S_1 *is used for multitarget detection.*

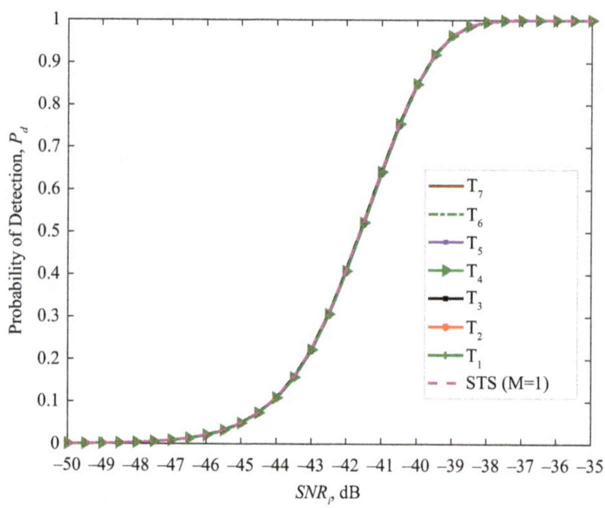

Figure 4.7 *Detection probability as a function of* SNR$_i$ *for the SW0 target model when* $P_{fa} = 10^{-6}$, *and the signal* S_2 *is used for multitarget detection.*

for detection, the DL is high due to the poor RR of the signal, which affects the delectability of targets that share the same velocity Doppler. However, when detection is performed using signal S_2, all targets follow the detection performance of the STS. In the case of using signal S_3 for detection, which has a high sidelobe level,

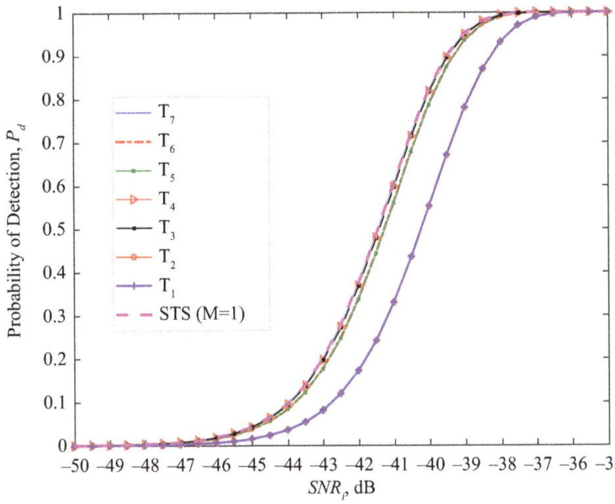

Figure 4.8 *Detection probability as a function of* SNR$_i$ *for the SW0 target model when* $P_{fa} = 10^{-6}$, *and the signal* S_3 *is used for multitarget detection.*

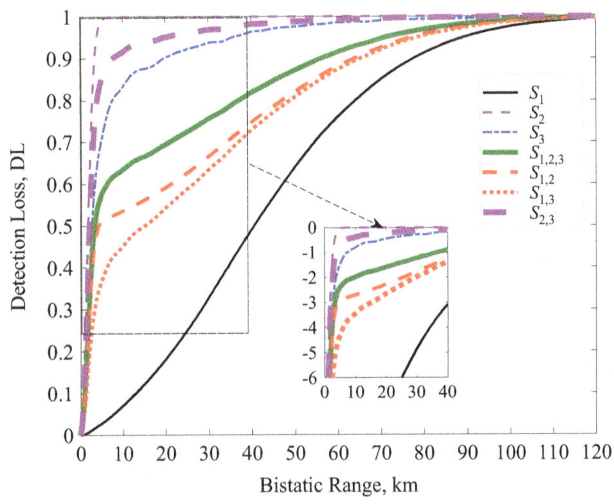

Figure 4.9 *Detection loss versus the bistatic range values when different combinations of signals* S_1, S_2, *and* S_3 *are applied to the proposed multiband detector.*

the target detection results are shown in Figure 4.8. From the results presented in Figures 4.6–4.8, it is clear that the performance of a single-band FBPBR system varies over time due to the changing program content of the broadcasted signal. However, by using three parallel frequency channels for detection, as shown in

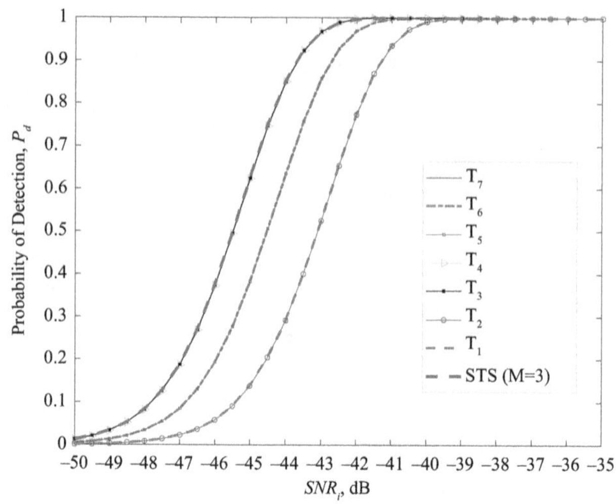

Figure 4.10 *Detection probability as a function of* SNR_i *for the SW0 target model for* $P_{fa} = 10^{-6}$*, and when three-channel detector are applied to a multitarget scenario.*

Figure 4.10, a significant improvement in the detection performance and robustness against time-varying conditions can be achieved. This is because using multiple channels for detection improves the quality of the resulting signal compared to using single-channel, as shown in Figure 4.9. To illustrate, the DL corresponding to single channel S_1, S_2 and S_3 and different multiple channels $S_{1,2}$, $S_{1,3}$, $S_{2,3}$ and $S_{1,2,3}$ are depicted in Figure 4.9.

To gain more insight into this, we define another parameter named processing gain. Here, we assume that there are m channels for detection. The processing gain in decibels is defined as the difference between the required SNR for an m -channel detector, denoted by $\mathrm{SNR}_i^{(m)}$, and the SNR of a single-channel detector in the presence of an STS, denoted as $\mathrm{SNR}_i^{(1)}$, i.e., $P_G^{(m)} = \mathrm{SNR}_i^{(1)} - \mathrm{SNR}_i^{(m)}$ for $m = 1, ..., M$. In Figures 4.11 and 4.12, the processing gain to achieve $P_d = 0.8$ against the range distance ΔR between an interfering and a testing targets is plotted for different target models SW0 (or SW1-C) and SW1-I. There are some points in order. First, the processing gain will always increase as the number of channels used increases, irrespective of the quality of the signals. Therefore, if possible, it is recommended to use more channels for detection to achieve greater diversity gain. Second, a robustness against the time-varying program content of FM radio signals can be obtained when using multiple channels for detection, especially in multitarget scenarios. Finally, it is worth noting that the processing gain demonstrated in Figure 4.12 is consistently higher than that in Figure 4.11 as the former shows diversity gain for the SW1-I target model, unlike the latter for the SW0 (or SW1-C) model.

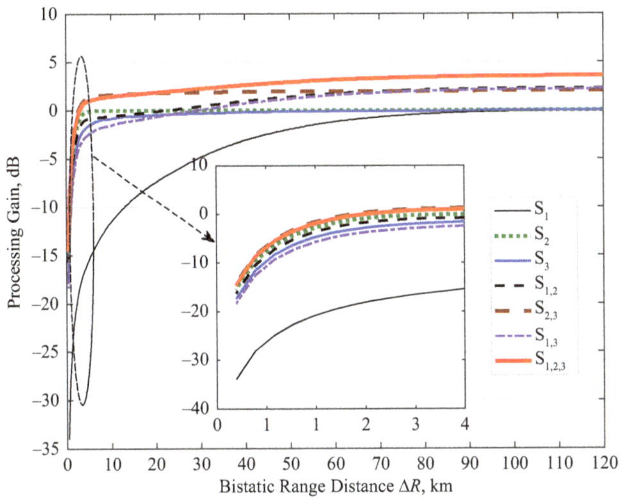

Figure 4.11 *Processing gain as a function of range distance* ΔR *between an interfering target and a testing target with the same velocity. Here, different combinations of signals* S_1, S_2, *and* S_3 *are used to detect the testing target with the SW0 (or SW1-C) model at* $P_d = 0.8$ *and* $P_{fa} = 10^{-6}$.

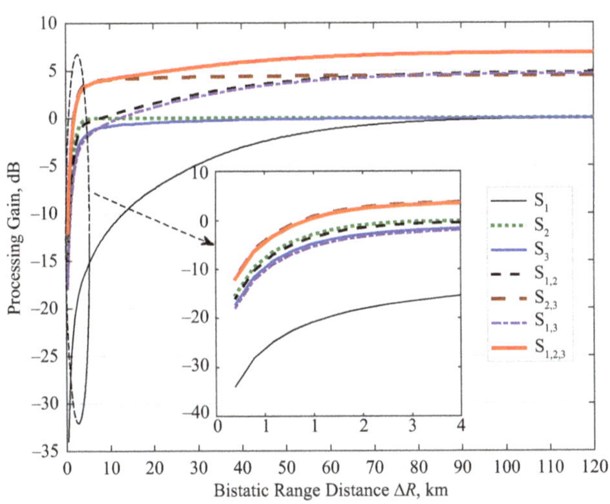

Figure 4.12 *Processing gain as a function of range distance* ΔR *between an interfering target and a testing target with the same velocity. Here, different combinations of signals* S_1, S_2, *and* S_3 *are used to detect the testing target with the SW1-I model at* $P_d = 0.8$ *and* $P_{fa} = 10^{-6}$.

4.6 Summary

This chapter addresses the problem of detecting multiband single-input multiple-output (SISO) passive radar targets in the presence of interference signals, such as receiver noise, direct signal, multipath/clutter echoes, and interfering targets. The chapter presents a comprehensive framework that includes signal modeling, detection methods, and statistical analysis. Multiband SISO passive radar target detection is formulated as a binary composite hypothesis-testing problem, which is solved using the UMPI classical method. The false alarm probability and detection probability in the presence of targets with SW0 and SW1 models are derived in closed forms for the multiband UMPI test. This approach not only demonstrates that the proposed UMPI test has a CFAR property against noise variance but also provides robust performance against the time-varying program content of FM radio signals. Additionally, the multiband UMPI test benefits from combined diversity gain due to a combination of multifrequency bands and frequency diversity gain due to independent returns from one target compared with the single-band UMPI (GLR) test presented in Chapter 3.

References

[1] Bongioanni C., Colone F., Lombardo P. 'Performance analysis of a multi-frequency FM-based passive bistatic radar'. *IEEE 2008 Radar Conference*; Rome, Italy, 2008. pp. 1–6.

[2] Tasdelen A.S., Koymen H. 'Range resolution improvement in passive coherent location radar systems using multiple FM radio channels'. *Proceedings of IET Forum on Radar and Sonar*; London, UK, 2006. pp. 23–31.

[3] Olsen K.E., Baker C.J. 'FM-based passive bistatic radar as a function of available bandwidth'. *IEEE 2008 Radar Conference*; 2008.

[4] Olsen K.E., Woodbridge K. 'FM-based passive bistatic radar target range improvement'. *Proceedings of International Radar Symposium*; 2009.

[5] Olsen K.E., Woodbridge K., Andersen I.A. 'FM-based passive bistatic radar target range improvement – Part II'. *Proceedings of International Radar Symposium*; 2010.

[6] Olsen K.E., Woodbridge K. 'Analysis of the performance of a multiband passive bistatic radar processing scheme'. *IEEE 2010 Radar Conference*; 2010.

[7] Zaimbashi A., Derakhtian M., Sheikhi A. 'Invariant target detection in multiband FM-based passive bistatic radar'. *IEEE Transactions on Aerospace and Electronic Systems*. 2014, vol. 50(1), pp. 720–36.

[8] Billingsley J.B. 'Low-angle land clutter measurements and empirical models' in William Andrew Publishing; 2002.

[9] Zaimbashi A., Derakhtian M., Sheikhi A. 'GLRT-based CFAR detection in passive bistatic radar'. *IEEE Transactions on Aerospace and Electronic Systems*. 2013, vol. 49(1), pp. 134–59.

[10] Colone F., O'Hagan D.W., Lombardo P., Baker C.J. 'A multistage processing algorithm for disturbance removal and target detection in passive bistatic radar'. *IEEE Transactions on Aerospace and Electronic Systems*. 2009, vol. 45(2), pp. 698–722.

[11] Kulpa K.S., Czekała Z. 'Masking effect and its removal in PCL radar'. *IEE Proceedings – Radar, Sonar and Navigation*. 2005, vol. 152(3), pp. 174–78.

[12] Lehman E.L., Romano J.P. 'Testing statistical hypothesis' in New York: Springer Verlag; 2005.

[13] Ramirez D., Via J., Santamaria I., Scharf L.L. 'Locally most powerful invariant tests for correlation and sphericity of Gaussian vectors'. *IEEE Transactions on Information Theory*. 2013, vol. 59(4), pp. 2128–41.

[14] Laub A.J. 'Matrix analysis for scientists and engineers' in Philadelphia, PA: Soc. for Industrial and Applied Math; 2005.

[15] Muirhead R.J. 'Aspect of multivariant statistical theory'. John Wiley and Sons; 2005.

[16] Several authors, Final acts. 'Regional administrative conference for the planning of VHF sound broadcasting (Region 1 and part of Region 3)' in Geneva: International Telecommunication Union; 1984.

[17] David H.A., Nagaraja H.N. 'Order statistics' in Third edition. New York: John Wiley and Sons; 2003.

Appendix 4A
Maximal invariant statistic distribution

In this appendix, the pdf of the MI z, represented in (4.24), is obtained under each hypothesis. To do this, we use a 1-1 function of z described as $t = \frac{N_e - P_e - K_t}{M}\frac{z}{1-z}$, where $K_t = K_e + M$. Note that, any 1-1 function of an MI is also an MI statistic. Notice that, t can be further written as follows:

$$\begin{aligned} t &= \frac{N_e - P_e - K_t}{M}\frac{L_n(\mathbf{r})}{L_d(\mathbf{r})} \\ &= \frac{N_e - P_e - K_t}{M}\frac{\sum_{m=1}^{M}(\frac{\mathbf{r}^{(m)}}{\sigma})^H(\Pi_{\mathbf{K}^{(m)}} - \Pi_{\mathbf{U}^{(m)}})(\frac{\mathbf{r}^{(m)}}{\sigma})}{\sum_{m=1}^{M}(\frac{\mathbf{r}^{(m)}}{\sigma})^H\Pi^{\perp}_{\mathbf{K}^{(m)}}(\frac{\mathbf{r}^{(m)}}{\sigma})} \end{aligned} \tag{4A.1}$$

where $\Pi^{\perp}_{\mathbf{K}^{(m)}} = \mathbf{I} - \mathbf{K}^{(m)}(\mathbf{K}^{(m)H}\mathbf{K}^{(m)})^{-1}\mathbf{K}^{(m)H}$ with $\mathbf{K}^{(m)} = [\mathbf{U}^{(m)}, \mathbf{s}^{(m)}]$. Extending the results of Chapter 3 to multiband case, it can be shown that

$$L_n(\mathbf{r}) \sim \begin{cases} \chi^2_M & \text{under} \quad \mathcal{H}_0 \\ \chi^2_M(\delta) & \text{under} \quad \mathcal{H}_1 \end{cases} \tag{4A.2}$$

where

$$\delta = \sum_{m=1}^{M}\frac{|\alpha_0{}^{(m)}|^2}{\sigma^2}\left(\mathbf{s}^{(m)\mathrm{H}}\Pi^{\perp}_{\mathbf{U}^{(m)}}\mathbf{s}^{(m)}\right) \tag{4A.3}$$

In (4A.2), $\chi^2_k(\delta)$ denotes an noncentral complex chi-squared distribution with k complex degrees of freedom and noncentrality parameter δ. Similarly, we can obtain

$$L_d(\mathbf{r}) \sim \begin{cases} \chi^2_{(N_e - P_e - K_t)} & \text{under} \quad \mathcal{H}_0 \\ \chi^2_{(N_e - P_e - K_t)} & \text{under} \quad \mathcal{H}_1 \end{cases} \tag{4A.4}$$

It is not difficult to show that the numerator and denominator of the statistic (4A.1) are independent. Hence, it is known to conclude that

$$t \sim \begin{cases} F_{M,(N_e - P_e - K_t)} & \text{under} \quad \mathcal{H}_0 \\ F_{M,(N_e - P_e - K_t)}(\delta) & \text{under} \quad \mathcal{H}_1 \end{cases} \tag{4A.5}$$

where $F_{M,(N_e - P_e - K_t)}$ denotes a central complex F distribution with M numerator complex degree of freedom and $(N_e - P_e - K_t)$ denominator complex degree of freedom, and $F_{M,(N_e - P_e - K_t)}(\delta)$ denotes a noncentral complex F distribution with M numerator complex degree of freedom and $(N_e - P_e - K_t)$ denominator complex degree of freedom and noncentrality parameter δ.

Appendix 4B

Detection probability formulas for the SW1-I and SW1-C target models

The SW1-I target model assumes that the amplitudes $\alpha_0^{(m)}$ (where $m = 1, ..., M$) are independent of each other, but they vary based on independent complex Gaussian variables with zero mean and a variance of $\sigma_s^{(m)^2}$. Based on this assumption, the detection probability can be calculated as

$$P_{d,1}^{I} = P(L_{\text{UMPI}}(\mathbf{r}) > \eta_{\text{UMPI}}|\mathscr{H}_1) = \int_0^{\infty} P_{d|\delta} f_\delta(\delta)d\delta \tag{4B.1}$$

where

$$P_{d|\delta} = e^{-(\delta+M\eta_t)} \sum_{k=0}^{\infty} \left(\frac{\delta^k}{k!} \sum_{n=0}^{M+k-1} \frac{(M\eta_t)^n}{n!} \right) \tag{4B.2}$$

and

$$\boldsymbol{\delta} = \sum_{m=1}^{M} \frac{|\alpha_0{}^{(m)}|^2}{\sigma^2} \| \Pi_{\text{U}}^{(m)\perp} \mathbf{s}^{(m)} \|^2 \tag{4B.3}$$

Here, $f_\delta(\boldsymbol{\delta})$ is the pdf of $\boldsymbol{\delta}$. To find $f_\delta(\boldsymbol{\delta})$, we resort to Theorem 2.

Theorem 2. *Assuming that z_m for $m = 1, \ldots, M$ are independent and follow a standard exponential distribution and that are positive and unique, it follows that*

$$P\left(\sum_{m=1}^{M} \beta_m z_m > z)\right) = \sum_{m=1}^{M} w_m e^{-\frac{z}{\beta_m}}, \qquad z > 0 \tag{4B.4}$$

where

$$w_m = \frac{1}{\prod_{s=1,s\neq m}^{M}\left(1 - \frac{\beta_s}{\beta_m}\right)} \tag{4B.5}$$

Proof. (see, e.g., [17], pp. 137).

Making use of Theorem 2, we can find the pdf of δ as follows:

$$f_\delta(\delta) = \sum_{m=1}^{M} \gamma_m e^{-\frac{\delta}{\beta_m}}, \qquad \delta > 0 \tag{4B.6}$$

where

$$\gamma_m = \frac{1}{\beta_m \prod_{s=1,s\neq m}^{M} \left(1 - \frac{\beta_s}{\beta_m}\right)} \tag{4B.7}$$

and

$$\beta_k = \frac{\sigma_s^{(k)2}}{\sigma^2} \| \Pi_{\mathbf{U}}^{(k)\perp} \mathbf{s}^{(k)} \|^2 \tag{4B.8}$$

By substituting (4B.2) and (4B.6) in (4B.1), and using $\int_0^\infty x^k e^{-cx} dx = \frac{k!}{c^{k+1}}$, the detection probability of the SW1-I target model can be obtained as

$$P_{d,1}^{I} = e^{-M\eta_t} \sum_{k=0}^{\infty} \sum_{n=0}^{M+k-1} \frac{(M\eta_t)^n}{n!} \sum_{m=1}^{M} \frac{1}{\beta_m (1 + \frac{1}{\beta_m})^{k+1} \prod_{s=1,s\neq m}^{M} \left(1 - \frac{\beta_s}{\beta_m}\right)} \tag{4B.9}$$

In the case of the SW1-C target model, it is assumed that the amplitudes $\alpha_0^{(m)}$, where $m = 1, \ldots, M$, are correlated and subject to complex Gaussian fluctuations with zero mean and a variance of $\sigma_s{}^2$. Given this assumption, the detection probability can be determined by

$$P_{d,1}^{C} = P(L_{\text{UMPI}}(\mathbf{r}) > \eta_{\text{UMPI}} | \mathcal{H}_1) = \int_0^\infty P_{d|x} f_x(x) dx \tag{4B.10}$$

where $f_x(x)$ is the pdf of random variable x defined as $x = |\alpha_0^{(m)}|^2 = |\alpha_0|^2$ with exponential distribution, given by

$$f_x(x) = \frac{1}{\sigma_s^2} e^{-\frac{x}{\sigma_s^2}} \tag{4B.11}$$

and

$$P_{d|x} = e^{-(\delta_c x + M\eta_t)} \sum_{k=0}^{\infty} \left(\frac{(\delta_c x)^k}{k!} \sum_{n=0}^{M+k-1} \frac{(M\eta_t)^n}{n!} \right) \tag{4B.12}$$

where

$$\delta_c = \frac{1}{\sigma^2} \sum_{m=1}^{M} \| \Pi_{\mathbf{U}}^{(m)\perp} \mathbf{s}^{(m)} \|^2 \tag{4B.13}$$

Plugging (4B.11) and (4B.12) into (4B.10), and using $\int_0^\infty x^k e^{-cx} dx = \frac{k!}{c^{k+1}}$, the detection probability of the SW1-C target model is computed as follows:

$$P_{d,1}^{C} = e^{-M\eta_t} \sum_{k=0}^{\infty} \frac{1}{\beta (1 + \frac{1}{\beta})^{k+1}} \sum_{n=0}^{M+k-1} \frac{(M\eta_t)^n}{n!} \tag{4B.14}$$

where $\beta = \frac{\sigma_s{}^2}{\sigma^2} \sum_{m=1}^{M} \| \Pi_{\mathbf{U}}^{(m)\perp} \mathbf{s}^{(m)} \|^2$.

Chapter 5

Multitarget detection in multiband FM-based PBR: target range resolution improvement

5.1 Introduction

A passive radar that operates on a single-band FM has limited target range resolution (TRR) due to low-modulation bandwidth and a high reliance on the content broadcasted on that specific FM band. To address this issue, multiple FM bands from the same illuminator of opportunity (IO) station can be utilized. In Reference 1, authors demonstrated that combining several FM radio bands can enhance target range resolution, but at the expense of some ambiguities in the output of a multiband autocorrelation function. Meanwhile, in Reference 2, it was shown that exploiting a lower channel correlation bandwidth can somewhat avoid range dimension ambiguities. Real data analysis in Reference 3 confirmed that this approach can improve range resolution, but the cross-correlation function in range may be very peaky [1, 2]. Furthermore, in References 4 and 5, the authors extended the concepts introduced in References 2 and 3 by providing a mathematical framework for dealing with non-regularly spaced carrier frequencies of FM and DVB-T signals, respectively. Two methods based on multiband and wideband autocorrelation have been introduced by References 6 and 7 to improve the range resolution of targets when exploiting multiple adjacent DVB-T channels of the same IO. The authors also conducted an analysis of the application of these techniques using real data. In Reference 8, the authors investigated how exploiting multiple broadcasted channels from a single transmitter can enhance the target range resolution in the range correlation of the matched filter processing. The algorithm proposed in Reference 8 is inspired by the high-range resolution techniques of active radar used in high-resolution radar systems. By combining multiple stepped frequency waveforms, high-range resolution can be achieved. The proposed algorithm is demonstrated to offer higher target range resolution using both simulation and real target data obtained from FM radio and DVB-T stations, provided that both Doppler and phase compensations have been accomplished.

Motivated by the above works, a systematic framework to improve target range resolution is presented in this chapter, which covers signal modeling, detection method, and statistical analysis. Unlike previous studies, the proposed algorithm in this chapter achieves asymptotic optimality. The problem of detecting multiple targets in a multiband FM-based PBR is formulated as a binary composite hypothesis testing problem. Interference signals such as noise, direct signal, multipath/clutter echoes, and interfering targets are considered in this formulation. Then, a UMPI-based detection algorithm as an optimum invariant test is derived. It should be noted that the proposed multiband detector in this chapter is different from the one presented in Chapter 4. While the focus of Chapter 4 was on improving the target detection quality, the aim of this chapter is to enhance the target range resolution through the coherent combination of multiple bands of an FM station. Therefore, the two chapters differ in signal modeling, detection method, and statistical analysis.

This chapter's content is based on our previous works in References 9 and 10, which are the first works directly related to improving target range resolution using a detection theory framework. First, we developed an optimal detection method that is invariant. We then demonstrated mathematically how to improve target range resolution. Our results have shown that increasing the bandwidth of each channel in a multiband FM-based PBR system's receiver is not sufficient for improving the target range resolution. Specifically, we showed that multiband FM-based PBR systems can improve range resolution, but this comes at the cost of ambiguities in the range dimension of the UMPI test statistic. Surprisingly, we also found that having a lower bandwidth for each channel in a multiband FM-based passive bistatic radar (PBR) system's receiver can result in smaller ambiguities and better target range resolution. We derived closed-form expressions for the false alarm probability and detection probability. Simulation results indicated that multiband FM-based PBR systems offer advantages over single-band systems in terms of coherent diversity gain and resolving nearby targets.

The following is the organization of this chapter: Section 5.2 presents the signal modeling of the multiband FM-based PBR, followed by section 5.3, which derives the UMPI detector as an invariant test. In section 5.4, analytical performance assessments are presented, which include closed-form formulas for detection and false alarm probabilities. Section 5.5 demonstrates how the target range resolution can be enhanced by exploring multiple FM bands from a single transmitter. Additionally, section 5.6 discusses the detection performance of the proposed detector in the presence of model mismatches. Simulation results and discussions are presented in section 5.7, and a summary is provided in section 5.8.

5.2 Signal modeling

Consider a single-input single-output (SISO) FM-based passive radar system that consists of a multichannel transmitter and a multiband receiver. The transmitter broadcasts signals in M different frequency bands or channels, and the receiver receives and processes these channels independently to form a wide bandwidth channel. The received surveillance signal due to the mth frequency channel at time delay index n can be represented by Reference 9

$$
\begin{aligned}
x^{(m)}[n] = {} & c_d^{(0,m)} y^{(m)}[n] + \sum_{i=1}^{N_c^{(m)}} c_n^{(i,m)} y^{(m)}[n - n_{c0}^{(i,m)}] \\
& + \sum_{k=0}^{K_m-1} \alpha_t^{(k,m)} y^{(m)}[n - n_t^{(k,m)}] e^{j2\pi(\frac{v_t^{(k)}}{\lambda_m}) n T_{sa}} \\
& + n_s^{(m)}[n], \quad n = 0, \ldots, N_m - 1, \quad m = 1, \ldots, M
\end{aligned}
\tag{5.1}
$$

where $\{\lambda_m\}_{m=1}^{M}$ are the wavelength of the transmitted radio channels; N_m is the data vector length corresponding to the integration time of the mth received surveillance channel denoted by T_m, i.e., $T_m = N_m T_{sa}$, where $f_{sa} = \frac{1}{T_{sa}}$ is the sampling frequency satisfying the Nyquist rate; $y^{(m)}[n]$ is the nth sample of the reference channel due to the mth broadcasted channel after accomplishing a proper signal conditioning to remove the effect of multipath and target echoes in the reference channel [11]; $\alpha_t^{(k,m)}$ represents the complex amplitude of the kth received target due to the mth broadcasted channel. It can be defined as $\alpha_t^{(k,m)} = a_t^{(k,m)} \sqrt{p^{(m,1)}} e^{-j2\pi f_c^{(m)} n_t^{(k,m)} T_{sa}}$, in which $f_c^{(m)}$ is the carrier frequency of mth broadcasting channel; $a_t^{(k,m)}$ for $k = 0, \ldots, K_m - 1$ takes into account the amplitudes attenuation of point targets. Assume that $y^{(m)}[n]$ has unit amplitude, so that the transmitted signal amplitude of mth broadcasted channel is represented by the term $\sqrt{p^{(m,1)}}$. In (5.1), $n_t^{(k,m)}$ and $v_t^{(k)}$ represent the time delay index and velocity of the kth target due to the mth broadcasted radio channel, respectively, where K_m is the number of targets to be detected in the mth frequency channel, and $k = 0, \ldots, K_m - 1$ is the target number; $c_d^{(0,m)}$ is the complex amplitude of the direct signal received by the sidelobe/backlobe antenna of the surveillance channel due to the mth broadcasted channel; $c_n^{(i,m)}$ and $n_{c0}^{(i,m)}$ represent the complex (time-varying) amplitude and time delay index of the ith clutter scatterers due to the mth broadcasted channel, respectively, for $i = 1, \ldots, N_c^{(m)}$ with $N_c^{(m)}$ being the number of clutter echoes resulted from nearby ground scatterers due to the mth broadcasted channel; $n_s^{(m)}[n]$ is also the nth noise sample at the mth received surveillance channel.

In this model, the time-varying amplitudes $c_n^{(i,m)}$ can be considered to be a slowly varying function of time, so that they can be represented by only a few frequency components around zero Doppler [12], i.e.,

$$
c_n^{(i,m)} = \sum_{q=-Q_i}^{Q_i} c^{(i,q,m)} e^{j2\pi f_c^{(i,q,m)} n T_{sa}}
\tag{5.2}
$$

where $c^{(i,q,m)}$ are considered as constant clutter amplitudes at the delay–Doppler coordinates of $(n_{c0}^{(i,m)}, f_c^{(i,q,m)})$, where $f_c^{(i,q,m)} = \dfrac{v_c^{(i,q)}}{\lambda_m}$ and $v_c^{(i,q)}$ are the Doppler velocities of clutter scatterers determined by the velocity spread in the clutter spectrum. For example, surface clutter usually has a small velocity spread due to the motion of the clutter scatterers, producing a small Doppler velocity spread for stationary radar. For terrain, the velocity spread ranges from near zero for bare terrain to about 0.33 m/s for wind-blown trees [13]. Alternatively, given an upper bound on the velocity spread of surface clutter, we can select $v_c^{(i,q)}$ uniformly spread across the clutter velocity bandwidth [12]. For example, we can choose $v_c^{(i,q)} = q\Delta v_c$ for $q = -Q_i, \ldots, Q_i$ with Q_i and Δv_c being two parameters that should be selected properly based on the velocity spread of the clutter spectrum [12, 13]. Henceforth, for simplicity in notation, we consider a maximum spread in velocity for all delay of clutter, i.e., $Q = \max_i\{Q_i\}$ for $i = 1, ..., N_c^{(m)}$ and $m = 1, ..., M$. In addition to approximate the spectrum of a clutter scatterer by a set of velocities denoted by $\{v_c^{(i,q)}\}_{q=-Q}^{Q}$, it is generally understood that the clutter echoes results from the nearby ground scatterers distributed from zero to maximum bistatic range of $R_c^{(m)}$ [12]. Taking these into consideration, we can consider a limited delay–velocity extent, say clutter region, for receiving possible clutter echoes. In the case at hand, we can consider P_m components with unknown amplitudes $\alpha_c^{(p,m)}$ at the delay–velocity coordinates of $(n_c^{(p,m)}, v_c^{(p)})$, including that of direct signal and clutter echoes, in the clutter region. Here, $P_m = (2Q + 1)N_c^{(m)} + 1$, where $N_c^{(m)}$ is the equivalent number of ground scatterers up to the bistatic clutter cut-off range of $R_c^{(m)}$, and $(2Q + 1)$ is the maximum number of Doppler velocity components of a clutter scatterer.

It is known that the range resolution in radar systems is inversely proportional to the bandwidth of the signal that is being used. This relation suggests that range resolution in PBR systems can be improved by increasing the total bandwidth of the signal used for detection. This may persuade one to use multiple frequency channels of FM radio in order to improve the bandwidth of the signal used for detection. To do this, without loss of generality, the desired received channels should be digitally up-sampled (UP-S), filtered at the selected bandwidth of B, and up-converted (UP-C) to f_m in order to be adjacent in frequency with no overlap or space between them. Figure 5.1 shows the workflow of the proposed algorithm to improve target

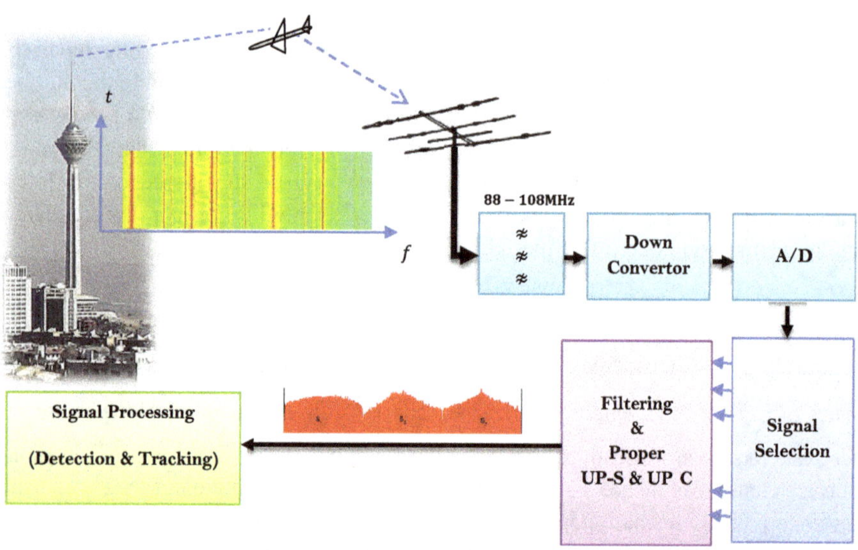

Figure 5.1 High target range resolution multiband FM-based PBR block diagram. Different frequency channels need to be digitally up-sampled (UP-S), filtered at the selected bandwidth of B, and up-converted (UP-C) to be adjacent in frequency.

range resolution when multiple frequency channels of FM radio are used. Now, the received signal model in the surveillance channel can be rewritten as

$$
\begin{aligned}
r[n] &= \sum_{m=1}^{M} x_f^{(m)}[n] e^{j2\pi f_m n T_s} \\
&= \sum_{m=1}^{M}\sum_{p=1}^{P_m} \alpha_c^{(p,m)} y_f^{(m)}[n - n_c^{(p,m)}] e^{j2\pi(\frac{v_c^{(p)}}{\lambda^{(m)}})nT_s} e^{j2\pi f_m n T_s} \\
&+ \sum_{m=1}^{M}\sum_{k=0}^{K_m-1} a_t^{(k,m)} \sqrt{p^{(m,1)}} e^{-j2\pi f_c^{(m)} n_t^{(k,m)} T_s} y_f^{(m)}[n - n_t^{(k,m)}] \\
&\times e^{j2\pi(\frac{v_t^{(k)}}{\lambda^{(m)}})nT_s} e^{j2\pi f_m n T_s} \\
&+ w[n]
\end{aligned}
\tag{5.3}
$$

where $x_f^{(m)}[n]$ and $y_f^{(m)}[n]$ are the filtered versions of $x^{(m)}[n]$ and $y^{(m)}[n]$ with selected bandwidth of B, respectively. In this case, we can consider $w[n]$ as a wideband noise with the bandwidth of MB, where the sampling frequency used in (5.3) is equal to $f_s = \frac{1}{T_s} = MB$. In the following sections, we obtain a necessary condition on B for improving target range resolution. For simplicity of notation, in (5.3), we use the same parameters as (5.1). Indeed, they can be converted to each other based on the different sampling frequency used in (5.1) and (5.3).

It is shown in Reference 14 that it is required to assume that $a_t^{(k,m)} = a_t^{(k)}$ for $m = 1, ..., M$ in hope of target range resolution improvement. This means that the received amplitudes of a target over different frequency channels of FM radio must be completely correlated. This can be achieved by considering that the broadcasted signal is reflected from a point target.* Besides, in practice, the target's radar cross section (RCS) may be expected to be frequency dependent. This will result in varying target amplitude over different frequency channels. In order to obtain correlated target echo signals over different frequency channels, the working frequency interval of two frequency channels, say Δf, should be selected in accordance with the criteria of $\Delta f \leq \frac{c}{2L}$, where c is the speed of light and L is the target size [17]. For example, we should have $\Delta f \leq 5$ MHz for the target size of 30 m. Based on these assumptions, the received signal (5.3) can be compactly written as

$$
\mathbf{r} = a_0\mathbf{s} + \mathbf{C}\mathbf{a}_c + \mathbf{T}_i\mathbf{a}_i + \mathbf{n}_s \tag{5.4}
$$

where $\mathbf{r}$ is the $N \times 1$ vector received by the surveillance channel, in which $N = \max_m\{N_m\}$. Here, $\mathbf{n}_s$ is the $N \times 1$ receiver thermal noise vector. In this case, we have $a_0 = a_t^{(0)}$ and the vector $\mathbf{s}$ is an $N \times 1$ vector defined as follows

$$
\mathbf{s} = \sum_{m=1}^{M} \sqrt{p^{(m,1)}} e^{j2\pi f_c^{(m)} n_0 T_s} \left(\mathbf{y}_f^{(n_0,m)} \odot \mathbf{e}^{(v_0,m)} \odot \mathbf{b}^{(f_m)}\right) \tag{5.5}
$$

where $\mathbf{y}_f^{(n_0,m)} = \mathbf{P}^{n_0}\mathbf{y}_f^{(m)}$, $[\mathbf{e}^{(v_0,m)}]_n = e^{j2\pi(\frac{v_0}{\lambda^{(m)}})nT_s}$, and $[\mathbf{b}^{(f_m)}]_n = e^{j2\pi f_m n T_s}$ with $n = 0, ..., N_m - 1$. In this case, $\mathbf{P}$ is an $N_m \times N_m$ permutation matrix defined as

*For narrow bandwidth waveforms, the point target model is often valid; see, e.g., References 15, 16.

$[\mathbf{P}]_{ij} = 1$ if $i = j + 1$ and 0 otherwise for $i = 0, ..., N_m - 1$; $j = 0, ..., N_m - 1$. Here, $\mathbf{y}_f^{(m)}$ is the reference signal received due to mth broadcasted channel.

The clutter subspace is described by the matrix $\mathbf{C}$, given by

$$\mathbf{C} \triangleq \left[\mathbf{C}^{(1)}, ..., \mathbf{C}^{(M)}\right] \tag{5.6}$$

where $\mathbf{C}^{(m)} = [\mathbf{c}_1^{(m)}, ..., \mathbf{c}_p^{(m)}, ..., \mathbf{c}_{P_m}^{(m)}]$ is an $N_m \times P_m$ matrix with $\mathbf{c}_p^{(m)} = (\mathbf{y}_f^{(n_c^{(p,m)}, m)} \odot \mathbf{e}^{(v_c^{(p)}, m)} \odot \mathbf{b}^{(f_m)})$, where $n = 0, ..., N_m - 1$, $m = 1, ..., M$, and $P_e = \sum_{m=1}^{M} P_m$. In this case, $n_c^{(p,m)}$ and $v_c^{(p,m)}$ are determined based on the delay–velocity coordinates of the clutter region for $p = 1, ..., P_m$. The vector $\mathbf{a}_c$ also contains the clutter amplitudes corresponding to the columns of the matrix $\mathbf{C}$. Since the true velocity-delay coordinate of the ground scatterers is unknown in practice and can take any velocity-delay coordinates within that of the clutter region, we assumed that we are faced with the worst-case condition of receiving clutter echoes. Hence, we considered all of the delay–velocity coordinates within the clutter region. However, one should note that clutter amplitudes $\alpha_c^{(p,m)}$ for any delay–velocity coordinate of the clutter region are considered as unknown complex parameters. In addition to this, three parameters of the clutter region introduced by Q, Δv_c and $R_c^{(m)}$ should be set properly in practice based on the maximum Doppler velocity spread of the clutter spectrum, and the radar location.

In (5.4), the interfering targets subspace can be described by the matrix $\mathbf{T}_i$, represented as

$$\mathbf{T}_i \triangleq \left[\mathbf{t}_1, ..., \mathbf{t}_{K-1}\right] \tag{5.7}$$

where

$$\mathbf{t}_k = \sum_{m=1}^{M} \sqrt{p^{(m,1)}} e^{j2\pi f_c^{(m)} n_t^{(k)} T_s} \left(\mathbf{y}_f^{(n_t^{(k)}, m)} \odot \mathbf{e}^{(v_t^{(k)}, m)} \odot \mathbf{b}^{(f_m)}\right) \tag{5.8}$$

where $k = 1, ..., K - 1$ with $K_m = K$ for $m = 1, ..., M$, and $\mathbf{a}_i = [a_1, ..., a_{K-1}]^T$ is the $(K - 1)$-dimensional column vector containing all amplitudes of the interfering targets. The interfering target amplitude vector, $\mathbf{a}_i$, is assumed to be deterministic and unknown. Moreover, the number of interfering targets $(K - 1)$, their corresponding delays $n_t^{(k)}$, and their bistatic velocities $v_t^{(k)}$ for $k = 1, ..., K - 1$ are assumed to be unknown.

Here, $\mathbf{n}_s$ is the surveillance noise vector, which is assumed to be zero-mean complex white Gaussian noise with the unknown covariance matrix of $E\{\mathbf{n}_s \mathbf{n}_s^H\} = \sigma^2 \mathbf{I}_N$.

5.3 Design of high-resolution UMPI test

The composite binary hypothesis-testing problem to detect a target with unknown amplitudes a_0 for a given delay–velocity coordinate of (v_0, n_0) can be formulated as

$$\begin{cases} \mathcal{H}_0: \mathbf{r} = \mathbf{C}\mathbf{a}_c + \mathbf{T}_i\mathbf{a}_i + \mathbf{n}_s \\ \mathcal{H}_1: \mathbf{r} = a_0\mathbf{s} + \mathbf{C}\mathbf{a}_c + \mathbf{T}_i\mathbf{a}_i + \mathbf{n}_s \end{cases} \tag{5.9}$$

The goal will be looking for a UMPI-based target detection algorithm in the presence of noise, clutter/multipath, and interfering targets to improve target range resolution. Thus, we will call the derived test a high-resolution UMPI detector. As such, the UMPI test cannot be straightforwardly obtained since there are lots of unknowns including $\mathbf{T}_i$, a_0, $\mathbf{a}_c$, $\mathbf{a}_t$, and σ^2. To address this, we obtained a two-stage UMPI-based detector rather than a one-stage detector. In this regard, it first assumes that the matrix $\mathbf{T}_i$ is known and drives the UMPI detector according to the classical approach over the remaining unknowns a_0, $\mathbf{a}_c$, $\mathbf{a}_t$, and σ^2. Then, the estimated matrix from the second stage is substituted in place of the true matrix $\mathbf{T}_i$ into the UMPI test. To do this, a multistage algorithm is proposed in the second stage of detector design. In what follows, the first and second stages are described in detail.

5.3.1 First stage of UMPI-based detector

In the first stage of the detector design, it is assumed that the matrix $\mathbf{T}_i$ is known, and the high-resolution UMPI detector is derived according to the classical approach over the remaining unknowns a_0, $\mathbf{a}_c$, $\mathbf{a}_t$, and σ^2. To derive the UMPI test, the classical approach introduced in Chapter 4 is applied. Before deriving the UMPI test, some transformations are applied to the observation data $\mathbf{r}$ to bring the detection problem to a simpler form. To do this, we first use the singular eigenvalue decomposition (SVD) of matrix $\mathbf{C}$ described as follows

$$\mathbf{C} = [\mathbf{U}_C^{(1)}, \mathbf{U}_C^{(2)}] \begin{bmatrix} \boldsymbol{\Sigma}_C^{(1)} & \mathbf{0} \\ \mathbf{0} & \mathbf{0} \end{bmatrix} \begin{bmatrix} \mathbf{V}_C^{(1)} \\ \mathbf{V}_C^{(2)} \end{bmatrix} \tag{5.10}$$

Here, $\mathbf{U}_C^{(1)}$ and $\mathbf{U}_C^{(2)}$ are the $N \times P_e$ and $N \times (N - P_e)$ orthogonal matrices that span the column space of matrix $\mathbf{C}$ and its orthogonal complement, respectively, i.e., $\mathbf{U}_C^{(2)^H}\mathbf{C} = \mathbf{0}$ [18]. Multiplying the observation vector $\mathbf{r}$ in (5.4) by $\mathbf{U}_C^{(2)^H}$, we come up with an observation vector as

$$\mathbf{r}_C \triangleq \mathbf{U}_C^{(2)^H}\mathbf{r} = a_0\mathbf{s}_C + \mathbf{W}_i\mathbf{a}_t + \mathbf{n}_C \tag{5.11}$$

where $\mathbf{r}_C$ is the $(N - P_e)$-dimensional column vector, vector $\mathbf{s}_C$ is defined as $\mathbf{s}_C \triangleq \mathbf{U}_C^{(2)^H}\mathbf{s}$, and $\mathbf{W}_i$ is an $(N - P_e) \times (K - 1)$ matrix defined as $\mathbf{W}_i \triangleq \mathbf{U}_C^{(2)^H}\mathbf{T}_i$. Since $\mathbf{U}_C^{(2)^H}\mathbf{U}_C^{(2)} = \mathbf{I}_{(N-P_e)}$ and $\mathbf{n}_s^{(m)}$ is zero-mean white Gaussian noise with the covariance matrix of $\sigma^2\mathbf{I}_N$, the transformed noise $\mathbf{n}_C$ is also zero-mean white Gaussian noise with the covariance matrix of $\sigma^2\mathbf{I}_{(N-P_e)}$.

Again, we employ another orthogonal transformation obtained using the SVD of $\mathbf{W}_i$ with the rank of $(K - 1)$, as follows

$$\mathbf{W}_i = [\mathbf{U}_{w_i}^{(1)}, \mathbf{U}_{w_i}^{(2)}] \begin{bmatrix} \boldsymbol{\Sigma}_{w_i}^{(1)} & \mathbf{0} \\ \mathbf{0} & \mathbf{0} \end{bmatrix} \begin{bmatrix} \mathbf{V}_{w_i}^{(1)} \\ \mathbf{V}_{w_i}^{(2)} \end{bmatrix} \tag{5.12}$$

Here, $\mathbf{U}_{w_i}^{(1)}$ and $\mathbf{U}_{w_i}^{(2)}$ are the $(N - P_e) \times (K - 1)$ and $(N - P_e) \times (N - P_e - K + 1)$ orthogonal matrices that span the column space of $\mathbf{W}_i$ and its orthogonal complement, respectively, i.e., $\mathbf{U}_{w_i}^{(2)^H}\mathbf{W}_i = \mathbf{0}$. Multiplying the observation data $\mathbf{x}_C$ by $\mathbf{U}_{w_i}^{(2)^H}$, we get

$$\mathbf{x} \triangleq \mathbf{U}_{w_i}^{(2)^H}\mathbf{r}_C = a_0\mathbf{s}_{w_i} + \mathbf{n}_{w_i} \tag{5.13}$$

where $\mathbf{x}$ is the $(N - P_e - K + 1)$-dimensional column vector, and vector $\mathbf{s}_{w_i}$ is defined as $\mathbf{s}_{w_i} \triangleq \mathbf{U}_{w_i}^{(2)^H}\mathbf{s}_c$. Since $\mathbf{U}_{w_i}^{(2)^H}\mathbf{U}_{w_i}^{(2)} = \mathbf{I}_{(N-P_e-K+1)}$ and $\mathbf{n}_c$ is zero-mean white Gaussian noise with the covariance matrix of $\sigma^2\mathbf{I}_{(N-P_e)}$, the transformed noise $\mathbf{n}_{w_i}$ is also zero mean, white, and Gaussian noise with the covariance matrix of $\sigma^2\mathbf{I}_{(N-P_e-K+1)}$.

To derive UMPI test, we first show that the detection problem in (5.9) is invariant under the composition of the transformation groups G_Q and G_d defined as follows

$$G_Q = \{g_Q : g_Q(\mathbf{x}) = \mathbf{Q}_{\mathbf{s}_{w_i}}^l\mathbf{x}\} \tag{5.14}$$

where l is an integer value and $\mathbf{Q}_{\mathbf{s}_{w_i}}$ is Householder matrices with Householder vectors $\mathbf{s}_{w_i}$ defined as $\mathbf{Q}_{\mathbf{s}_{w_i}} = \mathbf{I}_{(N-P_e-K+1)} - 2\dfrac{\mathbf{s}_{w_i}\mathbf{s}_{w_i}^H}{\|\mathbf{s}_{w_i}\|^2}$, and

$$G_d = \{g_d\text{: } g_d(\mathbf{x}) = d\mathbf{x},\ d \text{ is a complex scalar}\} \tag{5.15}$$

The above transformations, G_Q and G_d are groups, since they are closed sets and associative and contain the identity and inverse elements. In the following, we show that subject to any composition of the transformation groups in (5.14) and (5.15), the distribution of the observation and the parameter spaces remain unchanged under each hypothesis.

1. Under $\mathscr{H}_1$, we have $\mathbf{x} \sim \mathscr{CN}(a_0\mathbf{s}_{w_i}, \sigma^2\mathbf{I}_{(N-P_e-K+1)})$ and so, for the transformed data, we get $g_Q(\mathbf{x}) = \mathbf{Q}_{\mathbf{s}_{w_i}}\mathbf{x} \sim \mathscr{CN}((-1)^l a_0\mathbf{s}_{w_i}, \sigma^2\mathbf{I}_{(N-P_e-K+1)})$. Since a_0 is unknown complex value, the distribution family of the transformed data does not change. Under the null hypothesis, the proof is similar, except that $a_0 = 0$. Thus, the detection problem is invariant under $g_Q(\mathbf{x})$ transformation.
2. Under $\mathscr{H}_1$, from $\mathbf{x} \sim \mathscr{CN}(a_0\mathbf{s}_{w_i}, \sigma^2\mathbf{I}_{(N-P_e-K+1)})$, we get $g_d(\mathbf{x}) = d\mathbf{x} \sim \mathscr{CN}(d \times a_0\mathbf{s}_{w_i}, |d|^2\sigma^2\mathbf{I}_{(N-P_e-K+1)})$. Since a_0 and σ^2 are unknown complex and real values, respectively, the distribution family of the transformed data is not changed. Under the null hypothesis, the proof is similar, except that $a_0 = 0$. Therefore, the detection problem is invariant under $g_d(\mathbf{x})$ transformation.

In Appendix 5A, the maximum invariant statistic for the composite group G is derived. In term of original observation, the maximal invariant statistic can be written as

$$z = \frac{\dfrac{\left|\mathbf{s}^H\mathbf{U}_C^{(2)}\mathbf{U}_W^{(2)}\mathbf{U}_W^{(2)^H}\mathbf{U}_C^{(2)^H}\mathbf{r}\right|^2}{\|\mathbf{s}^H\mathbf{U}_C^{(2)}\mathbf{U}_W^{(2)}\mathbf{U}_W^{(2)^H}\mathbf{U}_C^{(2)^H}\mathbf{s}\|^2}}{\mathbf{r}^H\mathbf{U}_C^{(2)}\mathbf{U}_W^{(2)}\mathbf{U}_W^{(2)^H}\mathbf{U}_C^{(2)^H}\mathbf{r}} \tag{5.16}$$

After some matrix manipulations, it can be shown that

$$\mathbf{U}_C^{(2)}\mathbf{U}_w^{(2)}\mathbf{U}_w^{(2)^H}\mathbf{U}_C^{(2)^H} = \mathbf{\Pi}^{\perp}_{\mathbf{\Pi}_C^{\perp}\mathbf{T}_i}\mathbf{\Pi}^{\perp}_{\mathbf{C}} = \mathbf{\Pi}^{\perp}_{\mathbf{U}} \tag{5.17}$$

where $\mathbf{U} = [\mathbf{C}, \mathbf{T}_i]$, and $\mathbf{\Pi}^{\perp}_{\mathbf{X}}$ is the orthogonal projection onto the subspace span by columns of matrix $\mathbf{X}$. Using (5.17), we can rewrite the maximal invariant statistic in (5.16) as

$$z = \frac{\left|\mathbf{s}^H\mathbf{\Pi}^{\perp}_{\mathbf{U}}\mathbf{r}\right|^2}{(\mathbf{s}^H\mathbf{\Pi}^{\perp}_{\mathbf{U}}\mathbf{s})(\mathbf{r}^H\mathbf{\Pi}^{\perp}_{\mathbf{U}}\mathbf{r})} \tag{5.18}$$

In Appendix 5B, we also obtain the distribution of a 1–1 function of z described as $t = \frac{N-(P_e+K)}{1}\frac{z}{1-z}$ and show that t follows a central complex F distribution under the $\mathcal{H}_0$ hypothesis, whereas it follows a noncentral complex F distribution under the $\mathcal{H}_1$ hypothesis. In addition to this, it was shown in [19] that noncentral complex F distribution has a Monotone Likelihood Ratio (MLR) property of t, which implies that the threshold test on the likelihood ratio can be replaced by a threshold on the statistic t. Hence, the UMPI test is to reject $\mathcal{H}_0$ if $t > \eta_t$, or equivalently

$$L_U(\mathbf{r}) = \frac{\left|\mathbf{s}^H\mathbf{\Pi}^{\perp}_{\mathbf{U}}\mathbf{r}\right|^2}{(\mathbf{s}^H\mathbf{\Pi}^{\perp}_{\mathbf{U}}\mathbf{s})(\mathbf{r}^H\mathbf{\Pi}^{\perp}_{\mathbf{U}}\mathbf{r})} > \eta_U \tag{5.19}$$

where $\eta_U = \dfrac{\eta_t}{N-(P_e+K)+\eta_t}$. The threshold η_t is selected such that the desired false alarm probability (P_{fa}) requirement is satisfied in the test $t > \eta_t$.

It should be noted that the derived UMPI test in this chapter is different from that obtained in Chapter 4 for $M \neq 1$. More precisely, they coincide each other just for the single-band case, i.e., $M = 1$.

5.3.2 Second stage of UMPI-based detector

The first step in designing a UMPI-based detector assumes that the Interfering Target Matrix (ITM) $\mathbf{T}_i$ is known, which means that the target delay–velocity pairs are known at the receiver side. However, in practical situations, this information is not available and must be estimated. To solve this problem, a multistage algorithm called the High Resolution Imperative Target Positioning (HRITP) algorithm is introduced. This algorithm detects and localizes targets sequentially, by exploiting the target locations from the previous steps to update the ITM in the UMPI test for the detection and localization of a new imperative target. The HRITP algorithm assumes that the columns of ITM are sorted by target powers, with the first column corresponding to the strongest target echo. The algorithm detects the strongest target, cancels its effect from the received signal, and then proceeds to detect the next strongest imperative target signal, and so on. This iterative process continues until all the imperative targets are detected and localized.

The detailed steps of the HRITP algorithm can be described as follows:

1. Set $k \leftarrow 0$, $\mathbf{U}_0^{\perp} = \mathbf{C}$, $\mathbf{R}_0 = (\mathbf{C}^H\mathbf{C})^{-1}$ and $\mathbf{y}_{-1} = \mathbf{r}$.
2. Compute $\mathbf{y}_k = \mathbf{y}_{k-1} - \mathbf{U}_k^{\perp}\mathbf{R}_k\mathbf{U}_k^{\perp^H}\mathbf{y}_{k-1}$.

3. Compute the matrix $\mathbf{L}_a^{(k)}(n, l_v)$ using

$$\mathbf{L}_a^{(k)}(n, l_v) = \frac{\mathbf{f}(n) \odot \mathbf{f}^*(n)}{\| \sum_{m=1}^{M} \mathbf{y}_f^{(n,m)} \|^2 \|\mathbf{y}_k\|^2} \qquad n = 1, ..., N_r \tag{5.20}$$

where $\mathbf{f}(n) = \sum_{m=1}^{M} \sqrt{p^{(m,1)}}\mathbf{f}^{(m)}(n)$, in which $\mathbf{f}^{(m)}(n)$ is defined as

$$\mathbf{f}^{(m)}(n) = \mathcal{F}(e^{-j2\pi f_c^{(m)} nT_s}(\mathbf{y}_k \odot \mathbf{y}_f^{(n,m)} \odot \mathbf{b}^{(m)}), N_F^{(m)}) \tag{5.21}$$

Here, operator $\mathcal{F}(\mathbf{d}, N_F^{(m)})$ performs an $N_F^{(m)}$ point Fast Fourier Transform (FFT) of $\mathbf{d}$ such that the zero-frequency component is moved to the center of the $\mathbf{f}^{(m)}(n)$. Here, n and l_v are the time delay index and Doppler bin index, respectively. Note that in (5.23), N_r is the time delay index corresponding to the maximum bistatic range of interest. One should note that the target Doppler shift will be different when using broadcast signals at different carrier frequencies. To solve this in the HRITP algorithm, we use different number of FFT points over different frequency channels [9].
4. Compare the maximum value of $\mathbf{L}_a^{(k)}(n, l_v)$ over the time delay index n and Doppler bin index l_v, with the predetermined threshold η_a. If this maximum exceeds the threshold η_a, it means that the range-velocity coordinate corresponding to this maximum is the estimation of the bistatic range and velocity of an interfering target. In this case, we can continue the following steps to find other imperative targets. The algorithm stops when the obtained maximum is smaller than the threshold η_a. Note that, the threshold η_a should be selected according to the desired false alarm probability of the system (see section 5.5 for more details).
5. Set $k \leftarrow (k+1)$ and store the delay–velocity coordinate $(\hat{n}_t^{(k)}, \hat{l}_v f_s \frac{\lambda_r}{N_F^{(r)}})$ corresponding to the maximum value extracted at the previous step as the estimation of the kth interfering target position. Note that, this estimation may have an insufficient accuracy for a super interfering target cancelation. To circumvent this, a small area around the range-velocity coordinate of this maximum is considered for an effective cancelation. The velocity extent is selected to cover 7 velocity elements equal to $(\hat{l}_v \pm q)\frac{f_s}{6}\frac{\lambda_r}{N_F^{(r)}}$ for $q = 0, 1, 2, 3$ and spreads over 3 time delay index $\hat{n}_t^{(k)} - 1$, $\hat{n}_t^{(k)}$, and $\hat{n}_t^{(k)} + 1$, yielding a mask of dimension 7×3 or the total number of range-velocity bins equal to $N_X = 21$. Simulation results show that this extension is sufficient for a strong interfering target cancellation.
6. Compute and store the matrices $\mathbf{U}_k^{\perp}$ and $\mathbf{R}_k$ given by

$$\mathbf{U}_k^{\perp} = \mathbf{U}_k - \sum_{s=0}^{k-1} \mathbf{U}_s^{\perp}\mathbf{R}_s\mathbf{U}_s^{\perp H}\mathbf{U}_k \tag{5.22}$$

and

$$\mathbf{R}_s = (\mathbf{U}_s^{\perp H}\mathbf{U}_s^{\perp})^{-1} \tag{5.23}$$

where the ith column of $\mathbf{U}_k$, $[\mathbf{U}_k]_i = \mathbf{t}_i$, is defined based on the extended delay–velocity coordinates $(n_t^{(i)}, v_t^{(i)})$ for $i = 1, ..., N_x$ at step 5, given by

$$\mathbf{t}_i = \sum_{m=1}^{M} \sqrt{p^{(m,1)}} e^{j2\pi f_c^{(m)} n_t^{(i)} T_s} \left(\mathbf{y}_f^{(n_t^{(i)},m)} \odot \mathbf{e}^{(v_t^{(i)},m)} \odot \mathbf{b}^{(f_m)} \right) \tag{5.24}$$

7. Go back to step 2.

Step 4 provides the stopping criterion for the algorithm. The HRITP algorithm can detect targets above the noise level of the surveillance channel. Once the algorithm ends, the matrix $\mathbf{T}_i$ is obtained, which gives the number of strong targets detected. To confirm the detection of a target in the presence of other targets, the UMPI detector derived in (5.19) can be used. One of the strong targets is considered as the testing target, while the other $(K-1)$ targets are considered interfering targets. This step is called the confirmation step of the detection method. Although the algorithm requires several iterations to detect all possible targets and confirm their detection in the presence of other targets, it ensures excellent detection performance and regulates the probability of false alarms.

5.4 Analytical performance analysis

In this section, we derive closed-form expressions for both the detection probability and false alarm probability. Appendix 5B demonstrates that a 1–1 function of z, represented as $t = \frac{N-(P_e+K)}{1}\frac{z}{1-z}$, is distributed according to a central complex F distribution with 1 and $N-(P_e+K)$ complex degrees of freedom under the $\mathcal{H}_0$ hypothesis [20]. As a result, we can calculate the false alarm probability of the UMPI detector using the statistic t, i.e.,

$$P_{fa} = Q_{F_{1,N-(P_e+K)}}(\eta_t) \tag{5.25}$$

where $Q_{F_{1,N-(P_e+K)}}$ is the right-tail probability of central complex F distribution. To achieve the desired integration gain in the current problem, a large value is assigned to N. As a result, $F_{1,N-(P_e+K)}$ approaches the complex χ_1^2, as stated in Reference 20. This allows us to calculate P_{fa} as follows

$$P_{fa} = Q_{\chi_1^2}(\eta_t) = e^{-\eta_t} \tag{5.26}$$

For a predetermined value of P_{fa}, we should compute η_t from (5.26). Note that, in the UMPI test described in (5.19), the threshold η_U is set as $\eta_U = \frac{\eta_t}{(N-(P_e+K))+\eta_t}$. It is worthwhile to note that $N \gg (P_e+K)$, so an incorrect estimate of the number of targets cannot affect the UMPI test thresholding. As an important result, we can set threshold η_a in the UMPI algorithm, described in section 5.3, with an assumed maximum number of strong targets.

Similarly, the probability of detection based on the statistic t can be obtained as

$$P_d = Q_{F_{1,N-(P_e+K)}(\delta)}(\eta_t) \tag{5.27}$$

where $Q_{F_{1,N-(N_I+1)}(\delta)}$ is the right-tail probability of noncentral complex F distribution. Similarly, due to the large value of N in our detection problem, we can approximate $F_{1,N-(P_e+K)}(\delta)$ with $\chi_1^2(\delta)$. Hence, the detection probability of the UMPI test is given by

$$P_d = e^{-(\delta+\eta_t)} \sum_{k=0}^{\infty} \left(\frac{\delta^k}{k!} \sum_{n=0}^{k} \frac{\eta_t^n}{n!} \right) \tag{5.28}$$

Using (5.28), it is easy to show that P_d is an increasing function of $\delta = \frac{|\alpha_0|^2}{\sigma^2} \|\mathbf{\Pi}_{\mathrm{U}}^{\perp}\mathbf{s}\|^2$. If $\frac{|\alpha_0|^2}{\sigma^2}$ is considered as the input target signal-to-noise ratio (SNR), then the target SNR gain provided by the UMPI detector can be considered as $\|\Pi_{\mathrm{U}}^{\perp}\mathbf{s}\|^2$. Moreover, since $0 \leq \|\mathbf{\Pi}_{\mathrm{U}}^{\perp}\mathbf{s}\|^2 \leq \|\mathbf{s}\|^2$, we can define the detection loss as follows

$$\mathrm{DL} = \frac{\|\mathbf{\Pi}_{\mathrm{U}}^{\perp}\mathbf{s}\|^2}{\|\mathbf{s}\|^2} \tag{5.29}$$

5.5 Analytical assessment of improved range resolution

In this section, we demonstrated mathematically how multiple FM radio channels from a single transmitter can be used to improve range resolution in the range (delay) dimension of the UMPI detector. This improvement can provide more insight into how target range resolution can be improved in PBR systems. In what follows, we present the necessary conditions for improving target range resolution in PBR systems, which can be achieved by selecting the appropriate value of B. To further investigate the range (delay) dimension of the UMPI detector, it is useful to assume that the received signal in the surveillance channel consists of a testing target and noise. This is because our proposed detector does not require improvement in the range resolution of interference signals such as direct signal and clutter. Therefore, we can focus on improving target range resolution by simplifying the UMPI statistic as given by

$$L_{U_M}(\mathbf{x}) = \frac{|\mathbf{s}^H\mathbf{x}|^2}{(\mathbf{s}^H\mathbf{s})(\mathbf{x}^H\mathbf{x})} \tag{5.30}$$

The number of channels used for detection is denoted by subscript M. To focus on investigating range resolution, it is easier to assume that the Doppler frequency of the testing target is zero. In addition, we can ignore the noise term in the received signal, resulting in $\mathbf{x} = \sum_{m=1}^{M} \mathbf{x}^{(m)}$ with $\mathbf{x}^{(m)} = \sqrt{p^{(m,1)}}(\mathbf{y}^{(m)} \odot \mathbf{b}^{(f_m)})$. Furthermore, $\mathbf{s}$ can be expressed as $\mathbf{s} = \sum_{m=1}^{M} \sqrt{p^{(m,1)}} e^{j2\pi \frac{f_c^{(m)}}{f_S} n}(\mathbf{P}^n\mathbf{y}^{(m)} \odot \mathbf{b}^{(f_m)})$. The following Theorem is instrumental to proceed.

Theorem 1. Let **u** and **v** be two $N \times 1$ vectors, for large value of N, we have

$$\mathbf{u}^H\mathbf{v} = \int_0^1 U^*(f)V(f)df \tag{5.31}$$

where

$$U(f) = \sum_{i=0}^{N-1} u[i]e^{-j2\pi fi} \tag{5.32}$$

and

$$V(f) = \sum_{i=0}^{N-1} v[i]e^{-j2\pi fi} \tag{5.33}$$

Proof. [21].

Given the large value of N in the current problem, we can utilize Lemma 1 to express $\mathbf{s}^H\mathbf{x}$ as follows

$$\mathbf{s}^H\mathbf{x} = \int_0^1 S^*(f)X(f)df \tag{5.34}$$

where

$$X(f) = \sum_{i=1}^{M} \sqrt{p^{(m,1)}}Y^{(i)}(f - f^{(i)}) \tag{5.35}$$

with

$$Y^{(m)} = \sum_{i=0}^{N-1} y^{(m)}[i]e^{-j2\pi fi} \tag{5.36}$$

and

$$S(f) = \sum_{m=1}^{M} \sqrt{p^{(m,1)}}e^{-j2\pi(f-f^{(m)})n}e^{j2\pi\frac{f_c^{(m)}}{f_s}n}Y^{(m)}(f - f^{(m)}) \tag{5.37}$$

here, $f^{(m)} = \frac{f_m}{f_s}$ and we have $0 \leq f^{(m)} < 1$. By substituting (5.35) and (5.37) in (5.34), we find

$$\mathbf{s}^H\mathbf{x} = \sum_{m=1}^{M} p^{(m,1)}e^{-j2\pi\frac{f_c^{(m)}}{f_s}n} \int_0^1 G_m(f)e^{j2\pi fn}df \tag{5.38}$$

where

$$G_m(f) = \begin{cases} |Y^{(m)}(f)|^2 & -\frac{B}{2f_s} \leq f \leq \frac{B}{2f_s} \\ 0 & \text{otherwise} \end{cases} \tag{5.39}$$

By defining

$$g_m[n] \triangleq \int_0^1 G_m(f)e^{j2\pi fn}df \tag{5.40}$$

we can rewrite (5.38) as

$$\mathbf{s}^H\mathbf{x} = \sum_{m=1}^{M} p^{(m,1)} g_m[n] e^{-j2\pi \frac{f_c^{(m)}}{f_s} n} \tag{5.41}$$

We should observe that $g_m[n]$ represents the autocorrelation or zero Doppler ambiguity function for the mth channel of reception. We can apply a similar approach to find

$$\mathbf{s}^H\mathbf{s} = \sum_{m=1}^{M} p^{(m,1)} g_m[0] \tag{5.42}$$

and

$$\mathbf{x}^H\mathbf{x} = \sum_{m=1}^{M} p^{(m,1)} g_m[0] \tag{5.43}$$

Finally, we can rewrite (5.30) as

$$L_{U_M}[n] = \frac{u_M^{(2)}[n]}{v_m^{(2)}[n]} = \frac{\sum_{m=1}^{M}\sum_{l=1}^{M} p^{(m,1)} p^{(l,1)} g_m[n] g_l^*[n] e^{-j2\pi \frac{f_c^{(m)} - f_c^{(l)}}{f_s} n}}{\left|\sum_{m=1}^{M} p^{(m,1)} g_m[0]\right|^2} \tag{5.44}$$

The (5.44) indicates that the UMPI test's numerator relies on the value of n, which implies that it can identify the range resolution of the UMPI output. In order to analyze the range behavior of the UMPI detector within the maximum bistatic range of R_b, which is defined as $R_b = cN_bT_s$, we need to evaluate it for all values of n ranging from 0 to N_b. To accomplish this task, we can define the discrete Fourier transform of $u_M^{(2)}[n]$, given by

$$U_M^{(2)}(f) = \sum_{n=0}^{N_b-1} u_M^{(2)}[n] e^{-j2\pi fn} \tag{5.45}$$

We can obtain $U_M^{(2)}(f)$ as follows

$$U_M^{(2)}(f) = \sum_{m=1}^{M}\sum_{l=1}^{M} Z^{(m,l)}(f - \Delta f_c^{(m,l)}) \tag{5.46}$$

where

$$Z^{(m,l)}(f) = \sum_{n=0}^{N_b-1} \left(p^{(m,1)} p^{(l,1)} g_m[n] g_l^*[n]\right) e^{-j2\pi fn} \tag{5.47}$$

and

$$\Delta f_c^{(m,l)} = \frac{f_c^{(m)} - f_c^{(l)}}{f_s} \tag{5.48}$$

The conclusion can be drawn from (5.46) and (5.48) that the target range resolution cannot be enhanced when $f_c^{(m)} - f_c^{(l)} = kf_s$, where k is any arbitrary integer number and $m \neq l = 1, \ldots, M$. This implies that in order to improve the target range resolution, it is necessary to have at least two channels with carrier frequencies that do not satisfy $f_c^{(m)} - f_c^{(l)} = kf_s$. In PBR systems, it is not possible to control the transmitted waveforms or their carrier frequencies. Therefore, the only feasible option is to design B and, hence, $f_s = MB$. In general, if the condition mentioned below is met, the range resolution can be improved:

$$f_c^{(m)} - f_c^{(1)} = (m-1)B + kf_s, \qquad m = 2, \ldots, M \tag{5.49}$$

The process involves filtering the desired received channels with a suitable bandwidth of B, sampling them at a frequency of f_s, and up/down-converting them to ensure that they are adjacent in the frequency domain without overlapping or leaving gaps. To do so, the recommended value of B can be determined for (5.49), given by

$$B = \frac{2\sum_{m=1}^{M}(f_c^{(m)} - f_c^{(1)})}{M(M-1)(2k+1)} \tag{5.50}$$

It is important to mention that k must be an integer value chosen in a way that results in a bandwidth of less than 200 kHz (which is the standard FM radio channel bandwidth in many countries). It has been observed that the proposed algorithm provides better range resolution capabilities compared to the case where the signal bandwidth is used, which is supported by theoretical analysis. In the following section, we will examine the quality of the range resolution improvement achieved by the output of the UMPI detector. We will use the peak-to-sidelobe level ratio (PSLR) concept to investigate this. Based on the proposed test statistic, we can define PSLR as

$$\mathrm{PSLR}_M = \frac{L_{U_M}[0]}{\max_{n,n\neq 0}\{L_{U_M}[n]\}} \tag{5.51}$$

Making use of (5.44) in (5.51), we get

$$\mathrm{PSLR}_M = \frac{|\sum_{m=0}^{M-1} p^{(m+1,1)} g_{m+1}[0]|^2}{\max_{n,n\neq 0} |\sum_{m=0}^{M-1} p^{(m+1,1)} g_{m+1}[n] e^{-j2\pi \frac{B}{f_s} mn}|^2} \tag{5.52}$$

By defining

$$q_n[m+1] = \begin{cases} p^{(m+1,1)} g_{m+1}[n], & 0 \leq m \leq M-1 \\ 0, & \text{otherwise} \end{cases} \tag{5.53}$$

by noting that $f_s = MB$, we can find

$$\sum_{m=0}^{M-1} q_n[m+1] e^{-j2\pi \frac{B}{f_s} mn} = \sum_{m=-\infty}^{\infty} q_n[m+1] e^{-j2\pi \frac{n}{M} m} = Q_n\left(\frac{\mathrm{mod}(n, M)}{M}\right) \tag{5.54}$$

where $n = 0, \ldots, N_R$ and $Q_n(f)$ is the discrete Fourier transform of $q_n[m+1]$ with respect to m. Finally, we can rewrite (5.52) as follows

$$\mathrm{PSLR}_M = \frac{|Q_n(0)|^2}{\max_{n,n\neq 0} \left| Q_n\left(\frac{\mathrm{mod}(n, M)}{M}\right)\right|^2} \tag{5.55}$$

In a similar manner, we can recast (5.44) as

$$L_{U_M}[n] = \frac{\left| Q_n\left(\frac{\text{mod}(n, M)}{M}\right)\right|^2}{|Q_0(0)|^2} \tag{5.56}$$

In order to gain more understanding about how the target range resolution will be improved with the new detector, we present additional theoretical findings based on a simplified scenario. This simplified scenario assumes that $g_m[n] = g[n]$ and $p^{(m,1)} = 1$ for $m = 1, \ldots, M$. With this simplification, it becomes straightforward to demonstrate that

$$|Q_n(f)| = \left| Mg[n] \frac{\text{sinc}(Mf)}{\text{sinc}(f)} \right| \tag{5.57}$$

By substituting (5.57) into (5.56), we get

$$L^s_{U_M}[n] = \frac{\left| g[n] \dfrac{\text{sinc}(\text{mod}(n, M))}{\text{sinc}\left(\dfrac{\text{mod}(n, M)}{M}\right)} \right|^2}{|g[0]|^2} = \begin{cases} \dfrac{|g[n]|^2}{|g[0]|^2}, & n = \pm 0, \pm M, \pm 2M, \ldots \\ 0, & \text{otherwise} \end{cases} \tag{5.58}$$

Superscript s is used to denote the simplified case. Similarly, for the simplifying case, we can find the PSLR at the output of proposed detector given by

$$\text{PSLR}^s_M = \frac{|g[0]|^2}{\max_{k>0,k\in Z}\{|g[kM]|^2\}} \tag{5.59}$$

According to (5.58), we can provide a valid explanation for the peaky behavior observed in the output of the proposed test statistic, which is characterized by a high degree of ambiguity. This behavior has been empirically observed in practical results presented in References 1 and 2, but without any accompanying theoretical justification. Moreover, this explanation can shed light on the underlying reasons for why ambiguity peaks tend to be higher in the ambiguity function's range direction as the individual channel bandwidth increases. Specifically, as the bandwidth of each channel increases, the sampling frequency also increases, causing the time index kM to decrease and leading to larger values for $g_m[kM]$. This, in turn, results in lower values of PSLR^s_M and hence higher ambiguities in the output of a multiband test statistic or in the ambiguity function's range direction. Our simulation investigations, which we discuss later, confirm this behavior. Additionally, it is worth noting that the output of a multiband UMPI test statistic with M channels ($L_M[n]$) is weighted by that of a single-band UMPI detector ($L_1[n]$).

$$L^s_{U_M}[n] = L_{U_1}[n] \left| \frac{\text{sinc}(\text{mod}(n, M))}{\text{sinc}\left(\dfrac{\text{mod}(n, M)}{M}\right)} \right|^2 = \sum_{-\infty}^{\infty} L_{U_1}[n]\delta(n - kM) \tag{5.60}$$

Therefore, it appears that the ability to automatically choose high-quality channels could be extremely valuable. To achieve this goal, we utilize a simplified version of detection loss (DL) at the mth reference channel, given by

$$\mathrm{DL}^{(m)}[n] = 1 - \frac{|\mathbf{y}_f^{(0,m)^H}\mathbf{y}_f^{(n,m)}|^2}{\|\mathbf{y}_f^{(0,m)}\|^2\|\mathbf{y}_f^{(n,m)}\|^2} \tag{5.61}$$

If the mth reference channel signal meets the condition below, it can be considered a suitable signal for use in multiband detection:

$$\mathrm{argmax}_n\{\mathrm{DL}^{(m)}[n] < \mathrm{DL}_a\} < n_a \tag{5.62}$$

The value of n_a corresponds to the time index for a specific bistatic range of R_a, and DL_a represents the accepted detection loss for that range. Based on our simulation results for FM signals, it appears that choosing $\mathrm{DL}_a = 0.5$ and $R_a = 3$ km is a suitable option. However, it's important for radar designers to carefully select these parameters based on the specific needs and requirements of their system.

If we make the assumption that the received amplitudes of a target vary across frequency channels and create a detector accordingly, it is important to note that the target range resolution cannot be improved, as proven analytically in [14]. Therefore, simply building a multiband FM-based PBR receiver to generate a high bandwidth signal is not enough to improve the target range resolution.

5.6 Analytical performance analysis under amplitude mismatch

In this section, we aim to obtain the performance of the UMPI detector of (5.19) under complex amplitude mismatch (CAM). The results of this section was published in Reference 10. To do this, we assume that the received complex amplitudes of a target over different frequency channels (DFCs) are not the same, where the received signal vector can be described as

$$\mathbf{r}_2 = \mathbf{S}_0\mathbf{a}_0 + \mathbf{C}\mathbf{a}_c + \mathbf{T}\mathbf{a}_t + \mathbf{n}_s \tag{5.63}$$

In contrast to (5.19), matrices $\mathbf{T}$ and $\mathbf{S}$, respectively, with dimension $N \times K_e$ with $K_e = M(K-1)$ and $N \times M$ can be defined as

$$\mathbf{T} \triangleq [\mathbf{S}_1, \ldots, \mathbf{S}_{K-1}] \tag{5.64}$$

$$\mathbf{S}_k \triangleq [\mathbf{s}_k^{(1)}, \ldots, \mathbf{s}_k^{(M)}] \tag{5.65}$$

where $\mathbf{s}_k^{(m)} = (\mathbf{y}^{(n_k,m)} \odot \mathbf{e}^{(\nu_k,m)} \odot \mathbf{b}^{(f_m)})$ for $k = 0, \ldots, K-1$. In (5.63), $\mathbf{a}_t = [\mathbf{a}_1, \ldots, \mathbf{a}_K]$ where $\mathbf{a}_k \triangleq [\alpha_k^{(1)}, \ldots, \alpha_k^{(M)}]^T$. Note that the formulation in (5.4) becomes a special case of (5.63) with $\mathbf{a}_k = a_k\mathbf{1}$ for $k = 0, \ldots, K-1$, where $\mathbf{1}$ is the vector with all components 1. In the same way as [12], one can obtain the detection probability of the UMPI test in the presence of the CAM, given by

$$P_d = e^{-(\delta+\eta_t)} \sum_{k=0}^{\infty} \left(\frac{\delta^k}{k!} \sum_{n=0}^{k} \frac{\eta_t^n}{n!} \right) \tag{5.66}$$

where $\delta = \frac{1}{\sigma^2}\|(\mathbf{\Pi}_{\mathbf{U}}^{\perp} - \mathbf{\Pi}_{\mathbf{K}}^{\perp})(\mathbf{S}\mathbf{a}_0 + \mathbf{C}\mathbf{a}_c + \mathbf{T}\mathbf{a}_t)\|^2$. Here, $\mathbf{K} = [\mathbf{s}, \mathbf{U}]$. Since $\mathbf{\Pi}_{\mathbf{U}}^{\perp} = \mathbf{\Pi}_{\mathbf{\Pi}_{\mathbf{C}}^{\perp}\mathbf{T}_i}^{\perp}\mathbf{\Pi}_{\mathbf{C}}^{\perp}$, $\mathbf{\Pi}_{\mathbf{K}}^{\perp} = \mathbf{\Pi}_{\mathbf{\Pi}_{\mathbf{U}}^{\perp}\mathbf{s}}^{\perp}\mathbf{\Pi}_{\mathbf{\Pi}_{\mathbf{C}}^{\perp}\mathbf{T}_i}^{\perp}\mathbf{\Pi}_{\mathbf{C}}^{\perp}$, and $\mathbf{\Pi}_{\mathbf{C}}^{\perp}\mathbf{C} = 0$, we get $\delta = \frac{1}{\sigma^2}\|(\mathbf{\Pi}_{\mathbf{U}}^{\perp} - \mathbf{\Pi}_{\mathbf{K}}^{\perp})(\mathbf{S}\mathbf{a}_0 + \mathbf{T}\mathbf{a}_t)\|^2$. In the matched case, we can use of (5.66), but with $\delta = \frac{|a_0|^2}{\sigma^2}\|\Pi_{\mathbf{U}}^{\perp}\mathbf{s}\|^2$.

5.7 Simulation results

In this section, the performance of the suggested detector is verified by presenting simulation results that enhance target range resolution. The simulation includes direct signals with an input signal-to-noise ratio (DNR_i) ranging from approximately 58 dB to 63 dB across various channels. Additionally, ten clutter spikes with different input clutter-to-noise ratios (CNR_i) ranging from 5 dB to 35 dB are considered between bistatic ranges of 0 and 55 km. The clutter's complex amplitudes are assumed to follow an exponential power spectral density with a shape parameter of 7 and a ratio of 90 between dc and ac power. The detector's threshold is set analytically to achieve a false alarm probability of 10^{-3}, and parameters Δv_c and Q are set to suppress clutter for the proposed UMPI test. The simulation results indicate that to suppress clutter perfectly and attain the predetermined false alarm probability, clutter region parameters equal to $\Delta v_c = 0.75\ \text{ms}^{-1}$ and $Q = 4$ need to be set when the integration time is set to 0.5 s. Results with $\Delta v_c > 0.75\ \text{ms}^{-1}$ or $< 0.75\ \text{ms}^{-1}$ and $Q > 4$ or $= 4$ are also provided, resulting in a false alarm probability $> 10^{-3}$. The simulation results show that multiband PBRs do not have significant advantages over those of single-band cases in terms of clutter cancellation. Furthermore, the proposed UMPI test for $M = 1$ is equivalent to the GLRT presented in Chapter 3 for single-channel detection.

This section utilizes four simulated signals called S_1, S_2, S_3, and S_4, which have carrier frequencies of 92, 94, 96, and 98 MHz, respectively. These signals serve as the waveforms of opportunity. Figure 5.2 depicts the self-ambiguity function cutoff at zero velocity, corresponding to signals S_1, S_2, S_3, and S_4. The algorithm for selecting signals, as described in (5.62), only identifies signals S_1, S_2, and S_3 as high-quality signals for use in the multiband detector. Figure 5.3 illustrates the detection loss as a function of bistatic range for these signals, alongside a fixed curve for $\text{DL}_a = 0.5$. As can be seen, only the S_1, S_2, and S_3 signals satisfy the conditions outlined in (5.62) for $R_a = 3$ km. Accordingly, we will employ signals S_1, S_2, and S_3 in the subsequent analysis. Their integration times are 0.543, 0.531, and 0.520 s, respectively. These specific integration times were chosen because of their relevance to the implementation of the HRITP algorithm based on FFT, which is explained in Chapter 4.

To begin with, we confirm the accuracy of the theoretical findings stated in (5.60) by comparing them with the results obtained through simulations. The results of the simulation are exhibited in Figure 5.4. The weights of the delta functions indicated in (5.60) are illustrated by the dashed line in Figure 5.4. This line displays the output of the UMPI test statistic for a single-channel detection. This indicates that the theoretical results are in good agreement with the simulation outcomes.

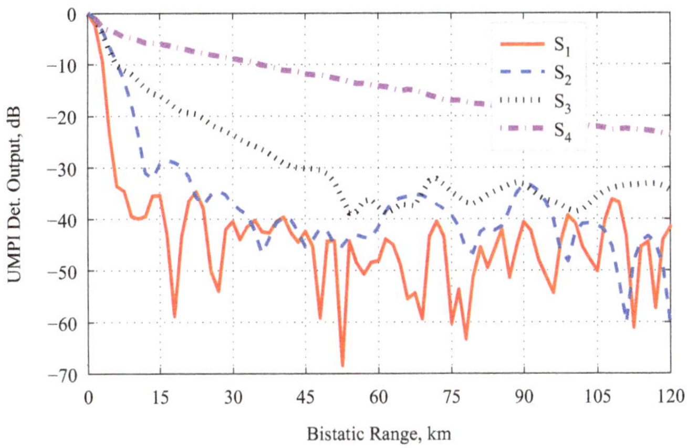

Figure 5.2 Zero-velocity cut of the UMPI test statistic for different FM signals S_1, S_2, S_3, and S_4

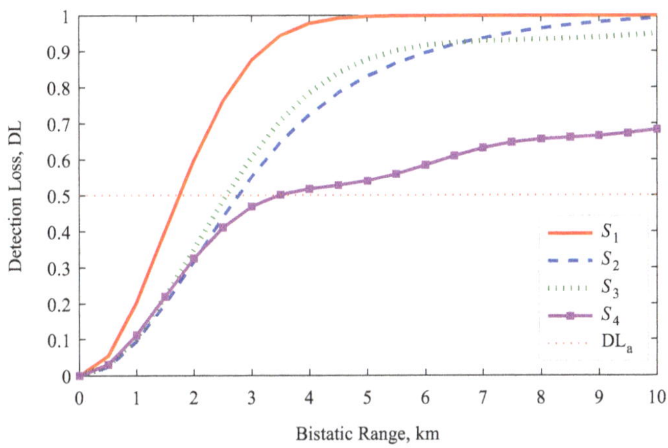

Figure 5.3 Detection loss as a function of bistatic range for different received signals S_1, S_2, S_3, and S_4

In Figure 5.5, we compare the range resolution of a single band detector to that of a multiband detector with $M = 2$ and different values of B, using signals S_1 and S_2. The results demonstrate that the proposed multiband detector has better range resolution than the single-band detector. However, increasing the bandwidth of each channel (and hence, the sampling frequency) leads to lower values of PSLR_M and higher ambiguities at the output of the multiband detector. These findings are consistent with the results obtained from real data in Reference 2, but are not supported by any mathematical explanation.

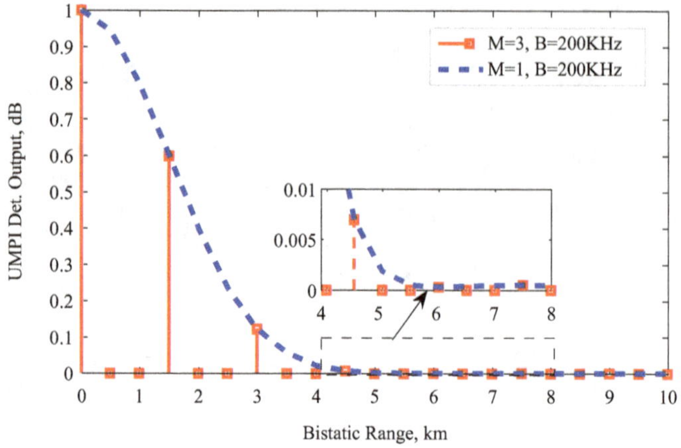

Figure 5.4 UMPI test statistic as a function of bistatic range for different number of channels (i.e., M) when B= 200 kHz

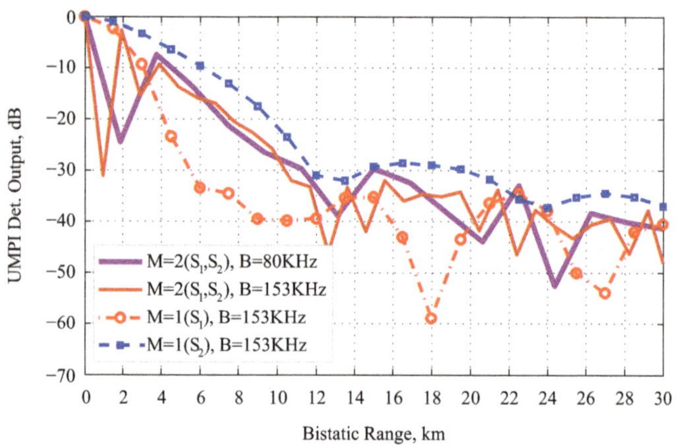

Figure 5.5 UMPI test statistic as a function of bistatic range for different combination of signals S_1 and S_2 and when bandwidth B is considered as parameter

To evaluate the detection performance of proposed multiband detector, the input signal-to-noise ratio (SNR_i) is defined as

$$\mathrm{SNR}_i = \frac{|\alpha_0|^2}{N_0 B} \tag{5.67}$$

where N_0 is the noise power per unit bandwidth, and B is the receiver bandwidth of a single channel. Throughout this chapter, the values of SNR_i are reported based on the bandwidth of $B = 80$ kHz.

Figure 5.6 shows the probability of detection as a function of the SNR_i for various values of M in the absence of any interfering target. Throughout subsequent simulations, the probability of false alarm P_{fa} is set to 10^{-6}. The solid lines in the figure represent the theoretical results, while the dashed lines show the results of the Monte Carlo simulations. These results confirm the accuracy of the theoretical results. As expected, it is observed that increasing the number of channels improves detection performance. For $P_d = 0.9$, this improvement in SNR is roughly equal to M, representing a coherent combined diversity gain. The simulation uses the same integration time of 0.5 s to demonstrate this. Consequently, our proposed detector outperforms a single-band PBR system in terms of coherent combined diversity gain. To test the effectiveness of the proposed HRITP algorithm, we define two scenarios. In the first scenario, including three target echoes with characteristics listed in Table 5.1, the results of HRITP algorithm for $M = 1$ and $M = 2$ are shown in Figures 5.7 and 5.8, respectively. In these figures, symbols □ and + are used to indicate the estimated and real position of the targets, respectively. Apparently, all

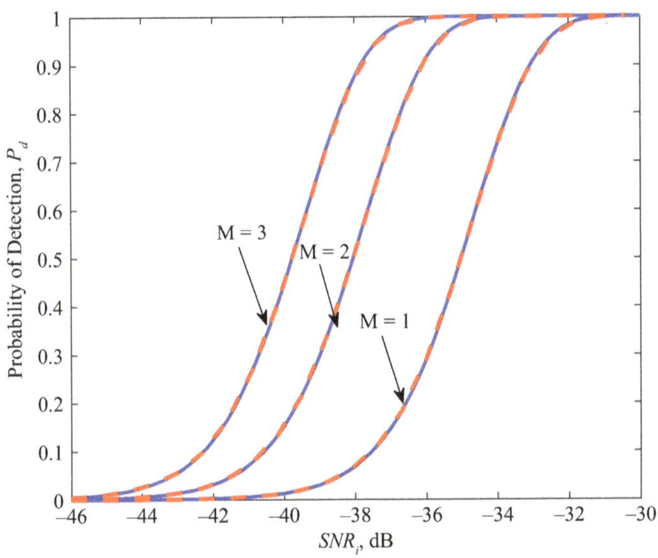

Figure 5.6 Detection probability as a function of SNR_i *for different values of M in the case of SW0 target model and* $P_{fa} = 10^{-6}$

Table 5.1 Characteristic of the targets in the first scenario

Target	T_1	T_2	T_3
Bistatic range (km)	15	18.75	150
Bistatic velocity (ms^{-1})	–69.4	–69.4	208.3
SNR_i (dB)	–30	–30	–30

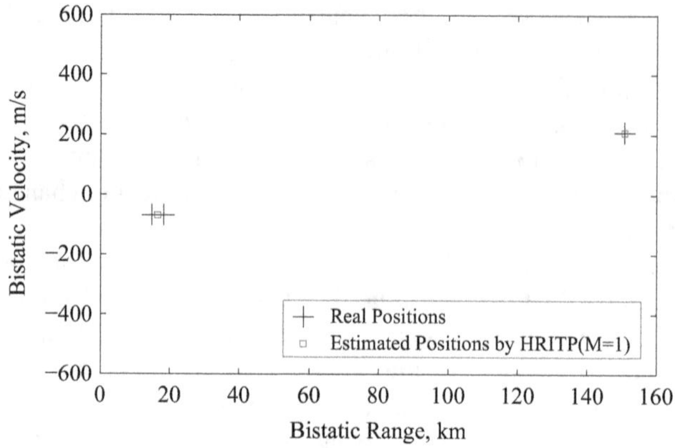

Figure 5.7 Result of applying the HRITP algorithm for the first scenario and when signal S_1 is used for detection

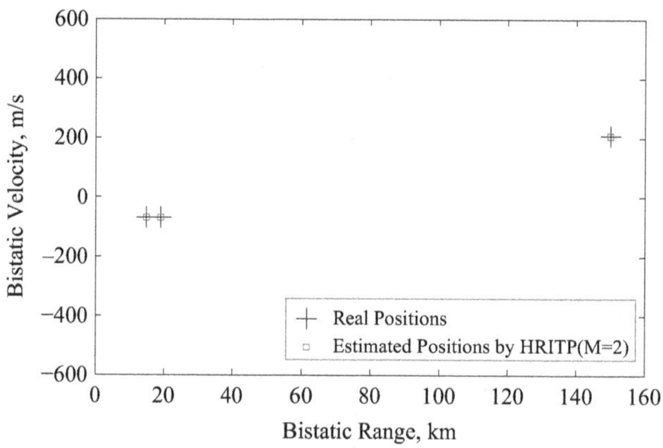

Figure 5.8 Result of applying the HRITP algorithm for the first scenario and when multiband channel (S_1, S_2) is used for detection

targets are correctly positioned when exploiting signals S_1 and S_2 simultaneously for detection, while in the case of single channel detection (based on signal S_1), HRITP algorithm cannot resolve targets T_1 and T_2. More precisely, to examine the ability of the proposed detector to resolve nearby targets as well as to see the effect of PSLR on the range resolution, we consider the second scenario in which there are two targets separated by ΔR in range dimension with $\text{SNR}_i = -30$ dB. In Figures 5.9 and 5.10, we plot the detection probability of distinguishing one target from other versus the range distance of ΔR for $M = 1$ and $M = 3$, respectively. For the case of single channel detector when signals S_1, S_2, or S_3 are exploited, it is observed that two targets can be resolved when they are separated by a distance of 6 km,

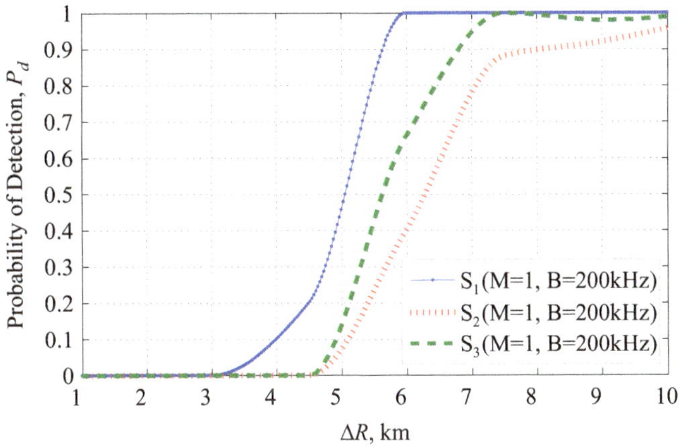

Figure 5.9 Target range resolution improvement when a single channel detector with signals S_1, S_2, and S_3 is used for detection. Here, the receiver bandwidth is fixed to B = *200 kHz.*

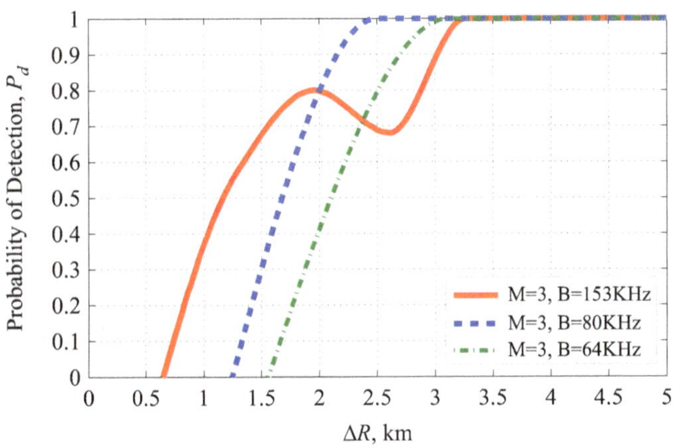

Figure 5.10 Target range resolution improvement for different values of receiver bandwidth B when multiband detector exploits signals S_1, S_2, and S_3 for detection

12 km, and 7.5 km, respectively, for $P_d = 0.9$. However, they are clearly resolved for distance of 2.5 km when exploiting three channels for detection for $P_d = 0.9$. To get more insight about this simulation, the range resolutions of the proposed detector are compared in Table 5.2 for different values of M and B, and for $P_d = 0.9$. From the results of Figure 5.9 and Table 5.2, it can be concluded that the range resolution

Table 5.2 Target range resolution improvement results for different values of M and B

Parameters	$M = 3$	$M = 2$	B
Range resolution, km	3.12	4.68	64 kHz
	2.50	3.75	80 kHz
	3.27	3.92	153 kHz

can be improved approximately from 6 km, for single target detection with nominal bandwidth of FM radio, to 3.75 km and 2.5 km by exploiting two and three FM channels, respectively. In addition to these results, it is seen from Figure 5.10 that the larger bandwidth of B cannot result in better target range resolution for $M \neq 1$. These simulations show that the range resolution in a multiband FM-based PBR not only depends on the bandwidth of the signal used for detection but also depends on the quality of resulted signal for detection. To qualify the quality of the resulted signal for detection, we use PSLR parameter presented in (5.55). Equation (5.55) shows that larger bandwidth of individual channels, B, results in lower values of PSLR, hence higher ambiguities in the peaky behavior of the range dimension of UMPI test. This means that in the multiband FM-based PBR, the range resolution is not inversely proportional to the bandwidth of signal used for detection. Our simulation results show that a bandwidth of approximately $B = 80$ kHz is a good choice for multiband detection in an FM-based PBR system.

To assess how well the UMPI-based detector performs in detecting targets when there are differences between the actual model and the assumed model, we examine a PBR system that uses multiband FM technology with two channels. These channels use carrier frequencies of 92 MHz and 94 MHz, and we use the same simulation parameters as those in this chapter. To begin our analysis for the two-channel target detection problem, we represent the complex amplitudes of the received target as $\alpha_0^{(m)} = \beta_0^{(m)} e^{-j\phi_0^{(m)}}$ for $m = 1, 2$. In order to investigate the impact of the CAM on the UMPI detector's detection performance and the degradation of the target range resolution, we examine three different cases, given by

Case1 (Angle Mismatch): $\beta_0^{(2)} = \beta_0^{(1)}$ and $\Delta\phi_0 = \phi_0^{(2)} - \phi_0^{(1)}$,

Case2 (Amplitude Mismatch): $\phi_0^{(2)} = \phi_0^{(1)}$ and $\Delta A = \frac{\beta_0^{(2)}}{\beta_0^{(1)}}$,

Case3 (Complex Amplitude Mismatch): $\Delta\phi_0 = \phi_0^{(2)} - \phi_0^{(1)}$ and $\Delta A = \frac{\beta_0^{(2)}}{\beta_0^{(1)}}$.

In order to quantify the reduction in detection performance caused by the CAM over different frequency channels, we introduce a measure called SNR loss. Specifically, we define SNR loss as the difference between the required SNRs for matched and mismatched cases when the desired detection probability and false alarm probability are achieved, denoted as $\text{SNR}^{(1)}$ and $\text{SNR}^{(2)}$, respectively. This measure allows us to compare the detection performance, and it is defined as

$$\text{SNRL} = \frac{\text{SNR}^{(2)}}{\text{SNR}^{(1)}} \tag{5.68}$$

Figures 5.11 and 5.12 evaluate the SNRL as a function of $\Delta\phi$ and ΔA, respectively. As expected, the larger $\Delta\phi$ or ΔA is, the more the performance degradation of the UMPI detector would be. However, under the case 3 of model mismatch, the worse-SNRL decreases as ΔA increases.

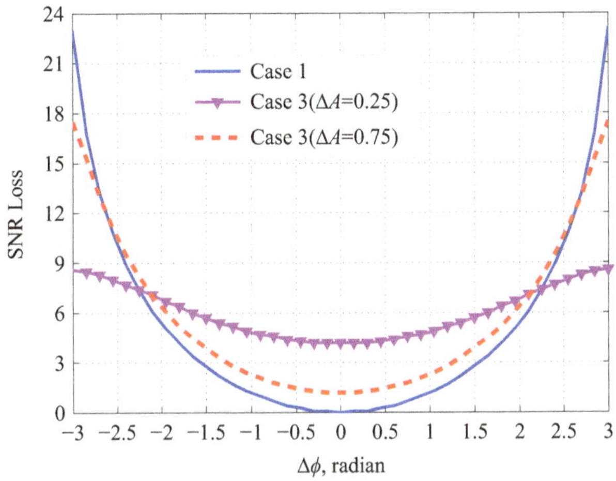

Figure 5.11 *SNR loss as a function of $\Delta\phi$ for $P_d = 0.9$ and $P_{fa} = 10^{-6}$ in the presence of the angle and complex amplitude mismatches*

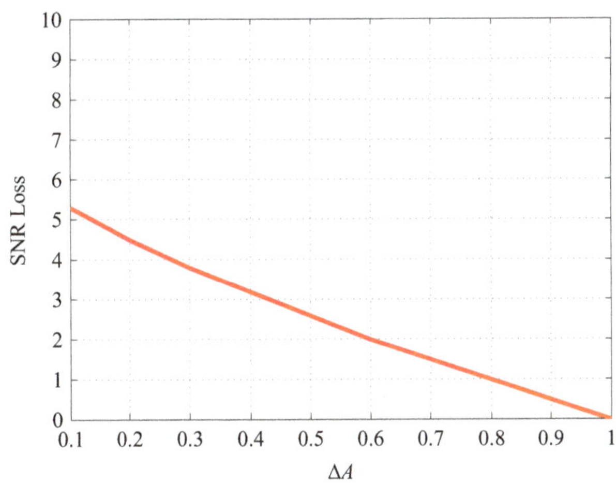

Figure 5.12 *SNR loss as a function of ΔA for $P_d = 0.9$ and $P_{fa} = 10^{-6}$ in the presence of amplitude mismatch*

In Figures 5.13–5.15, we investigate the effect of the CAM on the TRR of the UMPI detector. In this case, the M-UMPI detector output as a function of bistatic ranges is considered to evaluate the TRR degradation in the presence of the CAM. Note that narrower mainlobe at the output of the M-UMPI detector means better

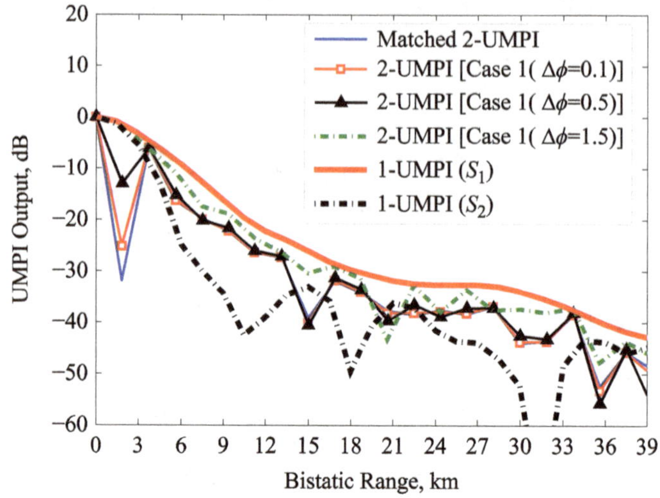

Figure 5.13 Range resolution comparison of the 1-UMPI detector with signals S_1 and S_2, and the 2-UMPI detector with signal (S_1,S_2) under the matched case with that of the 2-UMPI in the presence of the angle mismatch

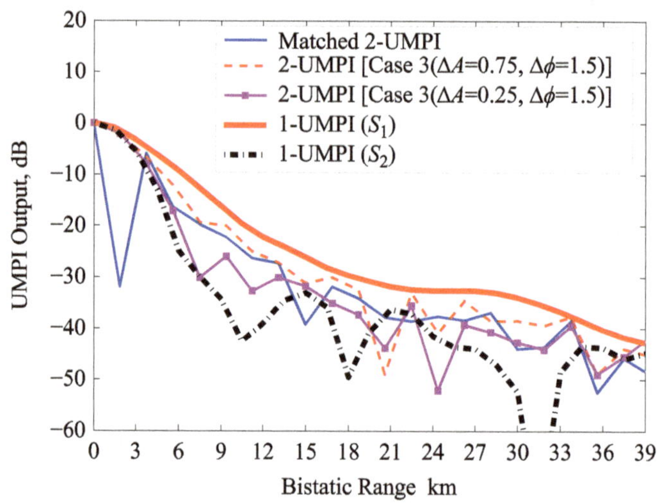

Figure 5.14 Range resolution comparison of the 1-UMPI detector with signals S_1 and S_2, and the 2-UMPI detector with signal (S_1,S_2) under the matched case with that of the 2-UMPI in the presence of the amplitude mismatch

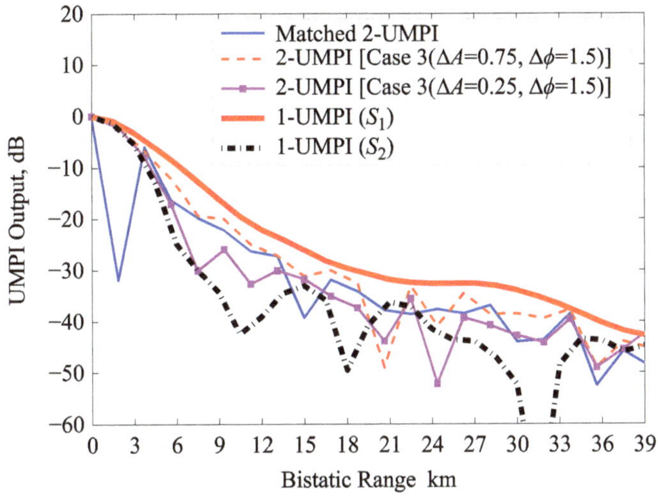

Figure 5.15 Range resolution comparison of the 1-UMPI detector with signals S_1 and S_2, and the 2-UMPI detector with signal (S_1,S_2) under the matched case with that of the 2-UMPI in the presence of the complex amplitude mismatch

target range resolution. For comparison, the output of the single channel UMPI (1-UMPI) detector and that of the two channels-UMPI (2-UMPI) detector under the matched case are also depicted in these figures. In the 1-UMPI detector, we employ two simulated signals named S_1 and S_2, while for the 2-UMPI detector, a combinations of signals S_1 and S_2 referred to as (S_1,S_2) is used. It is seen that the TRR of the 2-UMPI detector under any mismatched cases is better than that of the 1-UMPI, while it is worse than that of the 2-UMPI detector under the matched case. This means that the 2-UMPI detector of this chapter is capable of improving the TRR even under the CAM when compared to that of the 1-UMPI detector. However, increasing the values of $\Delta\phi$ and ΔA would worsen the matching TRR.

5.8 Summary

This chapter proposes a signal processing method to enhance target range resolution, which involves a systematic framework comprising signal modeling, UMPI-based detection, and statistical analysis. The UMPI-based detection method is found to offer superior range resolution capabilities compared to the single-band detectors discussed in Chapter 3, as supported by theoretical analysis and simulations. The amount of range resolution improvement depends on the bandwidth of the processed multiband signal, but the quality of this improvement is influenced by the quality of the channels used. In some cases, reducing the bandwidth of individual channels may be preferred to improve the quality of target range resolution improvement. Additionally, the impact of complex amplitude mismatch over different channels is examined, revealing that it can result in reduced detection performance and worsen target range resolution improvement.

References

[1] Tasdelen A.S., Koymen H. 'Range resolution improvement in passive coherent location radar systems using multiple FM radio channels'. *Proceedings of IET Forum on Radar and Sonar*; London, UK, 2006. pp. 23–31.

[2] Olsen K.E., Baker C.J. 'FM-based passive bistatic radar as a function of available bandwidth'. *IEEE 2008 Radar Conference*; Rome, Italy, 2008.

[3] Olsen K.E., Woodbridge K. 'FM based passive bistatic radar target range improvement'. *Proceedings of International Radar Symposium*; 2009.

[4] Olsen K.E., Woodbridge K., Andersen I.A. 'FM based passive bistatic radar target range improvement – Part II'. *Proceedings of International Radar Symposium*; Vilnius, Lithuania, 2010.

[5] Olsen K.E., Woodbridge K. 'Analysis of the performance of a multiband passive bistatic radar processing scheme'. *Presented at 2010 International Waveform Diversity and Design Conference (WDD)*; Niagara Falls, ON, Canada, 2010. pp. 142–49.

[6] Conti M., Berizzi F., Petri D., Capria A., Martorella M. 'High range resolution DVB-T passive radar'. *Radar Conference (EuRAD)*; 2010. pp. 109–12.

[7] Petri D., Capria A., Conti M., Berizzi F., Martorella M., Dalle Mese E. 'High-range resolution multiband DVB-T passive radar: Aerial target detection'. *International Journal of Microwave and Wireless Technologies*. 2012, vol. 4(2), pp. 147–53.

[8] Olsen K.E. 2011. 'Investigation of bandwidth utilisation methods to optimise performance in passive bistatic radar'. [PhD thesis]. University College London.

[9] Zaimbashi A. 'Multiband FM-based passive bistatic radar: Target range resolution improvement'. *IET Radar, Sonar and Navigation*. 2016, vol. 10(1), pp. 174–85.

[10] Zaimbashi A. 'Multiband FM-based PBR system in presence of model mismatch'. *Electronics Letters*. 2016, vol. 52(18), pp. 1563–65. Available from https://onlinelibrary.wiley.com/toc/1350911x/52/18

[11] Colone F., Cardinali R., Lombardo P, *et al*. 'Space–time constant modulus algorithm for multipath removal on the reference signal exploited by passive bistatic radar'. *IET Radar, Sonar & Navigation*. 2009, vol. 3(3), p. 253.

[12] Zaimbashi A., Derakhtian M., Sheikhi A. 'GLRT-based CFAR detection in passive bistatic radar'. *IEEE Transactions on Aerospace and Electronic Systems*. 2013, vol. 49(1), pp. 134–59.

[13] Billingsley J.B. 'Low-Angle land clutter measurements and empirical models' in William Andrew Publishing; 2002.

[14] Zaimbashi A. 2013. 'Target detection in passive radars based on commercial FM radio signals'. [PhD thesis]. Iran, Shiraz University.
[15] Van Trees H.L. 'Detection, estimation, and modulation theory' in Vol. III. Hoboken, NJ: Wiley; 1968.
[16] Skolnik M. 'Introduction to radar systems' in 3rd ed. Columbus, OH: McGraw-Hill; 2002.
[17] Barton D.K. 'Radar system analysis and modeling'. *IEEE Aerospace and Electronic Systems Magazine*. 2005, vol. 20(4), pp. 23–25.
[18] Laub A.J. 'Matrix analysis for scientists and engineers' in Philadelphia, PA: Soc. for Industrial and Applied Math; 2005.
[19] Lehman E.L., Romano J.P. 'Testing statistical hypothesis' in New York: Springer Verlag; 2005.
[20] Muirhead R.J. 'Aspect of multivariant statistical theory' in John Wiley and Sons; 2005.
[21] Kay S.M. 'Fundamentals of statistical signal processing: detection theory' in Vol. 2. Upper Saddle River, NJ: Prentice-Hall; 1998.

Appendix 5A
Maximal invariant derivation

In this appendix, we use Theorem 1 of Chapter 4 to derive the Maximal Invariant (MI). In our case, it can be shown that an MI for G_Q is

$$\mathbf{y} = M_Q(\mathbf{x}) = \left[\frac{|\mathbf{s}_{w_i}^H \mathbf{x}|^2}{\|\mathbf{s}_{w_i}\|^2}, \|\mathbf{x}\|^2 \right] \tag{5A.1}$$

This statistic is invariant, since for all $g_Q \in G_Q$, we have

$$M_Q(g_Q(\mathbf{x})) = \left[\frac{|\mathbf{s}_{w_i}^H \mathbf{Q}_{\mathbf{s}_{w_i}} \mathbf{x}|^2}{\|\mathbf{s}_{w_i}\|^2}, \|\mathbf{Q}_{\mathbf{s}_{w_i}} \mathbf{x}\|^2 \right] \tag{5A.2}$$

Since $\mathbf{Q}_{\mathbf{s}_{w_i}}^H = \mathbf{Q}_{\mathbf{s}_{w_i}}$, $\mathbf{Q}_{\mathbf{s}_{w_i}} \mathbf{Q}_{\mathbf{s}_{w_i}} = \mathbf{I}$, and $\mathbf{Q}^l{}_{\mathbf{s}_{w_i}} \mathbf{s}_{w_i} = (-1)^l \mathbf{s}_{w_i}$ for any integer value of l, it can be concluded that $M_Q(g_Q(\mathbf{x})) = M_Q(\mathbf{x})$. This show that the statistic $\mathbf{y}$ is invariant.

To see the maximality of $\mathbf{y}$, suppose that for given $\mathbf{x}_1$ and $\mathbf{x}_2$, we have $M_Q(\mathbf{x}_1) = M_Q(\mathbf{x}_2)$. Now, it follows from (5A.1) that

$$\begin{cases} \dfrac{|\mathbf{s}_{w_i}^H \mathbf{x}_1|^2}{\|\mathbf{s}_{w_i}\|^2} = \dfrac{|\mathbf{s}_{w_i}^H \mathbf{x}_2|^2}{\|\mathbf{s}_{w_i}\|^2} \\ \|\mathbf{x}_1\|^2 = \|\mathbf{x}_2\|^2 \end{cases} \tag{5A.3}$$

To proceed, it is required to find some $g_Q \in G_Q$ such that $\mathbf{x}_2 = g_Q(\mathbf{x}_1) = \mathbf{Q}_{\mathbf{s}_{w_i}}^l \mathbf{x}_1$. In other words, it is sufficient to find l based on the given vectors $\mathbf{x}_1$ and $\mathbf{x}_2$ to proof maximality. To do so, since $\mathbf{Q}_{\mathbf{s}_{w_i}}^H = \mathbf{Q}_{\mathbf{s}_{w_i}}$, $\mathbf{Q}_{\mathbf{s}_{w_i}}^{2k+1} \mathbf{s}_{w_i} = -\mathbf{s}_{w_i}$, and $\mathbf{Q}_{\mathbf{s}_{w_i}}^{2k} = \mathbf{I}$, it is easy to verify that for $\mathbf{x}_2 \neq \mathbf{x}_1$, l must take an odd integer value, i.e., $\mathbf{x}_2 = \mathbf{Q}_{\mathbf{s}_{w_i}}^{2k+1} \mathbf{x}_1$, whereas for $\mathbf{x}_2 = \mathbf{x}_1$, l must take an even integer value, i.e., $\mathbf{x}_2 = \mathbf{Q}_{\mathbf{s}_{w_i}}^{2k} \mathbf{x}_1 = \mathbf{x}_1$. In summary, it is shown that two given vectors $\mathbf{x}_1$ and $\mathbf{x}_2$ are related to each other by the transformation $g_Q \in G_Q$. Therefore, the statistic $\mathbf{y}$ is MI.

To drive a MI for second group, we have $M_Q(g_d(\mathbf{x}_1)) = M_Q(d\mathbf{x}_1) = |d|^2 M_Q(\mathbf{x}_1)$. From the assumption $M_Q(\mathbf{x}_1) = M_Q(\mathbf{x}_2)$, we obtain $M_Q(g_d(\mathbf{x}_1)) = M_Q(g_d(\mathbf{x}_2))$. Now, we should first find a group G_H that acts on $\mathbf{y}$ and then find an MI under that group. According to Theorem 1 of Chapter 4, the group is given by $g_H(\mathbf{y}) \triangleq M_Q(g_d(\mathbf{x})) = |d|^2 \mathbf{y}$. The MI for the group G_H is given by $z = M_H(\mathbf{y}) = M_H(y_1, y_2) = \frac{y_1}{y_2}$. This statistic is invariant, since $M_H(g_H(\mathbf{y})) = M_H(\mathbf{y})$. Furthermore, from $M_H(\mathbf{y}^{(1)}) = M_H(\mathbf{y}^{(2)})$ for given $\mathbf{y}^{(1)}$ and $\mathbf{y}^{(2)}$, we obtain $\frac{y_1^{(1)}}{y_2^{(1)}} = \frac{y_1^{(2)}}{y_2^{(2)}}$ and so $\frac{y_1^{(2)}}{y_1^{(1)}} = \frac{y_2^{(2)}}{y_2^{(1)}} = c$, where c is a positive scalar value. Therefore, $[y_1^{(2)}, y_2^{(2)}] = c[y_1^{(1)}, y_2^{(1)}]$ or $\mathbf{y}^{(2)} = g_H(\mathbf{y}^{(1)})$ for some $g_H \in G_H$, as was to be shown. Hence, $M_H(\mathbf{y})$ is the MI under group G_H. The MI under the composite group G is given by $z = M_H(M_Q(\mathbf{x})) = \frac{\frac{|\mathbf{s}_{w_i}{}^H \mathbf{x}|^2}{\|\mathbf{s}_{w_i}\|^2}}{\|\mathbf{x}\|^2}$.

Appendix 5B
Maximal invariant statistic pdfs

In this appendix, we obtain the pdf of z, represented in (5.19), under both hypotheses. To do so, we use a 1–1 function of MI statistic, given by

$$t = \frac{N - (P_e + K)}{1} \frac{L_n(\mathbf{r})}{L_d(\mathbf{r})} \tag{5B.1}$$

where

$$L_n(\mathbf{r}) = \left(\frac{\mathbf{r}}{\sigma}\right)^H (\mathbf{\Pi}_{\mathrm{K}} - \mathbf{\Pi}_{\mathrm{U}}) \left(\frac{\mathbf{r}}{\sigma}\right) = \left(\frac{\mathbf{r}}{\sigma}\right)^H (\mathbf{\Pi}_{\mathrm{U}}^{\perp} - \Pi_{\mathbf{K}}^{\perp}) \left(\frac{\mathbf{r}}{\sigma}\right) \tag{5B.2}$$

and

$$L_d(\mathbf{r}) = \left(\frac{\mathbf{r}}{\sigma}\right)^H \mathbf{\Pi}_{\mathrm{U}}^{\perp} \left(\frac{\mathbf{r}}{\sigma}\right) \tag{5B.3}$$

where $\mathbf{\Pi}_{\mathrm{U}}^{\perp} = \mathbf{U}(\mathbf{U}^H\mathbf{U})^{-1}\mathbf{U}^H$ is the projection onto the subspace span by columns of matrix $\mathbf{U}$, and $\mathbf{\Pi}_{\mathbf{K}}^{\perp} = \mathbf{I} - \mathbf{K}(\mathbf{K}^H\mathbf{K})^{-1}\mathbf{K}^H$ with $\mathbf{K} = [\mathbf{U}, \mathbf{s}]$ is the orthogonal projection onto the subspace span by columns of matrix $\mathbf{K}$. As can be seen from (5B.1), the test consists of two quadratic forms in the numerator and denominator. To find the pdf of the numerator and denominator of the test under either hypothesis, we resort to the Theorem 2.

Theorem 2. *Let* $\mathbf{z} \sim \mathcal{CN}(\mu, \mathbf{I})$ *be a complex* $N \times 1$ *vector, and let* $\mathbf{A}$ *be a Hermitian matrix of dimension* $N \times N$. *Then* $\mathbf{z}^H\mathbf{A}\mathbf{z}$ *has a noncentral complex chi-squared random variable with k complex degree of freedom and noncentrality parameters* δ *if and only if* $\mathbf{A}$ *is an idempotent matrix* ($\mathbf{A}^2 = \mathbf{A}$), *in which case the complex degree of freedom and the noncentrality parameters are, respectively,* $k = rank(\mathbf{A}) = tr(\mathbf{A})$ *and* $\boldsymbol{\delta} = \mu^{\mathrm{H}}\mathbf{A}\mu$.
Proof. See Reference 20.

Let k_n and k_d be the complex degree of freedom of the numerator and denominator of the test statistic (5B.1), respectively, that are given as $k_n = 1$ and $k_d = N - (P_e + K)$. Now, it can be shown that

$$L_n(\mathbf{r}) \sim \begin{cases} \chi_1^2 & \text{under } \mathcal{H}_0 \\ \chi_1^2(\delta) & \text{under } \mathcal{H}_1 \end{cases} \tag{5B.4}$$

From Theorem 2, the noncentrality parameters $\boldsymbol{\delta}$ can be obtained as

$$\boldsymbol{\delta} = \frac{|a_0|^2}{\sigma^2} \|\boldsymbol{\Pi}_{\mathrm{U}}^{\perp}\mathbf{s}\|^2 \tag{5B.5}$$

In (5B.4), $\chi_k^2(\boldsymbol{\delta})$ denotes an noncentral complex chi-squared distribution with k complex degree of freedom and noncentrality parameter $\boldsymbol{\delta}$. Similarly, it can be obtained that

$$L_d(\mathbf{x}) \sim \begin{cases} \chi^2_{N-(P_e+K)} & \text{under } \mathcal{H}_0 \\ \chi^2_{N-(P_e+K)} & \text{under } \mathcal{H}_1 \end{cases} \tag{5B.6}$$

Since $\boldsymbol{\Pi}_{\mathbf{K}}^{\perp}\left(\boldsymbol{\Pi}_{\mathbf{K}} - \boldsymbol{\Pi}_{\mathrm{U}}\right) = 0$, we can conclude that the numerator and denominator of the statistic (5B.1) are independent. Hence, it is known to conclude that

$$t \sim \begin{cases} F_{1,N-(P_e+K)} & \text{under } \mathcal{H}_0 \\ F'_{1,N-(P_e+K)}(\boldsymbol{\delta}) & \text{under } \mathcal{H}_1 \end{cases} \tag{5B.7}$$

where $F_{1,N-(P_e+K)}$ denotes a central complex F distribution with 1 numerator complex degree of freedom and $(N-(P_e+K))$ denominator complex degree of freedom, and $F'_{1,N-(P_e+K)}(\boldsymbol{\delta})$ denotes a noncentral complex F distribution with 1 numerator complex degree of freedom and $(N-(P_e+K))$ denominator complex degree of freedom and noncentrality parameter $\boldsymbol{\delta}$.

Chapter 6

Broadband target detection algorithm in FM-based passive bistatic radar systems

6.1 Introduction

The goal of this chapter is to jointly take advantage of the detectors proposed in Chapters 4 and 5. Chapter 4 introduced an optimal multiband detector that combined multiple FM radio channels in an incoherent way. This detector was shown to offer benefits in terms of noncoherent combining gain, frequency diversity gain, and robustness against time-varying characteristics of individual FM radio channels when compared to a single-band detector. However, while this detector enhanced detection performance quality, it did not improve target range resolution. In contrast, Chapter 5 proposed an optimal multiband detector that coherently combined multiple FM radio frequency channels, resulting in improved target range resolution. This second detector is called a high-range resolution (HRR) detector, while the first detector is called a high-quality performance (HQP) detector. The signal modeling in Chapter 5 revealed that the HRR detector requires some correlated channels to improve target range resolution, while the HQP detector can use multiple channels without any limitations. In practical scenarios, both HQP and HRR features are desirable simultaneously. This inspires us to formulate a new detection problem to take advantage of both HRR and HQP detectors.

In this chapter, a new broadband target detection problem is formulated as a binary composite hypothesis testing problem, and the uniformly most powerful invariant (UMPI) criterion is used to devise a new broadband detector. This detector is called UMPI(M,L), where M is the total number of channels, among which there are only L correlated frequency channels, i.e., $L \leq M$. The proposed algorithm improves upon the detectors presented in Chapters 4 and 5 and includes them as special cases of UMPI(M, 0) and UMPI(L, L), respectively. This chapter also provides analytical expressions for the probability of detection and the probability of false alarm. Additionally, it demonstrates how the target range resolution can be improved through this new detection algorithm. An integrated sidelobe level ratio (ISLR) is introduced to illustrate how the detection performance quality can be enhanced. Finally, simulation examples are presented to validate the theoretical analysis and demonstrate the effectiveness of the proposed detection algorithm. The material presented in this chapter was previously published in Reference 1.

The organization of this chapter is as follows: Section 6.2 presents the signal modeling. A two-stage broadband UMPI test is derived in section 6.3. In section 6.4, statistical analysis is derived in closed form. Target range resolution improvement is analytically discussed in section 6.5. Simulation results are provided in section 6.6 to verify the analytical findings and examine the detection performance of the proposed UMPI test. Finally, we summarize this chapter in section 6.8.

6.2 Signal modeling

Suppose we have a passive radar setup that includes a multichannel transmitter and a multiband receiver. The transmitter broadcasts M frequency channels simultaneously. In this setup, we assume that there are L correlated channels among the M channels. For these L correlated channels, we can consider the received target amplitudes as being completely correlated. The term "correlated channels" refers to a group of frequency channels transmitted by the same FM radio station, where the frequencies of the channels are closely spaced together. Specifically, if a set of L channels with carrier frequencies in the range of $f_c^{(i)} < f_c^{(i+1)} < \cdots < f_c^{(i+L)}$ have a frequency spacing of $f_c^{(i+L)} - f_c^{(i)} \ll \Delta f = \frac{c}{2D}$, where c is the speed of light and D is the effective target size, then they are considered correlated channels. On the other hand, if the frequency spacing between two channels with carrier frequencies of $f_c^{(i)}$ and $f_c^{(j)}$ is greater than Δf, then they are considered independent channels. For example, if we assume a target size of $D = 15$ m, then Δf would be 10 MHz. In the frequency range of 88–108 MHz for the FM radio standard, thus, it is possible to find both correlated and independent frequency channels.

In order to enhance the target resolution range of the target, it is necessary to increase the bandwidth of the detection signal. This can be accomplished by employing a technique similar to the one presented in Chapter 5, where each of the received channels is filtered at a selected bandwidth of $0.5B$, digitally up-sampled (UP-S) using a sampling frequency of $f_s = MB$, and up-converted (UP-C) to the center frequency of $\{f_m\}_{m=1}^{M}$. This will ensure that the channels are adjacent in frequency, with no overlap or gaps between them. A block diagram illustrating the system model of the broadband FM-based passive bistatic radar (PBR) system can be seen in Figure 6.1. Now, the received surveillance signal vector can be compactly written as

$$\mathbf{r} = \mathbf{S}_0\mathbf{a}_0 + \mathbf{Hc} + \mathbf{Tg} + \mathbf{n} \tag{6.1}$$

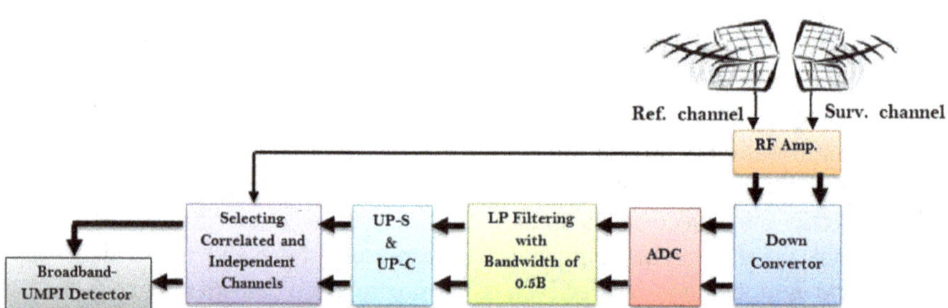

Figure 6.1 Broadband FM-based passive radar target detection scenario

- $\mathbf{r} \in \mathscr{C}^{N\times 1}$ is the $N \times 1$ vector received by the surveillance channel, in which $N = Tf_s$.
- $\mathbf{n} \in \mathscr{C}^{N\times 1}$ represents additive Gaussian noise in the broadband surveillance channel of the PBR system and has the distribution $\mathscr{CN}(0_{N\times 1}, \sigma^2\mathbf{I}_N)$ with unknown variance σ^2.
- Target under test signature matrix $\mathbf{S}_0$ is an $N \times M$ matrix defined as

$$\mathbf{S}_0 \triangleq [\mathbf{s}_1, \ldots, \mathbf{s}_M] \tag{6.2}$$

where

$$\mathbf{s}_m = \sqrt{p^{(m)}} e^{j2\pi f_c^{(m)} n_t^{(0)} T_s} (\mathbf{y}_f^{(n_t^{(0)},m)} \odot \mathbf{e}^{(v_t^{(k)},m)} \odot \mathbf{b}^{(f_m)}) \tag{6.3}$$

where $\mathbf{y}_f^{(m)}$ is the reference signal received due to mth broadcasted channel. In this case, subscript "f" is used to emphasize that $\mathbf{y}_f^{(m)}$ is a filtered version of the received signal by the bandwidth of $0.5B$ in the mth reference channel; transmitted signal amplitude of the mth broadcasted channel is represented by $p^{(m)}$; $(n_t^{(0)}, v_t^{(0)})$ represents the delay-velocity coordinate of the target under test; $\mathbf{y}^{(n_t^{(0)},m)} = \mathbf{P}^{n_t^{(0)}} \mathbf{y}_f^{(m)}$, $[\mathbf{e}^{(v_t^{(0)},m)}]_n = e^{j2\pi(\frac{v_t^{(0)}}{\lambda(m)})nT_s}$, and $[\mathbf{b}^{(f_m)}]_n = e^{j2\pi f_m n T_s}$ with $n = 0, \ldots, N_m - 1$. In this case, $\mathbf{P}$ is an $N_m \times N_m$ permutation matrix defined as $[\mathbf{P}]_{ij} = 1$ if $i = j + 1$ and 0 otherwise for $i = 0, \ldots, N_m - 1$; $j = 0, \ldots, N_m - 1$; and $\mathbf{a}_0 = [a_1, \ldots, a_M]^T$ is the M-dimensional column vector containing all amplitudes of the desired targets received over different channels. Throughout this chapter, it is assumed that there are only one set of correlated channels with L different frequency. For simplicity of notation, we assumed that the first L columns of matrix $\mathbf{S}_0$ are constructed based on the L correlated channels with carrier frequencies $\{f_c^{(1)} < f_c^{(i+1)} < \cdots < f_c^{(L)}\}$, respectively. By considering the fact that received target signal amplitudes over correlated channels may be fully correlated [2], i.e., $\{a_l\}_{l=1}^{l=L} = \alpha_1$, we can rewrite matrix $\mathbf{S}_0$ with lower $N \times R$ dimension as follows:

$$\mathbf{S}_k \triangleq \left[\mathbf{s}_1(n_t^{(k)}, v_t^{(k)}), \mathbf{s}_2(n_t^{(k)}, v_t^{(k)}), \ldots, \mathbf{s}_R(n_t^{(k)}, v_t^{(k)})\right], \tag{6.4}$$

for $k = 0$. Here, R is defined as $R = M - L + 1$ when $L \neq 0$, and $R = M$ otherwise. In (6.4), we have

$$\mathbf{s}_r(n, v) = \begin{cases} \sum_{l=1}^{L} \sqrt{p^{(l)}} e^{j2\pi f_c^{(l)} n T_s} (\mathbf{y}_f^{(n,l)} \odot \mathbf{e}^{(v,l)} \odot \mathbf{b}^{(f_l)}), & r = 1 \\ \sqrt{p^{(L-1+r)}} e^{j2\pi f_c^{(L-1+r)} n T_s} (\mathbf{y}_f^{(n,L-1+r)} \odot \mathbf{e}^{(v,L-1+r)} \odot \mathbf{b}^{(f_{L-1+r})}), & r = 2, \ldots, R \end{cases} \tag{6.5}$$

In a similar way, vector $\mathbf{a}_0$ changes to new vector $\mathbf{a}_0 = [\alpha_1^{(0)}, \alpha_2^{(0)}, \ldots, \alpha_R^{(0)}]^T$ with dimension $R \times 1$.

- Clutter signature matrix $\mathbf{H}$ is an $N \times P_e$ matrix defined as

$$\mathbf{H} \triangleq \left[\mathbf{H}^{(1)}, \ldots, \mathbf{H}^{(M)}\right] \tag{6.6}$$

where $\mathbf{H}^{(m)} = [\mathbf{h}_1^{(m)}, \ldots, \mathbf{h}_p^{(m)}, \ldots, \mathbf{h}_{P_m}^{(m)}]$ is an $N_m \times P_m$ matrix with $\mathbf{h}_p^{(m)} = (\mathbf{y}_f^{(n_c^{(p,m)},m)} \odot \mathbf{e}^{(v_c^{(p)},m)} \odot \mathbf{b}^{(f_m)})$, where $n = 0, \ldots, N_m - 1$, $m = 1, \ldots, M$, and $P_e = \sum_{m=1}^{M} P_m$. In this case, $n_c^{(p,m)}$ and $v_c^{(p,m)}$ are determined based on the delay-velocity coordinates of the clutter signature for $p = 1, \ldots, P_m$, and $\mathbf{c} = [c_1, \ldots, c_{P_e}]^T$ is the P_e-dimensional unknown column vector containing all amplitudes corresponding to any delay-velocity coordinates of the clutter signature over different channels. Here, we use a same clutter model as that of Chapter 3, and the interested reader is referred to it for more detail about clutter modeling.

- Interfering target signature matrix $\mathbf{T}$ is an $N \times KR$ matrix defined as

$$\mathbf{T} \triangleq \left[\mathbf{S}_1, \mathbf{S}_2, \ldots, \mathbf{S}_K\right], \tag{6.7}$$

where $\mathbf{S}_k$ for $k = 1, \ldots, K$, with dimension $N \times R$, is defined in (6.4); the KR-dimensional column vector $\mathbf{g}$ is defined as $\mathbf{g} = [\mathbf{a}_1, \mathbf{a}_2, \ldots, \mathbf{a}_K]^T$ in which the R-dimensional column vector $\mathbf{a}_k$ contains all amplitudes of the kth interfering targets.

6.3 Design of a broadband UMPI test

The composite binary hypothesis-testing problem can be compactly formulated as

$$\begin{cases} \mathcal{H}_0 : \mathbf{r} = \mathbf{Hc} + \mathbf{Tg} + \mathbf{n} \\ \mathcal{H}_1 : \mathbf{r} = \mathbf{Sa} + \mathbf{Hc} + \mathbf{Tg} + \mathbf{n} \end{cases} \tag{6.8}$$

where null ($\mathcal{H}_0$) and alternative ($\mathcal{H}_1$) hypotheses denote, respectively. For simplicity of notation, we replace $\mathbf{S}_0$ and $\mathbf{a}_0$ with $\mathbf{S}$ and $\mathbf{a}$, respectively.

The goal here is to devise a UMPI-based target detection algorithm in the presence of noise, clutter/multipath, and interfering targets to take advantages of the detectors introduced in Chapters 3, 4, and 5. In doing so, the UMPI test cannot be straightforwardly obtained since there are lots of unknowns, including $\mathbf{T}$, $\mathbf{a}$, $\mathbf{c}$, $\mathbf{g}$, and σ^2. To circumvent this, we obtain a two-stage UMPI-based detector. In the first stage, the UMPI detector is driven using the classical approach, assuming that the matrix $\mathbf{T}$ is known, and working with the remaining unknowns $\mathbf{a}$, $\mathbf{c}$, $\mathbf{g}$, and σ^2. In the second stage, a multistage algorithm is used to estimate the matrix $\mathbf{T}$ and substitute it for the true matrix used in the first stage. The first and second stages are elaborated in detail below.

6.3.1 First stage of a UMPI-based detector

In the first stage of the detector design, it is assumed that the matrix $\mathbf{T}$ is known, and a broadband UMPI detector is derived according to the classical approach over the remaining unknowns $\mathbf{a}$, $\mathbf{c}$, $\mathbf{g}$, and σ^2. To derive the UMPI test, we resort to the classical approach introduced in Chapter 4. Before deriving the UMPI test, some transformations are applied to the observation data $\mathbf{r}$ to bring the detection problem to a simpler form. To do this, we first use the singular eigenvalue decomposition (SVD) of matrix $\mathbf{H}$, given by

$$\mathbf{H} = [\mathbf{U}_H^{(1)}, \mathbf{U}_H^{(2)}] \begin{bmatrix} \Sigma_H^{(1)} & \mathbf{0} \\ \mathbf{0} & \mathbf{0} \end{bmatrix} \begin{bmatrix} \mathbf{V}_H^{(1)} \\ \mathbf{V}_H^{(2)} \end{bmatrix} \tag{6.9}$$

Here, $\mathbf{U}_H^{(1)}$ and $\mathbf{U}_H^{(2)}$ are the $N \times P_e$ and $N \times (N - P_e)$ orthogonal matrices that span the column space of matrix $\mathbf{H}$ and its orthogonal complement, respectively, i.e., $\mathbf{U}_H^{(2)^H}\mathbf{H} = \mathbf{0}$ [3]. Multiplying the observation vector $\mathbf{r}$ in (6.8) by $\mathbf{U}_H^{(2)^H}$, one can obtain the observation vector under full hypothesis ($\mathscr{H}_1$) as

$$\mathbf{r}_H \triangleq \mathbf{U}_H^{(2)^H}\mathbf{r} = \mathbf{S}_H\mathbf{a} + \mathbf{T}_H\mathbf{g} + \mathbf{n}_H \tag{6.10}$$

where $\mathbf{r}_H$ is the $(N - P_e)$-dimensional column vector, matrix $\mathbf{S}_H$ is defined as $\mathbf{S}_H \triangleq \mathbf{U}_H^{(2)^H}\mathbf{S}$, $\mathbf{T}_H$ is an $(N - P_e) \times KR$ matrix defined as $\mathbf{T}_H \triangleq \mathbf{U}_H^{(2)^H}\mathbf{T}$. Since $\mathbf{U}_H^{(2)^H}\mathbf{U}_H^{(2)} = \mathbf{I}_{(N-P_e)}$ and $\mathbf{n}$ is zero-mean white Gaussian noise with the covariance matrix of $\sigma^2\mathbf{I}_N$, the transformed noise $\mathbf{n}_H$ is also zero-mean white Gaussian noise with the covariance matrix of $\sigma^2\mathbf{I}_{(N-P_e)}$. Again, we employ another orthogonal transformation obtained using the SVD of $\mathbf{T}_H$ as follows:

$$\mathbf{T}_H = [\mathbf{U}_T^{(1)}, \mathbf{U}_T^{(2)}] \begin{bmatrix} \Sigma_T^{(1)} & \mathbf{0} \\ \mathbf{0} & \mathbf{0} \end{bmatrix} \begin{bmatrix} \mathbf{V}_T^{(1)} \\ \mathbf{V}_T^{(2)} \end{bmatrix} \tag{6.11}$$

Here, $\mathbf{U}_T^{(1)}$ and $\mathbf{U}_T^{(2)}$ are the $(N - P_e) \times KR$ and $(N - P_e) \times (N - P_e - KR)$ orthogonal matrices that span the column space of $\mathbf{T}_H$ and its orthogonal complement, respectively, i.e., $\mathbf{U}_T^{(2)^H}\mathbf{T}_H = \mathbf{0}$. Multiplying the observation data $\mathbf{r}_H$ in (6.10) by $\mathbf{U}_T^{(2)^H}$, one can obtain

$$\mathbf{x} \triangleq \mathbf{U}_T^{(2)^H}\mathbf{r}_H = \mathbf{G}\mathbf{a} + \mathbf{w} \tag{6.12}$$

where $\mathbf{x}$ is the $(N - P_e - KR)$-dimensional column vector, matrix $\mathbf{G}$ with dimension $(N - P_e - KR) \times R$ is defined as $\mathbf{G} \triangleq \mathbf{U}_T^{(2)^H}\mathbf{S}_H$. Since $\mathbf{U}_T^{(2)^H}\mathbf{U}_T^{(2)} = \mathbf{I}_{(N-P_e-KR)}$ and $\mathbf{n}_H$ is zero-mean white Gaussian noise with the covariance matrix of $\sigma^2\mathbf{I}_{(N-P_e)}$, the transformed noise $\mathbf{w}$ is also zero mean, white, and Gaussian noise with the covariance matrix of $\sigma^2\mathbf{I}_{(N-P_e-KR)}$. To simplify the problem at hand, we apply a unitary rotation matrix defined as

$$\mathbf{U}_G = [\mathbf{G}(\mathbf{G}^H\mathbf{G})^{-\frac{1}{2}}, \mathbf{U}_\perp], \tag{6.13}$$

where $\mathbf{U}_\perp^H\mathbf{G} = 0$. Here, $<\mathbf{G}>$ stands for R-dimensional signal subspace, and $<\mathbf{U}_\perp>$ is its corresponding $(N - P_e - KR - R)$-dimensional orthogonal subspace. We can then write the transformed received vector as

$$\mathbf{z} = \begin{bmatrix} \mathbf{z}_1 \\ \mathbf{z}_2 \end{bmatrix} \triangleq \mathbf{U}_G^H\mathbf{x} = \begin{bmatrix} (\mathbf{G}^H\mathbf{G})^{-\frac{1}{2}}\mathbf{G}^H\mathbf{x} \\ \mathbf{U}_\perp^H\mathbf{x} \end{bmatrix} \tag{6.14}$$

To derive an UMPI test, we first show that the transformed detection problem is invariant under the composition of the transformation groups $\mathscr{G}_Q$ and $\mathscr{G}_d$ defined as follows:

$$\mathscr{G}_Q = \left\{ g_Q : g_Q(\mathbf{z}) = \begin{bmatrix} \mathbf{U}_1 & \mathbf{0} \\ \mathbf{0} & \mathbf{U}_2 \end{bmatrix} \mathbf{z} \right\} \tag{6.15}$$

where $\mathbf{U}_1$ and $\mathbf{U}_2$ are both unitary matrices with dimension $R \times R$ and $(N - P_e - KR - R) \times (N - P_e - KR - R)$, respectively, and

$$\mathscr{G}_d = \{g_d : g_d(\mathbf{z}) = d\mathbf{z}; \quad d \text{ is a complex scalar}\} \tag{6.16}$$

The transformations $\mathscr{G}_Q$ and $\mathscr{G}_d$ satisfy the conditions of a group because they are closed sets, associative, and include both identity and inverse elements. The next step is to demonstrate that when we apply any combination of the transformation groups specified in (6.15) and (6.16), the distribution of the observation and parameter spaces remains unchanged under each hypothesis.

1. Under $\mathscr{H}_1$, we have $\mathbf{z}_1 \sim \mathscr{CN}((\mathbf{G}^H\mathbf{G})^{\frac{1}{2}}\mathbf{a}, \sigma^2\mathbf{I}_{(N-P_e-KR)})$ and $\mathbf{z}_2 \sim \mathscr{CN}(\mathbf{0}, \sigma^2\mathbf{I}_{(N-P_e-KR)})$, and so for the transformed data, we get $\mathbf{U}_1\mathbf{z}_1 \sim \mathscr{CN}(\mathbf{U}_1(\mathbf{G}^H\mathbf{G})^{\frac{1}{2}}\mathbf{a}, \sigma^2\mathbf{I}_R)$ and $\mathbf{U}_2\mathbf{z}_2 \sim \mathscr{CN}(\mathbf{0}, \sigma^2\mathbf{I}_{(N-P_e-KR-R)})$. Since $\mathbf{a}$ is an unknown complex vector, the distribution family of the transformed data does not change. Under the null hypothesis, the proof is similar, except that $\mathbf{a} = \mathbf{0}$. Thus, the detection problem is invariant under $g_Q(\mathbf{z})$ transformation.
2. Under $\mathscr{H}_1$, we have $g_d(\mathbf{z}_1) \sim \mathscr{CN}(d(\mathbf{G}^H\mathbf{G})^{\frac{1}{2}}\mathbf{a}, \sigma^2 d^2\mathbf{I}_{(N-P_e-KR)})$ and $g_d(\mathbf{z}_2) \sim \mathscr{CN}(\mathbf{0}, \sigma^2 d^2\mathbf{I}_{(N-P_e-KR)})$. Since $\mathbf{a}$ and σ^2 are unknown complex vector and real value, respectively, the distribution family of the transformed data is not changed. Under the null hypothesis, the proof is similar, except that $\mathbf{a} = \mathbf{0}$. Therefore, the detection problem is invariant under $g_d(\mathbf{z})$ transformation.

In Appendix 6A, the MI statistic for the composite group $\mathscr{G}$ is obtained as

$$t = \frac{\mathbf{x}^H \Pi_{\mathbf{G}} \mathbf{x}}{\mathbf{x}^H \Pi_G^{\perp} \mathbf{x}} \tag{6.17}$$

where $\Pi_{\mathbf{G}} = \mathbf{G}(\mathbf{G}^H\mathbf{G})^{-1}\mathbf{G}^H = \mathbf{I} - \Pi_{\mathbf{G}}^{\perp}$. Note that $\Pi_{\mathbf{G}}^{\perp}$ is the orthogonal projection onto the subspace span by columns of matrix $\mathbf{G}$. After some matrix manipulations, the maximal invariant statistic in terms of original observation can be written as

$$t = \frac{\mathbf{r}^H \Pi_{\Pi_{\mathbf{U}}^{\perp}\mathbf{S}} \mathbf{r}}{\mathbf{r}^H \Pi_{\mathbf{U}}^{\perp} \mathbf{r} - \mathbf{r}^H \Pi_{\Pi_{\mathbf{U}}^{\perp}\mathbf{S}}^{\perp} \mathbf{r}} \tag{6.18}$$

where $\mathbf{U} = [\mathbf{H}, \mathbf{T}]$. In (6.17), $\mathbf{x}$ is distributed as $\mathbf{x} \sim \mathscr{CN}(\mathbf{0}, \sigma^2\mathbf{I}_{(N-P_e-KR)})$ under the null hypothesis, and $\mathbf{x} \sim \mathscr{CN}(\mathbf{Ga}, \sigma^2\mathbf{I}_{(N-P_e-KR)})$ under the alternate hypothesis.

This gives rise to the following distribution for statistic $t' = \frac{N-P_e-KR-R}{R} t$:

$$t' \sim \begin{cases} F_{R,N-P_e-KR-R} & \text{under } \mathcal{H}_0 \\ F_{R,N-P_e-KR-R}(\delta) & \text{under } \mathcal{H}_1 \end{cases} \tag{6.19}$$

where $F_{R,N-P_e-KR-R}$ denotes a central complex F distribution with R numerator complex degree of freedom and $(N-P_e-KR-R)$ denominator complex degree of freedom, and $F_{R,N-P_e-KR-R}(\delta)$ denotes a noncentral complex F distribution with R numerator complex degree of freedom and $(N-P_e-KR-R)$ denominator complex degree of freedom and noncentrality parameter δ, given by

$$\delta = \frac{\|\Pi_{\mathbf{U}}^{\perp}\mathbf{Sa}\|^2}{\sigma^2} \tag{6.20}$$

It is not difficult to show that noncentral complex F distribution of maximal invariant has a monotone likelihood ratio (MLR) property of t', so UMPI test is to reject $\mathcal{H}_0$ if $t' > \eta_{t'}$, or equivalently

$$\frac{\mathbf{r}^H \Pi_{\Pi_{\mathbf{U}}^{\perp}\mathbf{S}}\mathbf{r}}{\mathbf{r}^H \Pi_{\mathbf{U}}^{\perp}\mathbf{r} - \mathbf{r}^H \Pi_{\Pi_{\mathbf{U}}^{\perp}\mathbf{S}}\mathbf{r}} > \eta_t \tag{6.21}$$

where threshold η_t is selected such that the desired false alarm probability (P_{fa}) requirement is satisfied in the test $t > \eta_t$.

6.3.2 Second stage of a UMPI-based detector

In the first stage of UMPI-based detector design, it was assumed that the delay-velocity coordinates of interfering targets were known. However, in practical situations, this information is unknown and needs to be estimated. To address this issue, we propose an algorithm that can estimate these unknown parameters. It is important to note that in FM-based PBR systems, strong targets can mask the detection of weaker targets even when they have large range-velocity separations. This observation motivated us to implement a multistage algorithm based on the sequential implementation of the B-UMPI test in the first step to remove the masking effect of strong targets. The proposed algorithm detects targets sequentially from the strongest to the weakest levels, and in each step, it cancels out the effect of targets detected in the previous step. Targets detected in each step are referred to as imperative targets. This algorithm is a broadband version of the imperative target positioning (ITP) algorithm presented in previous chapters and is called the broadband imperative target positioning algorithm (B-ITPA).

To implement B-UMPI sequentially, we use a 1-1 function of t, which can be shown to be also a UMPI test. In our detection problem, the function $t = \frac{z}{1-z}$ for $0 < z < 1$ is an increasing function of z, so the B-UMPI in (6.18) is statistically equivalent to

$$z = \frac{\mathbf{r}^H \Pi_{\Pi_{\mathbf{U}}^{\perp}\mathbf{S}}\mathbf{r}}{\mathbf{r}^H \Pi_{\mathbf{U}}^{\perp}\mathbf{r}} \tag{6.22}$$

In the B-UMPI test described in (6.22), the threshold η_z on statistic z can be set as $\eta_z = \dfrac{R\eta_{t'}}{(N - P_e - RK - R) + R\eta_{t'}}$. Also, after some matrix manipulations, we can rewrite the (6.22) as

$$z = \frac{\sum_{i=1}^{R} \dfrac{|\mathbf{s}_i^H \Pi_{\mathbf{U}}^{\perp} \mathbf{r}|^2}{\|\Pi_{\mathbf{U}}^{\perp} \mathbf{s}_i\|^2}}{\|\Pi_{\mathbf{U}}^{\perp} \mathbf{r}\|^2} \tag{6.23}$$

where (6.23) follows from the fact that $(\Pi_{\mathbf{U}}^{\perp}\mathbf{s}_i)^H(\Pi_{\mathbf{U}}^{\perp}\mathbf{s}_j) = 0$ for $i \neq j = 1, \ldots, R$. To compute the B-UMPI test for a desired delay-velocity plane, we can resort to

$$\mathbf{Z}_{n,v} = \frac{\sum_{i=1}^{R} \dfrac{|\mathbf{s}_i(n, v)^H \Pi_{\mathbf{U}}^{\perp} \mathbf{r}|^2}{\|\Pi_{\mathbf{U}}^{\perp} \mathbf{s}_i(n, v)\|^2}}{\|\Pi_{\mathbf{U}}^{\perp} \mathbf{r}\|^2} \tag{6.24}$$

for $n = 1, \ldots, N_B$ and $v < v_{\max}$. Here, N_B is the time delay index corresponding to the maximum bistatic range of interest, and $v_{\max}$ is also the maximum speed of interested targets. To reduce the computational complexity of the B-ITP algorithm, we use fast Fourier transform (FFT) to compute the numerator of (6.24) for each range (delay) of interest as follows:

$$\mathbf{Z}_{n,q}(\mathbf{p}) = \frac{\sum_{i=1}^{R} \dfrac{\mathbf{f}_n^{(i)}(\mathbf{p}) \odot \mathbf{f}_n^{(i)}(\mathbf{p})^*}{\|\mathbf{s}_i(n)\|^2}}{\|\Pi_{\mathbf{U}}^{\perp} \mathbf{r}\|^2} \tag{6.25}$$

where

$$\mathbf{f}_n^{(i)}(\mathbf{p}) = \begin{cases} \sum_{l=1}^{L} \mathrm{FFT}\Big(\big(\sqrt{p^{(l)}} e^{-j2\pi f_c^{(l)} nT_s} (\mathbf{y}_f^{(n,l)} \odot \mathbf{b}^{(f_l)})^* \odot \mathbf{p}\big), N_F^{(l)}\Big) & i = 1 \\ \mathrm{FFT}\Big(\big(\sqrt{p^{(L-1+i)}} e^{-j2\pi f_c^{(L-1+i)} nT_s} (\mathbf{y}_f^{(n,L-1+i)} \odot \mathbf{b}^{(f_{L-1+i})})^* \odot \mathbf{p}\big), N_F^{(L-1+i)}\Big), & i \neq 1 \end{cases} \tag{6.26}$$

and

$$\mathbf{s}_i(n) = \begin{cases} \sum_{l=1}^{L} \sqrt{p^{(l)}} e^{-j2\pi f_c^{(l)} nT_s} \mathbf{y}_f^{(n,l)} \odot \mathbf{b}^{(f_l)} & i = 1 \\ p^{(L-1+i)} e^{-j2\pi f_c^{(L-1+i)} nT_s} \mathbf{y}_f^{(n,L-1+i)} \odot \mathbf{b}^{(f_{L-1+i})}, & i \neq 1 \end{cases} \tag{6.27}$$

Here, $\mathbf{p} = \Pi_{\mathbf{U}}^{\perp}\mathbf{r}$ and transformation $\mathrm{FFT}(\mathbf{d}, N_F^{(l)})$ performs an $N_F^{(l)}$ point FFT of vector $\mathbf{d}$ such that the zero-frequency component is moved to the center of the $\mathbf{f}_i(n)$. In (6.25), n and q are the time delay index and Doppler bin index, respectively.

To benefit from the FFT implementation of B-UMPI test, it is necessary to use a different number of FFT points over the different frequency channels. This is due to the fact that we have different Doppler shifted frequencies $f_d^{(m)} = \frac{v}{\lambda_m}$ over the different FM channels with the wavelengths λ_m for $m = 1, \ldots, M$. In other words, by performing

$N_F^{(m)}$-point FFTs, the estimated Doppler frequencies are given by $\hat{f}_d^{(m)} = l_v^{(m)} \frac{f_s}{N_F^{(m)}}$, where f_s is the baseband sampling frequency of the system and $l_v^{(m)}$ are the frequency bins corresponding to the real Doppler frequencies $f_d^{(m)}$ for $m = 1, ..., M$. As a result, if the number of FFT points, $N_F^{(m)}$, are set by equal values, then $l_v^{(m)}$ take different values for the specific velocity of a target, or equivalently, the $\mathbf{f}^{(m)}(n)$ peak in different frequency bins for $m = 1, ..., M$. This is in conflict with the combination gain of the joint exploitation of the broadband detection of our proposed algorithm. To circumvent this, similar to Reference 4, it is necessary to have equal frequency bins for a specific velocity over different frequency channels, i.e., $l_v^{(1)} = \cdots = l_v^{(M)} = l_v$. This implies that we must have $\frac{N_F^{(1)}}{\lambda_1} = \cdots = \frac{N_F^{(m)}}{\lambda_m} = \cdots = \frac{N_F^{(M)}}{\lambda_M}$. For convenience, we can select a frequency in the FM-band as the reference frequency, and use $N_F^{(m)} = \frac{\lambda_m}{\lambda_r} N_F^{(r)}$, in which $N_F^{(r)}$ is the number of FFT-point assigned to the reference frequency. Note that $N_F^{(r)}$ can be set equal to $N_r = T_r f_s$, where N_r is the data vector length corresponding to the integration time of the reference frequency denoted by T_r. This means that the data vector length corresponding to the integration time of different channels can be computed as $N_m = \frac{\lambda_m}{\lambda_r} N_r$ for $m = 1, ..., M$.

The key rule of the proposed sequential algorithm is to compute the vector $\mathbf{p} = \Pi_{\mathbf{U}}^{\perp} \mathbf{r}$ sequentially. For example, in the first step of algorithm, the interference signature matrix (ISM) $\mathbf{U}$ contains only the clutter signature matrix, i.e., $\mathbf{U}_1 = \mathbf{H}$, while in the $(k+1)$th step the signatures of detected imperative targets are also added to ISM such that $\mathbf{U}_k = [\mathbf{H}, \mathbf{S}_1, ..., \mathbf{S}_{k-1}]$. The detailed steps of the B-ITP algorithm are described as follows:

Step 1: To initialize the B-ITP algorithm, set $k \leftarrow 0$, $\mathbf{U}_0^{\perp} = \mathbf{H}$, $\mathbf{R}_0 = (\mathbf{H}^H \mathbf{H})^{-1}$, and $\mathbf{p}_{-1} = \mathbf{r}$.
Step 2: Compute $\mathbf{p}_k = \mathbf{p}_{k-1} - \mathbf{U}_k^{\perp} \mathbf{R}_k \mathbf{U}_k^{\perp H} \mathbf{p}_{k-1}$, and then compute the matrix $\mathbf{Z}_{n,q}(\mathbf{p}_k)$ corresponding to the delay-Doppler plane of interest.
Step 3: Compare the maximum value of $\mathbf{Z}_{n,q}(\mathbf{p}_k)$ over the range-Doppler plane to the predetermined threshold level $\boldsymbol{\eta}_k = \frac{R\eta_{t'}}{(N - P_e - Rk - R) + R\eta_{t'}}$. If it exceeds the threshold η_k, the corresponding $(\hat{n}, \hat{q})$ should be preserved as an estimate of the kth imperative target coordinate; otherwise, there is no interfering target, and the algorithm stops. Note that, this estimate may have an insufficient accuracy for an efficient interfering target cancelation. To circumvent this, a mask with three velocity index bins $(\hat{q} \pm j)$ for $j = 0, 1, 2$, and three delay indexes $\hat{n}$ and $\hat{n} \pm 1$ are chosen, yielding a mask of dimension 3×3 or the total number of range-velocity bins equal to $N_M = 9$.
Step 4: To find new strong targets set $k \leftarrow k + 1$, then compute and preserve the matrices $\mathbf{U}_k^{\perp}$ and $\mathbf{R}_k$ given by

$$\mathbf{U}_k^{\perp} = \mathbf{S}_k - \left(\sum_{i=0}^{k-1} \mathbf{U}_i^{\perp} \mathbf{R}_i \mathbf{U}_i^{\perp H} \right) \mathbf{S}_k \tag{6.28}$$

and

$$\mathbf{R}_k = (\mathbf{U}_k^{\perp H}\mathbf{U}_k^{\perp})^{-1} \tag{6.29}$$

where $\mathbf{S}_k = [\mathbf{S}_k^{(1)}, ..., \mathbf{S}_k^{(N_M)}]$ with $\mathbf{S}_k^{(\zeta)} \triangleq \left[\mathbf{s}_1(n_t^{(\zeta)}, v_t^{(\zeta)}), ..., \mathbf{s}_R(n_t^{(\zeta)}, v_t^{(\zeta)})\right]$ for $\zeta = 1, ..., N_M$. In this case, $\mathbf{s}_r(n_t^{(\zeta)}, v_t^{(\zeta)})$ is defined in (6.5) in which $n_t^{(\zeta)} \in \{\hat{n}-1, \hat{n}, \hat{n}+1\}$ and $v_t^{(\zeta)} \in \{(\hat{q} \pm j)\frac{f_s\lambda_r}{N_F^{(r)}}; j = 0, 1, 2\}$.

Step 5: Iterate steps 2–4 until the algorithm stops.

By ending the B-ITP algorithm, the number of imperative targets (K+1) and their delay-velocity coordinates, which are required to construct matrices **T** and **S**, have been found.* After this, we can resort to the B-UMPI detector derived in (6.18) by assuming one of the imperative targets as the testing target and the other K imperative targets as interfering targets. This leads to a reduction in the probable false target corresponding to the B-ITP algorithm. To do this, if the B-UMPI detector does not confirm our assumptions, we consider the target under test as a false target; otherwise, the target will be considered as a real target.

6.4 Analytical performance evaluation

The probability of false alarm and the probability of detection for the proposed UMPI detector based on the statistic t' can be computed with the aid of (6.19), and they are given, respectively, by

$$P_{fa} = Q_{F_{R,N-P_e-RK-R}}(\eta_{t'}) \tag{6.30}$$

$$P_d = Q_{F_{R,N-P_e-RK-R}(\delta)}(\eta_{t'}) \tag{6.31}$$

where $Q_{F_{R,N-P_e-RK-R}}$ and $Q_{F_{1,N-(N_I+1)}(\delta)}$ are the right-tail probability of central and noncentral complex F distribution, respectively. For the detection problem at hand, N gets a large value to give the desired integration gain. As a consequence, the statistics of t' in (6.19) may be approximated as

$$t' \sim \begin{cases} \frac{\chi_R^2}{R} & \text{under } \mathcal{H}_0 \\ \frac{\chi_R^2}{R}(\delta) & \text{under } \mathcal{H}_1 \end{cases} \tag{6.32}$$

Now, the false alarm probability and the probability of detection may be approximated as

$$P_{fa} = e^{-R\eta_{t'}} \sum_{n=0}^{R-1} \frac{(R\eta_{t'})^n}{n!} \tag{6.33}$$

*This means that, in the detection problem at hand, it is not required to know delay-velocity coordinates $(n_t^{(0)}, v_t^{(0)})$ of the interested target.

and

$$P_d = e^{-(\delta + R\eta_{t'})} \sum_{k=0}^{\infty} \left(\frac{\delta^k}{k!} \sum_{n=0}^{R+k-1} \frac{(R\eta_{t'})^n}{n!} \right) \tag{6.34}$$

where the probability of detection is an increasing function of $\delta = \frac{\mathbf{a}^H \mathbf{S}^H \Pi_{\mathbf{U}}^{\perp} \mathbf{S}\mathbf{a}}{\sigma^2}$.

6.5 Enhancing range resolution with broadband detection

Understanding how the B-UMPI test statistic changes as a function of range (or time) is crucial in specifying the characteristics of the proposed broadband PBR signal, such as target range resolution (TRR), peak-to-sidelobe level ratio (PSLR), and ISLR. This is similar to studying the properties of a radar waveform by analyzing the output of a matched filter as a function of range. In cases where the signal-to-noise ratio is high, the output of the matched filter can be approximated by the auto-correlation function of the transmitted signal, with the noise and Doppler velocity ignored [5].

When there is no interference in the received signals over multiple FM channels (or after eliminating any interference signals from the received signals, as demonstrated in (6.12)), the output of the B-UMPI can be expressed as a function of the time delay index n', given by

$$\Lambda[n'] = \frac{\mathbf{a}^H \mathbf{S}[n_0]^H \Pi_{\mathbf{S}[n']} \mathbf{S}[n_0] \mathbf{a}}{\mathbf{a}^H \mathbf{S}[n_0]^H \mathbf{S}[n_0] \mathbf{a}} \tag{6.35}$$

where (6.35) follows from (6.17), and matrix $\mathbf{S}[n]$ can be obtained from (6.4) when there is no Doppler shift, that is

$$\mathbf{S}[n] \triangleq \left[\mathbf{s}_1(n, 0), \ldots, \mathbf{s}_R(n, 0) \right], \tag{6.36}$$

Here, n_0 and n' are the true time delay index and estimate of the time delay index (considered a variable) of the received target signal, respectively. For simplicity in understanding the output of the B-UMPI test as a function of time delay index, we take the origin to be the true time delay index; hence, $n_0 = 0$. Then, $n_0 - n' = -n' \triangleq n$. Also, it is assumed that received echo signals over multiple FM channels have equal amplitudes, i.e., ($\mathbf{a} = \alpha \mathbf{1}$). The output of the B-UMPI test is then

$$\Lambda[n] = \frac{\mathbf{x}^H \mathbf{h}[n]}{\mathbf{x}^H \mathbf{x}} \tag{6.37}$$

where $\mathbf{x} = \mathbf{S}[0]\mathbf{1}$ and so $\mathbf{h}[n] = \Pi_{\mathbf{S}[n]} \mathbf{x}$. The following Lemma is useful for developing metrics specifying any improvement in the target range resolution and the quality of detection at the output of the B-UMPI detector.

Theorem 1. *Let* **u** *and* **v** *be two* $N \times 1$ *vectors, for large value of N, we have*

$$\mathbf{u}^H\mathbf{v} = \int_0^1 U^*(f)V(f)df \tag{6.38}$$

where

$$U(f) = \sum_{i=0}^{N-1} u[i]e^{-j2\pi fi} \tag{6.39}$$

and

$$V(f) = \sum_{i=0}^{N-1} v[i]e^{-j2\pi fi} \tag{6.40}$$

where $u[i]$ *and* $v[i]$ *are the* ith *element of vectors* **u** *and* **v**, *respectively.*
Proof. [6].

In the problem at hand, N takes a large value, so we can apply Lemma 1 to find that

$$\mathbf{h}[n] = \sum_{r=1}^{R} \frac{\mathbf{s}_r(n,0)^H\mathbf{x}}{\|\mathbf{s}_r(n,0)\|^2}\mathbf{s}_r(n,0) \tag{6.41}$$

where (6.41) follows after using

$$\mathbf{S}[n]^H\mathbf{S}[n] = \text{Diag}\left(\|\mathbf{s}_1(\text{n},0)\|^2, \dots, \|\mathbf{s}_\text{R}(\text{n},0)\|^2\right) \tag{6.42}$$

The output of the B-UMPI test is then

$$\Lambda[n] = \frac{\sum_{r=1}^{R} \frac{|\mathbf{s}_r(n,0)^H\mathbf{x}|^2}{\|\mathbf{s}_r(n,0)\|^2}}{\|\mathbf{x}\|^2} \tag{6.43}$$

To more simplify (6.43), by invoking Lemma 1, we find that

$$\mathbf{s}_r(n,0)^H\mathbf{x} = \int_0^1 S_r^*(f)X(f)df \tag{6.44}$$

where

$$X(f) = \sum_{i=1}^{M} \sqrt{p^{(m)}}Y^{(i)}(f - f^{(i)}) \tag{6.45}$$

with

$$Y^{(m)} = \sum_{i=0}^{N-1} y_f^{(m)}[i]e^{-j2\pi fi} \tag{6.46}$$

In the problem at hand, we have

$$S_1(f) = \sum_{l=1}^{L} \sqrt{p^{(l)}}e^{-j2\pi(f-f^{(l)})n}e^{j2\pi\frac{f_c^{(l)}}{f_s}n}Y^{(l)}(f - f^{(l)}) \tag{6.47}$$

where $f^{(l)} = \frac{f_l}{f_s}$ and we have $0 \le f^{(l)} < 1$. By substituting (6.45) and (6.47) in (6.44), we find

$$\mathbf{s}_1(n,0)^H\mathbf{x} = \sum_{l=1}^{L} p^{(l)} e^{-j2\pi \frac{f_c^{(l)}}{f_S} n} \int_0^1 G_l(f) e^{j2\pi fn} df \tag{6.48}$$

where

$$G_l(f) = \begin{cases} |Y^{(l)}(f)|^2 & -\frac{B}{2f_S} \leq f \leq \frac{B}{2f_S} \\ 0 & \text{otherwise} \end{cases} \tag{6.49}$$

By defining

$$g_l[n] \triangleq \int_0^1 G_l(f) e^{j2\pi fn} df \tag{6.50}$$

Therefore, we get

$$\mathbf{s}_1(n,0)^H\mathbf{x} = \sum_{l=1}^{L} p^{(l)} g_l[n] e^{-j2\pi \frac{f_c^{(l)}}{f_S} n} \tag{6.51}$$

It should be noted that $g_l[n]$ is the zero Doppler ambiguity function (auto-correlation) for lth received correlated channels.

By following a similar line of argument as for $r = 1$, one can readily find that

$$\mathbf{S}_r(n,0)^H\mathbf{x} = p^{(L-1+r)} g_l[n] e^{-j2\pi \frac{f_c^{(L-1+r)}}{f_S} n}; \qquad r = 2, \ldots, R, \tag{6.52}$$

and

$$\|\mathbf{x}\|^2 = \sum_{l=1}^{M} p^{(l)} g_l[0] \tag{6.53}$$

and

$$\|\mathbf{S}_r(n,0)\|^2 = \begin{cases} \sum_{l=1}^{L} p^{(l)} g_l[0] & r = 1 \\ p^{(L-1+r)} g_{L-1+r}[0] & r = 2, \ldots, R \end{cases} \tag{6.54}$$

Finally, the output of B-UMPI test can be further simplified and is given by the following expression:

$$\Lambda[n] = \frac{u[n]}{v[n]} \tag{6.55}$$

where

$$u[n] = \frac{\sum_{k=1}^{L} \sum_{l=1}^{L} \left(p^{(k)} p^{(l)} g_k[n] g_l^*[n] e^{-j2\pi \left(\frac{f_c^{(k)} - f_c^{(l)}}{f_s} \right) n} \right)}{\sum_{l=1}^{L} p^{(l)} g_l[0]} + \sum_{r=2}^{R} p^{(L-1+r)} \frac{|g_{L-1+r}[n]|^2}{g_{L-1+r}[0]} \tag{6.56}$$

and

$$v[n] = \sum_{m=1}^{M} p^{(m)} g_m[0] \tag{6.57}$$

From (6.56), we can clearly see that the term defined by $u[n]$ depends on n. In the following, hence, we concentrate on it to evaluate the target range resolution over time delay index of $n = 0, \ldots, N_B$, where N_B is obtained based on the maximum bistatic range of R_B as $R_B = cN_BT_s$. To do so, by defining the discrete Fourier transform of $u[n]$ as

$$U(f) = \sum_{n=0}^{N_b-1} u[n]e^{-j2\pi fn} \tag{6.58}$$

one can find that

$$U(f) = \frac{1}{\sum_{l=1}^{L} p^{(l)} g_l[0]} \sum_{k=1}^{L}\sum_{l=1}^{L} Z^{(k,l)}(f - \Delta f_c^{(k,l)}) + \sum_{r=2}^{R} \frac{E^{(r)}(f)}{p^{(L-1+r)} g_{L-1+r}[0]} \tag{6.59}$$

where

$$Z^{(k,l)}(f) = \sum_{n=0}^{N_b-1} \left(p^{(k)}p^{(l)}g_k[n]g_l^*[n]\right)e^{-j2\pi fn} \tag{6.60}$$

and

$$\Delta f_c^{(k,l)} = \frac{f_c^{(k)} - f_c^{(l)}}{f_s} \tag{6.61}$$

and

$$E^{(r)}(f) = \sum_{n=0}^{N_R-1} p^{(L-1+r)^2} |g_l[n]|^2 e^{-j2\pi fn} \tag{6.62}$$

From (6.59) and (6.61), it is evident that coherent channels are the only way to enhance target range resolution when $f_c^{(k)} - f_c^{(l)} \neq gf_s$, where g is an arbitrary integer number for $k \neq l = 1, \ldots, L$. This implies that although independent channels increase the bandwidth of the received signal, they do not increase the equivalent bandwidth of the matched filter output. Therefore, it is necessary to have at least two correlated channels with carrier frequencies that satisfy $f_c^{(k)} - f_c^{(l)} \neq gf_s$ to improve target range resolution. In PBR systems, it is not possible to control the transmitted waveforms or their carrier frequencies. Hence, the range resolution can only be improved by designing the bandwidth B and consequently the sampling frequency of $f_s = MB$. In general, if the following condition is met, the range resolution can be enhanced, i.e.,

$$f_c^{(l)} - f_c^{(1)} = (l-1)B + gf_s, \qquad l = 2, \ldots, L \tag{6.63}$$

The method involves filtering the desired received signals, which have been sampled at frequency f_s, with a suitable bandwidth of B. This ensures that they are up-converted and positioned adjacent to each other in the frequency domain without any overlap or gaps. One way to determine the appropriate bandwidth B is to use (6.63) to arrive at

$$B = \frac{2\sum_{l=1}^{L}(f_c^{(l)} - f_c^{(1)})}{L(L-1) + 2gM} \tag{6.64}$$

It should be noted that g is an integer number and must be selected such that $B < 200$ kHz is obtained, where 200 kHz is the nominal bandwidth of FM radio channels in most countries.

6.6 Enhancing target performance quality with broadband detection

The variable and unpredictable nature of FM radio waveform signals, which is beyond the control of radar designers, often results in sidelobes with a time-varying structure and high levels at the output of a single-band detector. This issue can lead to target echoes being masked by echoes from other strong targets in a multiple-target scenario [7, 8]. In this section, we demonstrate analytically that the proposed broadband detector reduces the masking effect of strong targets. To assess the range-masking effect of the detector, we use the concept of broadband ISLR, denoted as ISLR(M,L) and given by

$$\mathrm{ISLR}(M, L) = \frac{\Lambda[0]}{\sum_{n=1}^{N_B} \Lambda[n]} \tag{6.65}$$

where the output of B-UMPI statistic $\Lambda[n]$ can be rewritten as

$$\Lambda[n] = \frac{\frac{\left|A_L\left(\frac{\mathrm{mod}(n,M)}{M}, n\right)\right|^2}{A_L(0,0)} + B_{R-1}(0,n)}{A_M(0,0)} \tag{6.66}$$

where $A_L(f, n)$ and $B_{R-1}(f, n)$ are the discrete Fourier transforms of $a_L[l+1, n]$ and $b_{R-1}[r+1, n]$ with respect to l and r, respectively, defined as

$$a_L[l+1, n] = \begin{cases} p^{(l+1)} g_{l+1}[n], & 0 \leq l \leq L-1 \\ 0, & otherwise \end{cases} \tag{6.67}$$

and

$$b_{R-1}[r+1, n] = \begin{cases} p^{(L+r+1)} \frac{|g_{L+r+1}[n]|^2}{g_{L+r+1}[0]}, & 0 \leq r \leq R-2 \\ 0, & otherwise \end{cases} \tag{6.68}$$

It should be noted that (6.66) comes from the fact that $f_s = MB$ and $0 < f < 1$.

To provide further insights into the improvement in the target detection quality as well as the target range resolution improvement of the proposed detector, we provide further theoretical results for a simplified case, assuming $g_m[n] = g[n]$ and $p^{(m)} = 1$. Under this simplified case, it is easy to show that

$$\left|A_L^{(s)}(f, n)\right| = \left|Lg[n]\frac{\mathrm{sinc}(Lf)}{\mathrm{sinc}(f)}\right| \tag{6.69}$$

and

$$B_{R-1}^{(s)}(0,n) = \frac{|g[n]|^2}{g^2[0]} \tag{6.70}$$

substituting (6.69) and (6.70) into (6.66), we get

$$\Lambda_s[n] = \frac{|g[n]|^2}{g^2[0]}\left(\frac{L}{M}\left|\frac{\mathrm{sinc}(\frac{L}{M}\mathrm{mod}(n,M))}{\mathrm{sinc}(\frac{1}{M}\mathrm{mod}(n,M))}\right|^2 + \frac{R-1}{M}\right) \tag{6.71}$$

In a same way, we can recast (6.65) as

$$\mathrm{ISLR}_s(M,L) = \frac{1}{\sum_{n=1}^{N_B}\frac{|g[n]|^2}{g^2[0]}\left(\frac{L}{M}\left|\frac{\mathrm{sinc}(\frac{L}{M}\mathrm{mod}(n,M))}{\mathrm{sinc}(\frac{1}{M}\mathrm{mod}(n,M))}\right|^2 + \frac{R-1}{M}\right)} \tag{6.72}$$

Subscript “s” is used to put emphasis on the simplified case. Again, by defining

$$\xi_m = \left|\frac{\mathrm{sinc}(\frac{L}{M}m)}{\mathrm{sinc}(\frac{1}{M}m)}\right|^2, \qquad m = 0,\ldots,M-1 \tag{6.73}$$

we can rewrite (6.71) and (6.72) as

$$\Lambda_s[n] = \Lambda_1[n]\left(1-\frac{L}{M}\left(1-\sum_{m=0}^{M-1}\sum_{k=-\infty}^{\infty}\xi_m\delta(n-kM-m)\right)\right) \tag{6.74}$$

and

$$\mathrm{ISLR}_s(M,L) = \frac{1}{\sum_{n=1}^{N_B}\Lambda_1[n]\left(1-\frac{L}{M}\left(1-\sum_{m=0}^{M-1}\sum_{K=-\infty}^{\infty}\xi_m\delta(n-kM-m)\right)\right)} \tag{6.75}$$

where $\Lambda_1[n] = \frac{|g[n]|^2}{g^2[0]}$ is the output of the single-band UMPI detector or B-UMPI(1,0). Now there are some remarks in order. First, one can find that the output of the B-UMPI(M, L) detector is weighted by that of the single-band UMPI detector. However, since $L \leq M$ and $\xi_m < 1$, we can conclude that the range-masking effect of the B-UMPI(M,L) for $M > 1$ is less than that of the B-UMPI(1,0) exploiting single channel for detection, i.e.,

$$\mathrm{ISLR}_s(M,L) \geq \frac{\Lambda_1[0]}{\sum_{n=1}^{N_B}\Lambda_1[n]\times 1} \triangleq \mathrm{ISLR}_s(1,0) \tag{6.76}$$

The simplified case assumes that the signal used in the B-UMPI(M, L) detector and B-UMPI(1,0) detector are similar, but the former produces a better ISLR than the latter. However, in a real-world scenario, signals received over different channels of a broadband FM-based PBR system have varying qualities, resulting in different ISLRs when using the B-UMPI(1,0) detector. Based on the aforementioned discussions, it can be concluded that the B-UMPI(M,L) detector with $M > 1$ offers a better ISLR than the inferior ISLR obtained by the B-UMPI(1,0) detector over different

FM radio channels. Simulation investigations conducted later confirm this result for some plausible cases that violate the simplified case.

6.7 Simulation results

This section includes simulation examples aimed at verifying our theoretical analysis, demonstrating the target detection quality and improvements in target range resolution achieved by our proposed broadband detection algorithm. We utilized the same simulation scenario as that presented in Chapter 3 for the direct-path and clutter signals. To ensure that the false alarm probability was controlled, we considered clutter parameters of $\Delta v_c = 0.75$ m/s and $Q = 4$ while using an integration time of around 0.5 seconds.

To conduct our evaluation, we utilized three FM signals labeled as S_1, S_2, and S_3 which have carrier frequencies of 92 MHz, 94 MHz, and 96 MHz, respectively. These signals were chosen because they contain different types of content such as music, speech, and speech with short pauses of silence. The zero velocity self-ambiguity function curve corresponding to these signals is shown in Figure 6.2. The integration time for each signal was set to 0.543, 0.531, and 0.520 seconds for S_1, S_2, and S_3, respectively, as discussed in Chapter 4. We use the notation B-UMPI(M,L) to represent a broadband FM-based PBR invariant detector with M total channels, in which L channels are correlated. For simplicity, we refer to this detector as B-UMPI. In comparison, S-UMPI(S_i) denotes a single channel detector that exploits signal S_i for detection. In other words, S-UMPI(S_i) is equivalent to B-UMPI(1,0) using signal S_i.

In order to confirm the theoretical results presented in (6.74) through simulation, the time waveform corresponding to signal S_2 was denoted as $g[n]$. Figure 6.3 illustrates the test statistic of B-UMPI as a function of bistatic range, revealing that the

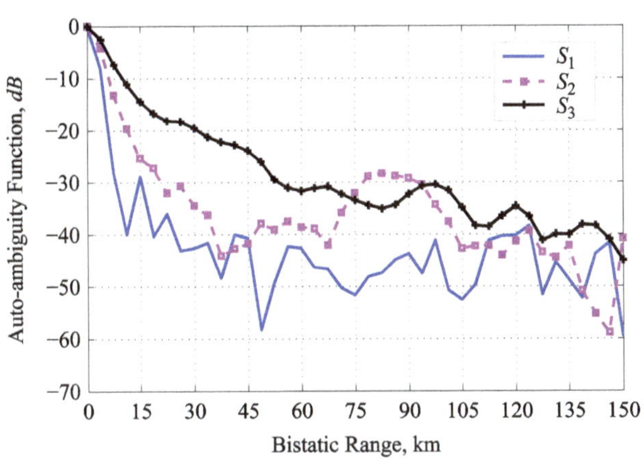

Figure 6.2 Auto-ambiguity function cut at zero velocity for received signals S_1, S_2, *and* S_3

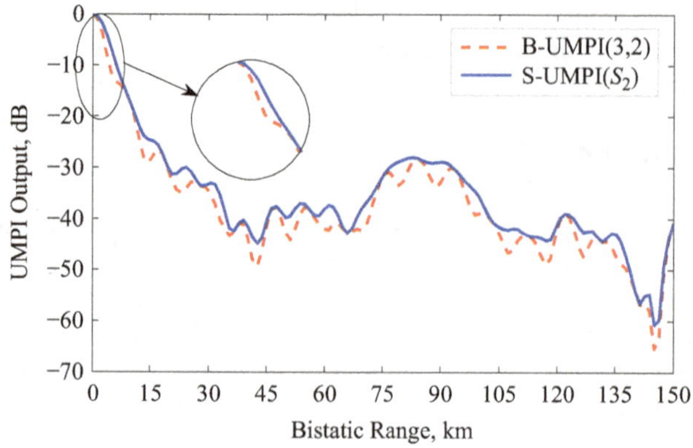

Figure 6.3 Comparison of B-UMPI(3,2) and S-UMPI(S_2) test statistics as a function of bistatic range

values of the B-UMPI test statistic are lower than those of the S-UMPI(S_2). The figure also indicates the weighting effects caused by coefficient $\xi_m < 1$ for $m = 0, \ldots, 3$ in the output of B-UMPI, which lead to a superior ISLR computed at the output of B-UMPI test. Specifically, the ISLR values provided by the B-UMPI(3,2) and the S-UMPI(S_2) are 37.58 dB and 34.18 dB, respectively, demonstrating an enhancement of around 3 dB offered by broadband detection. Additionally, the results demonstrate that the main lobe of the B-UMPI(3,2) output is narrower than that of the S-UMPI(S_2), indicating that the B-UMPI(3,2) improves the range resolution compared to the S-UMPI(S_2). Overall, these results confirm that the theoretical findings outlined in (6.65)–(6.76) are consistent with the simulation results obtained in this simplified scenario.

When cases arise that violate the simplified scenario, we compare the output of S-UMPI detector with that of the B-UMPI(3,0) one when working with signals S_1, S_2, and S_3. Figure 6.4 demonstrates that the B-UMPI(3,0) detector with $L = 0$ does not enhance target range resolution, but it is capable of achieving lower sidelobe levels when compared to the worst sidelobe level obtained from the S-UMPI detector with signals S_1, S_2, and S_3. The resulting ISLR values for the S-UMPI detectors using signals S_1, S_2, S_3, and B-UMPI(3,0) are 47.30, 38.88, 32.44, and 33.45 dB, respectively. Figure 6.5 explores the effect of increasing the number of correlated channels (L) on the output of the B-UMPI(3, L) detector and reveals that increasing L leads to improved target range resolution. The results of Figures 6.4 and 6.5 align with the analytical results discussed in sections 6.5 and 6.6. To compare the detection performance of the proposed B-UMPI(M,L) for different values of M and L, the input signal-to-noise ratio (SNR$_i$) is defined as

$$SNR_i = \frac{|\alpha_0|^2}{N_0 B} \tag{6.77}$$

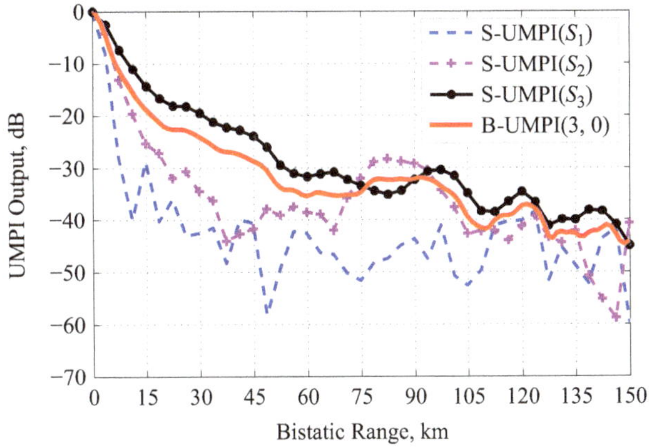

Figure 6.4 *Comparison of B-UMPI(3,0) and S-UMPI test statistics as a function of bistatic range when signals* S_1, S_2, *and* S_3 *are exploited for detection*

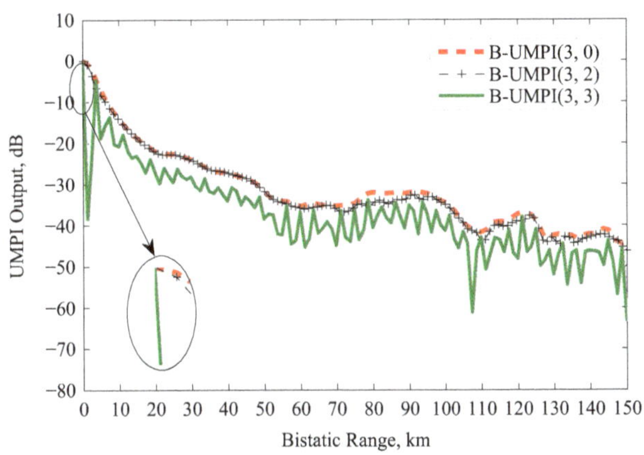

Figure 6.5 *Comparison of the B-UMPI(3, L) test statistic as a function of bistatic range for different number of correlated channels (L)*

where N_0 is the noise power per unit bandwidth, and B is the receiver bandwidth of a single channel. It should be noted that the value of B should be selected according to (6.63) and (6.64) to lead in the target range resolution improvement when exploiting correlated channels for detection. The corresponding bandwidth of the B-UMPI (3, L) is then computed using (6.64) to be equal to $B = 80, 82, 80$ Hz for $L = 0, 2, 3$, respectively.

Figure 6.6 illustrates the detection probability versus SNR_i for varying values of M, assuming no interfering targets. The probability of false alarm P_{fa} is maintained

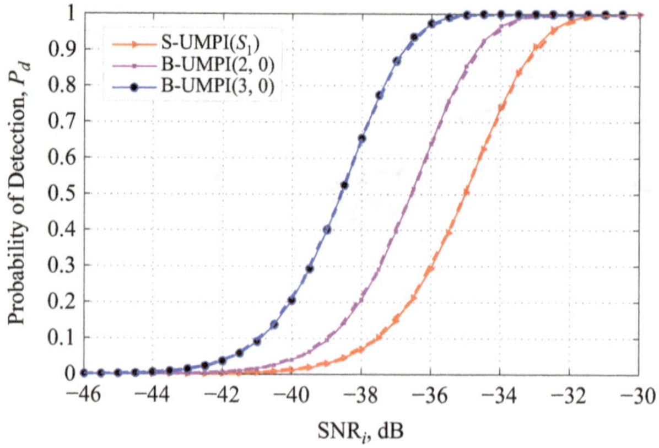

Figure 6.6 *Detection probability as a function of SNR$_i$ for the broadband UMPI(M,0) detector when different numbers of channels (M) are exploited. In this figure, solid lines denote the theoretical results, whereas the dashed lines denote the Monte Carlo simulation results.*

at 10^{-6} for all subsequent simulations. The solid lines in the figure correspond to theoretical results, while the dashed lines denote Monte Carlo (MC) simulation results. The figure confirms that the theoretical results align with the MC simulations, validating the theoretical model. As expected, the detection performance of the B-UMPI(M,0) algorithm improves as the number of channels used in the broadband system (M) increases. This enhancement is referred to as the non-coherent combined gain achieved by exploiting M channels for broadband detection. Figure 6.7

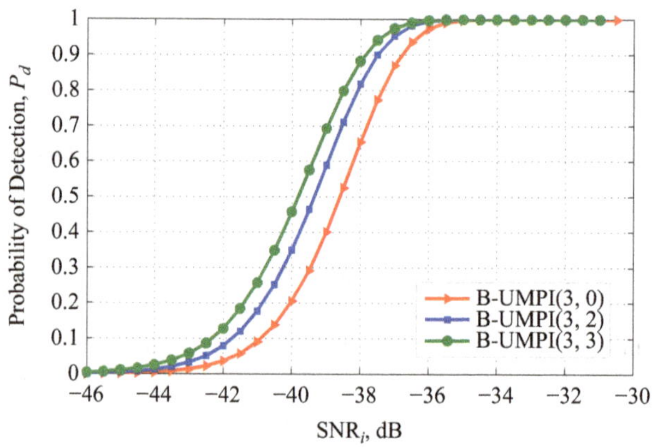

Figure 6.7 *Detection probability as a function of SNR$_i$ for the broadband UMPI(3,L) detector when different numbers of correlated channels (L) are used*

illustrates how the performance of B-UMPI(M,L) is affected by the number of correlated channels (L) for a fixed value of M. The results show that increasing the number of correlated channels in B-UMPI(3,L) improves its detection performance, which is referred to as coherent combined gain. Comparing the results of Figures 6.6 and 6.7, it is evident that enhancing the detection performance of the broadband detector B-UMPI(M, L) is more influenced by an increase in the number of correlated channels (L) rather than an increase in M. This is because an increase in L results in coherent gain in broadband detection, whereas an increase in M results in non-coherent gain, which is less effective than the coherent gain.

To evaluate the performance of the new broadband detector in comparison to the ones discussed in Chapters 4 and 5, a simulation is carried out. This simulation involves using an FM radio station that simultaneously broadcasts three frequency channels. In this case, we assume that there are two fully correlated, i.e., $M = 3$ and $L = 2$. Two detectors are evaluated, named B-UMPI(3,0) and B-UMPI(2,0), which are equivalent to the multiband detector described in Chapter 4 with three and two channels, respectively. Additionally, we use the B-UMPI(2,2) detector, which is equivalent to the detector presented in Chapter 5 with two correlated channels. According to Figure 6.8, the B-UMPI(3,2) detector is capable of utilizing all available broadcasted channels for detection. As a result, it is expected to outperform the detectors discussed in Chapters 4 and 5.

Up to now, we have investigated the range resolution improvement of the proposed broadband detector by comparing the output of the B-UMPI(M, L) test with different values of M and L. To get more insight, we consider a scenario with two targets separated by a range distance of ΔR, each with a signal-to-noise ratio of $\text{SNR}_i = -30$ dB. The detection probability of distinguishing one target from another ($P_{\text{res.}}$) is plotted against ΔR in Figure 6.9 for a broadband system with three channels

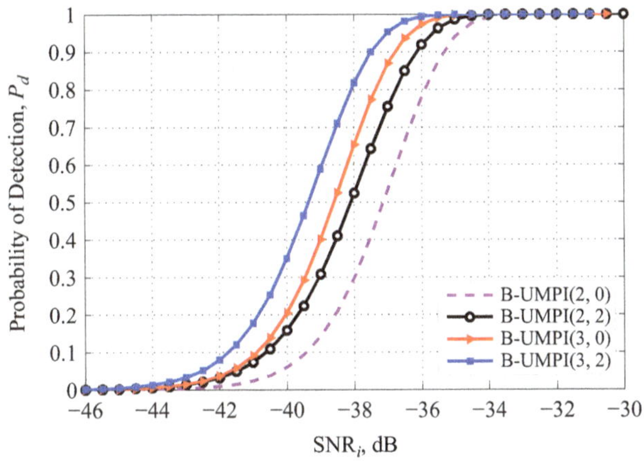

*Figure 6.8 Detection probability as a function of SNR$_i$ for the broadband UMPI(*M,L*) detector when different values of* M *and* L *are under* L≤*2*

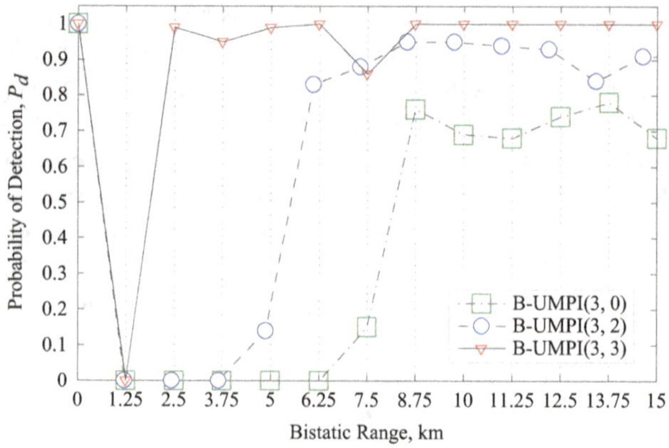

Figure 6.9 *Target range resolution improvement of the broadband UMPI(3,L) for different number of correlated channels (L)*

and various values of L. The plot shows the relationship between $P_{res.}$ and the distance between two targets, denoted by ΔR. The results indicate that if the two targets are separated by a distance of 8.75, 6.25, and 2.5 km, then they can be resolved with $P_{res.} > 0.75$ for $L = 0, 2, 3$, respectively. The range resolution of the system improves significantly as the number of correlated channels (L) increases. It is worth noting that the value of $P_{res.}$ not only depends on the number of correlated channels (L) but also on the quality of the broadband signal used in the B-UMPI(M,L) test. Moreover, the B-UMPI(3,3) detector's $P_{res.}$ degrades when the range distance between two targets is an integer multiple of 3.75 km. As such, the output of the B-UMPI(3,3) shows some peaks at these range distances, which degrades the quality of the broadband signal, leading to a decrease in $P_{res.}$. In contrast, Figure 6.5 shows that the broadband detector B-UMPI(3,3) has significantly lower sidelobe levels than the other B-UMPI(3,L) detectors for $L = 0, 2$, and therefore offers better overall performance. It is important to note that $\Delta R = 0$ in Figure 6.9 corresponds to the case where two targets have the same bistatic range position and can be detected as one target with probability 1.

In order to test the effectiveness of the proposed broadband detector in distinguishing closely spaced targets, a scenario was created that involved six targets with characteristics outlined in Table 6.1. The performance of the B-ITP algorithm and the B-UMPI(M,L) for different values of M and L were evaluated in this multi-target scenario. The results are presented in Figures 6.10–6.13. The original target positions, the estimated target positions using the B-ITP(M,L) algorithm, and the confirmed target positions using the B-UMPI(M,L) test were denoted by the symbols +, □, and ◯, respectively. The results indicated that all targets were correctly identified and positioned when using the B-UMPI(3,3)-based algorithm. However, when using the B-UMPI(3,2)-based algorithm, only two pairs of targets (T_{5-6} and T_{3-4}) were correctly identified and positioned. When using the B-UMPI(3,0)-based algorithm, only one pair of targets (T_{5-6}) was correctly positioned, although two

Table 6.1 Characteristic of the targets in the simulated scenario

Targets	T_1	T_2	T_3	T_4	T_5	T_6
Pair targets	T_{1-2}		T_{3-4}		T_{5-6}	
Bistatic range (km)	15	17.5	60	66.25	120	128.75
Bistatic velocity (m/s)	70	70	486	486	–208	–208
SNR_i (dB)	–28	–28	–30	–30	–20	–20

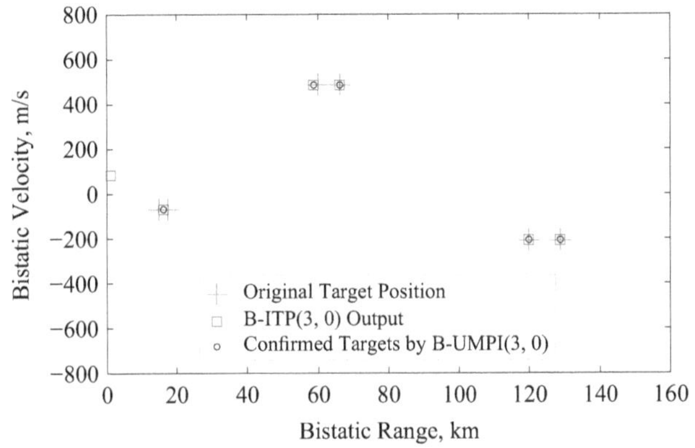

Figure 6.10 Results of applying the B-ITP(3,0) and its confirmation algorithm referred to as B-UMPI(3,0)

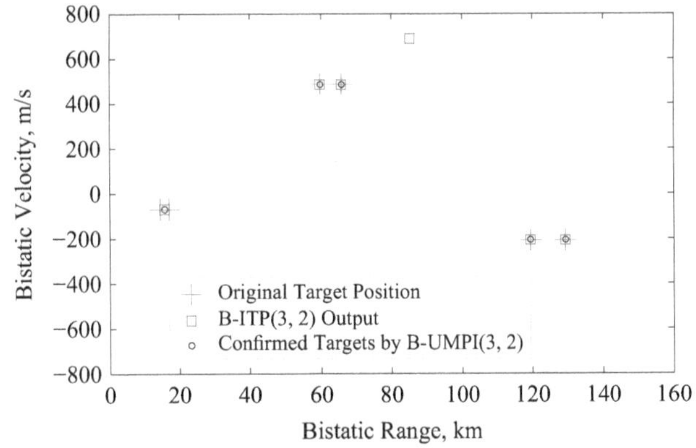

Figure 6.11 Results of applying the B-ITP(3,2) and its confirmation algorithm referred to as B-UMPI(3,2)

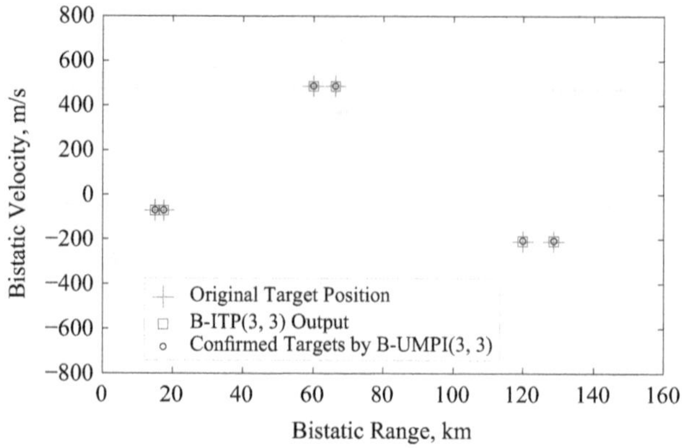

Figure 6.12 *Results of applying the B-ITP(3,3) and its confirmation algorithm referred to as B-UMPI(3,3)*

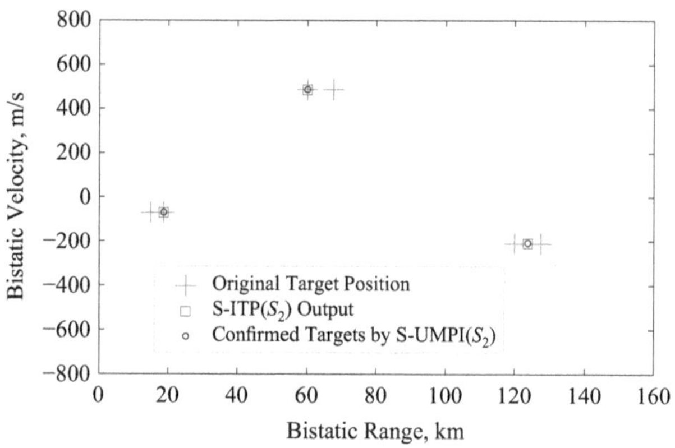

Figure 6.13 *Results of applying the S-ITP(S_2) and its confirmation algorithm referred to as S-UMPI(S_2)*

pairs of targets (T_{5-6} and T_{3-4}) were correctly identified. In the case of single-band detection, neither the B-UMPI(1,0) nor the S-UMPI(S_2)-based algorithm was able to correctly identify or position any of the targets of interest.

6.8 Summary

In this chapter, the focus was on addressing the issue of target detection in passive bistatic radar systems that use broadband FM. The chapter illustrates how the newly proposed broadband detection algorithm combines the advantages of the algorithms introduced in Chapters 4 and 5. The analytical framework developed in the chapter confirms that the proposed detector enhances both target range resolution and detection quality. Simulation results provide evidence that the algorithm is effective.

References

[1] Zaimbashi A. 'Broadband target detection algorithm in FM-based passive bistatic radar systems'. *IET Radar, Sonar & Navigation*. 2016, vol. 10(8), pp. 1485–99. Available from https://onlinelibrary.wiley.com/toc/17518792/10/8
[2] Zaimbashi A. 'Multiband FM-based passive bistatic radar: target range resolution improvement'. *IET Radar, Sonar & Navigation*. 2016, vol. 10(1), pp. 174–85. Available from https://onlinelibrary.wiley.com/toc/17518792/10/1
[3] Laub A.J. 'Matrix analysis for scientists and engineers' in Philadelphia, PA: Soc. for Industrial and Applied Math; 2005 Jan 1.
[4] Zaimbashi A., Derakhtian M., Sheikhi A. 'Invariant target detection in multiband FM-based passive bistatic radar'. *IEEE Transactions on Aerospace and Electronic Systems*. 2014, vol. 50(1), pp. 720–36.
[5] Skolnik M. Introduction to radar systems (3rd ed.). Columbus, OH: McGraw-Hill; 2002.
[6] Kay S. 'Fundamentals of statistical signal processing: Detection theory' in Upper Saddle River, NJ: Prentice-Hall; 1998.
[7] Colone F., O'Hagan D.W., Lombardo P., Baker C.J. 'A multistage processing algorithm for disturbance removal and target detection in passive bistatic radar'. *IEEE Transactions on Aerospace and Electronic Systems*. 2010, vol. 45(2), pp. 698–722.
[8] Bongioanni C. 2010. 'Multichannel passive radar: signal processing and experimental prototype development'. [PhD thesis]. Italy, Sapienza Università di Roma.Available from http://hdl.handle.net/10805/920

Appendix 6A

MI statistic derivation

This section presents the derivation of the maximal invariant (MI) statistic for the composite group $\mathcal{G}$. The transformation group $\mathcal{G}$ is formed by two subgroups, $\mathcal{G}_Q$ and $\mathcal{G}_d$, which are defined in (6.15) and (6.16). To derive the MI statistics, we utilize Theorem 1 from Chapter 4. According to Theorem 1, it can be shown that an MI for $\mathcal{G}_Q$ is a reduced-dimension function of the data $\mathbf{z} = [\mathbf{z}_1^T, \mathbf{z}_2^T]^T$ defined by

$$\mathbf{y} = M_Q(\mathbf{z}) = \left[\|\mathbf{z}_1\|^2, \|\mathbf{z}_2\|^2\right]^T \tag{6A.1}$$

This statistic is invariant, since for all $g_Q \in \mathcal{G}_Q$, we have

$$M_Q(g_Q(\mathbf{z})) = \left[\|\mathbf{U}_1\mathbf{z}_1\|^2, \|\mathbf{U}_2\mathbf{z}_2\|^2\right]^T \tag{6A.2}$$

Since $\mathbf{U}_1$ and $\mathbf{U}_2$ are both unitary matrices, it can be concluded that $M_Q(g_Q(\mathbf{z})) = M_Q(\mathbf{z})$. This shows that the statistic $\mathbf{y}$ is invariant.

To see the maximality of $\mathbf{y}$, it requires showing that given any pairs of data vectors $\mathbf{z}'$ and $\mathbf{z}$ such that $M_Q(\mathbf{z}') = M_Q(\mathbf{z})$, there exists a transformation $g_Q(.)$ of the form (6.15), such that $\mathbf{z}' = g_Q(\mathbf{z})$. Suppose, $M_Q(\mathbf{z}') = M_Q(\mathbf{z})$. Then, it follows from (6A.1) that

$$\begin{cases} \|\mathbf{z}_1'\|^2 = \|\mathbf{z}_1\|^2 \\ \|\mathbf{z}_2'\|^2 = \|\mathbf{z}_2\|^2 \end{cases} \tag{6A.3}$$

To proceed, we can construct the following unitary matrices from given any pairs of data vectors $\mathbf{z}'$ and $\mathbf{z}$

$$\mathbf{Q}_i = \left[\frac{\mathbf{z}_i'}{\sqrt{\mathbf{z}_i'^H\mathbf{z}_i'}} \middle| \mathbf{U}_i'\right] \left[\begin{array}{c} \dfrac{\mathbf{z}_i^H}{\sqrt{\mathbf{z}_i^H\mathbf{z}_i}} \\ \mathbf{U}_i^H \end{array}\right]. \tag{6A.4}$$

where $\mathbf{U}_i'^H\mathbf{z}_i' = \mathbf{0}$ and $\mathbf{U}_i^H\mathbf{z}_i = \mathbf{0}$ for $i = 1, 2$. Suppose $\|\mathbf{z}_i'\|^2 = \|\mathbf{z}_i\|^2$; then, we have $\mathbf{Q}_i\mathbf{z}_i = \mathbf{z}_i'$ for $i = 1, 2$. Substituting $\mathbf{Q}_1$ and $\mathbf{Q}_2$ instead of $\mathbf{U}_1$ and $\mathbf{U}_2$ in (6.15), we are able to construct transformation $g_Q(.)$ such that $\mathbf{z}' = g_Q(\mathbf{z})$.

To obtain an MI for the second group, we need to have $M_Q(g_d(\mathbf{z}')) = M_Q(d\mathbf{z}') = |d|^2 M_Q(\mathbf{z}')$. From the assumption $M_Q(\mathbf{z}') = M_Q(\mathbf{z})$, we obtain $M_Q(g_d(\mathbf{z}')) = M_Q(g_d(\mathbf{z}))$. Now, we should first find a group $\mathcal{G}_H$ that acts on $\mathbf{y}$ and then find an MI under that group. This group can be given by $g_H(\mathbf{y}) \triangleq M_Q(g_d(\mathbf{z})) = |d|^2\mathbf{y}$. The MI for the group $\mathcal{G}_H$ is given by $z = M_H(\mathbf{y}) = M_H(y_1, y_2) = \frac{y_1}{y_2}$. This statistic is invariant, since $M_H(g_H(\mathbf{y})) = M_H(\mathbf{y})$.

Furthermore, from $M_H(\mathbf{y}^{(1)}) = M_H(\mathbf{y}^{(2)})$ for given $\mathbf{y}^{(1)}$ and $\mathbf{y}^{(2)}$, we obtain $\frac{y_1^{(1)}}{y_2^{(1)}} = \frac{y_1^{(2)}}{y_2^{(2)}}$ and so $\frac{y_1^{(2)}}{y_1^{(1)}} = \frac{y_2^{(2)}}{y_2^{(1)}} = c$, where c is a positive scalar value. Therefore, $[y_1^{(2)}, y_2^{(2)}] = c[y_1^{(1)}, y_2^{(1)}]$ or $\mathbf{y}^{(2)} = g_H(\mathbf{y}^{(1)})$ for some $g_H \in \mathcal{G}_H$, as was to be shown. Hence, $M_H(\mathbf{y})$ is the MI under group $\mathcal{G}_H$. Thus, MI under the composite group $\mathcal{G}$ is given by $t = M_H(M_Q(\mathbf{z})) = \frac{\|\mathbf{z}_1\|^2}{\|\mathbf{z}_2\|^2}$.

Chapter 7

Multitarget detection in FM and digital TV-based passive radar: M-ary hypothesis testing framework

In the preceding chapters, we formulated the target detection problem in single- and multiband Frequency Modulation (FM)-based passive radar as binary hypothesis-testing problems. We tackled these problems with two-step Generalized Likelihood Ratio Test (GLRT) or Uniformly Most Powerful Invariant (UMPI) tests. In the second step of the detection methods, targets were sequentially detected, and previously detected targets were treated as interference and removed, resulting in the detection of weaker targets. This inspired us to approach the multitarget detection problem in this chapter as an M-ary hypothesis testing problem, rather than a binary one. Initially, it appeared that the detection problems in this chapter and the Chapters 3–6 shared some similarities. However, we will demonstrate that this new perspective will enable us to bypass the confirmation step of the detection methods used in the Chapters 3–6 by introducing a new threshold-setting strategy. Consequently, the computational complexity of the proposed detection algorithm in this chapter may be lower than that of the previous chapters.

A systematic framework for the passive radar target detection problem is presented in this chapter, which covers signal modeling, detection method, and statistical analysis. To solve this new detection problem, we consider the M-ary hypothesis testing problem as a series of binary hypothesis testing problems and propose a forward and sequential GLRT-based detector. The material in this chapter is based on our work published in Reference 1. The organization of this chapter is as follows. In section 7.1, the signal modeling is presented. Section 7.2 introduces M-ary hypothesis testing-based detection approach, where a forward and recursive Fast Fourier Transform (FFT)-based detection method is proposed. Section 7.3 provides our analytical analyses, while simulation results are provided in section 7.4. Finally, we draw some summaries in section 7.5.

7.1 Signal modeling

Consider a Ca.1-IR-based passive radar system with a transmitter and a single-band receiver. In a multitarget scenario with M targets, and in the presence of the system thermal noise, direct signal and clutter/multipath returns, the base-band received surveillance signal vector can be compactly represented by

$$\mathbf{x} = \mathbf{Sa} + \mathbf{Hc} + \mathbf{n} \tag{7.1}$$

where $\mathbf{x}$ is the $N \times 1$ complex vector received by the surveillance channel, in which $N = Tf_s$. Here, f_s and T are called the sampling frequency and the coherent processing interval (CPI), respectively. Vector $\mathbf{n}$ represents additive Gaussian noise in the surveillance channel of PBR system and has the distribution $\mathscr{CN}(0_{N\times 1}, \sigma^2 \mathbf{I}_N)$ with unknown variance σ^2. The clutter signature matrix $\mathbf{H}$ is an $N \times P$ matrix defined as:

$$\mathbf{H} \triangleq \left[\mathbf{h}_1, ..., \mathbf{h}_p, ..., \mathbf{h}_P\right] \tag{7.2}$$

where $\mathbf{h}_p = (\mathbf{y}^{n_c^{(p)}} \odot \mathbf{e}^{f_c^{(p)}})$ in which $[\mathbf{e}^{f_c^{(p)}}]_n = e^{j2\pi f_c^{(p)} nT_s}$ with $n = 0, ..., N-1$; $\mathbf{y}^{n_c^{(p)}} = \mathbf{p}^{n_c^{(p)}}\mathbf{y}$, where $\mathbf{y}$ is the reference signal after equalization techniques are applied to the reference channels to further isolate the direct path signals received; $\mathbf{P}$ is an $N \times N$ permutation matrix defined as $[\mathbf{P}]_{ij} = 1$ if $i = j + 1$ and 0 otherwise for $i = 0, ..., N-1$; $j = 0, ..., N-1$. Here, the clutters and direct path signal are modeled as multipaths characterized by the delays and Doppler frequencies in the low-frequency region. It is assumed that there are P significant paths where the delay and Doppler frequency of the pth path are denoted as $n_c^{(p)}$ and $f_c^{(p)}$, respectively. In this case, $n_c^{(p)}$ and $f_c^{(p)}$ are determined based on the delay-Doppler coordinates of a predefined clutter signature. In (7.1), $\mathbf{c} = [c_1, ..., c_P]^T$ is the P-dimensional unknown column vector containing all amplitudes corresponding to delay-Doppler coordinates of the clutter signature. Consequently, P will be much larger than the actual number of paths in the surveillance area:

Finally, the target signature matrix $\mathbf{S}$ is an $N \times K$ matrix defined as

$$\mathbf{S} \triangleq \left[\mathbf{s}_1, ..., \mathbf{s}_K\right] \tag{7.3}$$

where

$$\mathbf{s}_k = (\mathbf{y}^{(n_t^{(k)})} \odot \mathbf{e}^{(f_t^{(k)})}) \tag{7.4}$$

where $(n_t^{(k)}, f_t^{(k)})$ for $k = 1, ..., K$ represent the delay-Doppler coordinates of the interested targets. Here, $\mathbf{y}^{n_t^{(k)}} = \mathbf{P}^{n_t^{(k)}}\mathbf{y}$; $[\boldsymbol{e}^{(f_t^{(k)})}]_n = e^{j2\pi f_t^{(k)} nT_s}$ with $n = 0, ..., N-1$; $\mathbf{a} = [\alpha_1, ..., \alpha_K]^T$ is the K-dimensional column vector containing all complex amplitudes of the desired targets.

7.2 M-ary hypothesis testing-based detection approach

Passive radar systems exploit non-cooperative illuminators of opportunity. Although they offer some advantages, they need to cope with waveforms that are not tailored for radar applications. As such, it is likely that target echoes are masked by echoes from other strong targets even in the presence of large range Doppler separations in multitarget scenarios. To detect a target in a multitarget scenario, some iterative algorithms have been proposed in our previous chapters, which start to detect and remove the strongest target in order of descending signal strength. This inspires us to model the passive radar multitarget detection problem as a composite M-ary hypothesis problem and solve it based on the detection theory framework, given by

$$\begin{cases} \mathscr{H}_0 : \mathbf{x} = \mathbf{Hc} + \mathbf{n} \\ \mathscr{H}_1 : \mathbf{x} = \mathbf{s}_1\alpha_1 + \mathbf{Hc} + \mathbf{n} \\ \mathscr{H}_2 : \mathbf{x} = \mathbf{s}_2\alpha_2 + \mathbf{s}_1\alpha_1 + \mathbf{Hc} + \mathbf{n} \\ \mathscr{H}_3 : \mathbf{x} = \mathbf{s}_3\alpha_3 + [\mathbf{s}_1, \mathbf{s}_2][\alpha_1, \alpha_1]^T + \mathbf{Hc} + \mathbf{n} \\ \vdots \quad : \qquad \qquad \vdots \\ \mathscr{H}_m : \mathbf{x} = \mathbf{s}_m\alpha_m + [\mathbf{s}_1, ..., \mathbf{s}_{m-1}][\alpha_1, ..., \alpha_{m-1}]^T + \mathbf{Hc} + \mathbf{n} \\ \vdots \quad : \qquad \qquad \vdots \end{cases} \tag{7.5}$$

where hypothesis $\mathscr{H}_m$ denotes the presence of the mth targets in the presence of the system thermal noise, direct signal, and clutter/multipath returns. In (7.5), we have a set of hypotheses $\mathscr{H}_0 \subset \mathscr{H}_1 \subset \cdots \subset \mathscr{H}_m \subset \cdots$ and that each hypothesis specifies a probability density function (pdf) $f_m(\mathbf{x}; \boldsymbol{\theta}_m)$ with a parameter vector $\boldsymbol{\theta}_m$ of dimension $L_m = \dim(\boldsymbol{\theta}_m | \mathscr{H}_m)$ in the parameter space Θ_m. For the notational convenience, we, in the sequel, suppress the subscript m from $\boldsymbol{\theta}_m$ to Θ_m but the reader should keep in mind that these differ for each hypothesis. For the mth hypothesis, let

$$\ell_m(\boldsymbol{\theta}; \mathbf{x}) = ln(f_m(\mathbf{x}; \boldsymbol{\theta})) \tag{7.6}$$

denotes the log-likelihood function (LLF) of the observation under the mth hypothesis. In this chapter, we utilize a series of binary hypotheses ($\mathscr{H}_0$, $\mathscr{H}_1$), ($\mathscr{H}_1$, $\mathscr{H}_2$), …, ($\mathscr{H}_{m-1}$, $\mathscr{H}_m$), ($\mathscr{H}_m$, $\mathscr{H}_{m+1}$), …, ($\mathscr{H}_M$, $\mathscr{H}_{M+1}$) to sequentially detect targets. The GLRT criterion is applied to each of these binary hypothesis testing problems. Specifically, the GLRT statistic for testing $\mathscr{H}_m$ against $\mathscr{H}_{m-1}$ can be defined as

$$\Lambda_m(\mathbf{x}) = 2(\ell_m(\hat{\boldsymbol{\theta}}; \mathbf{x}) - \ell_{m-1}(\hat{\boldsymbol{\theta}}; \mathbf{x})) \underset{\mathscr{H}_{m-1}}{\overset{\mathscr{H}_m}{\gtrless}} \eta_m \tag{7.7}$$

where the threshold η_m is chosen to satisfy a specified probability of false alarm p_{fa}, and the natural point estimate of $\boldsymbol{\theta}$ is the maximum likelihood estimate (MLE) $\hat{\boldsymbol{\theta}} = \operatorname{argmax}_{\boldsymbol{\theta}_m} \ell_m(\boldsymbol{\theta}; \mathbf{x})$ which is assumed to be unique. By the general maximum likelihood theory, $\Lambda_m(\mathbf{x})$ has $\chi^2_{\mathrm{DOF}(m)}$ distribution asymptotically (i.e., chi-squared distribution with DOF(m) degrees of freedom), where

$$\mathrm{DOF}(m) = \dim(\boldsymbol{\theta} | \mathscr{H}_{\mathrm{m}}) - \dim(\boldsymbol{\theta} | \mathscr{H}_{\mathrm{m}-1}) \tag{7.8}$$

where DOF(m) is the number of scalar unknowns that take different values under $\mathscr{H}_m$ and $\mathscr{H}_{m-1}$ [2]. In the sequel, $\mathscr{H}_{m-1}$ ($\mathscr{H}_m$) is referred to as null (alternative) hypothesis, and we will derive binary hypothesis GLRT (BH-GLRT)-based detector when testing $\mathscr{H}_m$ against $\mathscr{H}_{m-1}$.

7.2.1 BH-GLRT detector design

The M-ary hypothesis testing problem (7.9) may be treated as a series of binary hypothesis testing problem. For example, to test $\mathscr{H}_m$ against $\mathscr{H}_{m-1}$, we get

$$\begin{cases} \mathscr{H}_{m-1} : \mathbf{x} = \mathbf{s}_{m-1}\alpha_{m-1} + [\mathbf{s}_1, ..., \mathbf{s}_{m-2}][\alpha_1, ..., \alpha_{m-2}]^T + \mathbf{Hc} + \mathbf{n} \\ \mathscr{H}_m : \mathbf{x} = \mathbf{s}_m\alpha_m + [\mathbf{s}_1, ..., \mathbf{s}_{m-1}][\alpha_1, ..., \alpha_{m-1}]^T + \mathbf{Hc} + \mathbf{n} \end{cases} \tag{7.9}$$

This detection problem can be compactly written as

$$\begin{cases} \mathscr{H}_{m-1} : \mathbf{x} = \mathbf{T}_{m-1}\mathbf{g}_{m-1} + \mathbf{Hc} + \mathbf{n} \\ \mathscr{H}_m : \mathbf{x} = \mathbf{s}_m\alpha_m + \mathbf{T}_{m-1}\mathbf{g}_{m-1} + \mathbf{Hc} + \mathbf{n} \end{cases} \tag{7.10}$$

where $\mathbf{T}_{m-1} = [\mathbf{s}_1, ..., \mathbf{s}_{m-1}]$ and $\mathbf{g}_{m-1} = [\alpha_1, ..., \alpha_{m-1}]^T$. In this new formulation, $\mathbf{s}_m$ represents the desired target to be detected, while columns of matrix $\mathbf{T}_{m-1}$ span the subspace of interfering targets to be detected.

The set of unknowns, given by $\Theta_m = \{(n_m, f_m), \alpha_m, \mathbf{g}_{m-1}, \mathbf{c}, \sigma^2\}$ and $\Theta_{m-1} = \{\mathbf{g}_{m-1}, \mathbf{c}, \sigma^2\}$ under the hypotheses $\mathscr{H}_m$ and $\mathscr{H}_{m-1}$, can be substituted by their MLEs to obtain the GLRT according to (7.7). It is more easier to apply a transformation to the observation data $\mathbf{x}$ to bring the detection problem to a simpler form. To begin, we first define the interference matrix $\mathbf{U}_{(P+m-1)}$ as $\mathbf{U}_{(P+m-1)} = [\mathbf{H}, \mathbf{T}_{m-1}]$ where $\mathbf{U}_{(P)} = \mathbf{H}^*$. Then, we use the full singular eigenvalue decomposition (SVD) of matrix $\mathbf{U}_{(P+m-1)}$ described as follows:

$$\mathbf{U}_{(P+m-1)} = [\mathbf{U}^{(1)}, \mathbf{U}^{(2)}] \begin{bmatrix} \Sigma^{(1)} & \mathbf{0} \\ \mathbf{0} & \mathbf{0} \end{bmatrix} \begin{bmatrix} \mathbf{V}^{(1)} \\ \mathbf{V}^{(2)} \end{bmatrix}^H \tag{7.11}$$

Here, $\mathbf{U}^{(1)}$ and $\mathbf{U}^{(2)}$ are the $N \times (P + m - 1)$ and $N \times (N - (P + m - 1))$ orthogonal matrices that span the column space of matrix $\mathbf{U}_{(P+m-1)}$ and its orthogonal complement, respectively, i.e., $\mathbf{U}^{(2)^H}\mathbf{U}_{(P+m-1)} = \mathbf{0}$. Multiplying the observation vector $\mathbf{x}$ in (7.10) by $\mathbf{U}^{(2)^H}$, we come up with a simpler hypothesis testing problem, given by

$$\begin{cases} \mathscr{H}_{m-1} : \mathbf{r} = \mathbf{w} \\ \mathscr{H}_m : \mathbf{r} = \mathbf{d}_m\alpha_m + \mathbf{w} \end{cases} \tag{7.12}$$

where $\mathbf{r}$ is the $(N - (P + m - 1))$-dimensional column vector, the vectors $\mathbf{d}_m$ and $\mathbf{w}$ are defined as $\mathbf{U}^{(2)^H}\mathbf{s}_m$ and $\mathbf{U}^{(2)^H}\mathbf{n}$, respectively. Since $\mathbf{U}^{(2)^H}\mathbf{U}^{(2)} = \mathbf{I}_{(N-(P+m-1))}$ and $\mathbf{n}$ is zero-mean white Gaussian noise with the covariance matrix of $\sigma^2\mathbf{I}_N$, the transformed noise $\mathbf{w}$ is also zero-mean white Gaussian noise with the covariance matrix of $\sigma^2\mathbf{I}_{(N-(P+m-1))}$. Now, the GLR test based on the transformed observation data $\mathbf{r}$ can be rewritten as

*Note that $(m-1)$ is the number of interfering targets to be detected.

$$\Lambda_m(\mathbf{r}) = 2(\ell_m(\hat{\boldsymbol{\theta}};\mathbf{r}) - \ell_{m-1}(\hat{\boldsymbol{\theta}};\mathbf{r})) \underset{\mathcal{H}_{m-1}}{\overset{\mathcal{H}_m}{\gtrless}} \eta_m \tag{7.13}$$

with the new set of unknowns, given by

$$\Theta_m = \{(n_m, f_m), \alpha_m, \sigma^2\} \tag{7.14}$$

$$\Theta_{m-1} = \{\sigma^2\} \tag{7.15}$$

One can show that

$$\mathcal{H}_{m-1}: \begin{cases} \ell_{m-1}(\hat{\sigma}^2_{m-1};\mathbf{r}) = -(N-(P+m-1))\ln\left(\pi e \hat{\sigma}^2_{m-1}\right) \\ \hat{\sigma}^2_{m-1} = \dfrac{1}{N-(P+m-1)}\|\mathbf{r}\|^2 \end{cases} \tag{7.16}$$

$$\mathcal{H}_m: \begin{cases} \ell_m((\hat{n}_m, \hat{f}_m), \hat{\alpha}_m, \hat{\sigma}^2_m;\mathbf{r}) = -(N-(P+m-1))\ln\left(\pi e \hat{\sigma}^2_m\right) \\ \hat{\sigma}^2_m = \dfrac{1}{N-(P+m-1)}\left(\|\mathbf{r}\|^2 - \max_{(n_m, f_m)} \dfrac{|\mathbf{d}_m^H \mathbf{r}|^2}{\|\mathbf{d}_m\|^2}\right) \\ (\hat{n}_m, \hat{f}_m) = \arg\max_{(n_m, f_m)} \dfrac{|\mathbf{d}_m^H \mathbf{r}|^2}{\|\mathbf{d}_m\|^2} \end{cases} \tag{7.17}$$

where $\hat{\sigma}^2_{m-1}$ and $\hat{\sigma}^2_m$ are the MLEs of the noise variance σ^2 under $\mathcal{H}_{m-1}$ and $\mathcal{H}_m$, respectively. Here, $\hat{\alpha}_m$ and $(\hat{n}_m, \hat{f}_m)$ are the MLEs of the complex amplitude and delay-Doppler of the mth target. By substituting (7.16) and (7.17) into (7.13), the GLRT based on the observation data $\mathbf{x}$ can be obtained as

$$\Lambda_m(\mathbf{x}) = -2N\ln\left(1 - \max_{(n_m, f_m)} i_m(n_m, f_m)\right) \underset{\mathcal{H}_{m-1}}{\overset{\mathcal{H}_m}{\gtrless}} \eta_m \tag{7.18}$$

where

$$i_m(n,f) = \frac{|\mathbf{s}(n,f)^H \Pi^{\perp}_{\mathbf{U}_{(P+m-1)}} \mathbf{x}|^2}{\|\Pi^{\perp}_{\mathbf{U}_{(P+m-1)}} \mathbf{s}(n,f)\|^2 \|\Pi^{\perp}_{\mathbf{U}_{(P+m-1)}} \mathbf{x}\|^2} \tag{7.19}$$

where

$$\Pi^{\perp}_{\mathbf{U}_{(P+m-1)}} = \mathbf{I} - \mathbf{U}_{(P+m-1)}(\mathbf{U}^H_{(P+m-1)}\mathbf{U}_{(P+m-1)})^{-1}\mathbf{U}^H_{(P+m-1)} \tag{7.20}$$

Here, the interference matrix $\mathbf{U}_{(P+m-1)}$ is defined as $\mathbf{U}_{(P+m-1)} = [\mathbf{H}, \mathbf{T}_{m-1}]$, where $\mathbf{U}_{(P)} = \mathbf{H}^{\dagger}$. It is observed that the direct signal, clutter, and the previously detected targets, namely, interfering targets, are treated as interference into the matrix $\mathbf{U}_{(P+m-1)}$ in order to be removed, thus allowing the detection of the mth target (if it exists). In order to determine the value of the threshold η_m, it is shown in Reference 3 that

†Note that $(m-1)$ is the number of interfering targets to be detected.

$$\Upsilon_m(\mathbf{x}) = \frac{\dfrac{i_m(n_m,f_m)}{1}}{\dfrac{1-i_m(n_m,f_m)}{N-(P+m)}} \sim \begin{cases} F_{1,N-(P+m)} & \text{under } \mathscr{H}_{m-1} \\ F_{1,N-(P+m)}(\delta_m) & \text{under } \mathscr{H}_m \end{cases} \tag{7.21}$$

where (n_m,f_m) is the true delay-Doppler coordinate of mth target under test, and $\delta_m = \frac{|\alpha_m|^2}{\sigma^2}\|\Pi^{\perp}_{\mathbf{U}_{(P+m-1)}}\mathbf{s}(n_m,f_m)\|^2$. In order to give a desired integration gain in PR systems, N takes a large value, so $F_{1,N-(P+m)}$ and $F_{1,N-(P+m)}(\delta_m)$ converge towards the complex χ_1^2 and $\chi_1^2(\delta_m)$, respectively. Note that $i_m(n_m,f_m)$ has a value close to zero under the $\mathscr{H}_{m-1}$ hypothesis, so we can approximate $-2N\ln\left(1-\max_{(n_m,f_m)} i_m(n_m,f_m)\right)$ with $2N\max_{(n_m,f_m)} i_m(n_m,f_m)$. Hence, the GLRT of (7.18) is asymptotically equivalent to

$$\lambda_m(\mathbf{x}) = 2N\max_{(n_m,f_m)} i_m(n_m,f_m) \underset{\mathscr{H}_{m-1}}{\overset{\mathscr{H}_m}{\gtrless}} \eta_m \tag{7.22}$$

Note that, for decision-making with the statistic $\Upsilon_m(\mathbf{x})$, we compare $\Upsilon_m(\mathbf{x})$ against a threshold, denoted here by η. From (7.21) to (7.22), one can obtain that $\eta_m = \frac{N\eta}{N-(P+m)-\eta}$. Under the large values of N, which is typical in PR systems, this threshold may be approximated by η. As one can notice, the threshold η_m does not depend on m. This means that the threshold of the proposed detector does not depend on the maximum number of targets to be detected under hypothesis $\mathscr{H}_m$, say m. Thus, in the sequel, we denote η_m by η.

In essence, the proposed detector decides that a new target with respect to that of hypothesis $\mathscr{H}_{m-1}$ is present, say the mth target, if the peak value of $i_m(n_m,f_m)$ exceeds a threshold, and if so, the delay-Doppler coordinate of that peak is the MLE of the target delay-Doppler coordinate denoted by $(\hat{n}_m,\hat{f}_m)$. When (n_m,f_m) is known, the GLRT of (7.22) becomes

$$\lambda'_m(\mathbf{x}) = 2Ni_m(n_m,f_m) \underset{\mathscr{H}_{m-1}}{\overset{\mathscr{H}_m}{\gtrless}} \eta' \tag{7.23}$$

It is not difficult to show that, for large values of N, $2Ni_m(n_m,f_m) \sim \mathscr{X}_2^2$ and $2Ni_m(n_m,f_m) \sim \mathscr{X}_2^2(2\delta_m)$ under $\mathscr{H}_{m-1}$ and $\mathscr{H}_m$ hypotheses, respectively. In this case, the threshold η' is chosen to satisfy

$$p'_{fa} = Pr\left\{2Ni_m(n_m,f_m) > \eta' | \mathscr{H}_{m-1}\right\} = e^{-0.5\eta'} \tag{7.24}$$

It should be noted that p'_{fa} is the probability of false alarm if we examine known delay-Doppler bin (n_m,f_m). In the case of unknown (n_m,f_m) and for large values of N, the only difference is that the false alarm probability increases approximately linearly with the number of delay-Doppler coordinates searched over [4]. Henceforth, the corresponding threshold, η, based on test statistic $\lambda_m(\mathbf{x})$ is derived as

$$p_{fa} = \zeta D_d e^{-0.5\eta} \tag{7.25}$$

where D_d is the number of delay-Doppler coordinates searched over, and ζ is a value less than one. Generally speaking, the parameter ζ takes different values according to the characteristics of the signals of opportunity used over different CPIs. In the simulation section, we will discuss the selection of parameter ζ in more detail.

7.2.2 Forward and recursive implementation of BH-GLRT detector

To check the presence of targets, we begin by testing $\mathscr{H}_1$ against $\mathscr{H}_0$ using the BH-GLRT detector. If the result is accepted, meaning that $\lambda_1(\mathbf{x}) > \eta$, we move on to test $\mathscr{H}_2$ against $\mathscr{H}_1$ and continue testing the sequence of hypotheses $(\mathscr{H}_0, \mathscr{H}_1)$, $(\mathscr{H}_1, \mathscr{H}_2)$, ..., $(\mathscr{H}_{m-1}, \mathscr{H}_m)$, $(\mathscr{H}_m, \mathscr{H}_{m+1})$, ..., $(\mathscr{H}_M, \mathscr{H}_{M+1})$ in a forward direction until we obtain the first rejection, which is when $\lambda_m(\mathbf{x}) < \eta$. This process is known as a forward M-ary hypothesis testing strategy and is used to detect new targets. For instance, if the hypothesis $\mathscr{H}_m$ is accepted when testing $\mathscr{H}_m$ against $\mathscr{H}_{m-1}$, we declare a new target (the mth target) and move on to the next hypothesis to decide $\mathscr{H}_{m+1}$ against $\mathscr{H}_m$. The algorithm stops when the hypothesis $\mathscr{H}_{m-1}$ is accepted, indicating that no further targets can be identified. To reduce the computational complexity, we will implement this target-detection strategy recursively. To do this, first, (7.19) can be rewritten as

$$i_m(n,f) = \frac{|\mathbf{s}(n,f)^H \mathbf{x}^{\perp(P+m-1)}|^2}{\|\mathbf{p}\|^2 \|\mathbf{x}^{\perp(P+m-1)}\|^2} \tag{7.26}$$

where

$$\mathbf{p} = \Pi^{\perp}_{\mathbf{U}(P+m-1)} \mathbf{s}(n,f) \tag{7.27}$$

where $\mathbf{x}^{\perp(P+m-1)} = \Pi^{\perp}_{\mathbf{U}(P+m-1)} \mathbf{x}$ represents the component of $\mathbf{x}$ orthogonal to the space spanned by the columns of $\mathbf{U}_{(P+m-1)}$. Let us write the interference matrix $\mathbf{U}_{(P+m-1)}$ as $\mathbf{U}_{(P+m-1)} = [\mathbf{U}_{(P)}, \mathbf{u}_{P+1}, ..., \mathbf{u}_{P+m-1}]$ with $\mathbf{U}_{(P)} = \mathbf{H}$ and $\mathbf{u}_{P+i} = \mathbf{s}_i$. By using (when $m \geq 2$) [5]

$$\Pi^{\perp}_{\mathbf{U}(P+m-1)} = \Pi^{\perp}_{\mathbf{U}(P+m-2)} - \frac{\Pi^{\perp}_{\mathbf{U}(P+m-2)} \mathbf{u}_{P+m-1} \mathbf{u}^H_{P+m-1} \Pi^{\perp}_{\mathbf{U}(P+m-2)}}{\mathbf{u}^H_{P+m-1} \Pi^{\perp}_{\mathbf{U}(P+m-2)} \mathbf{u}_{P+m-1}} \tag{7.28}$$

where $\mathbf{U}_{(P+m-1)} = [\mathbf{U}_{(P+m-2)}, \mathbf{u}_{P+m-1}]$, we can recursively update $\mathbf{x}^{\perp(P+m-1)}$ as

$$\mathbf{x}^{\perp(P+m-1)} = \mathbf{x}^{\perp(P+m-2)} - \frac{\mathbf{u}^{\perp(P+m-2)}_{P+m-1} \mathbf{u}^{\perp(P+m-2)H}_{P+m-1} \mathbf{x}^{\perp(P+m-2)}}{\|\mathbf{u}^{\perp(P+m-2)}_{P+m-1}\|^2} \tag{7.29}$$

where $\mathbf{x}^{\perp(P)} = \mathbf{x} - \mathbf{U}_{(P)} \mathbf{R}_P \mathbf{U}^H_{(P)} \mathbf{x}$ with $\mathbf{R}_P = (\mathbf{U}^H_{(P)} \mathbf{U}_{(P)})^{-1}$, and

$$\mathbf{u}^{\perp(P+m-2)}_{P+m-1} = \Pi^{\perp}_{\mathbf{U}(P+m-2)} \mathbf{u}_{P+m-1} \tag{7.30}$$

represents the component of $\mathbf{u}_{P+m-1}$ orthogonal to the space spanned by the columns of $\mathbf{U}_{(P+m-2)}$. In a similar manner, this may be recursively updated by using (for $m \geq 3$)

$$\mathbf{u}_{P+m-1}^{\perp(P+m-2)} = \mathbf{u}_{P+m-1}^{\perp(P+m-3)} - \frac{\mathbf{u}_{P+m-2}^{\perp(P+m-3)} \mathbf{u}_{P+m-2}^{\perp(P+m-3)^H} \mathbf{u}_{P+m-1}^{\perp(P+m-3)}}{\|\mathbf{u}_{P+m-2}^{\perp(P+m-3)}\|^2} \tag{7.31}$$

but

$$\mathbf{u}_{P+m-1}^{\perp(P)} = \mathbf{u}_{P+m-1} - \mathbf{U}_{(P)}\mathbf{R}_P\mathbf{U}_{(P)}^H\mathbf{u}_{P+m-1} \tag{7.32}$$

In general, to update $\mathbf{u}_{P+m-1}^{\perp(P+k)}$, we use

$$\mathbf{u}_{P+m-1}^{\perp(P+k)} = \begin{cases} \mathbf{u}_{P+m-1} - \mathbf{U}_{(P)}\mathbf{R}_P\mathbf{U}_{(P)}^H\mathbf{u}_{P+m-1} & k = 0 \\ \mathbf{u}_{P+m-1}^{\perp(P+k-1)} - \dfrac{\mathbf{u}_{P+m-2}^{\perp(P+k-1)} \mathbf{u}_{P+m-2}^{\perp(P+k-1)^H} \mathbf{u}_{P+m-1}^{\perp(P+k-1)}}{\|\mathbf{u}_{P+m-2}^{\perp(P+k-1)}\|^2} & k = 1, \ldots, m-2 \end{cases} \tag{7.33}$$

Now, using (7.29) and (7.33), we can recursively update the binary GLR statistic of (7.26) when testing $\mathscr{H}_m$ against $\mathscr{H}_{m-1}$. Based on this, it is referred to as forward and recursive BH-GLRT(FR-BH-GLRT)-based algorithm in the sequel.

7.2.3 Parallel implementation of FR-BH-GLR detector

In PR systems, a large value of N is needed to achieve the desired integration gain. However, the direct use of the FR-BH-GLR detector, which processes the entire signal simultaneously, is not practical due to the extremely high number of observation samples. As a solution, we will demonstrate that the FR-BH-GLR detector can be implemented in parallel on platforms such as Compute Unified Device Architecture (CUDA) graphic cards. This will enable its efficient execution in PR systems. To do so, we can rewrite (7.26) as

$$i_m(n,f) = \frac{|\sum_{r=1}^{R} \mathbf{s}_r^H \mathbf{x}_r^{\perp(P+m-1)}|^2}{\sum_{r=1}^{R} \|\mathbf{p}_r\|^2 \sum_{r=1}^{R} \|\mathbf{x}_r^{\perp(P+m-1)}\|^2} \tag{7.34}$$

where, the N-dimensional vectors $\mathbf{s}$, $\mathbf{p}$, and $\mathbf{x}^{\perp(P+m-1)}$ are partitioned into R subvectors with dimension $L \times 1$ as[‡] $\mathbf{s} = [\mathbf{s}_1^T, \ldots, \mathbf{s}_R^T]^T$, $\mathbf{p} = [\mathbf{p}_1^T, \ldots, \mathbf{p}_R^T]^T$, and $\mathbf{x}^{\perp(P+m-1)} = [\mathbf{x}_1^{\perp(P+m-1)^T}, \ldots, \mathbf{x}_R^{\perp(P+m-1)^T}]^T$. Using (7.33), we can come up with

$$\mathbf{x}_r^{\perp(P+m-1)} = \mathbf{x}_r^{\perp(P+m-2)} - \frac{\sum_{i=1}^{R} [\mathbf{u}_{P+m-1}^{\perp(P+m-2)}]_i^H \mathbf{x}_i^{\perp(P+m-2)}}{\sum_{i=1}^{R} [\mathbf{u}_{P+m-1}^{\perp(P+m-2)}]_i^H [\mathbf{u}_{P+m-1}^{\perp(P+m-2)}]_i} [\mathbf{u}_{P+m-1}^{\perp(P+m-2)}]_r \tag{7.35}$$

but

[‡]N is considered an integer multiple of L.

$$\mathbf{x}_r^{\perp(P)} = \mathbf{x}_r - [\mathbf{U}_{(P)}]_r \mathbf{R}_P \sum_{i=1}^{R} [\mathbf{U}_{(P)}]_i^H \mathbf{x}_i \tag{7.36}$$

where $\mathbf{x} = [\mathbf{x}_1^T, \ldots, \mathbf{x}_R^T]^T$, $\mathbf{u}_{P+m-1}^{\perp(P+m-2)} = \left[[\mathbf{u}_{P+m-1}^{\perp(P+m-2)}]_1^T, \ldots, [\mathbf{u}_{P+m-1}^{\perp(P+m-2)}]_R^T\right]^T$ and $[\mathbf{U}_{(P)}]_r$ is the rth submatrix with dimension $L \times P$ of matrix $\mathbf{U}_{(P)}$, i.e.,

$$\mathbf{U}_{(P)} = \begin{bmatrix} [\mathbf{U}_{(P)}]_1 \\ [\mathbf{U}_{(P)}]_2 \\ \vdots \\ [\mathbf{U}_{(P)}]_R \end{bmatrix} \tag{7.37}$$

Next, we use (7.33) to find

$$[\mathbf{u}_{P+m-1}^{\perp(P+k)}]_r = \begin{cases} [\mathbf{u}_{P+m-1}]_r - [\mathbf{U}_{(P)}]_r \mathbf{R}_P \sum_{i=1}^{R} [\mathbf{U}_{(P)}]_i^H [\mathbf{u}_{P+m-1}]_i \\ \qquad\qquad k = 0 \\ [\mathbf{u}_{P+m-1}^{\perp(P+k-1)}]_r - \dfrac{\sum_{i=1}^{R} [\mathbf{u}_{P+m-2}^{\perp(P+k-1)}]_i^H [\mathbf{u}_{P+m-1}^{\perp(P+k-1)}]_i}{\sum_{i=1}^{R} [\mathbf{u}_{P+m-2}^{\perp(P+k-1)}]_i^H [\mathbf{u}_{P+m-2}^{\perp(P+k-1)}]_i} [\mathbf{u}_{P+m-2}^{\perp(P+k-1)}]_r, \\ \qquad\qquad k = 1, \ldots, m-2 \end{cases} \tag{7.38}$$

where the vectors $\mathbf{u}_{P+m-j}^{\perp(P+k-q)}$ and $\mathbf{u}_{P+m-1}$ may be presented in the block form of

$$\mathbf{u}_{P+m-j}^{\perp(P+k-q)} = \left[\left[\mathbf{u}_{P+m-j}^{\perp(P+k-q)}\right]_1^T, \ldots, \left[\mathbf{u}_{P+m-j}^{\perp(P+k-q)}\right]_R^T\right]^T \tag{7.39}$$

$$\mathbf{u}_{P+m-1} = \left[\left[\mathbf{u}_{P+m-1}\right]_1^T, \ldots, \left[\mathbf{u}_{P+m-1}\right]_R^T\right]^T \tag{7.40}$$

From the representation of (7.34), it is clear that all vectors are partitioned into R non-overlapping sequences of length L, and the total length of each vector is equal to $N = RL$. In a similar way, matrix $\mathbf{U}_{(P)}$ is partitioned into R non-overlapping row matrices of dimension $L \times P$. As a result, it is possible to execute calculations for each segment on a separate processor. Afterward, the results of each segment are summed in one of the processors. Hence, the new binary hypothesis detector is a parallel implementation (PI) of FR-BH-GLR detector, named PI-FR-BH-GLRT.

7.2.4 FFT-based implementation of PI-FR-BH-GLR detector

Since the delay-Doppler coordinates of interested targets are unknown, we need to compute the test statistic $i_m(n, f)$ for a desired *delay-Doppler map*, in which $n = 1, \ldots, N_d$ with N_d being the time delay index corresponding to the maximum relative bistatic range of interest, and $|f| < f_{\max}$ with $f_{\max}$ being the maximum Doppler frequency of the interested targets. In practice, this can be implemented very efficiently by fast Fourier transform (FFT) as follows:

$$\mathbf{i}_a(\mathbf{x}^{\perp(P+m-1)}, n) = \frac{\mathbf{f}_n(\mathbf{x}^{\perp(P+m-1)}) \odot \mathbf{f}_n(\mathbf{x}^{\perp(P+m-1)})^*}{\sum_{r=1}^{R} \|\mathbf{y}_r^{(n)}\|^2 \sum_{r=1}^{R} \|\mathbf{x}_r^{\perp(P+m-1)}\|^2} \tag{7.41}$$

where $\mathbf{y}^{(n)} = [\mathbf{y}_1^{(n)^T}, \ldots, \mathbf{y}_R^{(n)^T}]^T$, and $\mathbf{f}_n(\mathbf{x}^{\perp(P+m-1)}) = \mathrm{FFT}\left(\mathbf{y}^{(n)^*} \odot \mathbf{x}^{\perp(P+m-1)}, N_f\right)$ denotes an N_f-point FFT of the vector $\mathbf{y}^{(n)^*} \odot \mathbf{x}^{\perp(P+m-1)}$. As a result, we can define matrix $\mathbf{I}_a(\mathbf{x}^{\perp(P+m-1)}) = [\mathbf{i}_a(\mathbf{x}^{\perp(P+m-1)}, 1), \ldots, \mathbf{i}_a(\mathbf{x}^{\perp(P+m-1)}, N_d)]$ with dimension $N_f \times N_d$ referred to as a range-Doppler map with N_f Doppler bins and N_d range bins. In this case, the kth Doppler bin corresponds to the Doppler frequency of $f_k = -\frac{f_s}{2} + (k-1)\frac{f_s}{N_f}$ for $k = 1, \ldots, N_f$, and the relative bistatic range corresponding to the nth range bin is $R_n = \frac{cn}{f_s}$, where c is the speed of light. In the denominator of (7.37), we also approximate vector $\mathbf{p}$ with $\mathbf{y}^{(n)}$ in order to further reduce the computational complexity of the proposed detection algorithm. In this case, subscript "a" is used to emphasize on this approximation, which is a reasonable approximate for target detection as shown in Chapter 3.

In actual practice, the Doppler frequencies of the targets of interest are much lower than the sampling frequency (f_s), so it is possible to employ a decimation technique to reduce the excessive processing load of calculating the N_f-point FFT with almost no loss in signal processing gain [6]. The cascaded integrator comb (CIC) is a very efficient implementation of a decimation filter [7, 8]. After applying a CIC filter, the decimated signal is low-pass filtered in order to remove the signal components out of the desired frequency band. Finally, the N_f-point FFT can be replaced by the N_F-point FFT, where $N_F = \frac{N_f}{D}$ and $D \geq 1$ is a decimation factor described in Reference 6. As such, the number of range-Doppler bins of the matrix $\mathbf{I}_a(\mathbf{x}^{\perp(P+m-1)})$ to be examined may be reduced to $D_d = N_F N_d$ rather than $D_d = N_f N_d$ [see (7.25)].

The proposed detection algorithm steps can be summarized as:

1. **Initialization:**
   - Set threshold η as (22), $m = 0$, and the partition vector $\mathbf{x}$ as $\mathbf{x} = [\mathbf{x}_1^T, \ldots, \mathbf{x}_R^T]^T$.
2. **Cancellation of direct signal and clutter:**
   - Calculate $\mathbf{R}_P = (\mathbf{U}_{(P)}{}^H \mathbf{U}_{(P)})^{-1}$, where $\mathbf{U}_{(P)} = \mathbf{H}$;
   - Calculate $\mathbf{x}_r^{\perp(P)} = \mathbf{x}_r - [\mathbf{U}_{(P)}]_r \mathbf{R}_P \sum_{i=1}^{R} [\mathbf{U}_{(P)}]_i{}^H \mathbf{x}_i$ for $r = 1, \ldots, R$.
3. **Successive target detection:** (if $\max_{n,k} \mathbf{I}_a(\mathbf{x}^{\perp(P+m)}) > \eta$ do)
   - Increase m by 1;
   - Find and save the MLE of delay-Doppler coordinate of the mth target, say (n_m, f_m), which is the location of the maximum of $\mathbf{I}_a(\mathbf{X}^{\perp(P+m)})$;
   - Construct $\mathbf{u}_{P+m} = (\mathbf{y}^{(nm)} \odot \mathbf{e}^{(fm)})$, and partition it as $\mathbf{u}_{P+m} = \left[[\mathbf{u}_{P+m}]_1^T, \ldots, [\mathbf{u}_{P+m}]_R^T\right]^T$,
   - Calculate $[\mathbf{u}_{P+m}^{\perp(P)}]_r = [\mathbf{u}_{P+m}]_r - [\mathbf{U}_{(P)}]_r \mathbf{R}_P \sum_{i=1}^{R} [\mathbf{U}_{(P)}]_i^H [\mathbf{u}_{P+m}]_i$
   - Calculate $[\mathbf{u}_{P+m}^{\perp(P+k)}]_r = [\mathbf{u}_{P+m}^{\perp(P+k-1)}]_r - \frac{\sum_{i=1}^{R} [\mathbf{u}_{P+m-1}^{\perp(P+k-1)}]_i^H [\mathbf{u}_{P+m}^{\perp(P+k-1)}]_i}{\sum_{i=1}^{R} [\mathbf{u}_{P+m-1}^{\perp(P+k-1)}]_i^H [\mathbf{u}_{P+m-1}^{\perp(P+k-1)}]_i} [\mathbf{u}_{P+m-1}^{\perp(P+k-1)}]_r$ for $k = 1 : m-1$,

- Calculate $\mathbf{x}_r^{\perp(P+m)} = \mathbf{x}_r^{\perp(P+m-1)} - \frac{\sum_{i=1}^{R} [\mathbf{u}_{P+m}^{\perp(P+m-1)}]_i^H \mathbf{x}_i^{\perp(P+m-1)}}{\sum_{i=1}^{R} [\mathbf{u}_{P+m}^{\perp(P+m-2)}]_i^H [\mathbf{u}_{P+m}^{\perp(P+m-1)}]_i} [\mathbf{u}_{P+m}^{\perp(P+m-1)}]_r,$ where $r = 1, \ldots, R$ and $\mathbf{X}^{\perp(P+m)} = [\mathbf{X}_1^{\perp(P+m)^T}, \ldots, \mathbf{X}_R^{\perp(P+m)^T}]^T$.

The algorithm proposed, which is based on PI-FR-BH-GLRT, indicates that if a target is identified, it will be considered as an interfering target in the following stage of target detection and must be eliminated in order to detect new targets. Once the PI-FR-BH-GLRT-based algorithm is completed, the number of targets is detected, say M, and their positions in both delay and Doppler are determined.

7.3 Analytical performance evaluation

The detection probability of the mth target can be computed from

$$p_d^{(m)'} = Pr\left\{2Ni_m(n_m, f_m) > \eta' | \mathcal{H}_m\right\} \tag{7.42}$$

Here, it can be shown that $2Ni_m(n_m, f_m) \sim \mathcal{X}_2^2$ or $Ni_m(n_m, f_m) \sim \chi_1^2$, resulting in

$$p_d^{(m)'} = e^{-(\delta_m + \frac{\eta'}{2})} \sum_{k=0}^{\infty} \left(\frac{\delta_m^k}{k!} \sum_{n=0}^{k} \frac{\left[\frac{\eta'}{2}\right]^n}{n!} \right) \tag{7.43}$$

where $\delta_m = \frac{|\alpha_m|^2}{\sigma^2} \| \Pi_{\mathbf{U}(P+m-1)}^{\perp} \mathbf{s}(n_m, f_m)\|^2$. Making use of (7.25), we finally find

$$p_d^{(m)} = e^{-\left(\delta_m + \ln\left(\frac{\zeta D_d}{P_{fa}}\right)\right)} \sum_{k=0}^{\infty} \left(\frac{\delta_m^k}{k!} \sum_{l=0}^{k} \frac{\left[\ln\left(\frac{\zeta D_d}{P_{fa}}\right)\right]^l}{l!} \right) \tag{7.44}$$

It is easy to show that the detection probability of the mth target is an increasing function of δ_m. Now, if we consider $\text{SNR}_m = \frac{|\alpha_m|^2}{\sigma^2}$ as the input SNR of the mth target, then $G_m = \| \Pi_{\mathbf{U}(P+m-1)}^{\perp} \mathbf{s}(n_m, f_m)\|^2$ can be regarded as its SNR gain. Since $\Pi_{\mathbf{U}(P+m-1)}^{\perp}$ is an idempotent matrix, it can be concluded that $0 \leq G_m \leq \|\mathbf{s}(n_m, f_m)\|^2$. This implies that the mth target may experience a big loss when its delay-Doppler coordinate is enough near to that of the interference signals (i.e., clutter or previously detected targets). In contrast, the maximum $\|\mathbf{s}(n_m, f_m)\|^2$, the so-called maximum integration gain, can be achieved when the delay-Doppler coordinate of the mth target is far from those of the interference signals. This inspires us to define detection loss (DL), experienced by the mth target, as

$$\text{DL}_m = \frac{\| \Pi_{\mathbf{U}(P+m-1)}^{\perp} \mathbf{s}(n_m, f_m)\|^2}{\|\mathbf{s}(n_m, f_m)\|^2} \tag{7.45}$$

Note that DL_m depends on the target under test position denoted by (n_m, f_m), the positions of previously detected targets, and the characteristics of opportunity signal used for PR detection.

7.4 Simulation results

In this section, first we evaluate the performance of the proposed PI-FR-BH-GLRT-based algorithm when exploited for both FM- and DVB-T-based bistatic PRs. Then, the detection performance of the proposed method is compared with other existing methods. For the FM-based bistatic PR, we consider an FM signal with 100 MHz carrier frequency, 100 kHz transmitter bandwidth, and 200 kHz sampling frequency for both the reference and surveillance channels. For DVB-T signal, we consider 8 k DVB-T signal with 600 MHz carrier frequency, 8 MHz bandwidth, 9.1429 MHz sampling frequency, code rate of 0.666, constellation of "64-QAM," and a cyclic prefix ratio of 1:8. For FM- and DVB-T based PR systems, an integration time of 0.8 seconds and 28 ms is considered, respectively. Other parameters such as R, N_f, N_F, D, N_d, and D_d are set, respectively, equal to 32, 524,288, 2,048, 256, 60, and 56,242 for the FM signal, while they are 32, 524,288, 256, 2,048, 2,743, and 373,184 for the DVB-T signal. Also, to see the effect of time-varying nature of FM signals on threshold setting and target detection performance, we consider two FM signals named as FM S_1 and S_2 signals with auto-ambiguity functions as shown in Figure 7.1.

In our simulations, we consider direct signals with input direct signal-to-noise ratio (DNR_i) of about 60. Also, ten clutter spikes with zero Doppler and relative bistatic ranges between 0 and 55 km are considered with different input clutter-to-noise ratios (CNR_i) over the interval 5 dB $\leq CNR_i \leq$ 45 dB. Note that the clutter cancellation capability of the detector of this chapter is similar to that of Chapter 3.

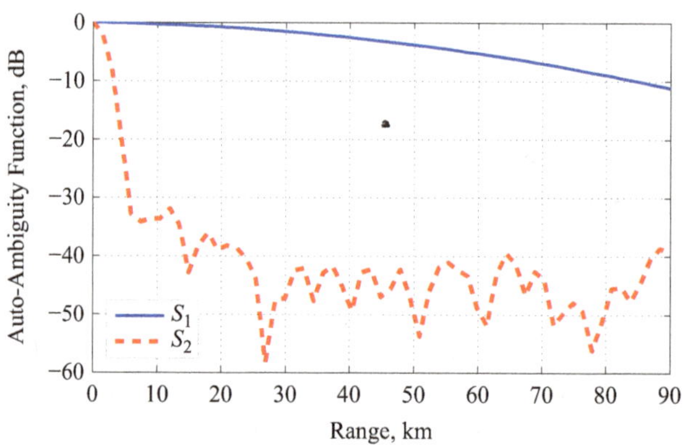

Figure 7.1 Auto-ambiguity function at zero velocity for the FM S_1 and S_2 signals corresponding to the silence and music voice track of an FM radio waveform, respectively

Thus, the interested reader is referred to Chapter 3 for more performance curves, analyses, and discussions.

7.4.1 The importance of optimal detectors for passive radar

As previously mentioned, passive radar uses existing broadcast or communication signals that were not specifically designed for radar sensing. Therefore, the focus in passive radar is on designing optimal signal processing algorithms for the receiver. Fortunately, modern digital transmitters for communications offer waveforms that are more suitable for radar use compared to those from a decade ago. Of these, it appears that an Orthogonal frequency division multiplexing (OFDM)-based waveform may be more effective for target detection than other radar signals, such as FM and analog TV (ATV). However, this is still not enough, especially when conducting target detection tasks in scenarios involving multiple targets. To demonstrate this, a simplified version of the proposed GLRT detector of this chapter is considered. Assume that the GLRT for target detection at hypothesized delay-Doppler (τ_H, f_H) is defined as

$$Ni_H(\tau_H, f_H) = N\frac{|\mathbf{s}^H(\tau_H, f_H)\mathbf{x}|^2}{\|\mathbf{s}(\tau_H, f_H)\|^2\|\mathbf{x}\|^2} \underset{\mathscr{H}_0}{\overset{\mathscr{H}_1}{\gtrless}} -\ln p_{fa} \tag{7.46}$$

This can be considered as the conventional detector of active radars especially when an optimal waveform is exploited for target detection purposes, and p_{fa} can be regarded as the desired value of false alarm probability.

Figure 7.2 shows only the portions of the statistic $Ni_H(n, f)$ that are greater than $-\ln p_{fa}$. In this simulation, we consider a desired target located at the range-velocity coordinate of (10 km, 65 m/s) with a signal-to-noise ratio of 0 dB. From the one-level contour plots shown in Figure 7.2, several remarks are now in order. For FM signal S_1, it is seen that high sidelobe level and poor range resolution of this signal can severely hamper target detection. For the DVB-T signal, several ambiguities, resulting from the pilot structure of the DVB-T signal, are seen aside from the large peak at the range-velocity coordinate of (10 km, 65 m/s). These regularly spaced undesired peaks might strongly limit the target detection capability of a DVB-T-based passive radar system and result in an excessive false alarm probability. The main idea of this examination is that the performance of passive radar detection is dependent on the transmitted waveform, which is not under the control of the passive radar designer. Consequently, even in a single-target scenario, using the statistical measure $Ni_H(\tau_H, f_H)$ with a threshold of $-\ln p_{fa}$ for target detection can lead to an excessively high false alarm probability due to inherent features of the communication waveform. In multitarget scenarios, if we continue to rely on detectors designed for active radars, weak targets may be obscured by strong ones. Therefore, the authors' concerns about the development of optimal detectors specifically designed for passive radar applications are justified.

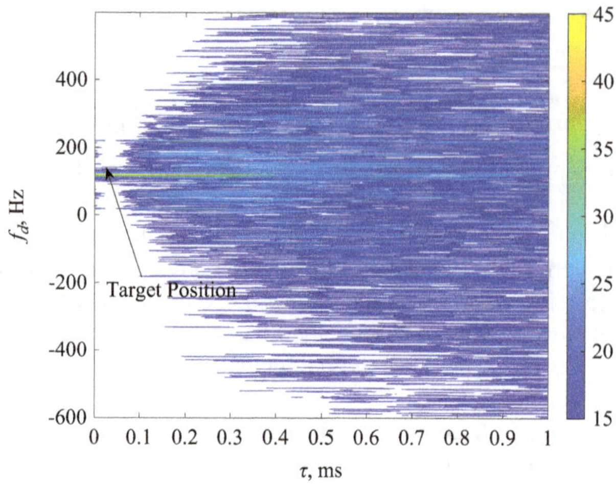

(a) FM signal (S_1)

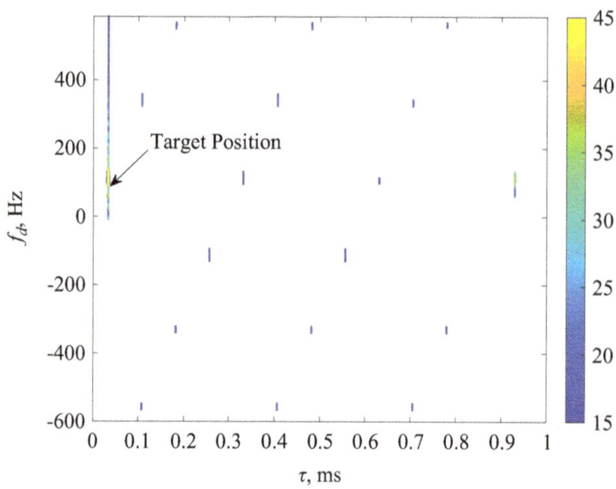

(b) DVB-T signal

Figure 7.2 *Bistatic range-Doppler map associated with the statistic* $Ni_H(n, f)$ *for a single-target scenario, where only portions greater than* $-\ln p_{fa}$ *for* $p_{fa} = 10^{-6}$ *are included*

7.4.2 Performance of the proposed detection algorithm

Extensive simulations are provided here to demonstrate the efficiency of the proposed detection algorithm for both FM and DVB-T passive radars. First, to demonstrate the validity of the asymptotic theoretical results presented in section 7.2, Figure 7.3 presents the estimated false alarm probability of the proposed detector against the predefined false alarm probability p_{fa} for different values of parameter ζ. The estimated false alarm probability, denoted here by P_{fa} is obtained from 10^8 Monte

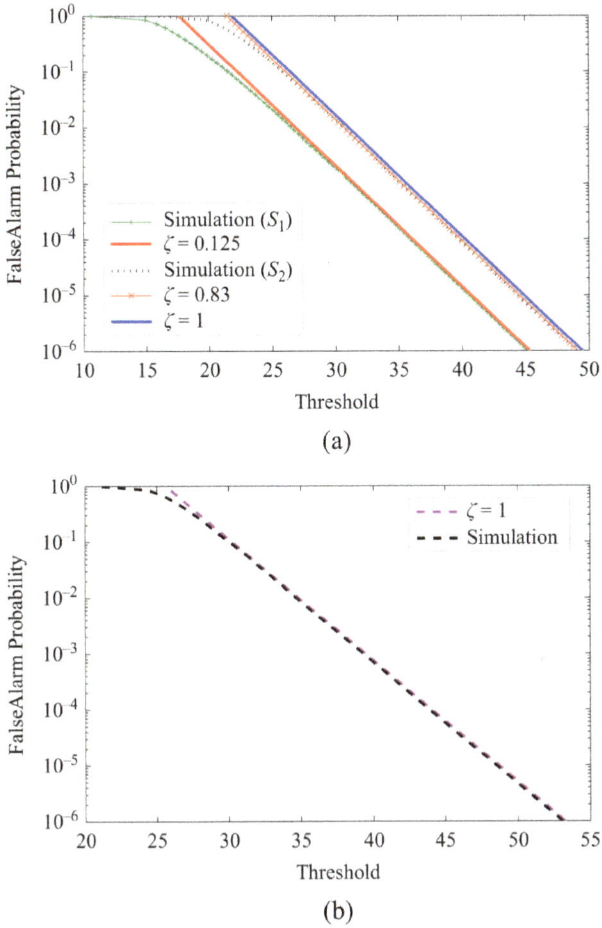

Figure 7.3 False alarm probability as a function of detection threshold to verify analytical solution (7.25) for (a) FM signals of S_1 and S_2 (b) DVB-T signal

Carlo simulation runs. We can observe that P_{fa} is very close to the asymptotic ones calculated from (7.25) when $P_{fa} < 10^{-2}$. For these results, we use $D_d = N_F N_d$, and the parameter ζ in (7.25) is set equal to 0.125 and 0.83 when using different FM S_1 and S_2 signals, respectively. In reality, since the transmitted waveforms are not within the control of the radar designer, it makes sense to calculate the threshold η based on the closed-form solution derived in (7.25) with $\zeta = 0.83$. As such, we could claim that we have designed a detector with the false alarm probability of level p_{fa} instead of size p_{fa}. Generally, a detector is said to have a false alarm probability of level (size) p_{fa} when the empirical false alarm probability (P_{fa}) satisfies $P_{fa} \leq p_{fa}$ ($P_{fa} = p_{fa}$) [9]. This means that, when using FM-based PRs with waveforms that change over time, it is challenging to create a constant false alarm rate (CFAR) detector that takes into account the opportunity signals received during different coherent processing intervals (CPIs). However, it is possible to create a detector with a controlled false alarm probability. Based on this finding, in the following paragraphs, the threshold η is set equal to 49.05 based on (7.25) with $\zeta = 0.83$ to attain $P_{fa} \leq 10^{-6}$, i.e., the proposed detector has a level of 10^{-6} for the FM signal S_1, while it has a size of 10^{-6} for the FM signal S_2, as shown in Figure 7.3. For DVB-T signal, we demonstrate both the asymptotic and the estimated false alarm probability in Figure 7.3(b). This figure supports our argument on the close relationship between asymptotic and actual false alarm probabilities of the proposed detector with parameter $\zeta = 1$ for $P_{fa} < 10^{-1}$. Due to the noise like and approximately the time-invariant nature of DVB-T signals, we can say that the proposed detector has CFAR property with respect to the different opportunity signals received over different CPIs of a DVB-T-based PR system.

In the sequel, we investigate the accuracy of the analytic formula for the detection probability when using the FM and DVB-T signals through numerical examples for $p_{fa} = 10^{-6}$. In this case, the numerical results are obtained from 10^4 Monte Carlo simulation trials. In this case, we demonstrate both the asymptotic and numerical detection probability of the proposed detector for the FM and DVB-T signals in Figure 7.4(a) and (b), respectively. Figure 7.4 shows the detection probability as a function of $\frac{|\alpha|^2}{\rho}$ with ρ being the noise power per unit bandwidth, set it equal to -197 dB/Hz for both FM- and DVB-T-based PRs. It can be seen that the estimated detection probability of the proposed test is very close to the asymptotic one, denoted as $p_d^{(1)}$ and calculated from (7.40) with $\mathbf{U}_{(P)} = \mathbf{H}$, in a single-target scenario (STS). By comparing the results of Figure 7.4(a) and (b), it is concluded that an FM-based PR system offers better detection performance (about 14.7 dB) as compared to that of DVB-T-based PR system thanks to the higher integration time used for FM-based PR system. It should be noted that, in the single-target scenario, the detection performance result using the FM S_1 signal is the same as that of the FM S_2 signal.

To examine the ability of the proposed algorithm in multitarget scenario (MTS), we define a scenario including five targets with characteristics listed in Table 7.1. In this multitarget scenario, the performance of the PI-FR-BH-GLRT algorithm for the FM and DVB-T signals are shown in Figures 7.5 and 7.6. In these figures, symbols + and □ are used to indicate the original target positions and estimate of target positions by the PI-FR-BH-GLRT algorithm, respectively. Using the FM S_1 signal, as shown in Figure 7.5(a), the targets T_2, T_3, T_4, and T_5 are correctly resolved and positioned. In this scenario, the targets T_1 and T_2 are placed at the same Doppler frequency, but in different relative bistatic ranges. We see that the poor range resolution of the considered FM signal yields into a big loss for detecting target T_1 in the presence of the target T_2 with higher power as compared to the target T_1. For the FM S_2 signal, we can observe that all the targets are correctly resolved and positioned as shown in Figure 7.5(b). From Figure 7.6, it is observed that for the DVB-T signal, the targets T_1, T_2, T_4, and T_5 are correctly resolved and positioned. In this case, poor frequency resolution of the DVB-T signal as compared to that of

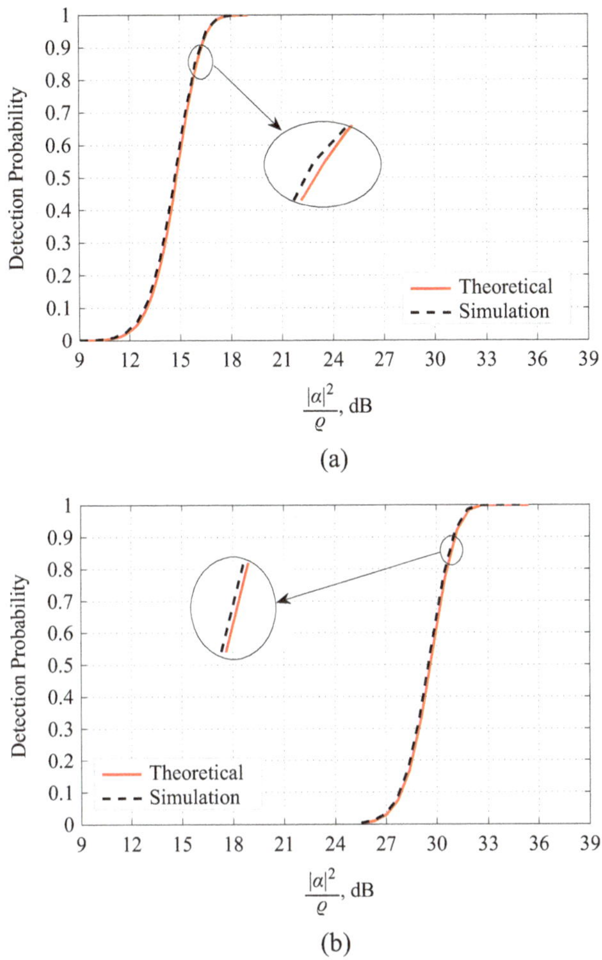

Figure 7.4 Detection performance evaluation of the proposed detector in a single-target scenario for (a) an FM signal and (b) DVB-T signal when $p_{fa} = 10^{-6}$

Table 7.1 Targets characteristics in the considered multitarget scenario

Targets	T_1	T_2	T_3	T_4	T_5
Bistatic range (km)	10	13	45	45	54.79
Bistatic velocity (m/s)	91.55	91.55	−117.7	−122.07	135.15
SNR_i (dB)	−27	−3	−28	−8	−17

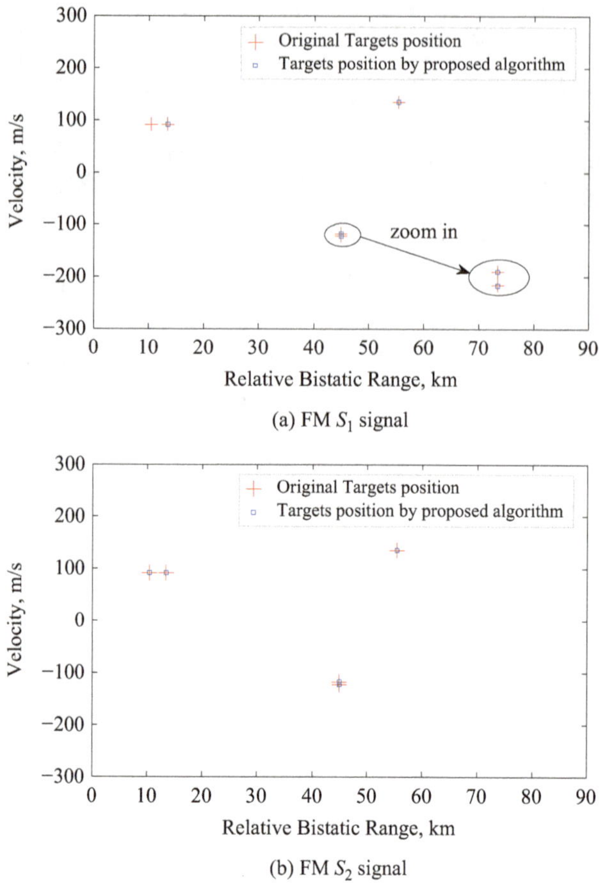

Figure 7.5 Results of the FM-based PI-FR-BH-GLRT-based detection algorithm when $p_{fa} = 10^{-6}$

FM signal leads to the missing of the target T_3, which is in the same relative bistatic range as T_4, but with lower power as compared to the target T_4.

In order to get further insight into target detection using the FM and DVB-T signals, we depict the detection probability $p_d^{(m)}$ for $m = 1, ..., 5$ as a function of $\frac{|\alpha|^2}{\rho}$ for all the targets in the considered multitarget scenario. To do this, we change the value of $\frac{|\alpha|^2}{\rho}$ corresponding to the mth testing target, say T_m, in the presence of the other targets known as interfering targets, whose $\frac{|\alpha|^2}{\rho}$ values and their positions are listed in Table 7.1. For this case, input SNR can be defined as $\frac{|\alpha|^2}{\rho}$ divided by the sampling frequency f_s, i.e., $\mathrm{SNR}_i = \frac{|\alpha|^2}{\rho f_s}$. The results of this simulation for FM and DVB-T signals are reported in Figures 7.7 and 7.8, respectively. From Figure 7.7, it is seen that different targets T_m for $m = 1, ..., 5$ have not the same performance due to the different characteristic of the considered FM signals. For the FM S_1 signal,

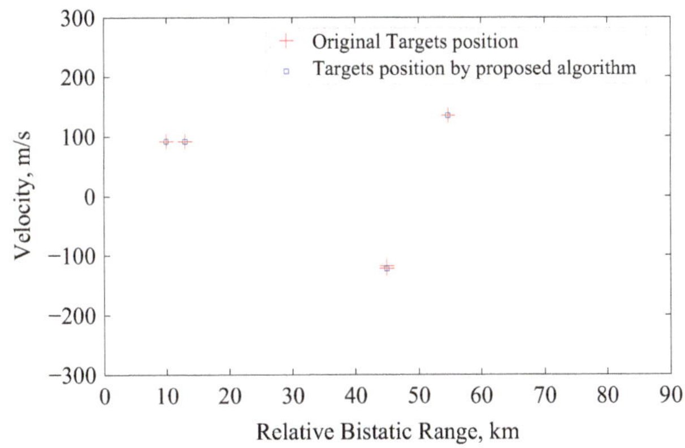

Figure 7.6 Results of the DVB-T-based PI-FR-BH-GLRT-based detection algorithm when $p_{fa} = 10^{-6}$

the targets T_3, T_4, and T_5 follow the performance of the proposed detector in the single-target scenario, while the target $T_1(T_2)$ experiences a performance loss of about 17.6 dB in the presence of the target $T_2(T_1)$ for the detection probability of 0.9 when compared with the results of single-target scenario shown in Figure 7.4(a). Using the FM signal S_2 reduces this loss to about 3 dB because of the better signal characteristic of the FM S_2 signal as compared to that of S_1 signal (see Figure 7.1). In the case of the DVB-T signal, as shown in Figure 7.8, the targets T_1, T_2, and T_5 follow the performance of the proposed detector in the single-target scenario, whereas target $T_3(T_4)$ experiences a performance loss of about 2.5 dB in the presence of the target $T_4(T_3)$. To explain this, it is interesting to evaluate the DL experienced by different targets as defined in (7.41). To do so, for the FM S_1 signal, we plot DL versus the relative bistatic range as shown in Figure 7.9, while for the DVB-T signal, we depict DL versus the bistatic velocity as shown in Figure 7.10. In both cases, we assume that an interfering target is placed in the zero relative bistatic range and the zero velocity for convenience. Generally speaking, the detection loss experienced by the *m*th target depends on the delay-Doppler coordinate of the *m*th target, the characteristics of the opportunity signal used for detection, and the delay-Doppler coordinates of the targets detected in previous steps of the algorithm. Thus, the curve of DL versus bistatic range (Doppler) can be referred to as range-dimension loss profile (Doppler-dimension loss profile) and may be used for justifying the losses experienced by different targets shown in Figure 7.7.

7.4.3 A comparative evaluation of passive radar detection algorithms

In Reference 10, a multistage processing algorithm for disturbance removal and strong targets detection based on projections of the received signal in a subspace orthogonal to both the disturbance and previously detected targets was presented. Then, a cell-averaging CFAR detector was applied to the cross-ambiguity function (CAF) of the cleaned signal to detect weak targets. Here, the detector proposed in Reference 10

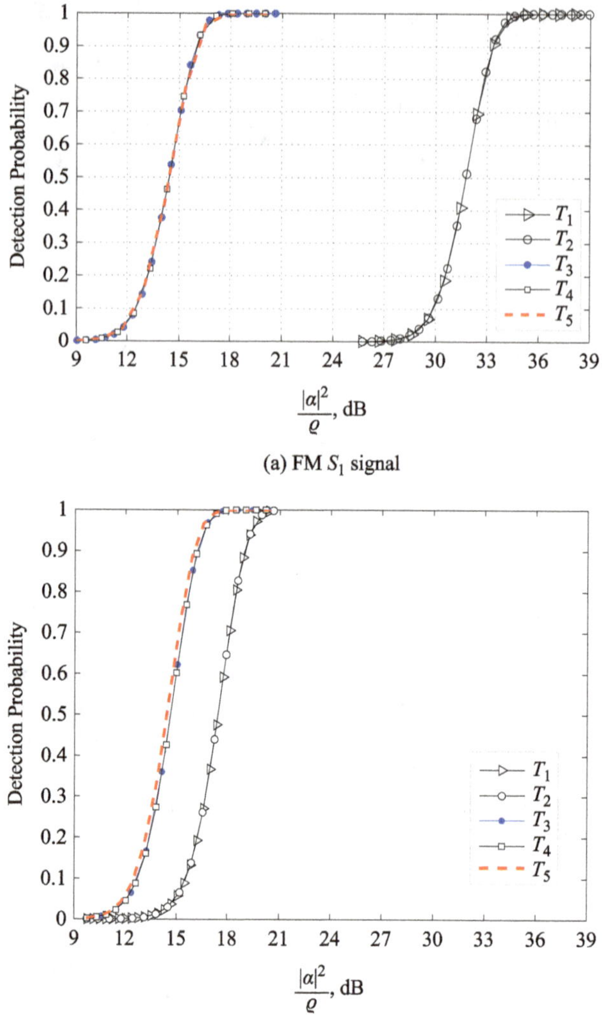

(b) FM S_2 signal

Figure 7.7 Detection performance evaluation of the PI-FR-BH-GLRT-based algorithm when the FM signals S_1 and S_2 are exploited for detection and $p_{fa} = 10^{-6}$

is referred to as the multistage processing (MP) and cell-averaging (CA) detection algorithm, abbreviated as MP-CA detector. Throughout the chapter, the CA detector is configured with a total number of training bins equal to 120, and two guard cells on each side of the test cell are also used to prevent from the self-masking effect [10]. Other parameters of the MP-CA detector are the same as those used in Reference 10 (e.g., $\varepsilon = 3$ dB and $\eta = 5$ dB). The detection performance of the MP-CA detector when using the FM S_2 signal is presented in Figure 7.11. By comparing Figures 7.7(b) and 7.11,

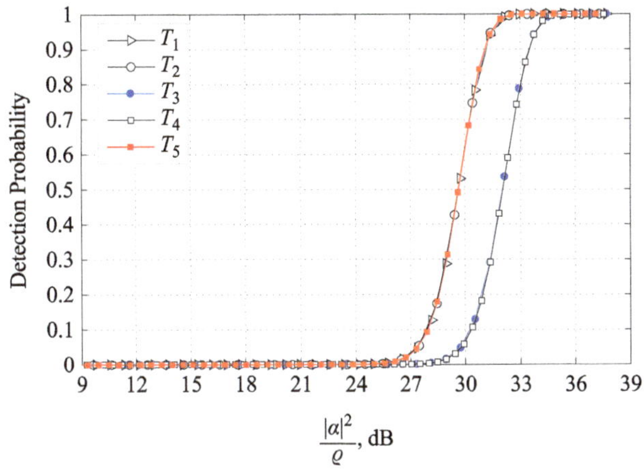

Figure 7.8 Detection performance evaluation of the PI-FR-BH-GLRT-based algorithm when the DVB-T is exploited for detection and $p_{fa} = 10^{-6}$

for the FM S_2 signal, we can observe that the detection performance of the proposed algorithm is superior to that of the MP-CA detector for all the targets in the considered scenario. Our simulation results show that the disturbance removal part of the MP-CA detector cannot detect target T_1 (T_3) when the target under test is T_2 (T_4), i.e., the MP-CA detector was proposed to only detect and remove strong targets (targets with high-input SNR). As a result, the interfering target T_1 (T_3) inevitably increases the threshold of the CA-CFAR detector for detecting the target T_2 (T_4); hence, a severe degradation of the MP-CA detector is seen. This target detection issue is generally known as the capture effect [11, 12], whereas our proposed algorithm does not have such problems. In this case, the MP-CA detector experiences an additional loss of about 9.7, 8.5, 16.1, and 4.5 dB

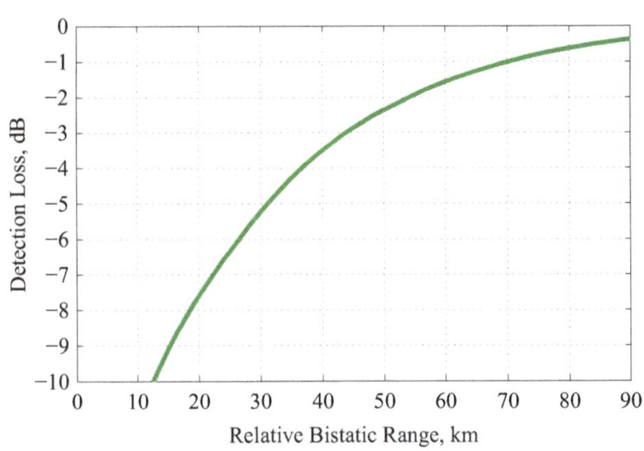

Figure 7.9 Detection loss as a function of relative bistatic range at the zero Doppler when the FM signal S_1 *is used for detection*

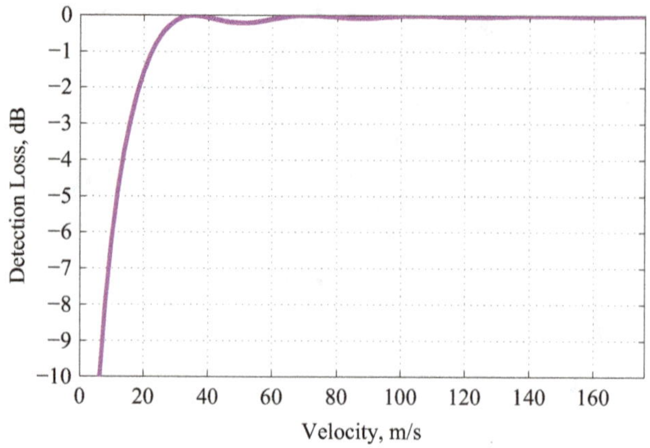

Figure 7.10 Detection loss as a function of bistatic Doppler frequency when the DVB-T signal is used for detection

as compared with that of the proposed detector for detecting targets T_2 to T_5 at detection probability of 0.9 and $p_{fa} = 10^{-6}$, respectively. Also, the MP-CA detector is unable to detect the target T_1, such as we obtain $p_d^{(1)} = 0$ for $\frac{|\alpha|^2}{\rho}$ values of between 9 and 39 dB. For the FM S_1 signal with the poor range resolution, the detection probabilities of zero have been obtained for all the targets in the considered scenario for $\frac{|\alpha|^2}{\rho}$ values between 9 and 39 dB when using the MP-CA detector. This significant detection loss of the MP-CA detector is related not only to the self-masking effect due to the

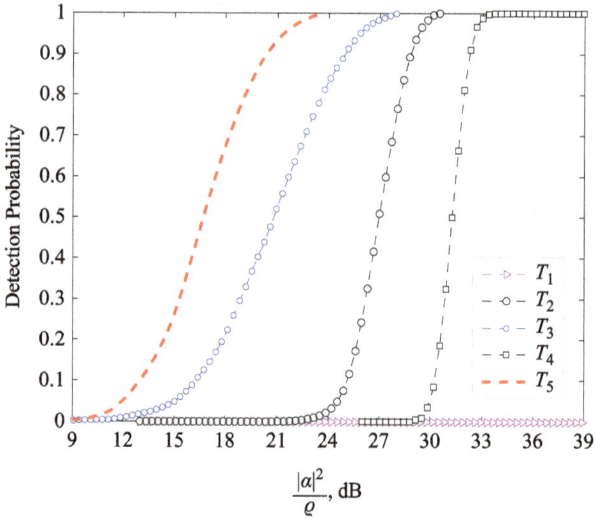

Figure 7.11 Detection performance evaluation of the MP-CA detector for $p_{fa} = 10^{-6}$ when using the FM signal S_2 for detection

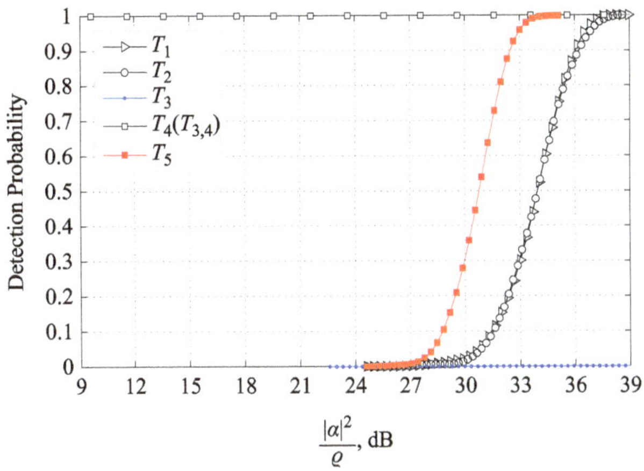

Figure 7.12 Detection performance evaluation of the ECA-CA detector for $p_{fa} = 10^{-6}$ *when using the DVB-T signal for detection*

poor range resolution of the FM S_1 signal but also to the heuristic algorithm used to estimate the noise power in its multistage processing algorithm. As a consequence of our simulations, it is concluded that most conventional PR detectors estimating the noise variance locally suffer from the self-masking and capture effect, especially in FM-based PR systems with poor ambiguity characteristics.

In Reference 13, the extensive cancellation algorithm (ECA) is used for multipath echoes removal, and the CA-CFAR threshold was applied for DVB-T target detection scenarios. Here, this detection strategy is abbreviated as ECA-CA detector. The detection performance of the ECA-CA-based detector is depicted in Figure 7.12. By comparing the results of Figures 7.8 and 7.12, the performance advantage of the proposed method as compared to that of the ECA-CA detector even for DVB-T signal can be observed . In this case, the target under test T_4 can be detected with probability of one since the targets T_3 and T_4 cannot be resolved by the ECA-CA detector, so they can be considered as a single target denoted by $T_{3,4}$ with high-input SNR yielding detection probability of one. In Figure 7.8, we see that in contrast to the proposed detector, the ECA-CA detector incurs losses of 4.8, 4.9, and 1.4 dB for detecting the targets T_1, T_2, and T_5 with detection probability of 0.9 and $p_{fa} = 10^{-6}$. More importantly, as mentioned before, exploiting the proposed detector makes it possible to detect the target T_3 in the presence of the strong interfering target T_4 with a loss of about 2.5 dB, whereas this is impossible with the ECA-CA detector due to the capture effect of the strong interfering target T_4.

In conclusion, we would like to comment on the computational complexity of the different detection methods discussed in this chapter in relation to the number of complex multiplications (CMs) required. Based on our proposed detection algorithm, the direct signal and clutter removal algorithm demands a total of $NP^2 + 3NP + \mathcal{O}(P^2 \log_2 P) + P^2$ CMs, where N is equal to LR. For the successive target detection part of the proposed algorithm with Q iterations, $3NN_dQ + 2NPQ + 4NQ + 1.5NQ$

$(Q-1) + QN_dN_F\log_2 N_F + QN_FN_d + QP^2 + QN_d + 0.5Q(Q-1) + Q$, CMs are required. Thus, the overall CMs during the Q iteration ($Q \geq M$) of the proposed detection algorithm is $C_1^{\times} = LRP^2 + 3LRP + \mathcal{O}(P^2\log_2 P) + P^2 + 3NN_dQ + 2NPQ + 4NQ + 1.5NQ(Q-1) + QN_dN_F\log_2 N_F + QN_FN_d + QP^2 + QN_d + 0.5Q(Q-1) + Q$. For the 2S-GLR-based detector, the overall complex multiplications is $C_2^{\times} = C_1^{\times} + 2NP + 2NQ + 2$. This shows a computational load saving of about $2NP + 2NQ + 2$ when using the proposed detector. Similarly, one can show that the ECA-CA detector has complexity $C_3^{\times} = NP^2 + 3NP + \mathcal{O}(P^2\log_2 P) + P^2 + NN_d + N_dN_F\log_2 N_F + O(N_FN_d)$. The overall CMs of the MP-CA detector after N_s stages ($N_s \leq Q$) is $C_4^{\times} = C_3^{\times} + N_s(N + N_F\log_2 N_F + N_F) + \sum_{i=1}^{N_s}(Nn_i^2 + 4Nn_i + n_i^2)$, where n_i is the number of Doppler-range bins used to update the cancellation mask at the ith stage ($n_i \geq 1$). This parameter depends on the number of Doppler-range bins at which a detection was declared, characteristics of opportunity signal used for detection, as well as the SNR levels of the detected targets. By comparing the computational complexity of the considered detection algorithms, it follows that the main computational complexity of these methods depends on term NP^2; hence, they have approximately the same computational complexity of order $\mathcal{O}(NP^2)$. The proposed detector has a slightly higher computational complexity than the MP-CA detector for FM-based PR systems, but it offers a 4–16 dB improvement in detection performance. On the contrary, for DVB-T-based PR systems, the ECA-CA detector has lower computational complexity than the proposed detector, but the proposed algorithm provides a 2–5 dB improvement in detection performance.

7.5 Summary

This chapter presents a model for detecting targets in PR systems using an M-ary hypothesis test for multitarget scenarios. The proposed approach is a forward and sequential GLR-based detector that detects targets one at a time and treats previously detected targets as interferences, enabling detection of even weak targets. This chapter also proposes a parallel implementation of the GLR-based detector to reduce memory requirements. Closed-form expressions for threshold and detection probability are derived. Simulation results show that the proposed algorithm outperforms existing methods without significant added complexity.

References

[1] Zaimbashi A. 'Forward M-ary hypothesis testing based detection approach for passive radar'. *IEEE Transactions on Signal Processing*. 2017, vol. 65(10), pp. 2659–71.

[2] Zoubir A.M., Viberg M., Chellappa R., Theodoridis S. 'Array and statistical signal processing' in academic press library in signal processing. Vol. 3; 2014. pp. 3–967.

[3] Zaimbashi A., Derakhtian M., Sheikhi A. 'GLRT-based CFAR detection in passive bistatic radar'. *IEEE Transactions on Aerospace and Electronic Systems*. 2013, vol. 49(1), pp. 134–59.
[4] Kay S.M. 'Fundamentals of statistical signal processing: detection theory'. *PTR Prentice-Hall, Englewood Cliff, NJ*. 1998.
[5] Kay S.M. 'Fundamentals of statistical signal processing: estimation theory'. *PTR Prentice-Hall, Englewood Cliffs, NJ*. 1993.
[6] Cherniakov M. 'Bistatic radar: emerging technology'. Wiley and Sons. 2007.
[7] Harris F.J. 'On the use of windows for harmonic analysis with the discrete Fourier transform'. *Proceedings of the IEEE*. 1978, vol. 66(1), pp. 51–83.
[8] Hogenauer E.B. 'An economical class of digital filters for decimation and interpolation'. *IEEE Trans. Acoustics, Speech, and Signal Processing*. 1981, vol. 29(2), pp. 155–62.
[9] Lehman E.L., Romano J.P. 'Testing statistical hypothesis'. *New York: Springer Verlag*. 2005.
[10] Colone F., O'Hagan D.W., Lombardo P., Baker C.J. 'A multistage processing algorithm for disturbance removal and target detection in passive bistatic radar'. *IEEE Transactions on Aerospace and Electronic Systems*. 2009, vol. 45(3), pp. 698–722.
[11] Zaimbashi A., Norouzi Y. 'Automatic dual censoring cell-averaging CFAR detector in non-homogenous environments'. *Signal Processing*. 2008, vol. 88(11), pp. 2611–21.
[12] Tao D., Doulgeris A.P., Brekke C. 'A segmentation-based CFAR detection algorithm using truncated statistics'. *IEEE Transactions on Geoscience and Remote Sensing*. 2016, vol. 54(5), pp. 2887–98.
[13] Colone F., Langellotti D., Lombardo P. 'DVB-T signal ambiguity function control for passive radars'. *IEEE Transactions on Aerospace and Electronic Systems*. 2014, vol. 50(1), pp. 329–47.

Chapter 8

Multitarget detection in analog TV-based passive radar systems

8.1 Introduction

Passive radars based on analog television (ATV) technology and phase alternating line (PAL) modulation format encode video information using two interlaced scans of 625 lines and a frame rate of 50 Hz. Each line begins with a sync pulse that lasts for 64 μs. However, due to the 64 μs line flyback time of the ATV signal, range ambiguity occurs at ranges corresponding to multiples of 19.2 km [1, 2]. As a result, determining the ranges of long-range targets is not feasible with ATV-based passive radar. Thus, most previous research has focused on detecting targets using Doppler information in the echoes of the television video carrier signal rather than range information [2–4].

The primary objective of this chapter is to develop specialized detection algorithms for ATV signals in the presence of significant range ambiguity. To achieve this goal, we begin by formulating the target detection problem in ATV-based passive bistatic radar (PBR) systems as an M-ary hypothesis testing problem for a multitarget scenario. We then propose a new sequential algorithm based on the generalized likelihood ratio (GLR) criterion and the chirp z-transform. This algorithm allows for the joint estimation of the bistatic range-Doppler coordinates of targets, even when range ambiguity is severe, and enables the detection of targets in sequence. Finally, we present simulation results that demonstrate the efficacy of our proposed detection algorithm in determining the range-Doppler coordinates of targets of interest. This approach is innovative in its application of M-ary hypothesis testing to ATV-based PBR systems with severe range ambiguity, and it differs from the approach taken in Chapter 7, where a different sequential detection method is used. In this chapter, we employ the chirp z-transform (CZT) instead of the fast Fourier transform (FFT) to implement the sequential detection algorithm. Additionally, we propose a robust sequential detector that effectively cancels out strong interfering targets and off-grid targets. This algorithm enables simultaneous range-Doppler processing in ATV-based PBR systems, which was previously considered impossible in the existing literature. The material in this chapter is based on our work published in Reference 5.

This chapter is structured as follows: in section 8.2, the detection problem is formulated. Section 8.3 introduces a multilayer GLR-based detection algorithm, which incorporates a chirp z-transform and recursive implementation to minimize computational complexity. Additionally, section 8.4 presents a robust variant of the proposed detector, capable of effectively mitigating the effects of strong interfering targets and off-grid targets. Section 8.5 provides our performance analyses, and in section 8.6, we draw summaries based on our findings.

8.2 Detection problem formulation

In the ATV-based passive radar system shown in Figure 8.1, a single transmitter is used to transmit a signal that illuminates the target of interest. The signal is then reflected off the target and detected by the receiver. The receiver has two channels, the reference channel and the surveillance channel. The system being discussed falls into the Ca.1-IR category. The main issues in the surveillance channel of the PBR system are direct path and multipath interference, as well as strong target echoes. These problems are particularly pronounced in ATV-based systems due to the 64 μs line flyback time of the ATV signal. In practical scenarios, the desired signal may be up to 60 dB weaker than the direct path interference. Consequently, there is a considerable masking effect of strong targets on weaker ones, making it challenging to estimate target ranges in a multitarget situation. Therefore, a specialized signal processing technique is required to address these limitations in the detection problem of an ATV-based passive radar. In this case, the detection problem can be expressed as a composite M-ary hypothesis testing problem, given by

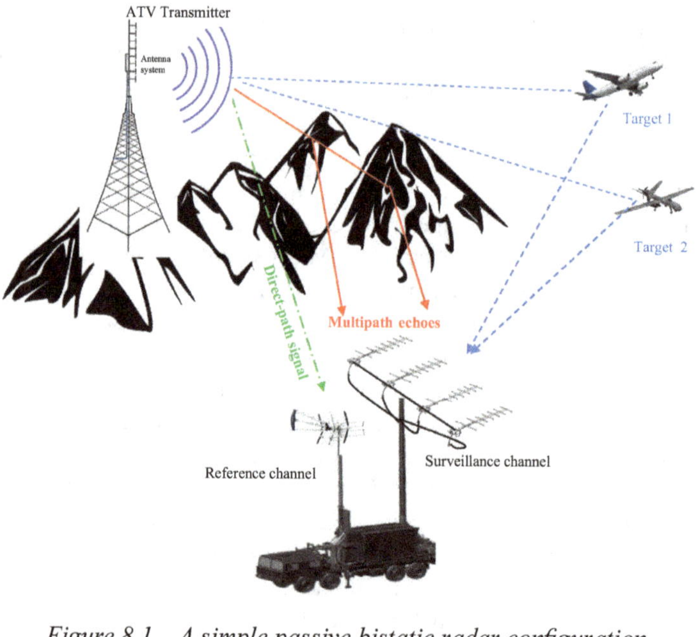

Figure 8.1 A simple passive bistatic radar configuration

$$\begin{cases} \mathcal{H}_0 : \mathbf{x} = \mathbf{Hc} + \mathbf{n} \\ \mathcal{H}_1 : \mathbf{x} = \mathbf{s}_1\alpha_1 + \mathbf{Hc} + \mathbf{n} \\ \mathcal{H}_2 : \mathbf{x} = \mathbf{s}_2\alpha_2 + \mathbf{s}_1\alpha_1 + \mathbf{Hc} + \mathbf{n} \\ \mathcal{H}_3 : \mathbf{x} = \mathbf{s}_3\alpha_3 + [\mathbf{s}_1, \mathbf{s}_2][\alpha_1, \alpha_1]^T + \mathbf{Hc} + \mathbf{n} \\ \vdots \quad \vdots \qquad \qquad \vdots \\ \mathcal{H}_m : \mathbf{x} = \mathbf{s}_m\alpha_m + [\mathbf{s}_1, ..., \mathbf{s}_{m-1}][\alpha_1, ..., \alpha_{m-1}]^T + \mathbf{Hc} + \mathbf{n} \\ \vdots \quad \vdots \qquad \qquad \vdots \end{cases} \tag{8.1}$$

For example, hypothesis $\mathcal{H}_K$ denotes the presence of the K targets in the presence of the system thermal noise, direct signal, and clutter/multipath returns, in which the received signal vector $\mathbf{x}$ can be compactly described by

$$\mathbf{x} = \mathbf{Sa} + \mathbf{Hc} + \mathbf{n} \tag{8.2}$$

where

- $\mathbf{x} \in \mathscr{C}^{N\times 1}$ is the $N \times 1$ vector obtained after sampling a window of the surveillance received signal, where $N = Tf_s$ with f_s denoting the sampling frequency, and T is the time duration of the window. The latter is called integration time and determines the Doppler resolution and the processing gain.
- $\mathbf{S}$ is termed as target signature matrix, represented by

  $$\mathbf{S} \triangleq [\mathbf{s}_1, ..., \mathbf{s}_M], \tag{8.3}$$

  where

  $$\mathbf{s}_m = (\mathbf{y}^{(nm)} \odot \mathbf{e}^{(fm)}) \tag{8.4}$$

  where (n_m, f_m) represents the delay-Doppler coordinates of the mth targets; $\mathbf{y}^{(nm)} = \mathbf{P}^{(nm)}\mathbf{y}$, where $\mathbf{y}$ is the reference signal vector; $\mathbf{P}$ is an $N \times N$ permutation matrix defined as $[\mathbf{P}]_{ij} = 1$ if $i = j + 1$ and 0 otherwise for $i = 0,\ldots, N - 1$; $j = 0,\ldots, N - 1$. Here, $[\mathbf{e}^{(fm)}]_n = e^{j2\pi f_m n T_s}$ with $n = 0,\ldots, N - 1$; $\mathbf{a} = [\alpha_1, \ldots, \alpha_M]^T$ is an M-dimensional column vector containing all the complex amplitudes of desired targets.
- $\mathbf{H}$ is referred to as clutter signature matrix, defined as

  $$\mathbf{H} \triangleq [\mathbf{h}_1, ..., \mathbf{h}_p, ..., \mathbf{h}_P], \tag{8.5}$$

  where $\mathbf{h}_p = (\mathbf{y}^{n_c^{(p)}} \odot \mathbf{e}^{f_c^{(p)}})$, where $n = 0, ..., N - 1$. In this case, $n_c^{(p)}$ and $f_c^{(p)}$ are specified by using the delay-Doppler coordinates of the clutter signature for $p = 1, ..., P$; $\mathbf{c} = [c_1, ..., c_P]^T$ denotes the P-dimensional unknown column vector containing all the clutter amplitudes corresponding to any delay-Doppler coordinates of the clutter signature.
- $\mathbf{n} \in \mathscr{C}^{N\times 1}$ models the additive Gaussian noise of the surveillance channel with the distribution of $\mathscr{CN}(0_{N\times 1}, \sigma^2\mathbf{I}_N)$, where the variance σ^2 is assumed to be unknown.

8.3 Multilayer GLR-based detection algorithm

A method similar to the one used in Chapter 7 can be applied to solve the M-ary hypothesis testing problem described in (8.1). In Chapter 7, we devised a sequential algorithm to test binary hypotheses such as $(\mathcal{H}_0, \mathcal{H}_1), (\mathcal{H}_1, \mathcal{H}_2), \ldots, (\mathcal{H}_{m-1}, \mathcal{H}_m), \ldots, (\mathcal{H}_K, \mathcal{H}_{K+1})$ with the goal of detecting K desired targets (if they exist). This strategy employs a multilayer detection algorithm, where a binary hypothesis testing problem is tackled in each layer. For instance, testing $\mathcal{H}_m$ against $\mathcal{H}_{m-1}$ requires addressing a binary composite hypothesis testing problem, given by

$$\begin{cases} \mathcal{H}_{m-1} : \mathbf{x} = \mathbf{T}_{m-1}\mathbf{g}_{m-1} + \mathbf{Hc} + \mathbf{n} \\ \mathcal{H}_m : \mathbf{x} = \mathbf{s}_m\alpha_m + \mathbf{T}_{m-1}\mathbf{g}_{m-1} + \mathbf{Hc} + \mathbf{n} \end{cases} \tag{8.6}$$

where $\mathbf{T}_{m-1} = [\mathbf{s}_1, \ldots, \mathbf{s}_{m-1}]$ and $\mathbf{g}_{m-1} = [\alpha_1, \ldots, \alpha_{m-1}]^T$. In this new formulation, $\mathbf{s}_m$ represents the desired target to be detected, while columns of matrix $\mathbf{T}_{m-1}$ spanned the subspace of interfering targets been detected.

By comparing the signal models (8.6) and (7.10) from Chapter 7, it is found that they have a same form. Therefore, we can use the same approach to derive the GLR test (GLRT), which can be expressed equivalently as follows:

$$\lambda_m(\mathbf{x}) = 2N \max_{(n_m, f_m)} i(n_m, f_m) \underset{\mathcal{H}_{m-1}}{\overset{\mathcal{H}_m}{\gtrless}} \eta_m \tag{8.7}$$

where

$$i_m(n,f) = \frac{|\mathbf{s}(n,f)^H \mathbf{x}^{\perp(P+m-1)}|^2}{\|\Pi^{\perp}_{\mathbf{U}(P+m-1)}\mathbf{s}(n,f)\|^2 \|\mathbf{x}^{\perp(P+m-1)}\|^2} \tag{8.8}$$

and

$$\eta_m = \frac{N\eta}{N - (P+m) - \eta} \tag{8.9}$$

where $\mathbf{x}^{\perp(P+m-1)} = \Pi^{\perp}_{\mathbf{U}(P+m-1)}\mathbf{x}$ represents the component of $\mathbf{x}$ orthogonal to the space spanned by the columns of $\mathbf{U}_{(P+m-1)}$, in which the projection matrix $\Pi^{\perp}_{\mathbf{U}(P+m-1)}$ is defined as

$$\Pi^{\perp}_{\mathbf{U}(P+m-1)} = \mathbf{U}^{(2)}\mathbf{U}^{(2)H} = \mathbf{I} - \mathbf{U}_{(P+m-1)}(\mathbf{U}^H_{(P+m-1)}\mathbf{U}_{(P+m-1)})^{-1}\mathbf{U}^H_{(P+m-1)} \tag{8.10}$$

It is worth noting that the statistic $i_m(n,f)$ can be regarded as a uniformly mostpower invariant (UMPI) test statistic when it is assumed that the delay-Doppler coordinate of interested target is known.

To set the threshold η in (8.9) according to a desired false alarm probability, the false alarm probability can be obtained in a similar manner to Chapter 7, given by

$$P_{fa} = \alpha L e^{-0.5\eta} \tag{8.11}$$

where L is the number of delay-Doppler coordinates searched over, and α is a value less than one. In the case of ATV signal, our simulation results show that the value of $\alpha = 0.77$ is a good choice (see Figure 8.4). In general, this parameter takes different values in various types of PBR systems based on different opportunity signals used for target detection. The detection probability can also be computed as

$$P_d^{(m)} = e^{-(\delta_m + \ln(\frac{\alpha L}{P_{fa}}))} \sum_{k=0}^{\infty} \left(\frac{\delta_m^k}{k!} \sum_{n=0}^{k-1} \frac{[\ln(\frac{\alpha L}{P_{fa}})]^n}{n!} \right) \tag{8.12}$$

where $\delta_m = \text{SNR}_m \times G_m$ with SNR_m being the input signal-to-noise ratio (SNR) associated to the mth target defined as $\text{SNR}_m = \frac{|\alpha_m|^2}{\sigma^2}$, while G_m can be regarded as the SNR gain of the mth target given by $G_m = \| \Pi^{\perp}_{\mathbf{U}(P+m-1)} \mathbf{s}(n_m,f_m) \|^2$. The detection loss (DL) experienced by the mth target can be defined in a similar manner to Chapter 7, given by

$$\text{DL}_m = \frac{\| \Pi^{\perp}_{\mathbf{U}(P+m-1)} \mathbf{s}(n_m,f_m) \|^2}{\| \mathbf{s}(n_m,f_m) \|^2} \tag{8.13}$$

In the MLa-GLR detection algorithm, as discussed before, we first test $\mathcal{H}_1$ against $\mathcal{H}_0$ if it is accepted by the binary GLRT statistic, i.e., $\lambda_1(\mathbf{x}) > \eta_1$ for some predetermined false alarm probability, then test $\mathcal{H}_2$ against $\mathcal{H}_1$, etc. until rejection, i.e., until $\lambda_m(\mathbf{x}) < \eta_m$. More precisely, in testing $\mathcal{H}_m$ against $\mathcal{H}_{m-1}$, if $\mathcal{H}_m$ is accepted, a new target is declared as detected, and the test procedure proceeds to the next hypothesis between $\mathcal{H}_{m+1}$ and $\mathcal{H}_m$. It stops when the null hypothesis ($\mathcal{H}_{m-1}$) is accepted, implying that no further target can be detected. It is worth to note that for large values of N, which is typical in PBR systems, the threshold η_m of (8.9) can be approximated by η, so the detection threshold does not depend on the number of targets been detected.

8.3.1 Chirp z-transform implementation of MLa-GLR detection algorithm

The presence of the mth target can be tested by computing $i(n,f)$* for a desired *delay-Doppler map*, in which $n = 1, \ldots, N_d$ with N_d being the time delay index corresponding to the maximum bistatic range of interest, and $|f| < f_{\max}$ with $f_{\max}$ being the maximum Doppler shift of interested targets. In line with the preceding chapters, the proposed detector can be implemented using the FFT. However, this approach produces a large number of bin frequencies that are not useful for our purposes. Specifically, in the present problem, we must disregard the bin frequencies that correspond to Doppler shifts from $f_{\max}$ to $\frac{f_s}{2}$, as well as from $-\frac{f_s}{2}$ to $-f_{\max}$, when analyzing the results of the FFT. Alternatively, we could use the Zoom-FFT method, as described in Reference 6. This method involves several additional steps, such as modulation, low-pass filtering, and decimation, which increase the computational cost of the algorithm [7]. To enhance computational efficiency in this scenario, we could use the chirp z-transform technique [8, 9]. This method is typically utilized for spectrum analysis when high-spectral resolution is required within a limited portion of a signal's frequency range.

*We suppress the subscript m from $i(n, f)$, but the reader should keep in mind that these differ for each hypothesis.

Compared with the conventional FFT, the chirp z-transform method is more efficient in such situations. For example, in our problem, we need to compute $i(n,f)$ for a limited range of desired Doppler shifts ranging from $-f_{\max}$ to $f_{\max}$, which is a small portion of overall frequencies resulted from the FFT transform (say f_s), i.e., $\zeta = \frac{2f_{\max}}{f_s} \ll 1$. Based on this, we implement the numerator of $i(n,f)$ for the desired frequency range and at each range (delay index n) of interest as

$$\mathbf{i}_a(\mathbf{r}, n) = \frac{\mathbf{z}_n(\mathbf{x}^{\perp(P+m-1)}) \odot \mathbf{z}_n(\mathbf{x}^{\perp(P+m-1)})^*}{||\mathbf{y}^{(n)}||^2||\mathbf{x}^{\perp(P+m-1)}||^2} \tag{8.14}$$

where

$$\mathbf{z}_n(\mathbf{x}^{\perp(P+m-1)}) = \text{CZT}\left(\mathbf{y}^{(n)*} \odot \mathbf{x}^{\perp(P+m-1)}, N_{cz}, W, A\right) \tag{8.15}$$

where $\text{CZT}(\mathbf{e}, N_{cz}, W, A)$ is the chirp z-transform of vector $\mathbf{e}$ along a spiral contour defined by W and A. In the problem at hands, W is the ratio between points along the z-plane spiral contour, and scalar A is the complex starting point on that contour. In this chapter, these parameters are set equal to $W = e^{j(\frac{2\pi}{N_{cz}})(\frac{2f_{\max}}{f_s})}$ and $A = e^{j2\pi\frac{f_{\min}}{f_s}}$, where N_{cz} is a scalar that specifies the length of the chirp z-transform.

To further reduce the computational complexity of the calculation (8.14) for the delay index n, we approximate $\Pi^{\perp}_{\mathbf{U}(P+m-1)}\mathbf{s}$ with $\mathbf{y}^{(n)}$ only in the denominator of (8.14). In this case, subscript "a" is used to emphasis this approximation, which is a reasonable approximate as shown in References 10, 11. Finally, we can define matrix $\mathbf{I}_a(\mathbf{x}^{\perp(P+m-1)}) = [\mathbf{i}_a(\mathbf{x}^{\perp(P+m-1)}, 1), \ldots, \mathbf{i}_a(\mathbf{x}^{\perp(P+m-1)}, N_d)]$ with dimension $N_{cz} \times N_d$ referred to as a range-Doppler matrix, in which $L = N_{cz}N_d$ is the number of Doppler and range bins examined (see (8.11)). In the following, this detector is referred to as the CZT-MLa-GLR.

In order to obtain the same Doppler frequency accuracy when using N'-point FFT[†] and N_{cz}-point CZT, it is required to set $\Delta f_{\text{FFT}} = \Delta f_{\text{CZT}}$, where $\Delta f_{\text{FFT}} = \frac{f_s}{N'}$ and $\Delta f_{\text{CZT}} = \frac{2f_{\max}}{N_{cz}}$, resulting in $N_{cz} = \zeta N'$, i.e., $N_{cz} \ll N'$. Thus, the Doppler frequency accuracy would be improved by increasing the values of N' and N_{cz}. However, this increase appears to be the most practically relevant one for the CZT as compared with that of the FFT method. Generally, when the CZT is used to evaluate for N_{cz} bin frequencies, the computational complexity of order $\mathcal{O}(N'\log_2 N_{cz})$ would be required, while it is about $\mathcal{O}(N'\log_2 N')$ when using N'-point FFT [9, 12]. This means that the proposed MLa-GLR detection algorithm implemented based on the chirp z-transform can increase both the calculation efficiency and frequency precision as compared with that introduced in Chapter 7.

8.3.2 Recursive implementation of the CZT-MLa-GLR detection algorithm

In this section, we aim to implement the CZT-MLa-GLR detector in a sequential manner over the sequence of binary hypotheses $(\mathcal{H}_0, \mathcal{H}_1), (\mathcal{H}_1, \mathcal{H}_2), \ldots, (\mathcal{H}_{m-1}, \mathcal{H}_m)$,

[†] N' is a power of 2 and $N' \geq N$.

..., $(\mathscr{H}_K, \mathscr{H}_{K+1})$. This will allow us to reduce the computational complexity of the CZT-MLa-GLR algorithm. Based on this, the new algorithm is referred to as sequential CZT-MLa-GLR (S-CZT-MLa-GLR) detection algorithm in the sequel. To implement the proposed algorithm in a sequential manner, we compute vector $\mathbf{x}^{\perp(P+m-1)}$ sequentially. For simplicity in notation, the interference matrix can be partitioned as $\mathbf{U}_{(P+m-1)} = [\mathbf{u}_1, ..., \mathbf{u}_P, \mathbf{u}_{P+1}, ..., \mathbf{u}_{P+m-1}]$, where $\mathbf{H} = [\mathbf{u}_1, ..., \mathbf{u}_P]$ and $\mathbf{u}_{P+i} = \mathbf{s}_i$.‡ To proceed, $\Pi^{\perp}_{\mathbf{U}_{(k+1)}}$ can be recursively written as follows [13]:

$$\Pi^{\perp}_{\mathbf{U}_{(k+1)}} = \Pi^{\perp}_{\mathbf{U}_{(k)}} - \frac{\mathbf{u}_{k+1}^{\perp(k)} \mathbf{u}_{k+1}^{\perp(k)^H}}{\|\mathbf{u}_{k+1}^{\perp(k)}\|^2} \tag{8.16}$$

where $\mathbf{U}_{(k+1)} = [\mathbf{U}_{(k)}, \mathbf{u}_{k+1}]$. Here, $\mathbf{u}_{k+1}^{\perp(l)} = \Pi^{\perp}_{\mathbf{U}_{(l)}} \mathbf{u}_{k+1}$ represents the component of $\mathbf{u}_{k+1}$ orthogonal to the space spanned by the columns of matrix $\mathbf{U}_{(l)}$. Therefore, by using (8.16), $\mathbf{u}_{k+1}^{\perp(l)}$ can be recursively computed as follows:

$$\mathbf{u}_{k+1}^{\perp(l)} = \begin{cases} \mathbf{u}_{k+1} - \dfrac{\mathbf{u}_1 \mathbf{u}_1^H \mathbf{u}_{k+1}}{\|\mathbf{u}_1\|^2} & l = 1 \\ \mathbf{u}_{k+1}^{\perp(l-1)} - \dfrac{\mathbf{u}_k^{\perp(l-1)} \mathbf{u}_k^{\perp(l-1)^H} \mathbf{u}_{k+1}^{\perp(l-1)}}{\|\mathbf{u}_k^{\perp(l-1)}\|^2} & l = 2, ..., k \end{cases} \tag{8.17}$$

Finally, $\mathbf{x}^{\perp(k+1)}$ defined as $\Pi^{\perp}_{\mathbf{U}_{(k+1)}} \mathbf{x}$ can be computed recursively as (for $k \geq 0$)

$$\mathbf{x}^{\perp(k+1)} = \mathbf{x}^{\perp(k)} - \frac{\mathbf{u}_{k+1}^{\perp(k)} \mathbf{u}_{k+1}^{\perp(k)^H} \mathbf{x}^{\perp(k)}}{\|\mathbf{u}_{k+1}^{\perp(k)}\|^2} \tag{8.18}$$

where $\mathbf{x}^{\perp(0)} = \mathbf{x}$ and $\mathbf{u}_1^{\perp(0)} = \mathbf{u}_1$. Using (8.17) and (8.18), the steps of the S-CZT-MLa-GLR detection algorithm for joint delay-Doppler estimation are summarized in Algorithm 8.1. The proposed detection algorithm operates in a layered manner, with one target being detected in each layer. The interference caused by the detected target in the first layer is estimated and removed from the received signal. This interference-canceled signal is then used to detect another target in the second layer, and so on, until all detectable targets are found. If the maximum value of $\max_{n,k} \mathbf{I}_a(\mathbf{x}^{\perp(P+m)})$ is below a certain threshold η, the algorithm terminates. It is important to note that the proposed detector detects targets from the strongest to the weakest. Once the algorithm ends, the number of detectable targets, say K, and their range-Doppler coordinates are determined. The algorithm comprises two main parts: removing multipath signals and detecting targets in multiple layers. In this chapter, the direct and clutter signals are removed without the need for constructing and inverting a gram matrix $(\mathbf{U}_{(P+m-1)}^H \mathbf{U}_{(P+m-1)})$, making the algorithm more efficient than the one presented in Chapter 7.

‡Note that $(P+m-1)$ is the number of interfering signals.

Algorithm 8.1 Steps of the S-CZT-MLa-GLR detection algorithm.

Input: Threshold η, received vector $\mathbf{x}$, and clutter signature matrix $\mathbf{H} = [\mathbf{u}_1, ..., \mathbf{u}_P]$;

1. **Initialization**
   - Set threshold η as (16) and $m = 0$.
2. **Part 1: Multi-path signals removal**
   - **for** $k = 1:1:P-1$, do
     for $l = 1:1:k$, calculate
     $$\mathbf{u}_{k+1}^{\perp(l)} = \begin{cases} \mathbf{u}_{k+1} - \frac{\mathbf{u}_1^H \mathbf{u}_{k+1}}{\|\mathbf{u}_1\|^2}\mathbf{u}_1 & l = 1 \\ \mathbf{u}_{k+1}^{\perp(l-1)} - \frac{\mathbf{u}_k^{\perp(l-1)\mathrm{H}} \mathbf{u}_{k+1}^{\perp(l-1)}}{\|\mathbf{u}_k^{\perp(l-1)}\|^2}\mathbf{u}_k^{\perp(l-1)} & l = 2, ..., k \end{cases}$$
     end
     Calculate
     $$\mathbf{x}^{\perp(k)} = \begin{cases} \mathbf{x} - \frac{\mathbf{u}_1^H \mathbf{x} \mathbf{u}_1}{\|\mathbf{u}_1\|^2} & k = 1 \\ \mathbf{x}^{\perp(k-1)} - \frac{\mathbf{u}_k^{\perp(k-1)\mathrm{H}} \mathbf{x}^{\perp(k-1)}}{\|\mathbf{u}_k^{\perp(k-1)}\|^2}\mathbf{u}_k^{\perp(k-1)} & k = 2, ..., P-1 \end{cases}$$
     end (save $\mathbf{x}^{\perp(P-1)}$ and $\mathbf{u}_P^{\perp(l)}$ for $l = 1, ..., P-1$)
   - Calculate and save
     $$\mathbf{x}^{\perp(P+m)} = \mathbf{x}^{\perp(P+m-1)} - \frac{\mathbf{u}_{P+m}^{\perp(P+m-1)\mathrm{H}} \mathbf{x}^{\perp(P+m-1)}}{\|\mathbf{u}_{P+m}^{\perp(P+m-1)}\|^2}\mathbf{u}_{P+m}^{\perp(P+m-1)}$$
3. **Part 2: Multi-layer target detection**
   While $\max_{n,k} \mathbf{I}_a(\mathbf{x}^{\perp(P+m)}) > \eta$; **then**
   - Increase m by 1;
   - Find and save the location of the maximum of $\mathbf{I}_a(\mathbf{x}^{\perp(P+m)})$ as the MLE of Delay-Doppler coordinate of the mth target, (n_m, f_m) say;
   - Construct vector $\mathbf{u}_{P+m} = (\mathbf{y}^{(n_m)} \odot \mathbf{e}^{(f_m)})$.
   - **for** $l = 1:1:P+m-1$
     $$\mathbf{u}_{P+m}^{\perp(l)} = \begin{cases} \mathbf{u}_{P+m} - \frac{\mathbf{u}_1^H \mathbf{u}_{P+m}}{\|\mathbf{u}_1\|^2}\mathbf{u}_1 & l = 1 \\ \mathbf{u}_{P+m}^{\perp(l-1)} - \frac{\mathbf{u}_{P+m-1}^{\perp(l-1)\mathrm{H}} \mathbf{u}_{P+m}^{\perp(l-1)}}{\|\mathbf{u}_{P+m-1}^{\perp(l-1)}\|^2}\mathbf{u}_{P+m-1}^{\perp(l-1)} & l = 2, ..., P+m-1 \end{cases}$$
     end (save $\mathbf{u}_{P+m}^{\perp(l)}$ for $l = 1, ..., P+m-1$)
   - Calculate and save
     $$\mathbf{x}^{\perp(P+m)} = \mathbf{x}^{\perp(P+m-1)} - \frac{\mathbf{u}_{P+m}^{\perp(P+m-1)\mathrm{H}} \mathbf{x}^{\perp(P+m-1)}}{\|\mathbf{u}_{P+m}^{\perp(P+m-1)}\|^2}\mathbf{u}_{P+m}^{\perp(P+m-1)}.$$

   end while

Output: Joint estimation of targets' delays and Doppler shifts.

8.4 Robust S-CZT-MLa-GLR detection algorithm

The proposed algorithm for detecting targets has two main parts: multipath signal removal and multilayer target detection. In this section, the focus is on the latter part. Once a target is detected, its signal is subtracted from the received signal before the next target is detected. While the targets' ranges and velocities are continuous, they are discretized into a number of grids for detection in the surveillance area. Some targets may not reside exactly on these grids, which are called off-grid targets. To ensure that these off-grid targets are effectively removed, a small area around the detected target's range and Doppler bins is considered for cancellation. For ATV-signal, our simulation results show that a cancellation area around the target position with an extension of $N_R = 3$ bins in the range dimension and $N_F = 3$ bins in Doppler dimension, yielding a mask of dimension $N_R \times N_F$. By considering this, the second part of the S-CZT-MLa-GLR detection algorithm can be modified as follows. Here, the vector $\mathbf{u}_{P+m}^{\perp(P+m-1)}$ must be replaced with $N \times L_M$-dimensional matrix $\mathbf{U}_{P+m}^{\perp(P+m-1)}$ whose ith column is constructed as $[\mathbf{U}_{P+m}]_i = (\mathbf{y}^{(n_m^{(i)})} \odot \mathbf{e}^{(f_m^{(i)})})$, where $n_m^{(i)} = n_m + \text{mod}(i, N_F) - 1$ and $f_m^{(i)} = f_m + (\text{mod}(i, N_R) - 1)\frac{\Delta f}{3}$ with $\Delta f = \frac{2f_{\max}}{N_{cz}}$ for $i = 1, \ldots, L_M$ with $L_M = N_R N_F$. It is important to mention that the first part of the proposed detection algorithm does not require any matrix inversion. However, the second part involves inverting matrices that have a dimension of $L_M \times L_M$. Similar to the first part, we can modify the second part to eliminate the need for matrix inversion. Therefore, this robust version of the proposed S-CZT-MLa-GLR is referred to as RS-CZT-MLa-GLR. The detail of the second part of this new algorithm can be seen in Algorithm 8.2.

Algorithm 8.2 Steps of the second part of RS-CZT-MLa-GLR detection algorithm

---

- **While** $\max_{n,k} \mathbf{I}_a(\mathbf{x}^{\perp(P+m)}) > \eta$; **then**
  - Increase m by 1;
  - Find and save the location of the maximum of $\mathbf{I}_a(\mathbf{x}^{\perp(P+m)})$ as the MLE of Delay-Doppler coordinate of mth target, (n_m, f_m) say;
  - Construct matrix $\mathbf{U}_{P+m}$ whose ith column is constructed as $[\mathbf{U}_{P+m}]_i = (\mathbf{y}^{(n_m^{(i)})} \odot \mathbf{e}^{(f_m^{(i)})})$ where $n_m^{(i)} = n_m + \text{mod}(i, N_F) - 1$ and $f_m^{(i)} = f_m + (\text{mod}(i, N_R) - 1)\frac{2f_{\max}}{3N_{cz}}$ for $i = 1, \ldots, L_M$ with $L_M = N_R N_F$.
  - **for** $l = 1:1:P+m-1$

    if (m == 1)

$$\mathbf{U}_{P+m}^{\perp(l)} = \begin{cases} \mathbf{U}_{P+m} - \mathbf{u}_1 \dfrac{\mathbf{u}_1^H \mathbf{U}_{P+m}}{\|\mathbf{u}_1\|^2} & l = 1 \\ \mathbf{U}_{P+m}^{\perp(l-1)} - \mathbf{u}_{P+m-1}^{\perp(l-1)} \dfrac{\mathbf{u}_{P+m-1}^{\perp(l-1)\,\mathrm{H}} \mathbf{U}_{P+m}^{\perp(l-1)}}{\|\mathbf{u}_{P+m-1}^{\perp(l-1)}\|^2} & l = 2, \ldots, P+m-1 \end{cases}$$

elseif

$$\mathbf{U}_{P+m}^{\perp(l)} = \begin{cases} \mathbf{U}_{P+m} - \mathbf{u}_1 \frac{\mathbf{u}_1^H \mathbf{U}_{P+m}}{\|\mathbf{u}_1\|^2} & l = 1 \\ \mathbf{U}_{P+m}^{\perp(l-1)} - \mathbf{U}_{P+m-1}^{\perp(l-1)} (\mathbf{U}_{P+m-1}^{\perp(l-1)\,\mathrm{H}} \mathbf{U}_{P+m-1}^{\perp(l-1)})^{-1} \mathbf{U}_{P+m-1}^{\perp(l-1)\,\mathrm{H}} \mathbf{U}_{P+m}^{\perp(l-1)} & \\ & l = 2, \ldots, P+m-1 \end{cases}$$

end end

– Calculate and save

$$\mathbf{x}^{\perp(P+m)} = \mathbf{x}^{\perp(P+m-1)} - \mathbf{U}_{P+m}^{\perp(P+m-1)} \times (\mathbf{U}_{P+m}^{\perp(P+m-1)\mathrm{H}} \mathbf{U}_{P+m}^{\perp(P+m-1)})^{-1} \mathbf{U}_{P+m}^{\perp(P+m-1)\mathrm{H}} \mathbf{x}^{\perp(P+m-1)}$$

- **end while**

8.5 Performance results

This section provides a performance analysis of the proposed detection algorithms, S-CZT-MLa-GLR and RS-CZT-MLa-GLR, when applied to PBR systems utilizing the PAL ATV signal. For this analysis, we assume there is a single transmitter emitting the ATV signal at a carrier frequency of $f_c = 479.25$ MHz, and we focus only on a bandwidth of ± 125 kHz around the carrier frequency. Although the ATV signal has a bandwidth of approximately 6 MHz, the majority of its energy is concentrated within this narrow frequency range. The integration time used for analysis is $T = 0.5$ *s*, and the sampling frequency is $f_s = 250$ kHz, resulting in a data length of $N = Tf_s$, which is 125,000.

To assess how well the proposed detectors perform, we create several scenarios that include both direct path signals and multiple targets and clutters within the surveillance channel. The noise power, which is represented by σ^2, is fixed at a level of –110 dBm. In these scenarios, we assume that the direct path signal, targets, and clutters are all point scatterers. We define clutter-to-noise ratio (CNR) or direct path-to-noise ratio (DNR) as $\frac{|c_p|^2}{\sigma^2}$ for $p = 1, \ldots, P$. Additionally, we assume that within the surveillance channel, there is a direct path signal with a DNR_s of approximately 30 dB and that its range and Doppler coordinate are both set to $(0, 0)$. Here, clutter scatterers are generated with CNRs ranging from –5 to 25 dB. The complex amplitudes of clutters are randomly generated according to a complex Gaussian distribution with zero means, and their variances are computed based on CNRs and the noise power. The velocities of the clutters are randomly generated between $-v_{c,\max}$ and $v_{c,\max}$ with $v_{c,\max} = 2$ m/s, while for targets, they are between -300 m/s and 300 m/s. We can compute the Doppler shift of a target based on its velocity. Specifically, if the target velocity is v, then its Doppler shift is $f = \frac{f_c v}{c}$, where f_c is the carrier frequency, and c is the light speed. Based on this simulation setup, other parameters such as N', N_{cz}, and $f_{\max}$ can be obtained as 524,288, 2,048, and 479.25 Hz, respectively. From this, we can find that the proposed CZT-based detector has slightly better frequency precision than that of the FFT-based one even with fewer points, i.e., $N_{cz} \ll N'$ leads to $\Delta f_{\mathrm{CZT}} < \Delta f_{\mathrm{FFT}}$. Also, the relative bistatic targets' ranges are uniformly considered between 10 and 150 km, while for clutters, they are generated from 0 km to $R_{c,\max}$, where the maximum range of clutter is $R_{c,\max} = 60$ km. Based on this simulation setup, the delays and Doppler frequencies of the clutter signature are set as $n_c^{(p)} = 0, \ldots, N_c - 1$ and $f_c^{(p)} = -g\frac{f_c v_{c,\max}}{c} : \frac{0.5}{T} : g\frac{f_c v_{c,\max}}{c}$, where N_c and g are set as $[\frac{R_{c,\max} f_s}{c}]$ and 2, respectively. In other words, the clutters can be eliminated by removing the paths around zero velocity region, while moving targets have higher Doppler shifts. In the reference channel, the received signal is regarded as a noisy and delayed version of the transmitted ATV signal with the DNR, denoted as DNR_r, ranging from 40 to 60 dB.[§]

In order to assess the effectiveness of the direct path and clutter removal part of our proposed detector, we conducted an analysis in the absence of any targets. Specifically, we evaluated the detector's performance in a surveillance channel scenario that included a direct path signal and two echoes of clutter. The direct path signal was assigned a range and velocity coordinate of (0,0), while the two clutter instances were assigned coordinates of (12 km, 0 m/s) and (36 km, 1 m/s). Here, the parameters DNR_s, CNRs are set as 30, 20, and 10 dB, respectively. To get an insight into the problem, we consider the amplitudes of clutters as deterministic but unknown parameters. For this scenario, Figure 8.2 shows the $2Ni(r, v)$ as a function of relative bistatic range and velocity coordinate (r, v) for $\text{DNR}_r = 50$ dB. To gain insight into this, the range cut associated to the range of clutters and zero velocity cut of $2Ni(r, v)$ for two values of DNR_r, which are set equal to 50 and 60 dB, are depicted in Figure 8.3(a) and (b), respectively. The proposed detector attempts to create a null in the region where clutter is present, as shown in Figures 8.2 and 8.3. However, the depth of these nulls is restricted by the value of DNR_r. Therefore, higher values of DNR_r lead to deeper nulls in the clutter region, resulting in greater clutter cancellation. It is evident from the simulation that having a clean reference channel is crucial, and this can be achieved through an appropriate signal conditioning algorithm. The detector's ability to eliminate clutter is similar to that of Chapter 3, and interested readers are referred to Chapter 3 for further information on performance curves, analyses, and discussions.

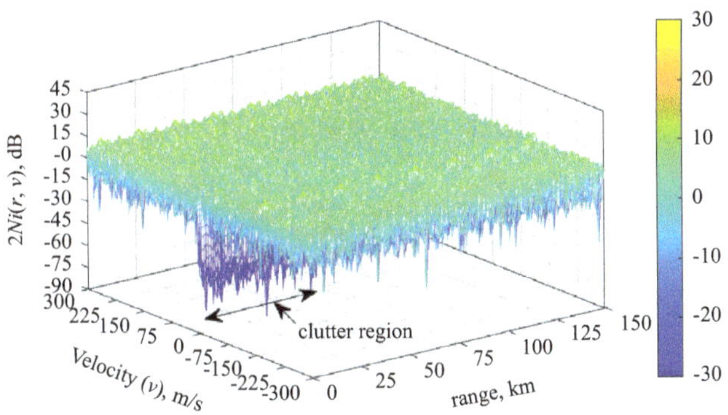

Figure 8.2 Range-velocity map of the S-CZT-MLa-GLR detector for $\text{DNR}_r = 50$ *dB*

[§]The reference channel is not so clean, so it is necessary to use a signal conditioning algorithm to accomplish this [14].

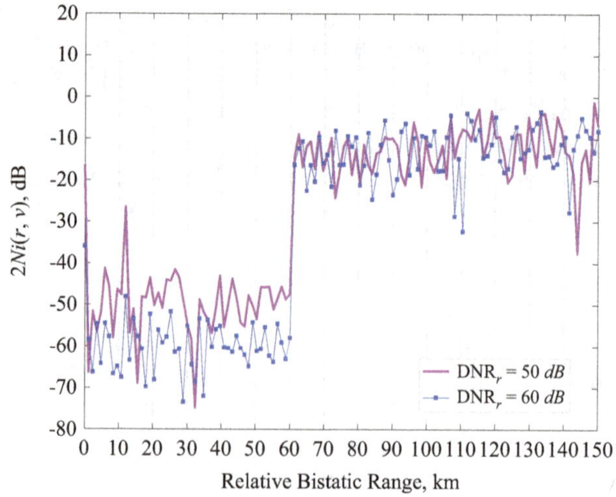

(a)

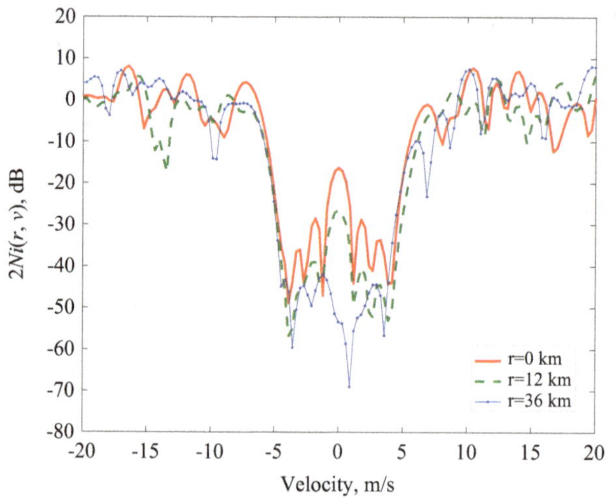

(b)

Figure 8.3 S-CZT-MLa-GLR test statistic (a) zero velocity cut of $2Ni(r, v)$ when DNR_r *are set as 50 and 60 dB and (b) range cut of $2Ni(r, v)$ for $r = 0$, 12, and 36 km*

Next, we compare the false alarm probability obtained through Monte Carlo (MC) simulation with the asymptotic probability of the detection algorithm discussed in section 8.3. For this, we conducted 10^7 MC simulation runs and obtained numerical results. As shown in Figure 8.4, we observed that the proposed test's actual false alarm probability is almost the same as the asymptotic probability when we set the value of α to 0.77 (as seen in (8.11)). Therefore, we can proceed with the proposed detection algorithm confidently.

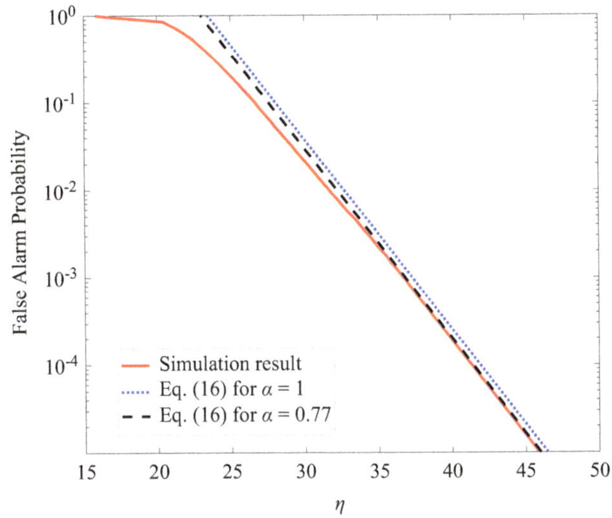

Figure 8.4 False alarm probability as a function of the detection threshold to verify the obtained analytical formula of (8.11) through MC simulation result

To better understand the significant ambiguity range sidelobe caused by the ATV signal's 64 μs line flyback time, we conducted a simulation of a single-target scenario (STS) and plotted a 2D representation on the range-velocity map of the S-CZT-MLa-GLR detector. We set the range and velocity of the simulated target at 12 km and 66.84 m/s, respectively, while the target SNR and DNR_r were –3 and 50 dB, respectively. The results are shown in Figures 8.5 and 8.6, where the velocity and range cuts of Figure 8.5 at velocity and range of 66.84 m/s and 12 km are presented in Figure 8.6(a) and (b), respectively. It is seen that the range dimension

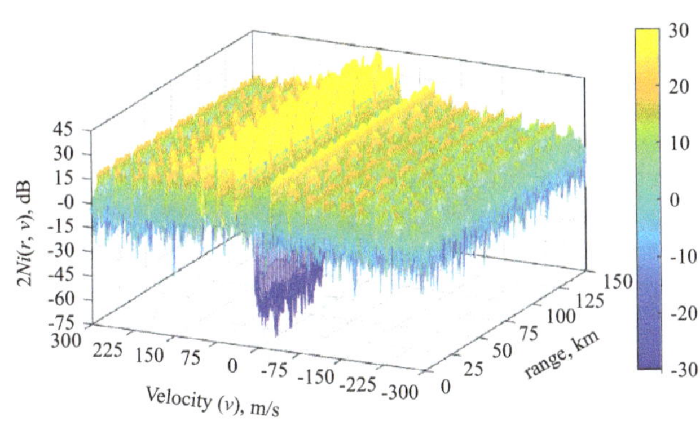

Figure 8.5 Two-dimensional representation of the S-CZT-MLa-GLR test statistic when there is a target with range and velocity of 12 km and 66.84 m/s, respectively. Here, we set $\text{DNR}_r = 50$ *dB.*

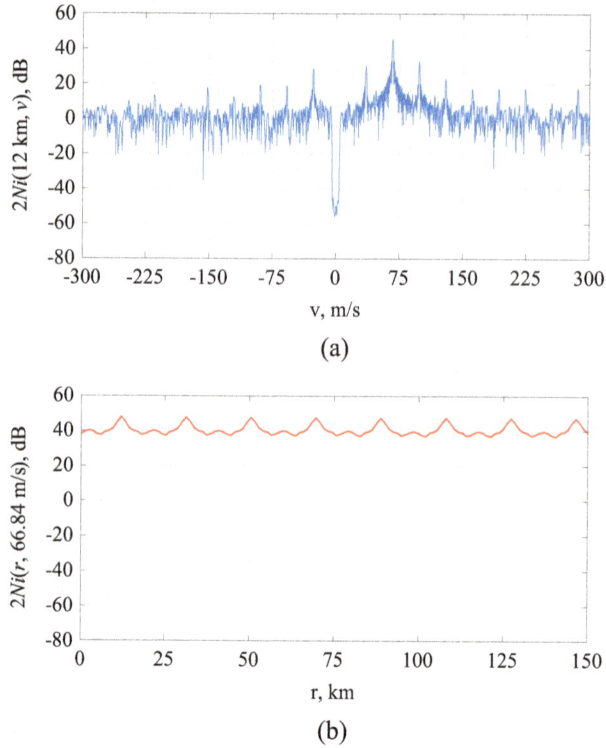

Figure 8.6 *S-CZT-MLa-GLR test statistic (a) velocity cut of* $Ni(n, v)$ *with* $v = 66.84$ *m/s and (b) range cut of* $Ni(r, v)$ *for* $r = 12$ *km*

sidelobe level due to the 64 μs line flyback time is almost equal to that of the main lobe, indicating the difficulty in estimating the exact range of a target using traditional methods. However, the proposed S-CZT-MLa-GLR detector was able to accurately estimate the joint range and velocity of the target without any false target, as shown in Figure 8.7.

Figure 8.8 illustrates how the target SNR impacts detection performance. Besides, it confirms the theoretical detection performance of the proposed detector, as presented in section 8.3, through Monte Carlo simulations. Specifically, the figure plots the detection probability as a function of the input SNR for the ATV signal in an STS. The simulation results demonstrate that the proposed detector's detection probability is almost identical to the asymptotic result of (8.12). To arrive at this conclusion, we conducted 10^4 MC simulation trials.

To evaluate the effectiveness of the proposed S-CZT-MLa-GLR and RS-CZT-MLa-GLR algorithms in determining the joint targets' ranges and velocities, two multitarget scenarios were examined. These scenarios are described in Tables 8.1 and 8.2. In the first scenario, referred to as MTS_1, there are five targets. Among these targets, T_1 and T_2 have similar velocities but are located at a range distance of

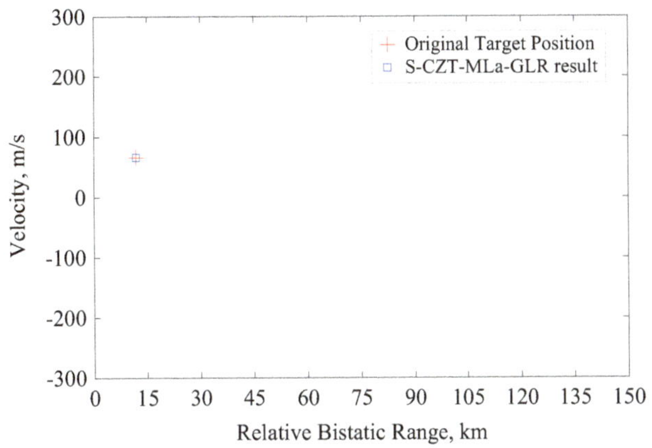

Figure 8.7 *Results of S-CZT-MLa-GLR algorithm for the simulated single-target scenario with the target's delay-Doppler coordinate of (12 km and 66.84 m/s) and when $P_{fa} = 10^{-6}$*

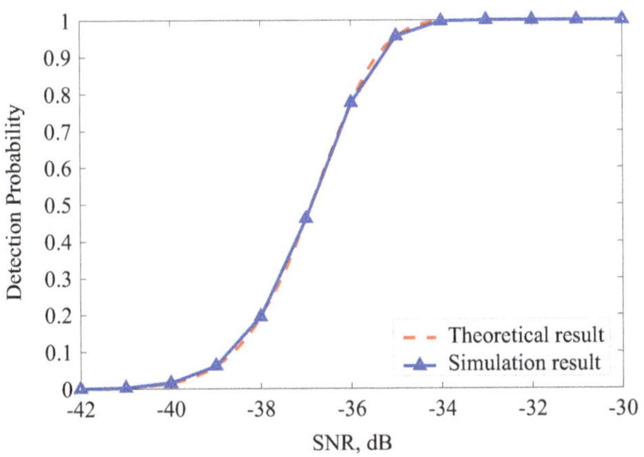

Figure 8.8 *Detection probability comparison of (8.12) with that obtained by Monte Carlo simulation for $P_{fa} = 10^{-6}$ in the single-target scenario*

Table 8.1 *Characteristics of simulated targets in scenario 1 (MTS_I)*

Targets	T_1	T_2	T_3	T_4	T_5
Relative bistatic range (km)	12	31.2	90	90	135.6
Velocity (m/s)	66.86	66.86	−182	−150.51	230
SNR_m (dB)	−3	−27	−11	−11	−26

approximately 19.5 km, which corresponds to the first range ambiguity for the ATV signal's 64 μs line flyback time. Additionally, T_2 and T_3 are in the same range cell but have different Doppler shifts of approximately 50 Hz, which is equivalent to the ATV signal's line frequency. The performance of two algorithms, S-CZT-MLa-GLR and RS-CZT-MLa-GLR, proposed for estimating the joint range-velocity of targets, was evaluated in Figure 8.9. The algorithms return delay-velocity pairs which can either correspond to targets or false alarms. A target cannot be identified if there is no delay-velocity pair near its position. The original target positions are denoted by symbols +, and the estimated positions by the proposed algorithms are denoted by □. For the RS-CZT-MLa-GLR algorithm, a cancellation area around the target position with an extension of three bins in the Doppler dimension is used. In this scenario, no extended range bins are required since all targets have ranges that are integer multiples of the sampling period. It can be observed from Figure 8.9 that both proposed algorithms correctly identified four targets. The S-CZT-MLa-GLR

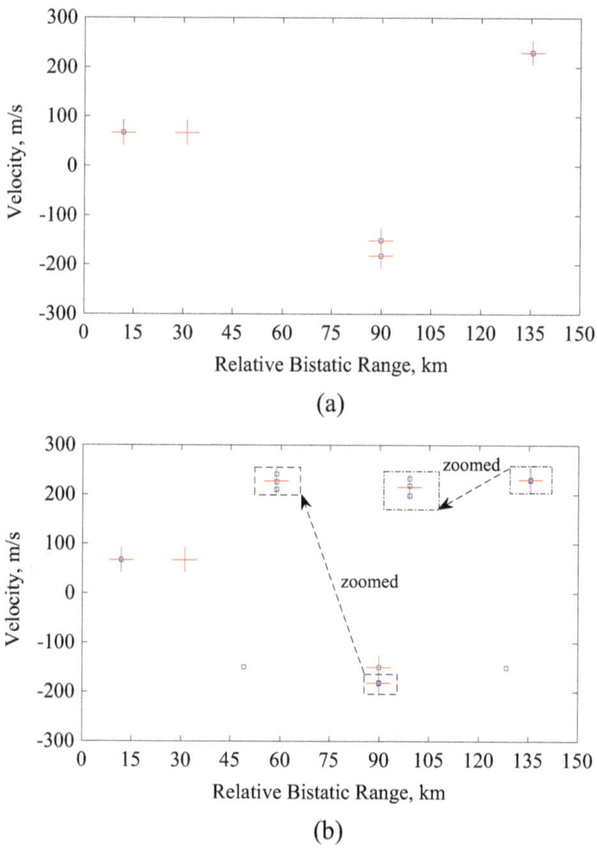

Figure 8.9 Results of (a) RS-CZT-MLa-GLR and (b) S-CZT-MLa-GLR algorithms for the multitarget of Table 8.1 and when $P_{fa} = 10^{-6}$

algorithm produced two false targets and split targets T_3 and T_5 into three scatterers, whereas the RS-CZT-MLa-GLR algorithm was able to detect four target positions with only four iterations, indicating its more robust performance. However, neither algorithm was able to accurately localize target T_2 in the presence of the strong target T_1. The reason for this will be explained below.

In order to examine the impact of off-grid targets on target detection, we analyze a new multitarget scenario known as MTS_2, which is defined in Table 8.2. In this scenario, targets T_2 and T_3 are off-grid targets, meaning that their ranges are not located on the range grids. To effectively eliminate the detected targets, particularly the off-grid ones, the RS-CZT-MLa-GLR algorithm employs three excess bins in the range dimension and three excess bins in the Doppler dimension. As demonstrated in Figure 8.10, both proposed algorithms accurately locate all of the targets. However, we noticed that the S-CZT-MLa-GLR algorithm generates two false targets and divides targets T_2 and T_3 into three scatterers. On the other hand, the RS-CZT-MLa-GLR algorithm discovers target positions with only three iterations and correctly identifies the number of targets. Therefore, the S-CZT-MLa-GLR algorithm is susceptible to the discretization of targets' ranges, which can result in erroneous targets and imprecise estimations of targets' ranges and velocities. In contrast, the RS-CZT-MLa-GLR algorithm can accurately detect the number of targets and is resistant to errors when determining targets' delay-Doppler coordinates, leading to superior target localization performance.

Figure 8.11 shows the $2N(r, v)$ versus bistatic range and velocity after removing the multipath/clutter signals and that of the strong targets T_1 and T_2 in the multitarget scenario characterized in Table 8.2. It should be noted that the results of Figure 8.11 are obtained after two iterations of the second part of the proposed RS-CZT-MLa-GLR detection algorithm. It is seen that the target T_3 is now visible, and it can be detected in the third iteration of this algorithm, as shown in Figure 8.10(a). In Figure 8.11, it is also observed that the proposed detector place deep nulls in the range-velocity coordinates of targets been detected and that of the clutter region to removing them.

In order to get further insight into target detection problem, we depict the detection probabilities $P_d^{(m)}$ for $m = 1, ..., 5$ as a function of SNR for all targets characterized in Tables 8.1 and 8.2. To do this, we change the SNR of the mth testing target, T_m say, in the presence of the other targets named interfering targets with the targets' SNR_m and delay-Doppler coordinates as listed in Tables 8.1 and 8.2.

Table 8.2 Characteristics of simulated targets in scenario 2 (MTS_2)

Targets	T_1	T_2	T_3
Relative bistatic range (km)	12	72.6	126.6
Velocity (m/s)	66.86	240	–210
SNR_m(dB)	–3	–13	–23

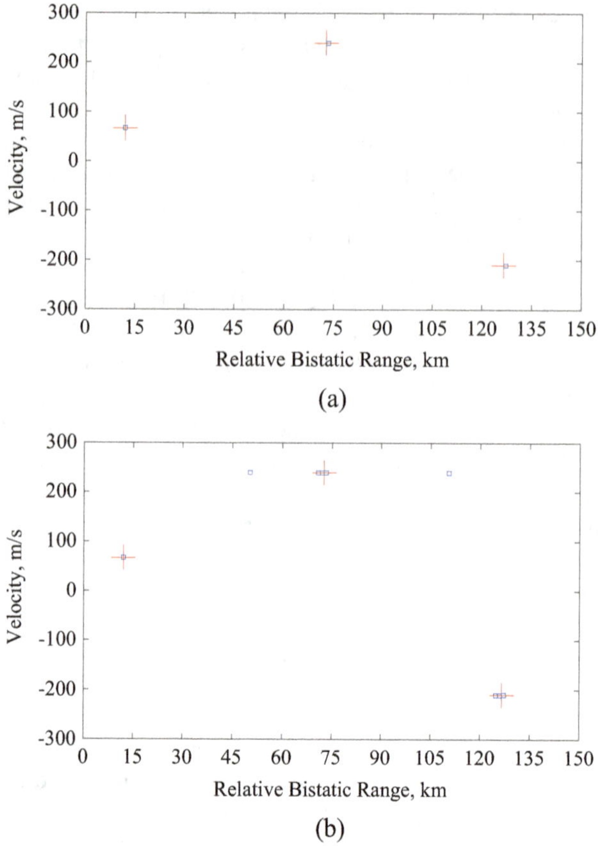

Figure 8.10 *Results of (a) RS-CZT-MLa-GLR and (b) S-CZT-MLa-GLR algorithms for the multitarget of Table 8.2 and when* $P_{fa} = 10^{-6}$

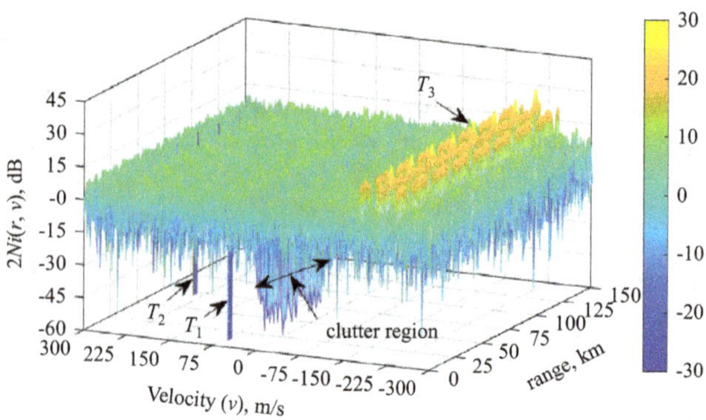

Figure 8.11 *Range-velocity map of the proposed RS-CZT-MLa-GLR detector after removing the multipath/clutter signals and that of the targets* T_1 *and* T_2 *in the multitarget scenario characterized in Table 8.2*

The results of this simulation are reported in Figures 8.12 and 8.13 The detection performance for the STS is also shown for comparison. Generally, it is seen that the target T_m for $m = 1, \ldots, 5$ does not experience the same detection performance due to the characteristics of the ATV signal. For the MTS_1, the targets 3, 4, and 5 follow the performance of the STS, but the target $T_1(T_2)$ experiences a performance loss of about 15 dB in the presence target $T_2(T_1)$ for the detection probability of 0.9 as compared with that of the STS. For the MTS_2, all targets approximately follow the performance of the STS. In this case, the off-grid targets experience slightly more losses as compared with the target T_1. To understand the decreased performance in detecting targets $8T_1$ and T_2 in the MTS_1, it is useful to analyze the DL experienced by various targets, as defined in (8.13). In order to do this, we plotted DL against the range distance ΔR between two targets, as shown in Figure 8.14. It is seen that

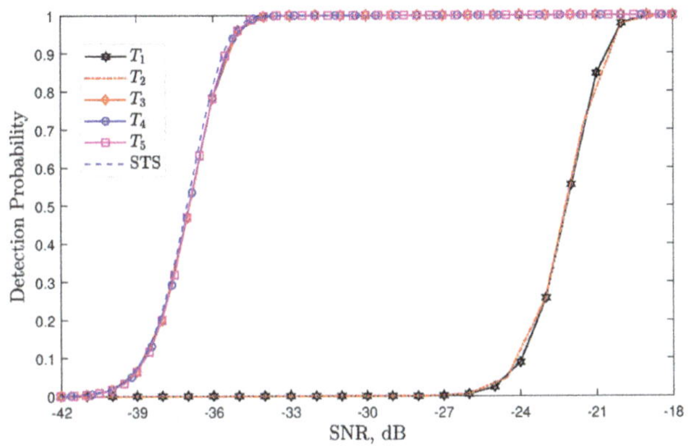

Figure 8.12 Detection probability comparison of different targets as a function of input SNR for the MTS_1 when $P_{fa} = 10^{-6}$

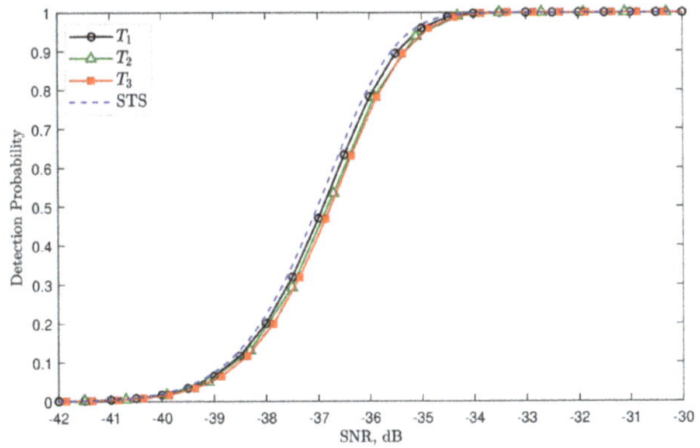

Figure 8.13 Detection probability comparison of different targets as a function of input SNR for the MTS_2 when $P_{fa} = 10^{-6}$

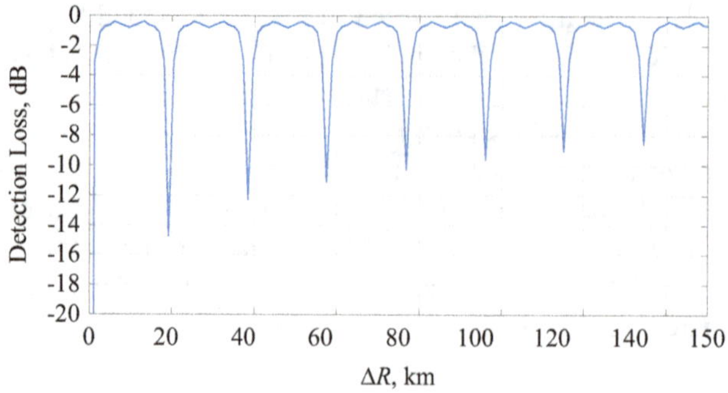

Figure 8.14 Range-dimension loss profile

the ATV signal's range ambiguities lead to significant losses at range distances corresponding to multiples of 19.2 km. These losses help to explain the reduced performance observed in Figures 8.9 and 8.12. Generally, this curve of DL versus the difference bistatic range (or Doppler) is known as the range-dimension loss profile (or Doppler-dimension loss profile), and it is useful for justifying the losses experienced by various targets in a multitarget scenario. By examining the range-dimension loss profile in Figure 8.14, we can conclude that the RS-CZT-MLa-GLR algorithm can detect the ranges of different targets but with some losses due to the range-dimension profile of the ATV signal.

In this chapter, a multitarget detection algorithm is introduced that can simultaneously determine both the range and Doppler coordinates of interesting targets. The proposed detector is distinct from previously existing ones in that it is designed to handle multitarget scenarios where the delay-Doppler coordinates of the targets are unknown. In contrast, existing detectors are only capable of working with the Doppler information of the received signal due to the difficulties posed by the range ambiguities of the ATV signal. It is important to note that the detector presented in Reference 15 assumes an STS and is not easily adapted for multitarget scenarios. As such, there is no previous research work that can be compared with the algorithm presented in this chapter. According to Algorithms 8.1 and 8.2, the number of complex multiplications (CMs) needed for part 1 of the proposed detectors is $\frac{3}{2}NP^2 + \frac{5}{2}NP + \frac{1}{2}P^2 + \frac{1}{2}P$. For the second part of the proposed robust detection algorithm with Q iterations ($Q \geq K$), the required number of CMs is approximately equal to $3NL_M^2(P-2) + NL_M(3Q+2P-2) + \frac{3}{2}NL_M^2Q^2 + \frac{5}{2}NL_M^2Q + QN'\log_2(N_{cz}) + 3NQ + (P+Q-1)L_M^2\log_2(L_M)$.

8.6 Summary

The signal used in ATV-based PBR is not under the control of radar system designers, so the only solution is to manage the signal's range ambiguities on the receiver side. This requires a sophisticated algorithm to prevent false alarms and target masking, particularly in the range dimension, and to estimate ranges in a multitarget scenario. To do this, a multilayer detector has been proposed in which one target is detected in each layer based on the GLR detector. Interference due to the detected target in the first layer was estimated and subtracted from the received signal. From the first layer's interference-canceled signal, another target (if it exists) was detected in the second layer. These detection and interference cancellation steps were carried out in each layer until all the targets were detected. To reduce the computational complexity, we use a chirp z-transform-based algorithm to compute the GLR-based detector over a limited frequency range. We also present a robust target detection algorithm to effectively cancel strong interference and off-grid targets. Our simulations demonstrate that the proposed detection algorithms can jointly estimate the number and delay-Doppler coordinates of desired targets.

References

[1] Griffiths H.D., Long N.R.W. 'Television-based bistatic radar'. *IEE Proceedings F Communications, Radar and Signal Processing*. 1986, vol. 133(7), pp. 649–57.

[2] Howland P.E. 'Target tracking using television-based bistatic radar'. *IEE Proceedings on Radar, Sonar and Navigation*. 1999, vol. 146(3), pp. 166–74.

[3] Baker C.J., Griffiths H.D., Papoutsis I. 'Passive coherent location radar systems. Part 2: Waveform properties'. *IEE Proceedings on Radar, Sonar and Navigation*. 2005, vol. 152(3), pp. 160–68.

[4] De Jong A.J., De Theije P.A.M., Gelsema S.J. 'NATO-RTO'. *Doppler-Bearing Tracking for Analog TV-Based Passive Radars*. 2007.

[5] Zaimbashi A. 'Target detection in analogue terrestrial TV-based passive radar sensor: Joint delay-Doppler estimation'. *IEEE Sensors Journal*. 2017, vol. 17(17), pp. 5569–80.

[6] Thrane N. *Zoom-FFT Tech*. Bruel-Kjaer; 1980.

[7] Lyons R. 'Understanding digital signal processing' in Prentice Hall PTR. 2nd edition. New Jersey (USA); 2004.

[8] Rabiner L., Schafer R., Rader C. 'The chirp z-transform algorithm'. *IEEE Transactions on Audio and Electroacoustics*. 1969, vol. 17(1), pp. 86–92.

[9] Rajmic P., Prusa Z., Wiesmeyr C. 'Computational cost of chirp z-transform and generalized Goertzel algorithm'. *22nd European Signal Processing Conference (EUSIPCO)*; 2014.

[10] Zaimbashi A., Derakhtian M., Sheikhi A. 'GLRT-based CFAR detection in passive bistatic radar'. *IEEE Transactions on Aerospace and Electronic Systems*. 2013, vol. 49(1), pp. 134–59.

[11] Zaimbashi A., Derakhtian M., Sheikhi A. 'Invariant target detection in multiband FM-based passive bistatic radar'. *IEEE Transactions on Aerospace and Electronic Systems*. 2014, vol. 50(1), pp. 720–36.

[12] Bailey D.H., Swarztrauber P.N. 'The fractional Fourier transform and applications'. *SIAM Review*. 1991, vol. 33(1), pp. 389–404.

[13] Kay, S.M. 'Fundamentals of statistical signal processing' in estimation theory. Englewood Cliffs, NJ: PTR Prentice-Hall; 1993.

[14] Zaimbashi A. 'Target detection in FM-based passive radars in the presence of interference signals in reference channel'. *Journal of Radar*. 2015, vol. 3(3), pp. 1–15.

[15] Wang H., Wang J., Zhong L. 'Mismatched filter for analogue TV-based passive bistatic radar'. *IET Radar, Sonar & Navigation*. 2011, vol. 5(5), pp. 573–81.

Chapter 9
Multi-accelerating-target detection in passive radar systems

9.1 Introduction

The range of passive radar can be enhanced by increasing the integration time, but this approach encounters limitations when attempting to detect accelerating targets with target detection algorithms presented in Chapters 2–8. Most passive radar target detection algorithms rely on a motion model that assumes constant target velocity. However, when attempting to extend integration time for accelerating targets, there is a loss in signal-to-noise ratio (SNR) due to the discrepancy between the assumed motion model and the actual one. Consequently, it becomes challenging to integrate the target's energy into a specific range-Doppler cell, resulting in significant detection losses due to range migration (RM) and Doppler frequency migration (DFM). These factors restrict any possible improvement in the SNR.

In Reference 1, a stretch processing algorithm has been introduced to address the issue of the RM caused by high-speed targets. However, this algorithm did not take into account the effects of target acceleration into the DFM. To address this limitation, References 2 and 3 introduced a modified cross-ambiguity function that considers the effect of the DFM and reduces SNR loss. Another approach proposed in Reference 4 involved a modified long-time coherent integration method based on short-time cross-correlation and two-step Doppler processing. The work in Reference 5 extended the findings of Reference 2 by introducing an extending cross-ambiguity function (ECAF) to estimate targets' range-Doppler-acceleration coordinates. Furthermore, a simplified algorithm was presented to reduce the computational complexity of the proposed ECAF. In Reference 6, a technique was suggested for DVB-T passive radar systems. The purpose of this method is to offset the range shift that occurs during integration time as a result of the target's non-zero bistatic velocity. Several approaches in both the time and frequency domains have been developed for compensating velocity and acceleration in active radar systems [7–9]. These approaches have also been applied to the PBR case, as seen in References 10–12. For instance, in Reference 10, the keystone transform (KT) was utilized to compensate for the range migration effect when using DVB-T signals. However, the authors only considered a constant velocity motion model and ignored the target acceleration. In Reference 11, compensation methods for range and Doppler migration based on KT and chirp-Fourier transform (CFT) were studied. Additionally, in Reference 12, a processing scheme was proposed that utilizes chirp-z transform (CZT) and fractional Fourier transform domain (FRFD) sharpness to compensate for range and Doppler migration effects.

This chapter presents a framework for accelerating target detection problem, consisting of signal modeling, detection methods, and statistical analysis. The received signal model is obtained, and the target detection problem is formulated as an M-ary hypothesis testing problem. The generalized likelihood ratio test (GLRT) principle is then applied to develop a new sequential detection algorithm that estimates the targets' range-Doppler-acceleration coordinates compared to the two- dimensional (2D) detectors presented in our previous chapters that only estimate range-Doppler coordinates. The proposed detection algorithm is implemented based on a modified chirp fast Fourier transform (MCFFT), which reduces the computational complexity of the detector. Additionally, a closed-form expression for the false alarm probability of the proposed detector is obtained to adjust the detection threshold. Simulation results demonstrate that the proposed three- dimensional (3D)-detection algorithm outperforms state-of-the-art classical target detection algorithms in the context of FM-based passive bistatic radar (PBR) systems. The material in this chapter is based on our previous works published in References 13 and 14.

This chapter is organized in the following manner: Section 9.2 introduces the formulation of the received signal model for multiple accelerating targets. The problem of detection is presented in section 9.3. A detection algorithm based on multilayer GLRT is proposed in section 9.4. The specific details of the proposed algorithm are discussed in section 9.5, and successive implementations aimed at reducing computational complexity are presented in section 9.6. Section 9.7 contains the performance analysis. Finally, the chapter concludes with a summary of the results in section 9.8.

9.2 Signal modeling

This section presents a general signal model for detecting multiple accelerating targets in a passive bistatic scenario. To begin, let us consider a single-target detection scenario, where the transmitter, receiver, and target's respective positions are denoted by $\mathbf{r}_T = [x_T, y_T, z_T]^T$, $\mathbf{r}_R = [x_R, y_R, z_R]^T$, and $\mathbf{r}_t = [x_t, y_t, z_t]^T$, respectively. The target is also assumed to have velocity and acceleration vectors of $\mathbf{v} = [v_x, v_y, v_z]^T$ and $\mathbf{a} = [a_x, a_y, a_z]^T$, respectively. As such, the baseband equivalent received signal can be described as

$$x(t) = \alpha s(t - \tau(t))e^{-j2\pi f_c \tau(t)} + w(t),\ 0 \leqslant t \leqslant T \tag{9.1}$$

where $s(t)$ and $w(t)$ are the baseband equivalent transmitted signal and the additive thermal noise received in the surveillance channel, respectively. Here, T is the integration time, α represents the unknown complex target reflectivity, $\tau(t)$ is the time-varying delay caused by velocity and acceleration of target, and f_c is the carrier frequency. In what follows, it is assumed that the integration time starts at time zero and ends at time T. According to the bistatic geometry shown in Figure 9.1, $\tau(t)$ can be computed as

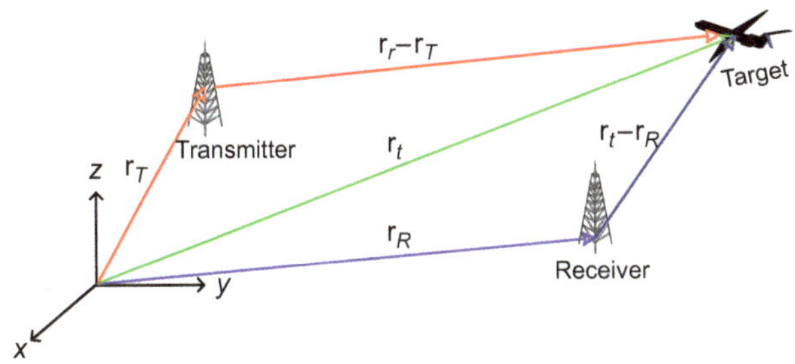

Figure 9.1 Passive bistatic radar configuration, where the transmitter, receiver, and target are located at $\mathbf{r}_T$, $\mathbf{r}_R$, *and* $\mathbf{r}_t$, *respectively*

$$\tau(t) = \frac{\| \mathbf{r}_t - \mathbf{r}_T \| + \| \mathbf{r}_t - \mathbf{r}_R \|}{c} \tag{9.2}$$

where c is the speed of light. In the following, we assume that the transmitter and receiver are stationary, while target moves with velocity of $\mathbf{v}$ and acceleration of $\mathbf{a}$ such that

$$\mathbf{r}_t = \mathbf{r}_0 + \mathbf{v}t + \frac{1}{2}\mathbf{a}t^2 \tag{9.3}$$

where $\mathbf{r}_0 = [x_0, y_0, z_0]^T$ is the location of the target at the beginning of the integration time. By substituting (9.3) in (9.2), we find

$$\tau(t) = \frac{\| \mathbf{r}_0 + \mathbf{v}t + \frac{1}{2}\mathbf{a}t^2 - \mathbf{r}_T \| + \| \mathbf{r}_0 + \mathbf{v}t + \frac{1}{2}\mathbf{a}t^2 - \mathbf{r}_R \|}{c} \tag{9.4}$$

By using the second-order Taylor approximation of $\tau(t)$ at $t = 0$, we obtain

$$\tau(t) = \tau(0) + \left.\frac{\partial \tau(t)}{\partial t}\right|_{t=0} \frac{(t-0)}{1!} + \left.\frac{\partial^2 \tau(t)}{\partial t^2}\right|_{t=0} \frac{(t-0)^2}{2!} \tag{9.5}$$

It can be shown that

$$\tau(0) = \frac{\| \mathbf{r}_0 - \mathbf{r}_T \| + \| \mathbf{r}_0 - \mathbf{r}_R \|}{c} \tag{9.6}$$

and

$$\left.\frac{\partial \tau(t)}{\partial t}\right|_{t=0} = \frac{1}{c}\left(\frac{\mathbf{v}^T(\mathbf{r}_0 - \mathbf{r}_T)}{\| \mathbf{r}_0 - \mathbf{r}_T \|} + \frac{\mathbf{v}^T(\mathbf{r}_0 - \mathbf{r}_R)}{\| \mathbf{r}_0 - \mathbf{r}_R \|}\right) \tag{9.7}$$

and

$$\left.\frac{\partial^2 \tau(t)}{\partial t^2}\right|_{t=0} = \frac{1}{c}\left(\frac{\mathbf{a}^T(\mathbf{r}_0 - \mathbf{r}_T)}{\|\mathbf{r}_0 - \mathbf{r}_T\|} + \frac{\|\mathbf{v}\|^2}{\|\mathbf{r}_0 - \mathbf{r}_T\|} + \frac{\mathbf{a}^T(\mathbf{r}_0 - \mathbf{r}_R)}{\|\mathbf{r}_0 - \mathbf{r}_R\|} + \frac{\|\mathbf{v}\|^2}{\|\mathbf{r}_0 - \mathbf{r}_R\|} - \frac{|\mathbf{v}^T(\mathbf{r}_0 - \mathbf{r}_T)|^2}{\|\mathbf{r}_0 - \mathbf{r}_T\|^3} - \frac{|\mathbf{v}^T(\mathbf{r}_0 - \mathbf{r}_R)|^2}{\|\mathbf{r}_0 - \mathbf{r}_R\|^3}\right) \tag{9.8}$$

By substituting (9.6), (9.7), and (9.8) into (9.5), we get

$$\tau(t) = \frac{R_0^{(1)}}{c} + \frac{v_b^{(1)}}{c} t + \frac{1}{2}\frac{a_b^{(1)}}{c} t^2 \tag{9.9}$$

where

$$R_0^{(1)} = \| \mathbf{r}_0 - \mathbf{r}_T \| + \| \mathbf{r}_0 - \mathbf{r}_R \| \tag{9.10}$$

$$v_b^{(1)} = \frac{\mathbf{v}^T(\mathbf{r}_0 - \mathbf{r}_T)}{\| \mathbf{r}_0 - \mathbf{r}_T \|} + \frac{\mathbf{v}^T(\mathbf{r}_0 - \mathbf{r}_R)}{\| \mathbf{r}_0 - \mathbf{r}_R \|} \tag{9.11}$$

and

$$a_b^{(1)} = \frac{\mathbf{a}^T(\mathbf{r}_0 - \mathbf{r}_T)}{\| \mathbf{r}_0 - \mathbf{r}_T \|} + \frac{\| \mathbf{v} \|^2}{\| \mathbf{r}_0 - \mathbf{r}_T \|} + \frac{\mathbf{a}^T(\mathbf{r}_0 - \mathbf{r}_R)}{\| \mathbf{r}_0 - \mathbf{r}_R \|} + \frac{\| \mathbf{v} \|^2}{\| \mathbf{r}_0 - \mathbf{r}_R \|} - \frac{| \mathbf{v}^T(\mathbf{r}_0 - \mathbf{r}_T) |^2}{\| \mathbf{r}_0 - \mathbf{r}_T \|^3} - \frac{| \mathbf{v}^T(\mathbf{r}_0 - \mathbf{r}_R) |^2}{\| \mathbf{r}_0 - \mathbf{r}_R \|^3} \tag{9.12}$$

where $v_b^{(1)}$ and $a_b^{(1)}$ are the bistatic velocity and bistatic acceleration of the considered target at the start of the integration time, respectively. Similarly, $R_0^{(1)}$ is the bistatic target range at the beginning of the integration time. Finally, by substituting (9.9) into (9.1), we obtain

$$x(t) = \alpha e^{-j2\pi \frac{R_0^{(1)}}{\lambda}} s\left(t - \frac{R_0^{(1)}}{c} - \frac{v_b^{(1)}}{c} t - \frac{1}{2}\frac{a_b^{(1)}}{c} t^2\right) \times e^{-j2\pi (f_b^{(1)} + \frac{1}{2}\frac{a_b^{(1)}}{\lambda} t)t} + w(t) \tag{9.13}$$

where $f_b^{(1)}$ is the bistatic Doppler frequency defined as $f_b^{(1)} = \frac{v_b^{(1)}}{\lambda}$, where λ is the carrier frequency wavelength, i.e., $\lambda = \frac{c}{fc}$. If the received signal $x(t)$ is sampled with a sampling frequency f_s at the time instants $t_n = \frac{n}{fs} = nT_s$, $n = 0, \ldots, N-1$, the nth sample of $x(t)$ can be described as

$$x[n] = \beta_1 s\left[n - \lceil \frac{R_0^{(1)} f_s}{c} + \frac{v_b^{(1)}}{c} n + \frac{1}{2}\frac{a_b^{(1)}}{c} n^2 T_s \rceil\right] \times e^{-j2\pi (f_b{}^{(1)} nT_s + \frac{1}{2}\frac{a_b{}^{(1)}}{\lambda}(nT_s)^2)} + w[n] \tag{9.14}$$

where $\beta_1 = \alpha e^{-j2\pi \frac{R_0^{(1)}}{\lambda}}$ is the complex received signal amplitude. Here, the assumption that the reference signal is delayed by integer number of samples is a reasonable approximation for sufficiently fast sampling. From (9.14), it is seen that the target acceleration leads to Doppler frequency migrations (DFMs) as well as range migrations (RMs), while very high-speed targets only cause RM. More precisely, the ratio of distance traveled by a target during the integration time to the range resolution length, denoted as ΔR, can be defined as

$$\delta R_T = \frac{v_b^{(1)} T + \frac{1}{2} a_b^{(1)} T^2}{\Delta R} \tag{9.15}$$

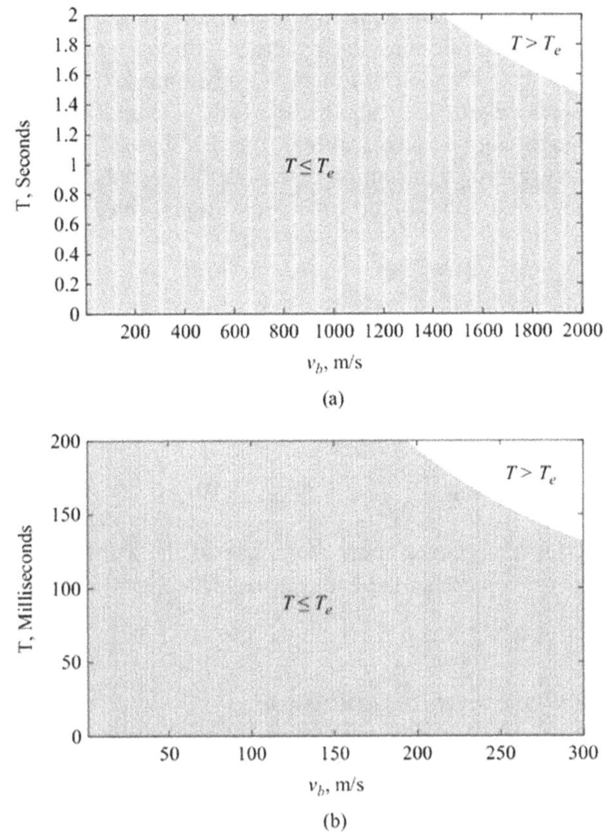

Figure 9.2 Limiting range migration condition versus target velocity when the target acceleration is equal to 60 m/s² and for (a) FM-based PBR with $\Delta R = 3$ *km (b) DVB-T-based PBR with* $\Delta R = 40$ *m*

The condition for limiting range migration is given by $\delta R_T \leq 1$. This is equivalent to the condition $T \leq T_e$, where

$$T_e = \frac{c}{a_b^{(1)}}\left(\sqrt{\left(\frac{v_b^{(1)}}{c}\right)^2 + \frac{2a_b^{(1)}\Delta R}{c^2}} - \frac{v_b^{(1)}}{c}\right) \tag{9.16}$$

The exit time T_e indicates the maximum duration of integration that can be conducted without experiencing range migration. It should be noted that the exit time is determined by various factors including the radar applications, such as the maximum speed and acceleration of the targets of interest, as well as the bandwidth of the exploited opportunity signals. Figure 9.2(a) and (b) depicts the evaluation of this condition against target velocity at a target acceleration of 60 m/s² for an FM-based PBR system with $\Delta R = 3$ km and a DVB-T-based PBR with $\Delta R = 40$ m, respectively. To avoid the range migration effect, the target acceleration, target velocity, and integration time must satisfy the condition given in (9.16), which is represented by the shaded region in Figure 9.2 where no RM occurs. Comparison between these two subfigures reveals that the DVB-T-based PBR system is more susceptible to range migration compared to the FM-based PBR system. This is mainly due to the high range resolution of the DVB-T signals, which amplifies the occurrence of range migration. In the radar applications with $T < T_e$, (9.14) reduces to

$$x[n] = \beta_1 s[n - n_0^{(1)}] e^{-j2\pi (f_b^{(1)} nT_s + \frac{1}{2}\frac{a_b^{(1)}}{\lambda}(nT_s)^2)} + w[n] \tag{9.17}$$

where $n_0^{(1)} = \lceil \frac{R_0^{(1)} f_S}{c} \rceil$. Similarly, in the presence of K accelerating targets, the received desired signal can be described as

$$x[n] = \sum_{k=1}^{K} \beta_k s[n - n_0^{(k)}] e^{-j2\pi (f_b^{(k)} nT_s + \frac{1}{2}\frac{a_b^{(k)}}{\lambda}(nT_s)^2)} + w[n] \tag{9.18}$$

where the delay-Doppler-acceleration coordinate of the kth target is denoted as $(n_0^{(k)}, f_b^{(k)}, a_b^{(k)})$, and β_k represents the complex received signal amplitude of the kth target.

9.3 Detection problem formulation

In this section, we formulate accelerating-target detection problem as a composite M-ary hypothesis testing problem, given by

$$\begin{cases} \mathscr{H}_0 : \mathbf{x} = \mathbf{Hc} + \mathbf{w} \\ \mathscr{H}_1 : \mathbf{x} = \mathbf{s}_1\beta_1 + \mathbf{Hc} + \mathbf{w} \\ \mathscr{H}_2 : \mathbf{x} = \mathbf{s}_2\beta_2 + \mathbf{s}_1\beta_1 + \mathbf{Hc} + \mathbf{w} \\ \mathscr{H}_3 : \mathbf{x} = \mathbf{s}_3\beta_3 + [\mathbf{s}_1, \mathbf{s}_2][\beta_1, \beta_2]^T + \mathbf{Hc} + \mathbf{w} \\ \quad \vdots \quad : \qquad\qquad \vdots \\ \mathscr{H}_m : \mathbf{x} = \mathbf{s}_m\beta_m + [\mathbf{s}_1, \ldots, \mathbf{s}_{m-1}][\beta_1, \ldots, \beta_{m-1}]^T + \mathbf{Hc} + \mathbf{w} \\ \quad \vdots \quad : \qquad\qquad \vdots \end{cases} \tag{9.19}$$

For example, hypothesis $\mathscr{H}_K$ denotes the presence of the K targets in the presence of the system thermal noise, direct signal and clutter/multipath returns, in which the received signal vector $\mathbf{x}$ can be compactly described by

$$\mathbf{x} = \mathbf{S}\beta + \mathbf{Hc} + \mathbf{w} \tag{9.20}$$

Here, we assume that N samples are collected for both reference and surveillance channels over the integration time T, i.e., $N = Tf_S$. In (9.20), $\mathbf{x}$ and $\mathbf{w}$ are formed from discrete samples of $x[n]$ and $w[n]$ for $n = 0, \ldots, N - 1$, respectively. In the following, the surveillance channel noise $\mathbf{w}$ is assumed to be a zero-mean white Gaussian noise with covariance matrix $\sigma^2\mathbf{I}_N$, where the noise variance σ^2 is an unknown parameter. Clutter signature matrix $\mathbf{H}$ is an $N \times P$ matrix defined as:

$$\mathbf{H} \triangleq \left[\mathbf{h}_1, ..., \mathbf{h}_p, ..., \mathbf{h}_P\right] \tag{9.21}$$

where $\mathbf{h}_p = (\mathbf{y}^{n_c^{(p)}} \odot \mathbf{e}^{f_c^{(p)}})$ in which $[\mathbf{e}^{f_c^{(p)}}]_n = e^{-j2\pi f_c^{(p)} nT_s}$ with $n = 0, ..., N-1$; $\mathbf{y}^{n_c^{(p)}} = \mathbf{P}^{n_c^{(p)}}\mathbf{y}$, where $\mathbf{y}$ is the cleaned reference signal obtained after exploiting some signal conditioning algorithms, such as beamforming and channel equalization in the reference channel [10, 11]; $\mathbf{P}$ is an $N \times N$ permutation matrix defined as $[\mathbf{P}]_{ij} = 1$ if $i = j + 1$ and 0 otherwise for $i = 0, ..., N-1; j = 0, ..., N-1$. Here, the clutters and direct path signal are modeled as multipaths characterized by the delays and Doppler frequencies in the low-frequency region. It is assumed that there are P significant paths where $(n_c^{(p)}, f_c^{(p)})$ represent the delay-Doppler coordinates of the pth clutter signature, determined based on the delay-Doppler coordinates of a predefined clutter signature [15, 16]. In (9.20), $\mathbf{c} = [c_1, ..., c_P]^T$ contains the unknown complex amplitudes of all P clutter signatures. Consequently, P will be much larger than the actual number of multi-paths in the surveillance channel.

Finally, target signature matrix $\mathbf{S}$ is an $N \times K$ matrix defined as

$$\mathbf{S} \triangleq \left[\mathbf{s}_1, ..., \mathbf{s}_K\right] \tag{9.22}$$

where

$$\mathbf{s}_k = (\mathbf{y}^{(n_r^{(k)})} \odot \mathbf{e}^{(f_b^{(k)})} \odot \mathbf{e}^{(a_b^{(k)})}) \tag{9.23}$$

where $(n_r^{(k)}, f_b^{(k)}, a_b^{(k)})$s for $k = 1, ..., K$ represent the delay-Doppler-acceleration coordinates of the interested targets at the beginning of the integration time. Here, $\mathbf{y}^{n_r^{(k)}} = \mathbf{P}^{n_r^{(k)}}\mathbf{y}$; $[\mathbf{e}^{(f_b^{(k)})}]_n = e^{-j2\pi f_b^{(k)} nT_s}$, and $[\mathbf{e}^{(a_b^{(k)})}]_n = e^{-j2\pi \frac{1}{2} \frac{a_b^{(k)}}{\lambda}(nT_s)^2}$ with $n = 0, ..., N-1$; is the K-dimensional column vector containing all complex amplitudes of the targets under test. In (9.23), $n_r^{(k)}$ is the relative time delay index with respect to the direct signal received in the reference channel.

9.4 Multilayer GLR-based detection algorithm

To solve the M-ary hypothesis testing problem (9.19), we adopt a strategy of detecting targets in a layered manner. To do this, the M-ary hypothesis testing problem is considered as a sequence of multiple binary hypothesis $(\mathcal{H}_0, \mathcal{H}_1)$, $(\mathcal{H}_1, \mathcal{H}_2)$, ..., $(\mathcal{H}_{m-1}, \mathcal{H}_m)$, ..., $(\mathcal{H}_K, \mathcal{H}_{K+1})$ to find K desired targets (if exist). Thus, it is called a multilayer detection algorithm. In each layer, we have a binary hypothesis testing problem to detect one target. For example, to test $\mathcal{H}_m$ against $\mathcal{H}_{m-1}$, we have the following binary composite hypothesis testing problem:

$$\begin{cases} \mathcal{H}_{m-1} : \mathbf{x} = \mathbf{T}_{m-1}\mathbf{g}_{m-1} + \mathbf{Hc} + \mathbf{w} \\ \mathcal{H}_m : \mathbf{x} = \mathbf{s}_m\beta_m + \mathbf{T}_{m-1}\mathbf{g}_{m-1} + \mathbf{Hc} + \mathbf{w} \end{cases} \tag{9.24}$$

where $\mathbf{T}_{m-1} = [\mathbf{s}_1, ..., \mathbf{s}_{m-1}]$ and $\mathbf{g}_{m-1} = [\beta_1, ..., \beta_{m-1}]^T$. In this formulation, $\mathbf{s}_m$ represents the mth target under test vector to be detected (if exist), while columns of matrix $\mathbf{T}_{m-1}$ spanned the subspace of interfering targets detected in previous layers.

The GLRT principle is employed to create a detector for each layer in this chapter. As a result, the detection method is referred to as 3-dimensional multi-layer GLRT, abbreviated to as 3D-MLa-GLRT. In the problem at hand, the set of unknown parameters is denoted by $\Theta_m = \{(n_r^{(m)}, f_b^{(m)}, a_b^{(m)}), \beta_m, \mathbf{g}_{m-1}, \mathbf{c}, \sigma^2\}$ and $\Theta_{m-1} = \{\mathbf{g}_{m-1}, \mathbf{c}, \sigma^2\}$ under the hypotheses $\mathscr{H}_m$ and $\mathscr{H}_{m-1}$, respectively. Similar to Chapter 7, but for accelerating targets, the 3D-MLa-GLRT test can be obtained as

$$\mathscr{L}_m(\mathbf{x}) = 2N \max_{(n_r^{(m)}, f_b^{(m)}, a_b^{(m)})} d(n, f, a) \underset{\mathscr{H}_{m-1}}{\overset{\mathscr{H}_m}{\gtrless}} \eta_m \tag{9.25}$$

where

$$d(n, f, a) = \frac{|\mathbf{s}(n, f, a)^H \mathbf{x}^{\perp(P+m-1)}|^2}{\|\Pi^{\perp}_{\mathbf{U}(P+m-1)} \mathbf{s}(n, f, a)\|^2 \|\mathbf{x}^{\perp(P+m-1)}\|^2} \tag{9.26}$$

and

$$\eta_m = \frac{N\eta}{N - (P + m) - \eta} \tag{9.27}$$

where $\mathbf{x}^{\perp(P+m-1)} = \Pi^{\perp}_{\mathbf{U}(P+m-1)} \mathbf{x}$ represents the component of $\mathbf{x}$ orthogonal to the space spanned by the columns of interference matrix $\mathbf{U}_{(P+m-1)} = [\mathbf{H}, \mathbf{T}_{m-1}]$, where $\mathbf{U}_{(P)} = \mathbf{H}$, in which $(m-1)$ is the number of interfering targets that have been detected. Here, the projection matrix $\Pi^{\perp}_{\mathbf{U}(P+m-1)}$ is given by $\Pi^{\perp}_{\mathbf{U}(P+m-1)} = \mathbf{I} - \Pi_{\mathbf{U}(P+m-1)}$ and $\Pi_{\mathbf{U}(P+m-1)} = \mathbf{U}_{(P+m-1)}(\mathbf{U}^H_{(P+m-1)} \mathbf{U}_{(P+m-1)})^{-1} \mathbf{U}^H_{(P+m-1)}$. In (9.27), the threshold η should be selected according to a predetermined false alarm probability (p_{fa}) such that

$$p_{fa} = \zeta L_{3D} e^{-0.5\eta} \tag{9.28}$$

where L_{3D} is the number of delay-Doppler-acceleration coordinates searched over, and ζ is a value less than one. More precisely, the probability of false alarm, p_{fa}, is dependent on the real number of resolution cells, which is lower than L_{3D} and is indicated by ζL_{3D}.

As previously discussed, the 3D-MLa-GLR detection algorithm tests $\mathscr{H}_1$ against $\mathscr{H}_0$ first. If the binary GLRT statistic accepts $\mathscr{H}_1$, meaning that $\mathscr{L}_1(\mathbf{x}) > \eta_1$ for a predetermined false alarm probability, then it proceeds to test $\mathscr{H}_2$ against $\mathscr{H}_1$, and so on, until a hypothesis is rejected, i.e., $\mathscr{L}_m(\mathbf{x}) < \eta_m$. When testing $\mathscr{H}_m$ against $\mathscr{H}_{m-1}$ and $\mathscr{H}_m$ is accepted, a new target is declared as detected, and the test procedure moves on to the next hypothesis between $\mathscr{H}_{m+1}$ and $\mathscr{H}_m$. The procedure stops when the null hypothesis ($\mathscr{H}_{m-1}$) is accepted, indicating that no further targets can be detected. It is important to note that for large values of N, typical in PBR systems, the threshold η_m of (9.27) can be approximated by η, so the detection threshold is not affected by the number of targets detected.

9.5 Chirp-FFT implementation of 3D-MLa-GLRT

Unlike the detection algorithms discussed in Chapters 7 and 8, the 3D-MLa-GLR detection algorithm proposed here relies on the range-Doppler-acceleration coordinates of the targets. This means that we need to calculate $d(n,f,a)$ to create a delay-Doppler-acceleration map, where n ranges from 1 to N_d, with N_d representing the number of time delay indexes corresponding to the maximum relative bistatic range of interest, $R_{\max}$. Additionally, $|f_b| \le f_{\max}$, where $f_{\max}$ is the maximum bistatic Doppler frequency, and $|a_b| \le a_{\max}$, where $a_{\max}$ is the maximum bistatic acceleration of the targets of interest. To implement the proposed detection algorithm, this section utilizes a computationally efficient algorithm based on a modified version of the discrete chirp-fast Fourier transform (DCFT). Before proceeding with this, let us first introduce the DCFT.

Definition 1. *The N-point discrete chirp-Fourier transform (DCFT) of discrete signal* $b[n]$ *for* $n = 0, ..., N-1$ *with length* N *can be defined as* [17–19]

$$[\mathscr{P}_0(\mathbf{b}; N)]_{k,l} = \sum_{n=0}^{N-1} b[n] W_N^{\frac{l}{N}n^2+kn}, \quad 0 \le k, l \le N-1 \tag{9.29}$$

where $\mathbf{b} = [b[0], ..., b[N-1]]^T$ *and* $W_N = e^{-j2\pi/N}$. *In* (9.29), $\mathscr{P}_0(\mathbf{b}; N)$ *is an* $N \times N$*-dimensional matrix whose* (k, l)*th element is denoted by* $[\mathscr{P}_0(\mathbf{b}; N)]_{k,l}$, *where k and l represent the bin frequencies and the chirp rates, respectively. Interestingly, the lth column of matrix* $\mathscr{P}_0(\mathbf{b}; N)$, *denoted as* $[\mathscr{P}_0(\mathbf{b}; N)]_{*l}$, *can be obtained from the discrete Fourier Transform (DFT) of vector* $[b[0]W_N^0, ..., b[n]W_N^{l\frac{(N-1)^2}{N}}]^T$.

Based on the definition 1, the numerator of the proposed GLR detector in (9.26) can be computed as follows:

$$\begin{aligned}(n_r^{(m)}, f_b^{(m)}, a_b^{(m)})^H \mathbf{x}^{\perp(P+m-1)} &= \sum_{n=0}^{N-1} x^{\perp(P+m-1)}[n] s^*[n - n_r^{(m)}] e^{-j2\pi f_b^{(m)} nT_s} \\ &\quad \times e^{-j2\pi \frac{1}{2} \frac{a_b^{(m)}}{\lambda} (nT_s)^2}\end{aligned} \tag{9.30}$$

It is worth noting that this term of the proposed detector is the discrete version of the modified complex cross-ambiguity function defined in [5]. Let us introduce $f_b^{(m)} = (k_m - \frac{N}{2})\frac{f_s}{N}$ and $a_b^{(m)} = (l_m - \frac{N}{2})\frac{2\lambda}{N^2 T_s^2}$. Now, (9.30) can be rewritten as

$$\begin{aligned}\mathbf{s}(n_r^{(m)}, f_b^{(m)}, a_b^{(m)})^H \mathbf{x}^{\perp(P+m-1)} &= \sum_{n=0}^{N-1} x^{\perp(P+m-1)}[n] s^*[n - n_r^{(m)}] W_N^{\frac{l_m}{N}n^2+k_m n} \\ &= [\mathscr{P}_{n_r^{(m)}}(\mathbf{b}_{n_r^{(m)}}; N)]_{k_m, l_m}\end{aligned} \tag{9.31}$$

where $\mathbf{b}_{n_r^{(m)}} = \mathbf{x}^{\perp(P+m-1)} \odot \mathbf{P}^{n_r^{(m)}} \mathbf{s}^*$. To efficiently compute (9.31), we may use N_f-point fast Fourier transform (FFT) with $N_f = 2^\gamma \ge N$ instead of an N-point DFT, i.e.,

$$\begin{aligned}\mathbf{s}(n_r^{(m)}, f_b^{(m)}, a_b^{(m)})^H \mathbf{x}^{\perp(P+m-1)} &= \sum_{n=0}^{N_f-1} [\mathbf{z}_{n_r^{(m)}}]_n W_{N_f}^{\frac{l_m}{N_f}n^2+k_m n} \\ &= [\mathscr{P}_{n_r^{(m)}}(\mathbf{z}_{n_r^{(m)}}; N_f)]_{k_m, l_m}\end{aligned} \tag{9.32}$$

where $\mathbf{z}_{n_r^{(m)}} = [\mathbf{b}^T_{n_r^{(m)}}, \mathbf{0}^T_{N_f-N}]^T$, where $\mathbf{0}_{N_f-N}$ is the vector of $N_f - N$ zeros. In this case, we have $f_b^{(m)} = (k_m - \frac{N_f}{2})\frac{f_s}{N_f}$ and $a_b^{(m)} = (l_m - \frac{N_f}{2})\Delta a$ with $\Delta a = 2\lambda\left(\frac{N}{N_f}\right)^2\frac{1}{T^2}$. It should be noted that the width of a Doppler bin is given by $\Delta f = \frac{1}{T}$ satisfying $\Delta f \geq \frac{f_s}{N_f}$, and the acceleration bin width is equal to Δa. In the sequel, this new transformation is called MCFFT. For notational simplicity, let us define tensors $\mathscr{D}$ and $\mathscr{C}$ whose elements are $[\mathscr{D}]_{n_r^{(m)},k_m,l_m} = d(n_r^{(m)}, f_b^{(m)}, a_b^{(m)})$ and $[\mathscr{C}]_{n_r^{(m)},k_m,l_m} = [\mathscr{P}_{n_r^{(m)}}]_{k_m,l_m}$, respectively. Thus, a 3D-data matrix is created for a desired delay-Doppler-acceleration map of dimension $N_d \times N_f' \times N_a$, in which N_d, N_f', and N_a are the number of time delay index, Doppler frequency bins, and acceleration bins, respectively. Here, the N_d, N_f', and N_a can be defined by $N_d = \lceil \frac{R_{\max}}{\Delta r} \rceil$ with $\Delta r = cT_s$, $N_f' = \lceil \frac{2f_{\max}}{\Delta f} \rceil$ with $\Delta f = \frac{f_s}{N_f}$, and $N_a = \lceil \frac{2a_{\max}}{\Delta a} \rceil$, respectively. Finally, the test statistic $d(n, f, a)$ can be given by

$$[\mathscr{D}]_{n,f,a} = \frac{[\mathscr{C} \odot \mathscr{C}^*]_{n,f,a}}{\|\Pi^{\perp}_{\mathbf{U}(P+m-1)}\mathbf{s}(n,f,a)\|^2\|\mathbf{x}^{\perp(P+m-1)}\|^2} \tag{9.33}$$

In the sequel, the detector with the test statistic of $2N\max_{(n_r^{(m)}, f_b^{(m)}, a_b^{(m)})}[\mathscr{D}]_{n,f,a}$ is referred to as 3D-MCFFT-MLa-GLR detector.

9.6 Successive implementation of 3D-MCFFT-MLa-GLRT

The proposed method for detecting targets involves a layered approach, in which a binary GLRT is used to detect new targets in each layer. Once a target is detected, the interference matrix is updated using that information, allowing for the detection of weaker targets in subsequent layers. This means that the size of the interference matrix increases with each layer, with an additional column added in each subsequent layer. For example, the interference matrix is $\mathbf{U}_{(P)} = \mathbf{H}$ in the first layer, but becomes $\mathbf{U}_{(P+1)} = [\mathbf{H}, \mathbf{s}_1]$ in the second layer. As a result, the computational complexity of the detection process also increases. Using the recursive approach for the orthogonal projection matrix, $\Pi^{\perp}_{\mathbf{U}(P+m-1)}$ can simplify the complexity. For instance, the implementation of $\Pi^{\perp}_{\mathbf{U}(P+1)}$ in the second layer can be achieved by using $\Pi^{\perp}_{\mathbf{U}(P+1)} = \Pi^{\perp}_{\mathbf{U}(P)} - \frac{\Pi^{\perp}_{\mathbf{U}(P)}\mathbf{s}_1\mathbf{s}_1^H\Pi^{\perp}_{\mathbf{U}(P)}}{\mathbf{s}_1^H\Pi^{\perp}_{\mathbf{U}(P)}\mathbf{s}_1}$ and so on for the successive layers. This approach enables us to recursively implement $\Pi^{\perp}_{\mathbf{U}(P+1)}\mathbf{x}$ as follows:

$$\Pi^{\perp}_{\mathbf{U}(P+1)}\mathbf{x} = \Pi^{\perp}_{\mathbf{U}(P)}\mathbf{x} - \frac{\Pi^{\perp}_{\mathbf{U}(P)}\mathbf{s}_1\mathbf{s}_1^H\Pi^{\perp}_{\mathbf{U}(P)}}{\mathbf{s}_1^H\Pi^{\perp}_{\mathbf{U}(P)}\mathbf{s}_1}\mathbf{x} \tag{9.34}$$

where $\Pi^{\perp}_{\mathbf{U}(P)} = \mathbf{I} - \mathbf{U}_{(P)}(\mathbf{U}^H_{(P)}\mathbf{U}_{(P)})^{-1}\mathbf{U}^H_{(P)}$. In (9.34), the first term is a cancellation operation that removes the interference due to direct and clutter signals.

The detection algorithm based on the above principle is referred to as 3D-successive and recursive (SaR) detection algorithm, referred to as 3D-SaR, and is described below:

Step 1: Set the threshold η according to (9.28), $\mathbf{U}_{(P)} = \mathbf{H}$ and set the layer index equal to $m := 0$.

Step 2: Remove the direct signal and clutter by computing $\mathbf{x}^{\perp(P)} = \mathbf{x} - \mathbf{U}_{(P)}\mathbf{R}_P\mathbf{U}_{(P)}^H\mathbf{x}$, where $\mathbf{R}_P = (\mathbf{U}_{(P)}^H\mathbf{U}_{(P)})^{-1}$.

Step 3: Compute tensor $\mathscr{D}$.

Step 4: If $\max_{n,k,l}\mathscr{D} > \eta$ (i.e., there is a new target), then increase m by 1 ($m := m+1$) and continue; otherwise terminate.

Step 5: Find and save the location of maximum $\mathscr{D}$ as the MLE of delay-Doppler-acceleration coordinate of mth target denoted as $(n_r^{(m)}, f_b^{(m)}, a_b^{(m)})$.

Step 6: Construct $\mathbf{u}_{P+m} = (\mathbf{y}^{(n_r^{(m)})} \odot \mathbf{e}^{(f_b^{(m)})} \odot \mathbf{e}^{(a_b^{(m)})})$. Here, $\mathbf{u}_{P+m}$ is the $(P+m)$th column of the interference matrix $\mathbf{U}_{(P+m)}$.

Step 7: Calculate $\mathbf{u}_{P+m}^{\perp(P)} = \mathbf{u}_{P+m} - \mathbf{U}_{(P)}\mathbf{R}_P\mathbf{U}_{(P)}^H\mathbf{u}_{P+m}$.

Step 8: For $k = 1, ..., m-1$, calculate

$$\mathbf{u}_{P+m}^{\perp(P+k)} = \mathbf{u}_{P+m}^{\perp(P+k-1)} - \frac{\mathbf{u}_{P+m-1}^{\perp(P+k-1)^H}\mathbf{u}_{P+m}^{\perp(P+k-1)}}{\mathbf{u}_{P+m-1}^{\perp(P+k-1)^H}\mathbf{u}_{P+m-1}^{\perp(P+k-1)}}\mathbf{u}_{P+m-1}^{\perp(P+k-1)}.$$

Step 9: Calculate $\mathbf{x}^{\perp(P+m)} = \mathbf{x}^{\perp(P+m-1)} - \dfrac{\mathbf{u}_{P+m}^{\perp(P+m-1)^H}\mathbf{x}^{\perp(P+m-1)}}{\mathbf{u}_{P+m}^{\perp(P+m-2)^H}\mathbf{u}_{P+m}^{\perp(P+m-1)}}\mathbf{u}_{P+m}^{\perp(P+m-1)}.$

Step 10: Return to Step 3.

To implement the proposed SaR detection algorithm, an approximate calculation indicates that a total of $NP^2 + 3NP + \mathscr{O}(P^2\log_2 P) + P^2 + 3NN_dN_aQ + QN_dN_aN_f\log_2(N_f) + 2NPQ + 10NQ + P^2Q + N_f'N_aN_d$ complex multiplications are required. This high level of computational complexity indicates that the proposed detectors will need to calculate the desired range-velocity-acceleration map numerous times, equivalent to the number of targets present. However, since the transmitted waveforms used by passive radar systems are beyond the control of passive radar system designers, all limitations of the opportunity signals must be managed at the receiver. This emphasizes the importance of a sophisticated algorithm implemented in the receiver of passive radar systems.

9.7 Performance results

This section is split into two subsections. The first subsection assesses the detection performance of the UMPI test of Chapter 4 for single-channel scenarios and in the presence of accelerating targets. This UMPI detector is referred to as the 2D-SaR detection algorithm. The second subsection employs Monte Carlo simulations to evaluate the effectiveness of the proposed 3D-SaR detection algorithm and compare it with the traditional detection algorithm.

9.7.1 Detection performance of the 2D-SaR detection algorithm

The aim is to conduct an analytical investigation on the deterioration of detection performance of the UMPI test outlined in Chapter 4 for single-channel scenarios when a maneuvering target is present. This detector was originally developed based on a target model with constant velocity. When dealing with a target that is maneuvering, the signal received by the surveillance channel can be described as follows:

$$\mathbf{x}_a = \mathbf{s}_a\alpha + \mathbf{C}\mathbf{a}_c + \mathbf{n}_s \tag{9.35}$$

where $\mathbf{s}_a = (\mathbf{r}_r^{(n_0,(1-\frac{v_b}{c}))} \odot \mathbf{e}^{(\Omega_0)} \odot \mathbf{a})$ in which the nth element of vector $\mathbf{a}$ is given by $[\mathbf{a}]_n = e^{j\frac{\pi}{\lambda}a_b(n/f_s)^2}$ with a_b and λ being target bistatic acceleration and the carrier frequency wavelength, respectively, and $\mathbf{r}_r^{(n_0,(1-\frac{v_b}{c}))}$ is a time-delayed and time-scaled version of vector $\mathbf{r}_r$ by n_0 and $(1-\frac{v_b}{c})$, respectively.

Let us define

$$\zeta = \left(\frac{\mathbf{x}_a}{\sigma}\right)^H \left(\mathbf{P}_C^\perp - \mathbf{P}_H^\perp\right)\left(\frac{\mathbf{x}_a}{\sigma}\right)^H \tag{9.36}$$

$$\varsigma = \left(\frac{\mathbf{x}_a}{\sigma}\right)^H \mathbf{P}_H^\perp \left(\frac{\mathbf{x}_a}{\sigma}\right) \tag{9.37}$$

to write the UMPI test as

$$\Lambda(\mathbf{x}_a) = \frac{N-(P+1)}{1}\frac{\zeta}{\varsigma} \tag{9.38}$$

The statistic ζ can be written under $\mathcal{H}_0$ as follows:

$$\zeta = \left(\frac{\mathbf{C}\mathbf{a}_c + \mathbf{n}_s}{\sigma}\right)^H (\mathbf{P}_C^\perp - \mathbf{P}_H^\perp)\left(\frac{\mathbf{C}\mathbf{a}_c + \mathbf{n}_s}{\sigma}\right) \tag{9.39}$$

Since $\mathbf{P}_C^\perp\mathbf{C} = 0$ and $\mathbf{P}_H^\perp\mathbf{C} = 0$, (9.39) reduces to $\zeta = (\frac{\mathbf{n}_s}{\sigma})^H(\mathbf{P}_C^\perp - \mathbf{P}_H^\perp)(\frac{\mathbf{n}_s}{\sigma})$. Under $\mathcal{H}_0$, we have $(\frac{\mathbf{n}_s}{\sigma}) \sim \mathcal{CN}(0, \mathbf{I})$, it follows that $\zeta \sim \chi^2_{k_n}$, where it is easy to show that $k_n = \text{rank}(\mathbf{P}_C^\perp - \mathbf{P}_H^\perp) = 1$. Under $\mathcal{H}_1$, we have $(\frac{\mathbf{x}_a}{\sigma}) \sim \mathcal{CN}(\frac{\mathbf{s}_a\alpha + \mathbf{C}\mathbf{a}_c}{\sigma}, \mathbf{I})$ and, in turn, this results in $\zeta \sim \chi_1^2(\delta_a)$, where the noncentrality parameters δ_a can be obtained from $\delta_a = \left(\frac{\alpha^* \mathbf{s}_a^H + \mathbf{a}_c^H\mathbf{C}^H}{\sigma}\right)(\mathbf{P}_C^\perp - \mathbf{P}_H^\perp)\left(\frac{\mathbf{s}_a\alpha + \mathbf{C}\mathbf{a}_c}{\sigma}\right)$. Since $\mathbf{P}_C^\perp\mathbf{C} = 0$ and $\mathbf{P}_H^\perp\mathbf{C} = 0$, and making use of $(\mathbf{P}_C^\perp - \mathbf{P}_H^\perp) = \frac{\mathbf{P}_C^\perp \mathbf{s}\mathbf{s}^H \mathbf{P}_C^\perp}{\mathbf{s}^H\mathbf{P}_C\mathbf{s}}$ [15] helps us to obtain $\delta_a = \frac{|\alpha|^2}{\sigma^2}\frac{|\mathbf{s}_a^H\mathbf{P}_C^\perp\mathbf{s}|^2}{\mathbf{s}^H\mathbf{P}_C^\perp\mathbf{s}^H}$. We can proceed along the same line to obtain

$$\varsigma \sim \begin{cases} \chi^2_{k_d} & \text{under } \mathcal{H}_0 \\ \chi^2_{k_d} & \text{under } \mathcal{H}_1 \end{cases} \tag{9.40}$$

where it is easy to show that $k_d = \text{tr}(\mathbf{I}_N - (\mathbf{H}^H\mathbf{H})^{-1}\mathbf{H}^H\mathbf{H}) = N-(P+1)$. Finally, since $\mathbf{P}_H^\perp(\mathbf{P}_C^\perp - \mathbf{P}_H^\perp) = \mathbf{P}_H^\perp(\mathbf{P}_H - \mathbf{P}_C) = 0$, it can be concluded that the numerator and the denominator of (9.38) are independent. Hence, the distribution of the UMPI detector statistic in the presence of highly maneuvering target becomes

$$\Lambda(\mathbf{x}_a) \sim \begin{cases} F_{1,N-(P+1)} & \text{under } \mathcal{H}_0 \\ F_{1,N-(P+1)}(\delta_a) & \text{under } \mathcal{H}_1 \end{cases} \tag{9.41}$$

where $F_{1,N-(P+1)}(\delta_a)$ denotes a noncentral complex F distribution with 1 numerator complex degree of freedom and $N-(P+1)$ denominator complex degrees of freedom and the noncentrality parameter δ_a. As a result, the detection probability of the UMPI test of Chapter 4 in the presence of highly maneuvering target can be obtained as

$$P_d^{(a)} = e^{-(\delta_a+\eta)} \sum_{k=0}^{\infty} \left(\frac{\delta_a^{\ k}}{k!} \sum_{n=0}^{k-1} \frac{\eta^n}{n!} \right) \tag{9.42}$$

where

$$\delta_a = \frac{|\alpha|^2}{\sigma^2} \frac{|\mathbf{s}_a^H \mathbf{P}_C^{\perp} \mathbf{s}|^2}{\mathbf{s}^H \mathbf{P}_C^{\perp} \mathbf{s}^H} \tag{9.43}$$

It can be observed that the detection performance of the UMPI test is affected by a maneuvering target through the noncentrality parameter δ_a. In this case, the maximum SNR gain (or maximum integration gain) provided by the UMPI detector is $G_m^{(a)} = \frac{|\mathbf{s}_a^H \mathbf{s}|^2}{\mathbf{s}^H \mathbf{s}}$ instead of $G_m = \mathbf{s}^H\mathbf{s}$. In a simplified case where $\mathbf{r}_r^{(n_0,(1-\frac{v_b}{c}))} \approx \mathbf{r}_r^{(n_0)}$, we can write $\mathbf{s}_a \approx \mathbf{s} \odot \mathbf{a}$. As a result, we obtain $\mathbf{s}_a^H \mathbf{s}_a \approx (\mathbf{a}^H \odot \mathbf{a})(\mathbf{s}^H\mathbf{s}) = \mathbf{s}^H\mathbf{s}$. Using the Cauchy-Schwarz inequality, we have $|\mathbf{s}_a^H\mathbf{s}|^2 \leq (\mathbf{s}^H\mathbf{s})(\mathbf{s}_a^H\mathbf{s}_a)$, where equality holds for a target with no acceleration. Thus, we can conclude that $G_m^{(a)} < G_m$ or $\delta_a < \delta$ when $a_b \neq 0$. Moreover, P_d and $P_d^{(a)}$ in (3.69) and (9.42) have the same form, but they are monotonically increasing functions of different non-centrality parameters δ and δ_a, respectively. Since $\delta_a < \delta$, we can conclude that $P_d^{(a)} < P_d$. This means that the detection performance of the UMPI detector degrades in the presence of an accelerating target. Taking this into account, we can establish a factor that measures the reduction in detection for maneuvering targets, given by

$$L_f = \frac{G_m^{(a)}}{G_m} = \frac{|\mathbf{s}_a^H\mathbf{s}|^2}{|\mathbf{s}^H\mathbf{s}|^2} \tag{9.44}$$

In general, we have $L_f \leq 1$, which means that the UMPI detector may experience a detection loss factor in the presence of maneuvering targets.

To evaluate the performance of the UMPI test, we use an FM radio signal with the carrier frequency of 100 MHz as waveform of opportunity which is sampled by sampling frequency of 200 kHz. For the UMPI detector, a comparison of detection performance and loss factor for different values of bistatic velocity and acceleration is carried out by evaluating (9.42) and (9.44), respectively.

Figure 9.3 depicts the relationship between the loss factor and integration time for various bistatic velocity values, when no acceleration target is present. When the target's velocity is increased to approximately Mach 9, a loss factor of 0.95 dB is observed for an integration time of 1 s, compared to a target velocity of 330 m/s with no loss factor. Furthermore, it is noted that the loss factor increases more quickly as the integration time is increased. This is due to the target echo's range migration, which results in the target's energy being distributed over a larger number of range cells as the integration time is increased.

In Figure 9.4, a comparison is made between the UMPI detector's performance for $v_b = 0$ m/s and $v_b = 2{,}970$ m/s with $a_b = 0$ and $P_{fa} = 10^{-6}$. To achieve $P_d = 0.9$

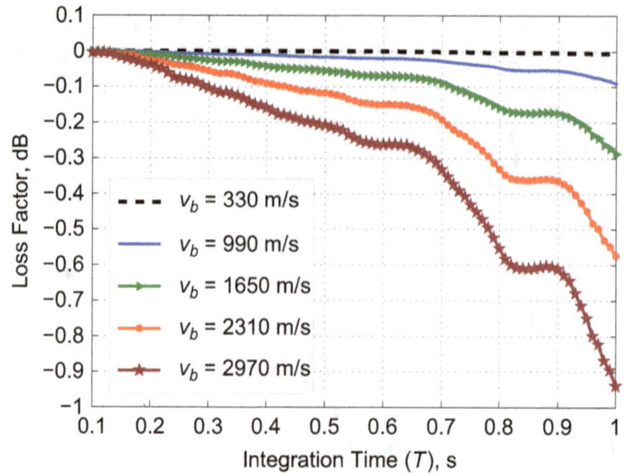

Figure 9.3 Loss factor, defined in (9.44), as a function of integration time for different values of target velocity with zero acceleration

at $P_{fa} = 10^{-6}$ for a 2,970 m/s target, an additional 0.95 dB of input SNR is required when compared to a 0 m/s target.

Figure 9.5 shows the loss factor as a function of integration times for various acceleration values, with a bistatic target velocity of 660 m/s. The loss factor due to Doppler frequency migration becomes more prominent as acceleration values increase. If a radar designer uses the uncompensated UMPI detector for a target with

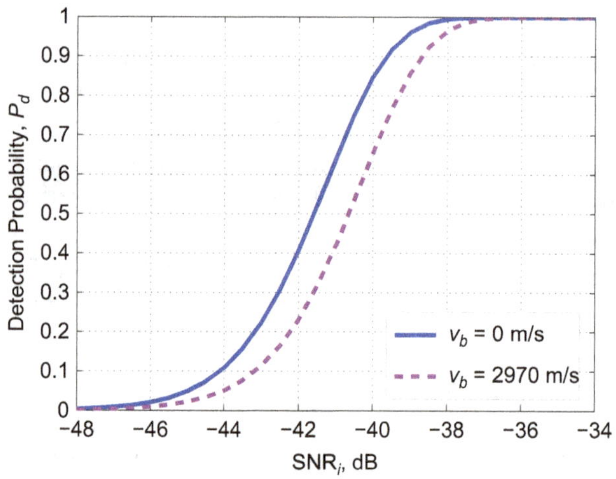

Figure 9.4 Detection performance comparison of the UMPI detector for different target velocity values of $v_b = 0$ m/s and $v_b = 2,970$ m/s, and when $a_b = 0$ and $P_{fa} = 10^{-6}$

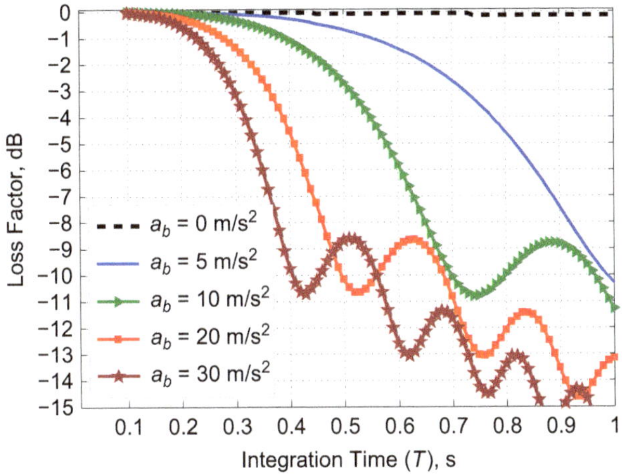

Figure 9.5 *Loss factor, defined in (9.44), as a function of integration time for different values bistatic acceleration and when v_b = 660 m/s*

20 m/s² acceleration and an integration time of 1 s, they should consider a loss factor of approximately 13.3 dB compared to a target with no acceleration. Figure 9.6 illustrates the corresponding UMPI detection probability comparison as a function of the input SNR for $a_b = 0\,\text{m/s}^2$ and $a_b = 20\,\text{m/s}^2$ with $v_b = 660\,\text{m/s}$. It is evident from Figure 9.6 that an SNR loss of about 13.3 dB is necessary to achieve $P_d = 0.9$ at $P_{fa} = 10^{-6}$ for a target with 20 m/s² acceleration, as compared to a target with no bistatic acceleration.

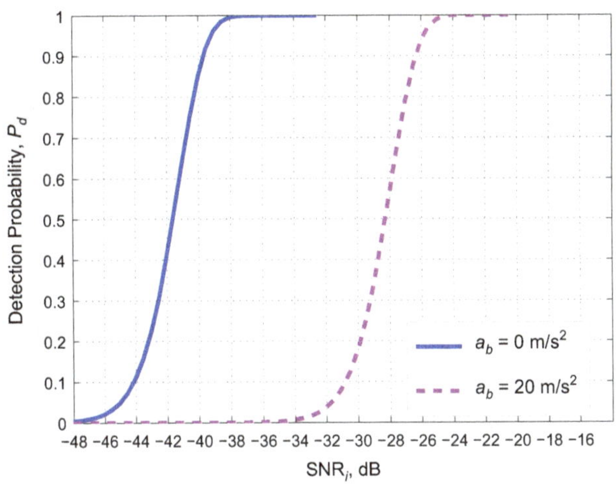

Figure 9.6 *Detection performance comparison of UMPI detector for two target acceleration values of $a_b = 0\,m/s^2$ and $a_b = 20\,m/s^2$, and when $v_b = 660\,m/s$ and $P_{fa} = 10^{-6}$*

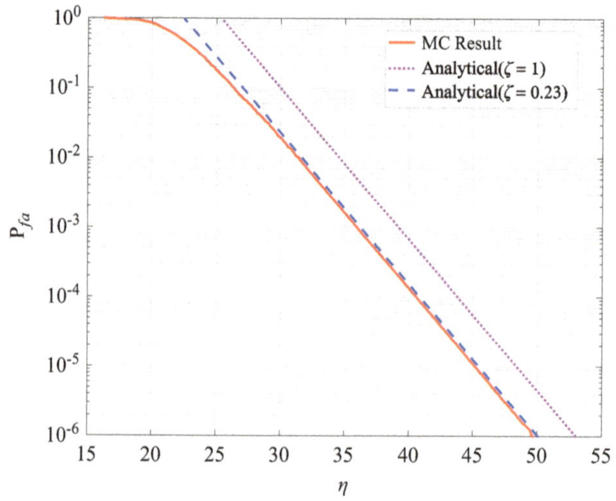

Figure 9.7 False alarm probability as a function of detection threshold for the proposed detector when values $\zeta = 1$ and 0.23 are considered to verify the analytical solution (9.28)

9.7.2 Performance evaluation of the 3D-SaR detection algorithm

Here, we consider single-channel detection problem in an FM-based PBR system with carrier frequency of $f_c = 88$ MHz. Also, an integration time of $T = 1$ s and the sampling frequency of $f_s = 200$ kHz are considered. The desired bistatic targets' ranges, velocities, and accelerations are considered within the range of [0, 70] km, [−600, 600] m/s, and [−20, 20] m/s^2, respectively. Based on this simulation setup, other parameters such as N, N_d N_f, N_f', N_a, $L_{3D} = N_d N_f' N_a$, Δf, and Δa can be obtained as 200,000, 47, 524,288, 923, 42, 1,822,002, 0.381, and 0.992, respectively.

Figure 9.7 depicts false alarm probability versus the detection threshold η to verify the obtained closed-form expression for false alarm probability of (9.28) for different values of ζ. Here, each simulated point has been obtained from 10^8 MC simulation runs. It can be observed that the obtained false alarm probability of the proposed test is very close to the asymptotic closed-form expression when the parameter ζ is set equal to 0.23. In general, for the FM-based PBR, the value of ζ depends on the signals of opportunity received over different integration times. In practice, since the transmitted signals are not within the control of the radar designer, we can adjust the detection threshold with $\zeta = 1$, resulting in a detector with the level of desired false alarm probability p_{fa}, i.e., $P_{fa} \leq p_{fa}$. In such cases, P_{fa} is called empirical false alarm probability obtained through Monte Carlo simulation or a practical situation, while p_{fa} is the desired false alarm probability set according to the closed-form expression in (9.28). In the following, for the considered signal, ζ was set to 0.23 to achieve $P_{fa} = p_{fa}$.

In Figure 9.8, we simulate a single-target scenario and plot 3-dimensional representation of the 3D-MCFFT-MLa-GLR test statistic over the desired

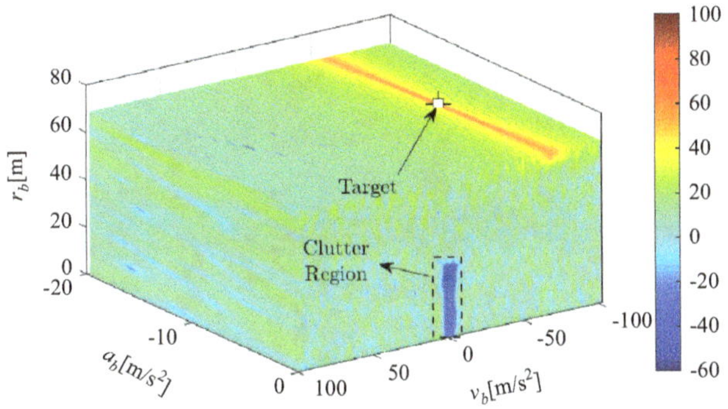

Figure 9.8 *Three-dimensional representation of the 3D-MCFFT-MLa-GLR test statistic in the presence of clutter echoes with the maximum range of 30 km and the velocities in the range of [–2, 2] m/s when the interested target is located at* (69 *km*, −63.72 *m/s*, −9.92 *m/s*^{2}). *Here, symbols* + *and* □ *indicate the actual target position and its estimate, respectively.*

range-velocity-acceleration map, where three axes represent targets' range, velocity, and acceleration. Here, the range, velocity, and acceleration of the simulated target are set equal to 69 km, –63.72 m/s, and −9.92 m/s^2, respectively. The maximum range of clutter is also considered equal to 30 km with the velocities in the range of [–2, 2] m/s. For clutter echoes, input clutter-to-noise ratios (CNR_i) are chosen over the interval 5 dB $\leq \mathrm{CNR}_i \leq$ 45 dB. From Figure 9.8, it is observed that the proposed detector places deep nulls in the range-Doppler coordinates of the clutter region in order to able the detection of the target located in (69 km, −63.72 m/s, −9.92 m/s^2). The multipath removal part of this new 3D-SaR detection algorithm is similar to that of Chapter 3; thus, an interested reader is referred to this chapter.

In Figure 9.9(a)–(c), the range, Doppler, and acceleration slices corresponding to the test statistic $d(n_r^{(1)}, f_b^{(1)}, a_b^{(1)})$ around a target located in (69 km, −63.72 m/s, −9.92 m/s^2) is depicted. For example, Figure 9.9(c) shows the values for constant bistatic range and Doppler versus bistatic acceleration for different integration times. As can be seen, increasing integration time leads to tighten the main lobe of Doppler and acceleration slices, while the main lobe of range slice remains constant for the considered signal. As explained before, the acceleration bin width Δa is proportional to $\frac{1}{T^2}$, while it is $\frac{1}{T}$ for the Doppler bin width. Thus, the broadening of the acceleration bin width is twice as that in the Doppler when integration time is varied from 1 s to 0.5 s.

Figure 9.10 compares the detection performance of the proposed 3D-SaR detection algorithm with that of classical 2D-SaR detection algorithm presented infor a single-target scenario, where a target is located at (69 km, −63.72 m/s, −9.92 m/s^2). In this figure, the input SNR is defined as $\mathrm{SNR}_i^{(m)} = \frac{|\beta_m|^2}{N_0 f_s}$ with $N_0 = -197$ dB/Hz being the noise power per unit bandwidth. In Figure 9.10, we also consider 2D-detection algorithm with the known acceleration of −9.92 m/s^2 as benchmark detector, which is called

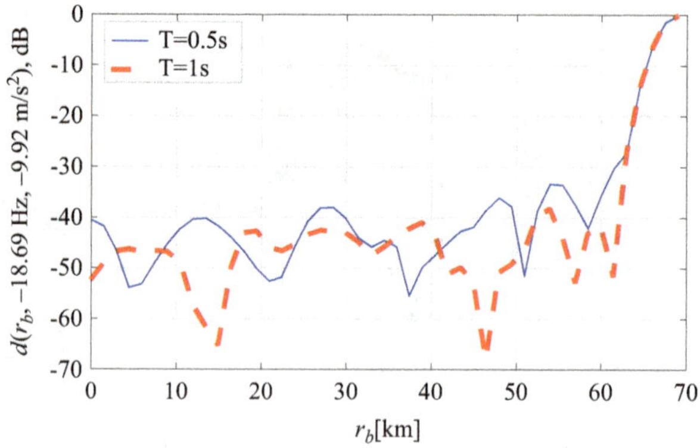

(a) Range slice

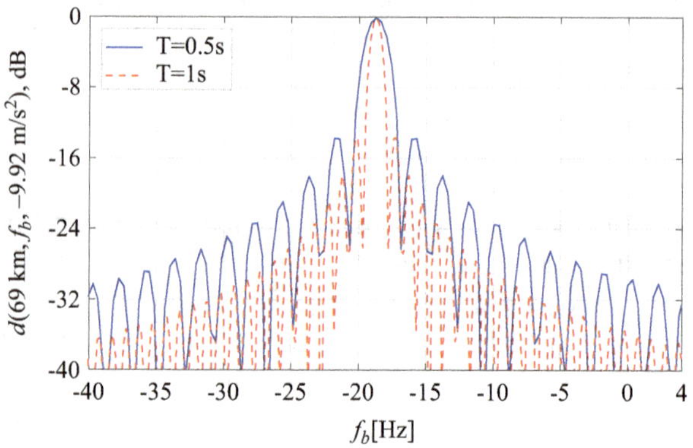

(b) Doppler slice

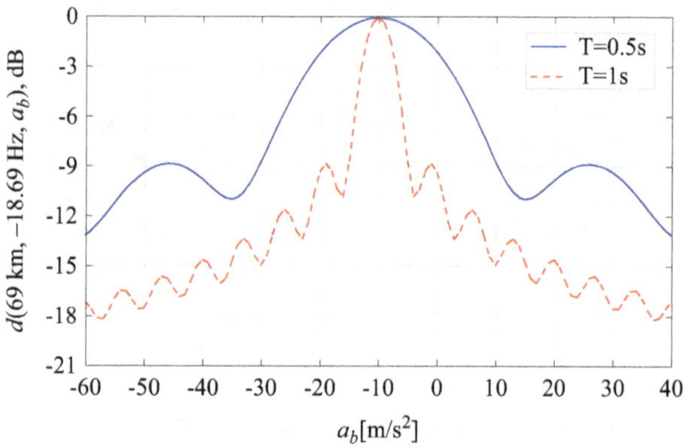

(c) Accerelation slice

Figure 9.9 *Range, Doppler, and acceleration slices corresponding to the test statistic $d(n,f,a)$ around a target located in* $(69\ km, -63.72\ m/s, -9.92\ m/s^2)$

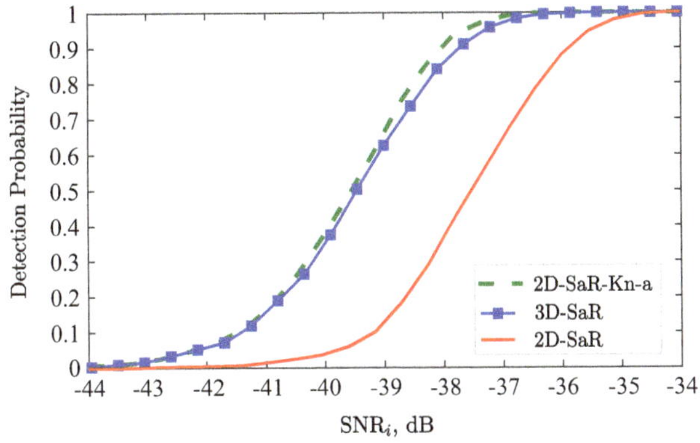

Figure 9.10 *Detection probability comparison of proposed 3D-SaR detector with those of the 2D-SaR and 2D-SaR-Kn-a (known acceleration* −9.92 m/s^2*) for the single-target scenario when* $P_{fa} = 10^{-6}$

2D-SaR-Kn-a. It is seen that the 3D-SaR detection algorithm follows the detection performance of the benchmark detector with a loss of 0.25 dB at $P_d = 0.9$ and $P_{fa} = 10^{-6}$. In this case, the 2D-SaR detection algorithm experiences a loss of about 2.1 dB due to the DFM of the considered accelerating target. To see the severity of this DFM, which depends on the magnitude of the acceleration of the target, we compare the detection performance of the proposed 3D-SaR detection algorithm with that of classical 2D-SaR detection algorithm when the target accelerations are varied from 0 to 20 m/s^2. In contrast to proposed 3D-SaR detection algorithm, as shown in Figure 9.11, the detection

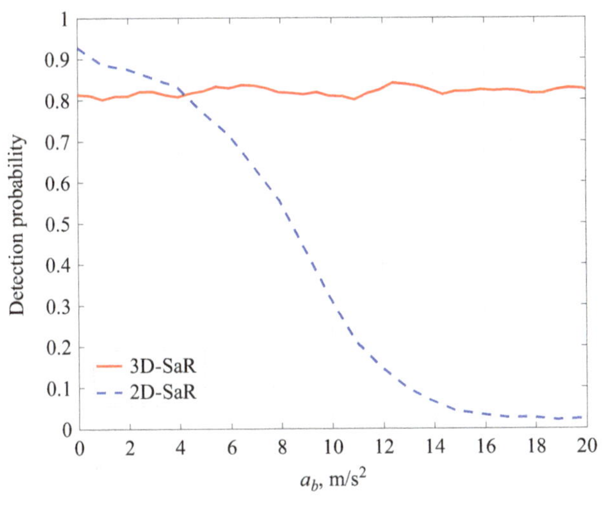

Figure 9.11 *Detection probability as a function of target acceleration for* $SNR_i^{(1)} = -38$ *dB and* $P_{fa} = 10^{-6}$

Table 9.1 *Characteristics of the simulated multitarget scenario*

Targets	T_1	T_2	T_3	T_4
Relative bistatic range (km)	10.5	30	69	66
Velocity (m/s)	50.72	–269.19	161.26	546.19
Acceleration (m/s^2)	0	9.92	16.86	11.90
$SNR_i^{(m)}$(dB)	–3	–17	–13	–10

performance of the 2D-SaR detection algorithm is reduced since the effective SNR in the range-Doppler bin associated with the target location at the start of the integration time is reduced. This clearly shows that the proposed detection algorithm significantly outperforms the 2D-SaR detector that ignores the targets' acceleration, especially in the high acceleration region. However, the 2D-SaR detector performs better than that of the 3D-SaR detector for small target acceleration values, inducing negligible DFMs. This is mainly due to the fact that the 3D-SaR detector needs to estimate more unknown parameters, i.e., delay, velocity, and acceleration of a target.

In order to assess the ability of the proposed 3D-SaR detection algorithm to determine the ranges, velocities, and accelerations of targets, we will examine a multitarget scenario (MTS) with the specific characteristics outlined in Table 9.1. This MTS will include four targets, referred to as T_1, T_2, T_3, and T_4, with T_1 having a zero acceleration, while the other targets have positive accelerations. We will evaluate the accuracy of both the 3D-SaR and 2D-SaR detectors in estimating the target positions, as shown in Figures 9.12 and 9.13, respectively. The proposed 3D-SaR detection algorithm successfully determined the positions of inserted targets, whereas the 2D-SaR detector only accurately determined the

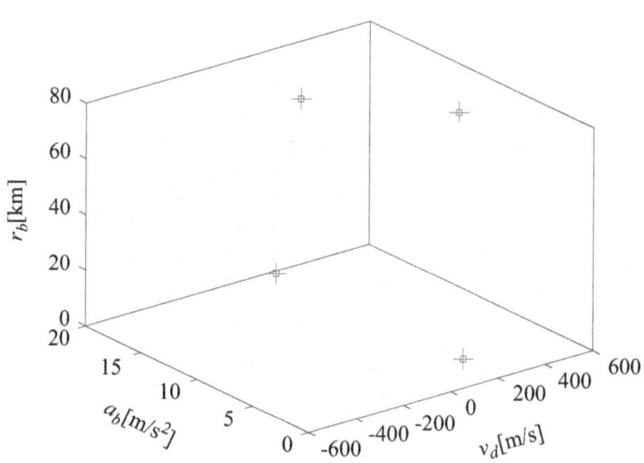

Figure 9.12 *Results of the 3D-SaR detection algorithm for the simulated multitarget scenario characterized in Table 9.1 at* $P_{fa} = 10^{-6}$

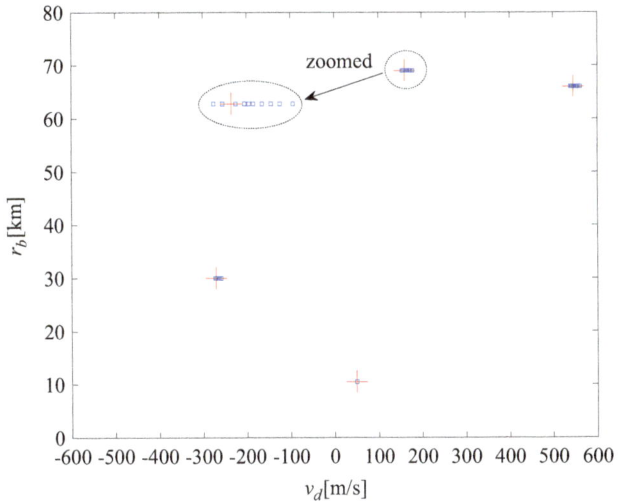

Figure 9.13 *Results of the 2D-SaR detection algorithm for the simulated multitarget scenario characterized in Table 9.1 when* $P_{fa} = 10^{-6}$

position of one target, T_1. The 2D-SaR algorithm produced a high number of false targets and spread the energy of T_2, T_3, and T_4 over several Doppler bins due to DFMs. The 3D-SaR algorithm obtained the positions of four targets after four iterations, while the 2D-SaR algorithm produced 28 positions. It is important to mention that the 2D-SaR detection method does not take into account the acceleration of targets, resulting in reduced effectiveness when facing accelerating targets.

Figure 9.14 illustrates the assessment of the detection capabilities of targets within the MTS being analyzed. Specifically, the SNR_i of a particular target, referred to as T_m, is changed while taking into account the presence of other interfering targets with their respective $\mathrm{SNR}_i^{(m)}$ values and delay-Doppler-acceleration coordinates, as indicated in Table 9.1. The proposed 3D-SaR detection algorithm shows similar performance for all the targets. However, when the 2D-SaR detection algorithm is applied, different SNR losses are observed for these targets. This is due to the Doppler frequency migration caused by the acceleration of the targets, which causes the energy of targets T_3, T_4, and T_2 to be, respectively, spread over 12, 10, and 5 Doppler bins. As a result, the 2D-SaR detection algorithm performs better in detecting the target T_1 with zero acceleration, as expected. However, the detection performance of the 2D-SaR algorithm is degraded for the other targets with acceleration, resulting in additional SNR losses of approximately 1.75 dB, 4.05 dB, and 2.25 dB for targets T2, T3, and T4, respectively, as compared to the proposed 3D-SaR detection algorithm.

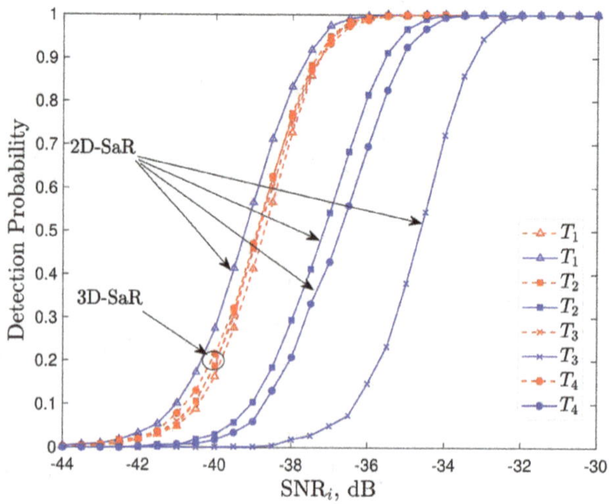

Figure 9.14 *Detection performance comparison of the 3D-SaR and 2D-SaR detection algorithms when $P_{fa} = 10^{-6}$*

9.8 Summary

First, we have investigated the problem of detecting a maneuvering target using the UMPI detector designed in Chapter 4. The closed-form expression for the detection probability of the UMPI detector has been obtained in the presence of a maneuvering target. Simulation results show that the bistatic velocity and acceleration of a maneuvering target induce range migration and Doppler frequency migration of echo signal, which degrade the performance of the UMPI detector in the integration time as long as seconds. To solve this, then, we proposed a new 3-dimensional detector in which delay-Doppler and acceleration of interesting targets are estimated, so it was called a 3-dimensional detection algorithm. To reduce the computational complexity of the proposed GLR-based detector, it was implemented based on a modified version of the chirp fast Fourier transform. To reduce its computational complexity, a 3-dimensional sequential algorithm, in which targets have been detected in a layered manner, was proposed for multitarget scenarios. A closed-form formula for the false alarm probability of the proposed detector has been derived which is very instructive to adjust the detection threshold. Our simulation results indicated that the 3-dimensional detection algorithm is required to prevent degraded detection probability as well as excessive false alarm probability of the 2-dimensional classical target detection algorithm, especially for highly accelerating targets.

References

[1] Kulpa K.S., Misiurewicz J. ‘Stretch processing for long integration time passive covert radar’. *2006 CIE International Conference on Radar*; Shanghai, China, 2006. pp. 1–4.

[2] Malanowski M., Kulpa K., Olsen K.E. 'Extending the integration time in DVB-T-based passive radar'. *2011 8th European Radar Conference*; Manchester, UK, IEEE, 2011. pp. 190–93.

[3] Borowiec K., Malanowski M. 'Accelerating rocket detection using passive bistatic radar'. *2016 17th International Radar Symposium (IRS)*; Krakow, Poland, 2016. pp. 1–5.

[4] Hoshino T., Suwa K., Nakamura S, *et al.* 'Long-time integration by short-time cross-correlation and two-step Doppler processing for passive bistatic radar'. *2013 European Radar Conference*; Nuremberg, Germany, 2013. pp. 451–54.

[5] Malanowski M. 'Detection and parameter estimation of manoeuvring targets with passive bistatic radar'. *IET Radar, Sonar & Navigation*. 2012, vol. 6(8), pp. 739–45.

[6] Christiansen J.M., Olsen K.E. 'Range and Doppler walk in DVB-T based passive bistatic radar'. *2010 IEEE Radar Conference*; Arlington, VA, USA, 2010. pp. 620–26.

[7] Raney R.K., Runge H., Bamler R., Cumming I.G., Wong F.H. 'Precision SAR processing using chirp scaling'. *IEEE Transactions on Geoscience and Remote Sensing*. 1994, vol. 32(4), pp. 786–99.

[8] Almeida L.B. 'The fractional Fourier transform and time-frequency representations'. *IEEE Transactions on Signal Processing*. 1994, vol. 42(11), pp. 3084–91.

[9] Zhu D., Li Y., Zhu Z. 'A keystone transform without interpolation for SAR ground moving-target imaging'. *IEEE Geoscience and Remote Sensing Letters*. 2007, vol. 4(1), pp. 18–22.

[10] Pignol F., Colone F., Martelli T. 'Lagrange-polynomial-interpolation-based keystone transform for a passive radar'. *IEEE Transactions on Aerospace and Electronic Systems*. 2017, vol. 54(3), p. 99.

[11] Feng Y., Shan T., Zhuo Z., Tao R. 'The migration compensation methods for DTV based passive radar'. *2013 IEEE Radar Conference*; Ottawa, ON, Canada, 2013. pp. 1–4.

[12] Bai X., Feng Y., Zhao J. 'A processing scheme for long integration time passive radar based on CZT and FRFD-sharpness'. *2016 IEEE International Conference on Signal Processing, Communications and Computing (ICSPCC)*; Hong Kong, IEEE, 2016. pp. 1–4.

[13] Zaimbashi A. 'Integration gain limitations in passive coherent location radars for manoeuvring target detection'. *Electronics Letters*. 2016, vol. 52(21), pp. 1801–04.

[14] Solatzadeh Z., Zaimbashi A. 'Accelerating target detection in passive radar sensors: delay-Doppler-acceleration estimation'. *IEEE Sensors Journal*. 2018, vol. 18(13), pp. 5445–54.

[15] Zaimbashi A., Derakhtian M., Sheikhi A. 'GLRT-based CFAR detection in passive bistatic radar'. *IEEE Transactions on Aerospace and Electronic Systems*. 2013, vol. 49(1), pp. 134–59.

[16] Palmer J.E., Searle S.J. 'Evaluation of adaptive filter algorithms for clutter cancellation in passive bistatic radar'. *2012 IEEE Radar Conference*; Atlanta, GA, USA, 2012. pp. 493–98.

[17] Xia X.G. 'Discrete Chirp-Fourier transform and its application to Chirp rate estimation'. *IEEE Transactions on Signal Processing*. 2000, vol. 48(11), pp. 3122–33.

[18] Guo X., Sun H.B, Wang S.L, Liu G.S. 'Comments on discrete chirp-Fourier transform and its application to chirp rate estimation [with reply]'. *IEEE Transactions on Signal Processing*. 2002, vol. 50(12), pp. 3115–16.

[19] Xia X.G. 'Response to comments on discrete chirp-Fourier transform and its application to chirp rate estimation'. *IEEE Transactions on Signal Processing*. 2002, vol. 50, pp. 3115–16.

Part III

Multistatic passive radar target detection under noisy reference channels

Passive radars utilize signals from non-cooperative sources like radio, television, and cellular signals instead of having their own dedicated transmitter to detect targets of interest. Traditional matched filter-based detectors used in active radars may not be suitable for passive radars because the transmitted signals are typically unknown to the passive radar receiver. One solution is to exploit an additional channel, known as a reference channel, to collect a delayed version of the unknown transmitted signal. However, many existing methods assume a high direct-path signal power-to-noise ratio (DNR) in the reference channel and ignore the noise. These methods are categorized as category 1 with an ideal reference channel (Ca.1-IR) and have been extensively studied in Part 2. However, their performance can be significantly impacted, especially when the DNR of the reference channel is low. In this part, we explore situations in which it is not possible to ignore the effects of noise in the reference channel. Consequently, we encounter a challenge known as passive radar target detection with a noisy reference (NR) channel. Similar to the Ca.1-IR methods, here, we employ two receive channels for target detection: one for surveillance and the other for reference. This similarity leads us to categorize the target detection techniques developed for the NR channel as Ca.1-NR methods. Although both Ca.1-IR and Ca.1-NR detection strategies fall under the same category of passive radar target detection methods, they differ in hypothesis-testing problem formulations. Firstly, Ca.1-NR approaches incorporate both reference and surveillance channels when formulating the passive radar target detection problem, while Ca.1-IR methods only use the surveillance channel for this purpose. Secondly, in Ca.1-NR approaches, it is not feasible to treat multipath signals as interference; thus, these interference signals are ignored to obtain closed-form detectors. However, it is possible to utilize the reference channel separately to estimate the unknown transmitted signal and exploit the subspace-based interference removal algorithm of Chapter 7 to cancel the effects of multipath signals.

In this part, there are two chapters. Chapter 10 assumes a single-input single-output (SISO) configuration for addressing an NR channel detection problem, while Chapter 11 considers a single-input multi-output (SIMO) configuration.

In chapter 10 with SISO configuration, also known as passive bistatic radar, a detection problem is formulated as a composite hypothesis-testing problem and solved using the likelihood ratio test (LRT) principle. It is shown that uncertainty in the DNR of the reference channel (DNR_r) can lead to excessive false alarm probability for the proposed noisy reference channel-based detector in the low-DNR_r regime. To address this issue, a new threshold-setting strategy is proposed to adjust the level of the detector to operate efficiently under uncertain DNR_r conditions. This introduces a new concept for setting a detection threshold in uncertain situations. The proposed strategy is evaluated through extensive Monte Carlo simulations. Additionally, a framework for exploring passive radar target detection in a non-linear feature space is introduced using kernel theory. The kernel theory is utilized to replace the inner products of test statistics with those of nonlinear mapped data in the feature space or their kernel tricks to achieve better detection performance. New detection algorithms for the target detection problem of noisy and ideal reference channel passive radar are proposed using this framework.

In Chapter 11, with a calibrated SIMO configuration, we present two target detection problems, which are formulated as composite hypothesis testing problems. The difference between the two problems lies in the availability of noisy reference channels. The chapter uses the LRT principle to develop two detection methods, one with a fixed level and the other with a fixed size (CFAR property) to deal with uncertainty in DNR_r. The chapter provides several simulation results to compare and demonstrate the effectiveness of the proposed detection methods.

Chapter 10

Noisy RC-based bistatic passive target detection

10.1 Introduction

Two-channel passive radar target detection, which takes into account noisy reference channels, has been studied in References 1–6. However, it can be demonstrated that the detection thresholds of these methods are highly dependent on the direct-path signal power-to-noise ratio (DNR) values of the reference channel. Despite this crucial influence, most previous studies [1–6 and references therein] have overlooked the impact of the DNR of the reference channel on the adjustment of detection thresholds. To address this gap, this chapter aims to devise new target detection algorithms and provide a more comprehensive investigation into this aspect.

This chapter focuses on the issue of detecting moving targets in a noisy reference channel passive radar system with a single-input and single-output (SISO) configuration. Specifically, it addresses the challenge of detecting a single moving target with an unknown delay-Doppler coordinate in the presence of a noisy reference channel. A new target detection algorithm is proposed based on the likelihood ratio test (LRT) principle. The chapter then introduces the concept of kernel theory and applies it to improve the efficiency of the proposed detection algorithm. Kernel-based methods have been found to be highly effective in nonlinear decision problems, as they can leverage the nonlinear characteristics of input data through the use of nonlinear kernel functions. To this end, two types of kernelized-likelihood ratio tests are derived, using polynomial kernel functions, for both ideal and noisy reference channel detection problems. As such, the tests' inner products are replaced with appropriate polynomial kernel functions, resulting in improved detection performance. This chapter further exploits the principle of invariance and the maximization of detection performance to set the polynomial kernel parameters of the proposed kernelized detectors. Moreover, a novel approach to threshold setting in the field of passive radar target-detection is introduced. The study reveals that when dealing with noisy reference channel detectors in low DNR scenarios, any uncertainty in the DNR values of the reference channel, namely DNR_r, can result in an increased number of false alarms. An approach is suggested to tackle the problem of reference channel uncertainty (RCU) by introducing a new threshold-setting strategy. This strategy aims to control the levels of the proposed detectors rather than their sizes. Extensive Monte Carlo simulations are conducted to investigate these problems and to evaluate the effectiveness of the proposed strategy. By combining detection theory and kernel theory frameworks, better detection performance is achieved. This chapter builds upon the previous works of References 7 and 8.

In this chapter, we have arranged the content in the following manner. First, in section 10.2, we establish the formulation of the detection problem. Then, our proposed LRT-based detector is presented in section 10.3. Next, we introduce and investigate two novel kernelized LRT-based detectors in section 10.4.2. In section 10.5, we provide extensive Monte Carlo simulation results and examine them. Additionally, we introduce a new strategy for setting the threshold level of the proposed detectors to control the level of the proposed detector when working with noisy reference channels. Finally, in section 10.6, we summarize the chapter.

10.2 Target detection problem formulation

The situation depicted in Figure 10.1 involves detecting a target, in which a passive bistatic radar (PBR) makes use of an reference channel (RC) to gather direct-path signal and an surveillance channel (SC) to receive possible echoes from an interested target. In the case of detecting a single target, the observations received by the RC and SC can be expressed as follows:

$$x_r(t) = \alpha s(t) + n_r(t) \tag{10.1}$$

$$x_s(t) = \beta s(t-\tau_t)e^{j2\pi f_t t} + n_s(t) \tag{10.2}$$

where $0 \leq t < T$ with T being the integration time; $s(t)$ is the deterministic but unknown transmitted signal; α is deterministic but unknown complex amplitude scaling factor accounting for propagation loss, transmit antenna gain, and antenna gain of the RC; β is a deterministic but unknown complex coefficient factor considering attenuation due to transmission from illuminator opportunity (IO) to SC and target reflectivity; (τ_t, f_t) is the unknown delay-Doppler coordinate of a desired target; and $n_r(t)$ and $n_s(t)$ are zero-mean additive white Gaussian noise at the RC and SC with unknown variances σ_r^2 and σ_s^2, respectively. It should be pointed out that τ_t and f_t denote bistatic relative

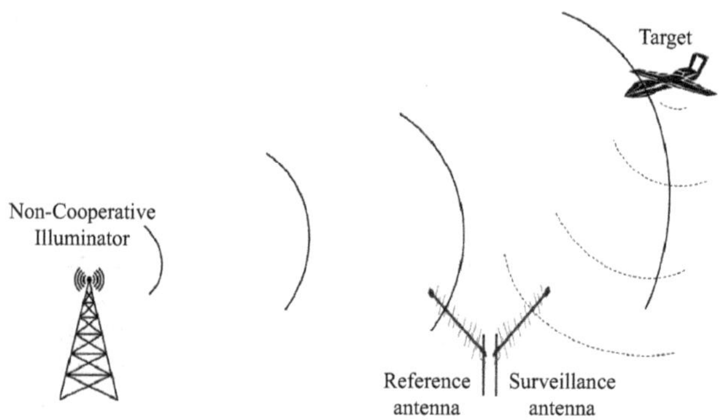

Figure 10.1 Configuration of a typical SISO passive radar system

delay and Doppler frequency shift of the target. After sampling observations in both channels with proper sampling frequency f_s, a discretized and vectorized models of the RC and SC data can be described by

$$\mathbf{x}_r = \alpha\, \mathbf{s} + \mathbf{n}_r \tag{10.3}$$

$$\mathbf{x}_s = \beta\, \Phi(\tau_t, f_t)\mathbf{s} + \mathbf{n}_s \tag{10.4}$$

where $\mathbf{x}_r, \mathbf{x}_s, \mathbf{s}, \mathbf{n}_r, \mathbf{n}_s \in \mathcal{C}^{N\times 1}$ are column vectors formed by N adjacent samples of $x_r(t)$, $x_s(t)$, $s(t)$, $n_r(t)$, and $n_s(t)$, respectively. Here, $N = Tf_s$ is the number of samples during integration time. By using N-point discrete Fourier transform (DFT) matrix $\mathbf{F}$ with elements $[\mathbf{F}]_{i,k} = e^{-j2\pi(i-1)\Delta f(k-1)T_S}/\sqrt{N}$, $i, k = 1, \ldots, N$, where $T_s = \frac{1}{f_S}$ is sampling interval and $\Delta f = \frac{1}{T} = \frac{1}{NT_S}$ is frequency bin of N-point DFT. The delay-Doppler compensator matrix $\Phi(\tau_t, f_t)$ can be defined as

$$\Phi(\tau_t, f_t) = \mathbf{D}(f_t T_s)\mathbf{F}^H \mathbf{D}(-\tau_t \Delta f)\mathbf{F} \tag{10.5}$$

where $\mathbf{D}(a)$ is a diagonal matrix with diagonal elements: $[\mathbf{D}(a)]_{ii} = e^{j2\pi(i-1)a}$, $i = 1, \ldots, N$. Henceforth, for notational simplicity, we use Φ instead of $\Phi(\tau_t, f_t)$.

The moving target detection problem can be expressed as the following binary hypothesis testing:

$$\mathcal{H}_0 : \begin{cases} \mathbf{x}_r = \alpha\, \mathbf{s} + \mathbf{n}_r \\ \mathbf{x}_s = \mathbf{n}_s \end{cases} \tag{10.6a}$$

$$\mathcal{H}_1 : \begin{cases} \mathbf{x}_r = \alpha\, \mathbf{s} + \mathbf{n}_r \\ \mathbf{x}_s = \beta \Phi \mathbf{s} + \mathbf{n}_s \end{cases} \tag{10.6b}$$

where the unknowns are the transmitted signal waveform $\mathbf{s}$, α, β, delay-Doppler (τ_t, f_t) coordinate of target and noise powers σ_r^2 and σ_s^2 in both RC and SC, respectively.

10.3 LRT-based detector

In the following, we deploy the LRT principle to develop a detector for the problem (10.6). The LRT principle is given as follows.

Definition 1: *The LRT for testing $\mathcal{H}_0 : \theta \in \Theta_0$ versus $\mathcal{H}_1 : \theta \in \Theta_0^c$ is [9]*

$$\Lambda(\mathbf{x}) = \sup_{\mathbf{\Theta}_0} \mathcal{L}(\mathbf{x}) - \sup_{\mathbf{\Theta}} \mathcal{L}(\mathbf{x}) \underset{\mathcal{H}_1}{\overset{\mathcal{H}_0}{\gtrless}} \eta_l \tag{10.7}$$

The first term of $\Lambda(\mathbf{x})$ is the supremum of the log-likelihood function (LLF) of observation vector, the supremum should be computed over the parameters in the null hypothesis denoted as Θ_0, while for the second term the supremum should be computed over all possible parameters, defined as $\Theta = \Theta_0 \cup \Theta_0^c$. Here, η_l is any number satisfying $\eta_l \leq 0$. It is important to note that if the maximums are used instead of supremums, and if Θ in the second component of (10.7) is replaced with Θ_0^c, then the LRT becomes the generalized LRT (GLRT). In order to use the GLRT, it is *customary to obtain the maximum likelihood estimates of the unknown parameters under both hypotheses. However, for the LRT, it is necessary to find the supremum.*

In our case, we have $\Theta_0 = \{\sigma_r^2, \sigma_s^2, \mathbf{s}, \alpha\}$, $\Theta_0^c = \{(\tau_t, f_t), \sigma_r^2, \sigma_s^2, \mathbf{s}, \alpha, \beta\}$, and $\Theta = \Theta_0 \cup \Theta_0^c$ and $\mathbf{x} = [\mathbf{x}_r^T, \mathbf{x}_s^T]^T$. To solve the problem, we need to assume that $\sigma_r^2 = \sigma_s^2 = \sigma^2$. In such case, it is readily found that the LLF of the general hypothesis can be expressed as

$$\mathscr{L}(\mathbf{x}) = -2N \ln(\pi\sigma^2) - \frac{1}{\sigma^2}\left(\|\mathbf{x}_r - \alpha\mathbf{s}\|^2 + \|\mathbf{x}_s - \beta\Phi\mathbf{s}\|^2\right) \tag{10.8}$$

Let us consider the first term in the left-hand side (lhs) of (10.7), where it can be rewritten as

$$\sup_{\sigma^2}\sup_{\mathbf{s}}\sup_{\alpha}\left\{-2N\ln(\pi\sigma^2) - \frac{1}{\sigma^2}\left(\|\mathbf{x}_r - \alpha\mathbf{s}\|^2 + \|\mathbf{x}_s\|^2\right)\right\} \tag{10.9}$$

The supremum with respect to (w.r.t.) the unknown parameter α is further obtained as

$$\sup_\alpha \mathscr{L}(\mathbf{x}) = -2N\ln(\pi\sigma^2) - \frac{1}{\sigma^2}\left(\|\mathbf{x}_s\|^2 - \frac{|\mathbf{s}^H\mathbf{x}_r|^2}{\|\mathbf{s}\|^2} + \|\mathbf{x}_s\|^2\right) \tag{10.10}$$

Then, the supremum w.r.t. the unknown vector $\mathbf{s}$ can be computed as

$$\sup_{\mathbf{s}}\sup_{\alpha}\mathscr{L}(\mathbf{x}) = -2N\ln(\pi\sigma^2) - \frac{1}{\sigma^2}\left(\|\mathbf{x}_r\|^2 - \lambda_{\max}(\mathbf{x}_r\mathbf{x}_r^H) + \|\mathbf{x}_s\|^2\right) \tag{10.11}$$

It is readily found that $\lambda_{\max}(\mathbf{x}_r\mathbf{x}_r^H) = \|\mathbf{x}_r\|^2$, resulting in

$$\sup_{\mathbf{s}}\sup_{\alpha}\mathscr{L}(\mathbf{x}) = -2N\ln(\pi\sigma^2) - \frac{\|\mathbf{x}_s\|^2}{\sigma^2} \tag{10.12}$$

Finally, the first term in the lhs of (10.7), can be further expressed as

$$\sup_{\sigma^2}\sup_{\mathbf{s}}\sup_{\alpha}\mathscr{L}(\mathbf{x}) = -2N\ln(\pi) - 2N\ln\left(\frac{\|\mathbf{x}_s\|^2}{2N}\right) - 2N \tag{10.13}$$

In a similar manner, the second term of the LRT in (10.7) can be obtained from

$$\sup_{\Theta}\mathscr{L}(\mathbf{x}) = \sup_{\sigma^2}\sup_{(\tau_t,f_t)}\sup_{\mathbf{s}}\sup_{\alpha,\beta}\{\mathscr{L}(\mathbf{x})\} \tag{10.14}$$

Similar to the computation of the first term of (10.7), the supremum w.r.t. to both unknown parameters α and β gives

$$\sup_{\alpha,\beta}\mathscr{L}(\mathbf{x}) = -2N\ln(\pi\sigma^2) - \frac{1}{\sigma^2}\left(\|\mathbf{x}_r\|^2 - \frac{\mathbf{s}^H\mathbf{R}\,\mathbf{s}}{\|\mathbf{s}\|^2} + \|\mathbf{x}_s\|^2\right) \tag{10.15}$$

where $\mathbf{R} = \mathbf{x}_r\mathbf{x}_r^H + \Phi^H\mathbf{x}_s\mathbf{x}_s{}^H\Phi$. Similar to (10.11), the supremum w.r.t. unknown vector $\mathbf{s}$ can be given by

$$\sup_{\mathbf{s}}\sup_{\alpha,\beta}\mathscr{L}(\mathbf{x}) = -2N\ln(\pi\sigma^2) - \frac{1}{\sigma^2}\left(\|\mathbf{x}_r\|^2 + \|\mathbf{x}_s\|^2 - \lambda_{\max}(\mathbf{R})\right) \tag{10.16}$$

Let us define $\mathbf{Z} = [\Phi^H \mathbf{x}_s, \mathbf{x}_r]$ and use the fact that $\lambda_{\max}(\mathbf{R}) = \lambda_{\max}(\mathbf{Z}\mathbf{Z}^H) = \lambda_{\max}(\mathbf{Z}^H\mathbf{Z})$ to obtain

$$\lambda_{\max}(\mathbf{R}) = \lambda_{\max}\left(\begin{bmatrix} \|\mathbf{x}_s\|^2 & (\Phi^H\mathbf{x}_s)^H\mathbf{x}_r \\ \mathbf{x}_r{}^H\Phi^H\mathbf{x}_s & \|\mathbf{x}_r\|^2 \end{bmatrix}\right) \tag{10.17}$$

After some simple computations, we find

$$\lambda_{\max}(\mathbf{R}) = \frac{\|\mathbf{x}_r\|^2 + \|\mathbf{x}_s\|^2}{2} + \frac{\sqrt{(\|\mathbf{x}_s\|^2 - \|\mathbf{x}_r\|^2)^2 + 4\left|\mathbf{x}_r^H\Phi^H\mathbf{x}_s\right|^2}}{2} \tag{10.18}$$

By substituting (10.18) into (10.16), after some algebraic manipulations, we find the supremum w.r.t. the delay-Doppler coordinate (τ_t, f_t), given by

$$\begin{aligned}\sup_{(\tau_t,f_t)}\sup_{s}\sup_{\alpha,\beta}\{\mathscr{L}(\mathbf{x})\} &= -2N\ln(\pi\sigma^2) - \frac{1}{\sigma^2}\frac{\|\mathbf{x}_r\|^2 + \|\mathbf{x}_s\|^2}{2} \\ &+ \frac{1}{\sigma^2}\frac{\sqrt{(\|\mathbf{x}_s\|^2 - \|\mathbf{x}_r\|^2)^2 + 4\,\max_{(\tau_t,f_t)}\{\left|\mathbf{x}_r^H\Phi^H\mathbf{x}_s\right|^2\}}}{2}\end{aligned} \tag{10.19}$$

Finally, the second term of (10.7) can be written as

$$\begin{aligned}&\sup_{\sigma^2}\ \sup_{(\tau_t,f_t)}\sup_{s}\sup_{\alpha,\beta}\{\mathscr{L}(\mathbf{x})\} = -2N\ln(\pi) - 2N \\ &-2N\ln\left(\frac{\|\mathbf{x}_r\|^2 + \|\mathbf{x}_s\|^2 - \sqrt{(\|\mathbf{x}_s\|^2 - \|\mathbf{x}_r\|^2)^2 + 4\,\max_{(\tau_t,f_t)}\{\left|\mathbf{x}_r^H\Phi^H\mathbf{x}_s\right|^2\}}}{2N}\right)\end{aligned} \tag{10.20}$$

By substituting (10.13) and (10.20) into (10.7), and after some algebraic manipulations, we obtain the LRT, given by

$$\begin{aligned}\Gamma_{\text{NR-LRT}}\quad(\mathbf{x}_s, \mathbf{x}_r) &= \frac{1}{2}\left(1 - \frac{\|\mathbf{x}_r\|^2}{\|\mathbf{x}_s\|^2}\right) \\ &+ \frac{1}{2}\sqrt{\left(1 - \frac{\|\mathbf{x}_r\|^2}{\|\mathbf{x}_s\|^2}\right)^2 + 4\frac{\|\mathbf{x}_r\|^2}{\|\mathbf{x}_s\|^2}\max_{(\tau_t,f_t)}\left|\left(\frac{\mathbf{x}_r}{\|\mathbf{x}_r\|}\right)^H\Phi^H\left(\frac{\mathbf{x}_s}{\|\mathbf{x}_s\|}\right)\right|^2}\ \underset{\mathscr{H}_0}{\overset{\mathscr{H}_1}{\gtrless}}\ \eta\end{aligned} \tag{10.21}$$

This new detector is referred to as the noisy reference channel-based detector, abbreviated as the NR-LRT detector.

In the case of a cleaned RC and single-target detection problem, the simplified IR-LRT-based detector can be obtained in a similar manner of Chapter 3, given by

$$\Lambda_{\text{IR-LRT}}(\mathbf{x}_s, \mathbf{x}_r) = \frac{\max_{(\tau_t,f_t)}\left|\mathbf{x}_r^H\Phi^H\mathbf{x}_s\right|^2}{\|\mathbf{x}_r\|^2\|\mathbf{x}_s\|^2}\ \underset{\mathscr{H}_0}{\overset{\mathscr{H}_1}{\gtrless}}\ \zeta \tag{10.22}$$

or, equivalently,

$$\Lambda_{\text{IR-LRT}}(\mathbf{x}_s, \mathbf{x}_r) = \max_{(\tau_l, f_l)} \left|\mathbf{z}_r^H \Phi^H \mathbf{z}_s\right|^2 \underset{\mathscr{H}_0}{\overset{\mathscr{H}_1}{\gtrless}} \zeta \tag{10.23}$$

where

$$\mathbf{z}_s = \frac{\mathbf{x}_s}{\|\mathbf{x}_s\|} \tag{10.24}$$

$$\mathbf{z}_r = \frac{\mathbf{x}_r}{\|\mathbf{x}_r\|} \tag{10.25}$$

In (10.23), ζ is the detection threshold to be set to attain the desired false alarm probability. This detector is called a simplified ideal reference LRT-based detector, abbreviated to as SIR-LRT. For the purpose of comparison, we will be using the SIR-LRT detector to demonstrate the advantages of the newly proposed detectors.

10.4 Kernel-based detectors

This section will start with an explanation of the basics of kernel theory, followed by an introduction of the kernel-based detectors that are expected to outperform the current state-of-the-art detectors in terms of detection accuracy.

10.4.1 Preliminaries

In most detection statistics we encounter with, the inner product, e.g., $\langle \mathbf{a}, \mathbf{g} \rangle$, which can provide a proper measure of similarity between vectors **a** and **g** if the representation space of the vectors **a** and **g** (i.e., original space) is rich enough. In many problems, the original space is limited in resolution and expressive power, and thus, it does not necessarily provide the best support for simple inner product type of similarity measure [10] and [11]. In kernel-based methods, in general, the original input data belonging to the data space $\mathscr{X}$ are mapped to a new space $\mathscr{H}$, known as the feature space [10]. In general, this mapping can be represented by a nonlinear mapping function φ_i, i.e.,

$$\varphi_i : \mathscr{X} \rightarrow \mathscr{H}, \qquad \mathbf{a} \rightarrow \varphi_i(\mathbf{a}), \qquad \mathbf{g} \rightarrow \varphi_i(\mathbf{g}) \tag{10.26}$$

where **a** and **g** are input vectors in $\mathscr{X}$, which are mapped into a potentially much higher-dimensional feature space $\mathscr{H}$. Then, a given similarity measure (i.e., the inner product) in the original space can be replaced with that of the feature space using the corresponding inner product $\langle \varphi_i(\mathbf{a}), \varphi_i(\mathbf{g}) \rangle$. To this end, we build on the idea that the original space does not necessarily provide the best support for simple inner product type of similarity measure; thus, the original data space is mapped into the feature space, aiming to obtain a richer space. It should be noted that computing $\langle \varphi_i(\mathbf{a}), \varphi_i(\mathbf{g}) \rangle$ would be impossible unless the feature mapping $\varphi_i(.)$ is explicit. In addition, implementing $\langle \varphi_i(\mathbf{a}), \varphi_i(\mathbf{g}) \rangle$ directly in the feature space is not necessarily straightforward or computationally efficient due to the high dimensionality of vectors $\varphi_i(\mathbf{a})$ and $\varphi_i(\mathbf{g})$. For example, if φ_i is selected to be a polynomial of order d, then the length of the vectors $\varphi_i(\mathbf{a})$ and $\varphi_i(\mathbf{g})$ is equal to $\frac{(N+d-1)!}{d!\,(N-1)!}$, which is a large value if either N or d is large [10]. For the most common real-valued Gaussian kernel, the dimensionality of the associated feature space is infinite. Fortunately, in the kernel method, this mapping function can be defined implicitly; rather, we need to define a function, often called a kernel, such that [11]

$$\mathscr{K}_i(\mathbf{a}, \mathbf{g}) = \langle \varphi_i(\mathbf{a}), \varphi_i(\mathbf{g}) \rangle \tag{10.27}$$

Interestingly, it is seen that the kernel function computes the similarity using the kernel function $\mathscr{K}_i$ in the input space $\mathscr{X}$ instead of the feature space $\mathscr{H}$. In such cases, we do not require to know the explicit form of the feature map φ_i, but instead, it is implicitly defined through the corresponding kernel function. This technique is known as *kernelization* or a *kernel trick* in the kernel theory. A function that takes as its input vectors in the original space and returns the inner product of the vectors in the feature space is called a kernel function. Thus, using kernels, we never need the coordinates of the data in the feature space, because the detection algorithms only require the inner products between data in the feature space, which can be obtained by kernel function. To illustrate this, consider second-order real polynomial (SORP) kernel defined as $\mathscr{K}_i(\mathbf{a}, \mathbf{g}) = (\mathbf{a}^T\mathbf{g} + \delta_i)^{d_i}$ with $d_i = 2$, where we consider $\mathbf{a} = [a_1, a_2]^T$ and $\mathbf{g} = [g_1, g_2]^T$ as a simplified example. In this case, the SORP kernel can be expanded as

$$\mathscr{K}_i(\mathbf{a}, \mathbf{g}) = (a_1 g_1 + a_2 g_2 + \delta_i)^2 = \langle \varphi_i(\mathbf{a}), \varphi_i(\mathbf{g}) \rangle = \varphi_i(\mathbf{a})^T \varphi_i(\mathbf{g}) \tag{10.28}$$

where

$$\varphi_i(\mathbf{a}) = [a_1^2, a_2^2, \sqrt{2\delta_i}a_1, \sqrt{2\delta_i}a_2, \sqrt{2}a_1 a_2, \delta_i]^T \tag{10.29}$$

$$\varphi_i(\mathbf{g}) = [g_1^2, g_2^2, \sqrt{2\delta_i}g_1, \sqrt{2\delta_i}g_2, \sqrt{2}g_1 g_2, \delta_i]^T \tag{10.30}$$

Similarly, in the case of $\delta_i = 0$, we find that

$$\varphi_i(\mathbf{a}) = [a_1^2, a_2^2, \sqrt{2}a_1 a_2]^T \tag{10.31}$$

$$\varphi_i(\mathbf{g}) = [g_1^2, g_2^2, \sqrt{2}g_1 g_2]^T \tag{10.32}$$

As can be seen from (10.29) and (10.30), the two-dimensional input space via nonlinear mapping φ_i is mapped onto a six (or three)-dimensional feature space, aiming to obtain a richer space. To see this, a binary classification example is considered in Figure 10.2, where a two-dimensional original space $\mathscr{X}$, represented by $\mathbf{x} = [x_1, x_2]^T$, is mapped onto a three-dimensional feature space $\mathscr{H}$, represented by $\varphi_i(\mathbf{x}) = [z_1, z_2, z_3] = [x_1^2, x_2^2, \sqrt{2}x_1 x_2]$. It is seen that the mapped data become linearly separable, i.e., the ellipse decision boundary of the input space of Figure 10.2(a) becomes a hyperplane of Figure 10.2(b) since ellipses can be written as linear equations in the entries of $[z_1, z_2, z_3]$ [10].

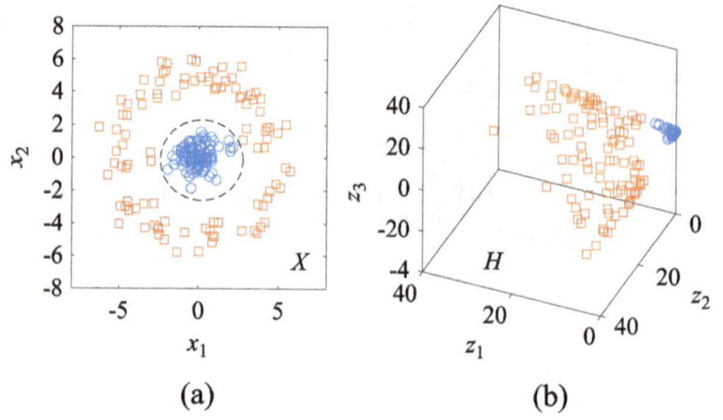

Figure 10.2 *A binary classification example, (a) data of two-dimensional input space $\mathscr{O}$, represented by $[x_1, x_2]$, (b) mapped data onto a three-dimensional feature space $\mathscr{F}$, represented by $\varphi_i(\mathbf{x}) = [z_1, z_2, z_3] = [x_1^2, x_2^2, \sqrt{2}x_1x_2]$, where data become linearly separable in the feature space $\mathscr{F}$*

Remark 1. *In general, there are different real-value kernel functions used in the literature, and the most common ones are the real polynomial $\mathscr{K}_i(\mathbf{x}_1, \mathbf{x}_2) = (\mathbf{x}_1^T\mathbf{x}_2 + \delta_i)^{d_i}$ and the real Gaussian radial basis function (RBF) $\mathscr{K}_i(\mathbf{x}_1, \mathbf{x}_2) = \exp(-(\mathbf{x}_1 - \mathbf{x}_2)^2/2\sigma_i^2)$, where the first belongs to the projective kernels and the latter is radial ones [10, 12]. Here, δ_i and d_i are, respectively, bias and order parameters of the polynomial kernel, while the single parameter $\sigma_i > 0$ is called the width of Gaussian RBF. Although the Gaussian RBF is the most popular, this radial kernel can deal with real-valued data sequences only. While the complex Gaussian RBF kernel with application in the complex kernel LMS algorithm has been introduced in [13], its applications to the target detection problems remain in obscurity, especially when the Doppler frequency of interested target is assumed to be unknown [14]. In contrast, the polynomial kernel can also be used in complex domains. To do so, it is enough to end up with a complex inner product space. Based on this, for the first time, we use complex-valued polynomial kernel function for radar target detection problems, defined as $\mathscr{K}_i(\mathbf{x}_1, \mathbf{x}_2) = (\mathbf{x}_1^\dagger\mathbf{x}_2 + \delta_i)^{d_i}$ with $d_i > 0$ [10]. This new complex-valued kernel can be served as a kernel function (in the sense of the kernel trick) if it satisfies two key properties: (1) positive semi-definiteness (i.e., $\mathscr{K}_i(\mathbf{x}_i, \mathbf{x}_i) \geq 0$ for $i = 1, 2$) and (2) symmetry (i.e., $\mathscr{K}_i(\mathbf{x}_1, \mathbf{x}_2) = \mathscr{K}_i^*(\mathbf{x}_2, \mathbf{x}_1)$). In our case, the positive semi-definiteness implies that $\mathscr{K}_i(\mathbf{x}_i, \mathbf{x}_i) = (\langle \mathbf{x}_i, \mathbf{x}_i \rangle + \delta_i)_i^d = (\|\mathbf{x}_i\|^2 + \delta_i)^{d_i} \geq 0$ for $i = 1, 2$, which holds true for $\delta_i \geq 0$. The symmetry property requires that $(\mathbf{x}_1^\dagger\mathbf{x}_2 + \delta_i)^{d_i} = (\mathbf{x}_2^T\mathbf{x}_1^* + \delta_i^*)^{d_i}$, implying that δ_i must be a real value. All of these mean that $\mathscr{K}_i(\mathbf{x}_1, \mathbf{x}_2) = (\mathbf{x}_1^\dagger\mathbf{x}_2 + \delta_i)^{d_i}$ with $d_i > 0$ is a valid complex-valued kernel function provided that $\delta_i \geq 0$.*

10.4.2 Kernelized LRT-based proposed detectors

The research in the area of capitalizing the idea of kernel theory and detection theory in order to develop new target detection algorithms is at a very incipient stage, and only a few detection-theory-based works are available in the literature [8] and [15]. In general, the kernelization procedure of a detector consists of two basic steps. First, we need to find an expression of the considered detector as a function of Euclidean inner products between the original (input) data only. This representation is called the dual representation in contrast to the primal representation [10]. In our case, we can re-express the test statistic (10.23) as

$$\Lambda_{\text{IR-LRT}}(\mathbf{x}_s, \mathbf{x}_r) = \max_{(\tau_t, f_t)} \left|\langle \Phi \mathbf{z}_r, \mathbf{z}_s \rangle\right|^2 \underset{\mathcal{H}_0}{\overset{\mathcal{H}_1}{\gtrless}} \zeta \tag{10.33}$$

and that of (10.21) can be written as

$$\begin{aligned} \Gamma_{\text{NR-LRT}}(\mathbf{x}_s, \mathbf{x}_r) &= \frac{1}{2}\left(1 - \frac{\langle \mathbf{x}_r, \mathbf{x}_r \rangle}{\langle \mathbf{x}_s, \mathbf{x}_s \rangle}\right) \\ &+ \frac{1}{2}\sqrt{\left(1 - \frac{\langle \mathbf{x}_r, \mathbf{x}_r \rangle}{\langle \mathbf{x}_s, \mathbf{x}_s \rangle}\right)^2 + 4\frac{\langle \mathbf{x}_r, \mathbf{x}_r \rangle}{\langle \mathbf{x}_s, \mathbf{x}_s \rangle} \max_{(\tau_t, f_t)} \left|\langle \Phi \mathbf{z}_r, \mathbf{z}_s \rangle\right|^2} \underset{\mathcal{H}_0}{\overset{\mathcal{H}_1}{\gtrless}} \eta \end{aligned} \tag{10.34}$$

From (10.33) to (10.34), it becomes evident that the similarity measure utilized in the proposed detectors can be expressed using the Euclidean inner products of the original data. In order to enhance the detection performance, the second step involves substituting the Euclidean inner products with appropriate kernel functions in a high-resolution space. For our situation, where we are dealing with complex-valued data, we substitute the Euclidean inner products with suitable polynomial kernel functions. This substitution is equivalent to transforming the original data space into a richer space, which is commonly referred to as the feature space. This is because the elements of the vectors in the original space can be viewed as monomial features with degree one, whereas the elements in the feature space correspond to monomials with degrees up to d when a polynomial kernel function of order d is employed [8, 15]. As a result, the SIR-LRT detector can be represented by a polynomial kernel-based detector with order d_1 and bias parameter c_1, as follows:

$$\Lambda_{\text{KSIR-LRT}}(\mathbf{x}_s, \mathbf{x}_r) = \max_{(\tau_t, f_t)} \left|\left(\langle \Phi \mathbf{z}_r, \mathbf{z}_s \rangle + c_1\right)^{d_1}\right|^2 \underset{\mathcal{H}_0}{\overset{\mathcal{H}_1}{\gtrless}} \zeta_1 \tag{10.35}$$

or, equivalently,

$$\Lambda'_{\text{KSIR-LRT}}(\mathbf{x}_s, \mathbf{x}_r) = \max_{(\tau_t, f_t)} \left|\langle \Phi \mathbf{z}_r, \mathbf{z}_s \rangle + c_1\right|^2 \underset{\mathcal{H}_0}{\overset{\mathcal{H}_1}{\gtrless}} \zeta_1^{\frac{1}{d_1}} = \zeta_1' \tag{10.36}$$

Similarly, the corresponding kernelized detector of the NR-LRT of (10.34) can be obtained as

$$\Gamma_{\text{KNR-LRT}}(\mathbf{x}_s, \mathbf{x}_r) = \frac{1}{2}\left(1 - \frac{(\langle \mathbf{x}_r, \mathbf{x}_r\rangle + c_2)^{d_2}}{(\langle \mathbf{x}_s, \mathbf{x}_s\rangle + c_3)^{d_3}}\right)$$

$$+\frac{1}{2}\sqrt{\left(1 - \frac{(\langle \mathbf{x}_r, \mathbf{x}_r\rangle + c_2)^{d_2}}{(\langle \mathbf{x}_s, \mathbf{x}_s\rangle + c_3)^{d_3}}\right)^2 + 4\frac{(\langle \mathbf{x}_r, \mathbf{x}_r\rangle + c_2)^{d_2}}{(\langle \mathbf{x}_s, \mathbf{x}_s\rangle + c_3)^{d_3}} \max_{(\tau_l, f_l)} \left|(\langle \Phi \mathbf{z}_r, \mathbf{z}_s\rangle + c_4)^{d_4}\right|^2} \underset{\mathcal{H}_0}{\overset{\mathcal{H}_1}{\gtrless}} \zeta_2' \tag{10.37}$$

In the sequel, the former detector is called kernelized SIR (KSIR)-LRT detector, while the latter is referred to as kernelized NR (KNR)-LRT detector.

It is worth to note that the main NR-LRT detector possesses constant false alarm rate (CFAR) property against the noise variances in the reference and surveillance channels, that is,

$$\Gamma_{\text{NR-LRT}}(\xi \mathbf{x}_r, \xi \mathbf{x}_s) = \Gamma_{\text{NR-LRT}}(\mathbf{x}_r, \mathbf{x}_s) \tag{10.38}$$

where $\xi > 0$. By requiring that

$$\Gamma_{\text{KNR-LRT}}(\xi \mathbf{x}_r, \xi \mathbf{x}_s) = \Gamma_{\text{KNR-LRT}}(\mathbf{x}_r, \mathbf{x}_s) \tag{10.39}$$

we obtain $c_2 = c_3 = 0$ and $d_2 = d_3 = d_{2,3}$ in (10.37), which guarantees to preserve the CFAR property against the noise variances in the reference and surveillance channels for the proposed KNR-LRT detector. Thus, under this condition, the proposed KNR-LRT detector can be simplified as

$$\Gamma_{\text{KNR-LRT}}(\mathbf{x}_s, \mathbf{x}_r) = \frac{1}{2}\left(1 - \left(\frac{\langle \mathbf{x}_r, \mathbf{x}_r\rangle}{\langle \mathbf{x}_s, \mathbf{x}_s\rangle}\right)^{d_{2,3}}\right)$$

$$+\frac{1}{2}\sqrt{\left(1 - \left(\frac{\langle \mathbf{x}_r, \mathbf{x}_r\rangle}{\langle \mathbf{x}_s, \mathbf{x}_s\rangle}\right)^{d_{2,3}}\right)^2 + 4\left(\frac{\langle \mathbf{x}_r, \mathbf{x}_r\rangle}{\langle \mathbf{x}_s, \mathbf{x}_s\rangle}\right)^{d_{2,3}} \max_{(\tau_l, f_l)} \left|(\langle \Phi \mathbf{z}_r, \mathbf{z}_s\rangle + c_4)^{d_4}\right|^2} \underset{\mathcal{H}_0}{\overset{\mathcal{H}_1}{\gtrless}} \zeta \tag{10.40}$$

where ζ_2 is a suitably chosen threshold.

The key issue in kernelized detectors is determining the appropriate values for the parameters c_1 and d_1 in (10.35) (or c_1 in [10.36]), as well as $d_{2,3}$, c_4, and d_4 in (10.40). To achieve maximum detection probabilities, it is important to select these parameters thoughtfully. Section 10.5 elaborates on this topic and presents numerical results.

Finally, it is seen from (10.33), (10.34), (10.36), and (10.40) that they have the same term $\langle \Phi \mathbf{z}_r, \mathbf{z}_s\rangle$, denoted by $\mathcal{D}(\tau_l, f_l)$ hereafter. It was shown in Reference 16 that $\mathcal{D}(\tau_l, f_l)$ can be efficiently implemented by using fast Fourier transform (FFT) and its inverse (IFFT). In our case, this implementation is used and illustrated in Figure 10.3.

10.5 Performance results

In this section, we utilized numerical simulations to assess the effectiveness of the proposed SIR-LRT, KSIR-LRT, NR-LRT, and KNR-LRT, or simply SIR, KSIR, NR,

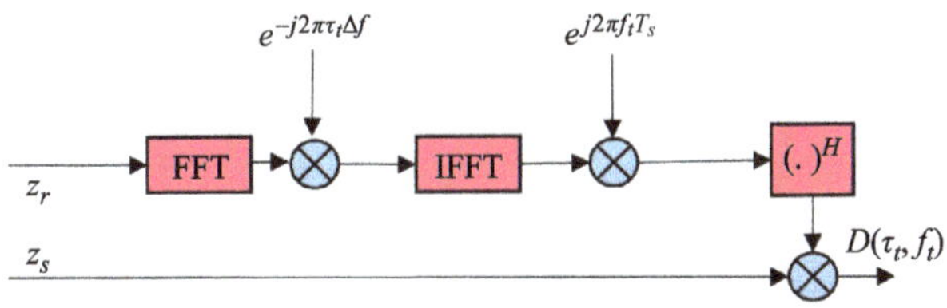

Figure 10.3 *Efficient implementation of* $\mathscr{D}(\tau_t, f_t) = \langle \Phi \mathbf{z}_r, \mathbf{z}_s \rangle$ *in the detectors of (10.23), (10.34), (10.36), and (10.40)*

and KNR detectors. The SIR detector was used as a benchmark for comparison. The SNR_s and DNR_r were defined as the signal power-to-noise ratio in the SC and the direct-path signal power-to-noise ratio in the RC, respectively, given by

$$\text{SNR}_s = \frac{|\beta|^2}{\sigma^2} \tag{10.41}$$

$$\text{DNR}_r = \frac{|\alpha|^2}{\sigma^2} \tag{10.42}$$

It is challenging to derive explicit formulas for the detection and false alarm probabilities, denoted by p_d and p_{fa}, respectively. Therefore, we turn to Monte Carlo (MC) simulations to investigate the performance of our proposed detectors in terms of false alarm and detection probabilities. It is widely accepted that for reliable MC evaluations, we should perform a minimum of $\frac{100}{p_{fa}}$ simulations to obtain the detection threshold accurately. For example, for $p_{fa} = 10^{-3}$, this equates to 10^5 simulations. To ensure the accuracy of our results, we conduct 2×10^6 MC simulations, which provides a safety margin. Our simulations employ an integration time of $T = 0.00128$ s and a sampling frequency of $f_s = 200$ kHz, resulting in a total of $N = 256$ samples.

In order to utilize a detector effectively, it is necessary to establish its detection threshold in accordance with a desired false alarm probability. Figure 10.4 presents false alarm probabilities as a function of detection thresholds for the NR, KNR, SIR, and KSIR detectors across a range of DNR_r values from $[-10, 20]$ dB. The results indicate that the NR and KNR-LRT detectors exhibit a high degree of variability in detection thresholds when working with small DNR_r values, while the SIR-LRT and KSIR-LRT detectors show less variation. This suggests that precise knowledge of the DNR_r value is essential when setting detection thresholds for the proposed NR-LRT and KNR-LRT detectors, particularly in practical scenarios where DNR_r values are small. The result of this is the need to explore alternative methods for establishing detection thresholds, beyond what is shown in Figure 10.4. This is crucial because in practical scenarios, the accuracy of the value of DNR_r may be uncertain, which is referred to as reference channel uncertainty (RCU). Fortunately, there is a viable and effective approach for determining the detection thresholds using the concept of test level.

To begin, we will use $P_{fa}^{(e)}$ to represent the empirical false alarm probability and p_{fa} to represent the desired false alarm probability. A detector is said to have an

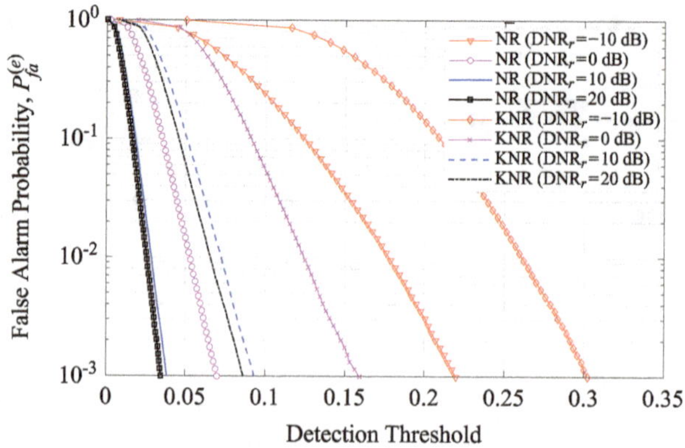

(a) NR-LRT and KNR-LRT detectors

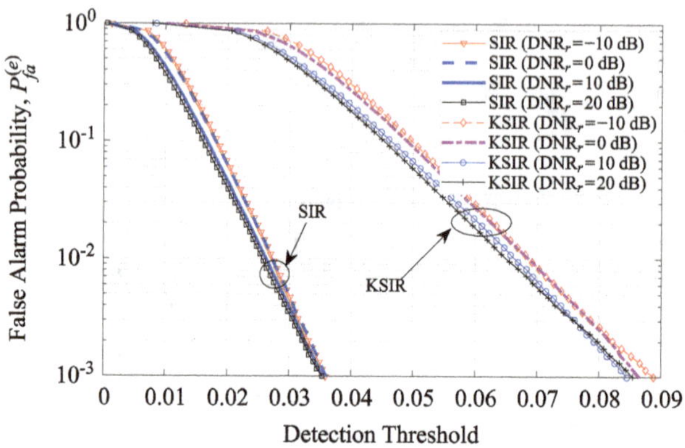

(b) SIR-LRT and KSIR-LRT detectors

Figure 10.4 *Empirical probability of false alarm as a function of detection threshold for different values of* DNR$_r$

empirical false alarm probability of p_{fa} if $P_{fa}^{(e)} = p_{fa}$, which is also known as a CFAR detector. A detector has an empirical false alarm probability at the level of p_{fa} if $P_{fa}^{(e)} \leq p_{fa}$. The former is called a size-p_{fa} detector, while the latter is referred to as a level-p_{fa} test. In practical scenarios, it is crucial for a detector to ensure that the false alarm rate does not exceed the required level of p_{fa}. This approach is a new concept for determining a detection threshold in the field of radar target detection. In order to utilize this concept, we will introduce some new parameters. The term DNR$_r$ will denote the real DNR$_r$, while DNR$_r^{\eta}$ will represent the value of DNR$_r$ that serves as the basis for establishing the detection thresholds of the designed detectors. As we will demonstrate, it is imperative to determine or acquire the minimum received DNR$_r$ value in order to create or devise a detector with a predetermined level. To gain a better understanding of this new concept, we will examine two scenarios, as described below.

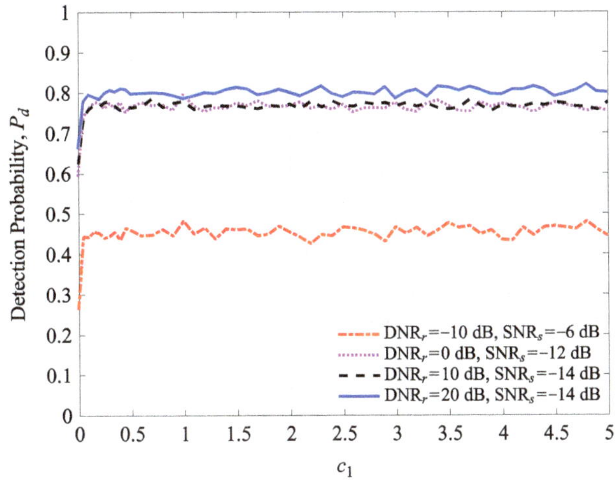

Figure 10.5 *Detection probability as a function of polynomial kernel parameter c_1 of the proposed KSIR detector for different values of* DNR_r *and* SNR_s

Case A1: We assume that the received DNR_r for different radar-setup scenarios is in the range of −10 dB to 10 dB. Here, we assume that the radar designer (or radar operator) does not use the fact (does not know) that the detection threshold of the considered detector is strongly dependent on the values of the received DNR_r and use a specific value of DNR_r to set the threshold. In such a case, we are interested to investigate the effect of this threshold-setting strategy on the false alarm regulation as well as detection performance.

Case A2: We have some knowledge about the received DNR_r with some uncertainties. Here, we assume that the minimum values of the received DNR_r, denoted by $\min\{\mathrm{DNR}_r\}$, and the amount of the RCU, denoted by ρ, are known based on the accuracy of the knowledge of DNR_r. This means that we know the range of the received DNR_r, that is, $[\min\{\mathrm{DNR}_r[\mathrm{dB}]\}, \max\{\mathrm{DNR}_r[\mathrm{dB}]\}]$ with $\max\{\mathrm{DNR}_r[\mathrm{dB}]\} = \min\{\mathrm{DNR}_r[\mathrm{dB}]\} + \rho_{\mathrm{dB}}$ where $\mathrm{DNR}_r[\mathrm{dB}] = 10\log_{10}\mathrm{DNR}_r$ and $\rho_{\mathrm{dB}} = 10\log_{10}\rho$. In practice, ρ_{dB} reflects how accurate the estimate of DNR_r is.

Before proceeding with the novel threshold-setting approach, it is necessary to select the polynomial parameters for the proposed KSIR-LRT and KNR-LRT detectors. Within the context of detection theory, our focus is on tests that achieve the highest possible detection probability while maintaining a specified false alarm probability. Consequently, we will seek to determine the kernel parameters that optimize the detection probabilities of these detectors according to these criteria. In Figure 10.5, the detection probability of the proposed KSIR-LRT detector against c_1 is plotted when we use different values of DNR_r and SNR_s, and when $\mathrm{DNR}_r^{\eta} = \mathrm{DNR}_r$. In the case of $c_1 = 0$, the KSIR detector corresponds to that of the SIR detector; thus, the SIR detector is a special case of the KSIR one. From Figure 10.5, it is observed that the value of $c_1 > 0.1$ results in the best detection performance. In the

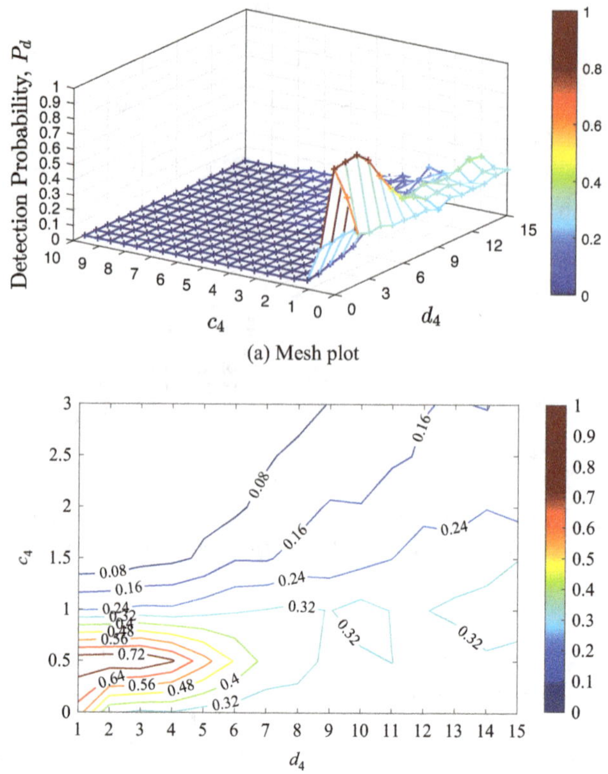

(a) Mesh plot

(b) Colormap plot

Figure 10.6 *Detection probability as a function of polynomial kernel parameters d_4 and c_4 of the proposed KNR detector when* DNR_r = *−10 dB,* SNR_s = *−6 dB, and* $d_{2,3} = 1$

following, we use $c_1 = 0.5$ to compare the detection performance of the proposed KSIR detector with that of its counterparts. For the proposed KNR-LRT detector, the above simulation is repeated, and the results are shown in Figures 10.6–10.9 for several values of parameters c_4 and d_4 and when $d_{2,3} = 1$. The results show that selecting appropriate values for c_4 and d_4 is crucial for maximizing the detection probability. Additionally, the impact of kernel parameter $d_{2,3}$ on the performance of the KNR-LRT detector was examined in Figure 10.10, where $c_4 = 0.5$ and $d_4 = 3$. The results indicate that increasing $d_{2,3}$ leads to a detection probability of zero. Based on the results of Figures 10.6–10.10, it is evident that the optimal kernel parameters are dependent on DNR_r and SNR_s. Nevertheless, for further analysis, we set $c_4 = 0.5$, $d_4 = 3$, and $d_{2,3} = 1$ as suitable kernel parameters for the proposed KNR-LRT detector across various DNR_r and SNR_s ranges.

We will now examine how effective the proposed method of setting thresholds is in two scenarios, A1 and A2. For A1, we assume that the detection thresholds are

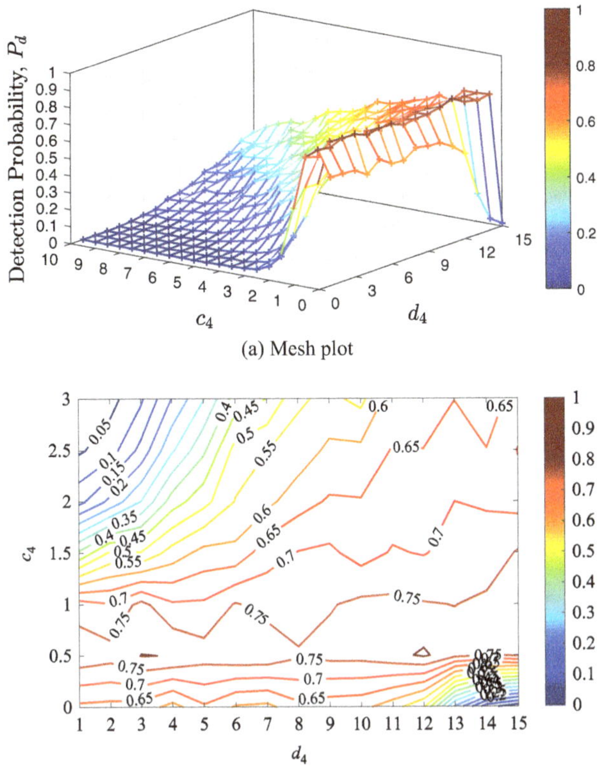

(a) Mesh plot

(b) Colormap plot

Figure 10.7 *Detection probability as a function of polynomial kernel parameters d_4 and c_4 of the proposed KNR detector when* DNR_r = 0 *dB,* $\text{SNR}_s = -12$ *dB, and* $d_{2,3} = 1$

established based on two different values of DNR_r^η, specifically, $\text{DNR}_r^\eta = -10$ dB and $\text{DNR}_r^\eta = 0$ dB, while we encounter with situations in which $\text{DNR}_r \in [-10, 6]$ dB. The results of this simulation are reported in Figure 10.11. For the proposed NR-LRT and KNR-LRT detectors, it is seen that empirical false alarm probability is limited to desired false alarm $p_{fa} = 10^{-3}$ for $\text{DNR}_r \geq \text{DNR}_r^\eta$ when the detection thresholds are set according to DNR_r^η. For example, when we set the detection threshold according to the value $\text{DNR}_r^\eta = 0$ dB, we get excessive false alarm probability with respect to the desired false alarm $p_{fa} = 10^{-3}$ for $\text{DNR}_r < 0$ dB, while it is bounded for $\text{DNR}_r > 0$ dB. This implies that it is required to know the minimum value of the received DNR_r to achieve a detector with a fixed level, i.e., $P_{fa}^{(e)} \leq p_{fa} = 10^{-3}$, for the proposed NR-LRT and KNR-LRT detectors. Also, for the SIR-LRT and KSIR-LRT detectors, it is observed from Figure 10.11 that the empirical false alarm probability is equal to the desired false alarm probability of $p_{fa} = 10^{-3}$ for different values of DNR_r and DNR_r^η.

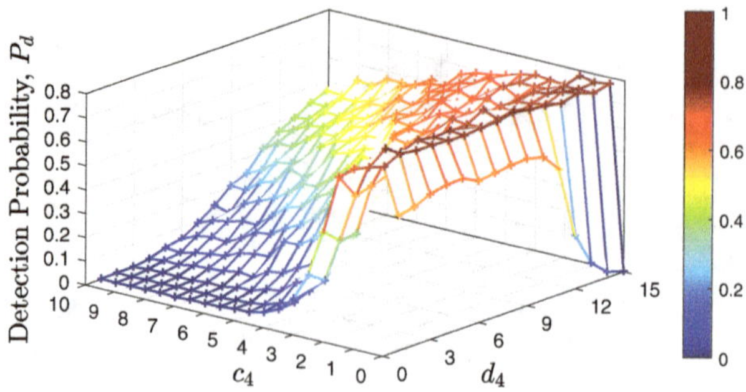

a) Mesh plot

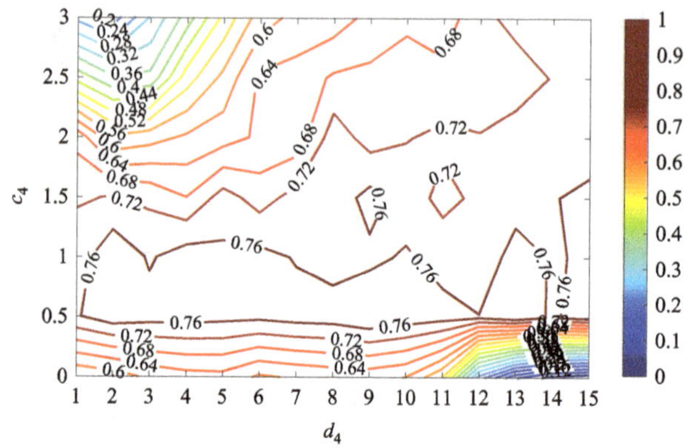

(b) Colormap plot

Figure 10.8 *Detection probability as a function of polynomial kernel parameters d_4 and c_4 of the proposed KNR detector when* DNR_r = 10 dB, $\mathrm{SNR}_s = -14$ dB, *and* $d_{2,3} = 1$

It may not always be feasible to determine the minimum value of received DNR in practical scenarios. Nevertheless, we can establish a detection threshold based on a worst-case scenario for received DNR. For this study, we have considered two cases, where DNR_r is equal to 10 dB and –10 dB. However, we have used a DNR_r^η of –10 dB for threshold-setting purposes. As such, the detection performances of the proposed detectors are compared and shown in Figure 10.12. Interestingly, it is seen that by increasing the DNR_r, the detection probability of the proposed NR-LRT detector is not improved, while that of the proposed KNR-LRT detector is improved. In other words, by increasing the DNR_r from –10 dB to 10 dB, a 2.5 dB performance gain is obtained for the KNR-LRT detector. The reason of this detection performance limitation is that fixing the level of the NR-LRT detector increases its detection threshold substantially, resulting in $P_{fa}^{(e)} \leq 10^{-3}$, where $p_{fa} = 10^{-3}$ in Figure 10.11. In other words, the adjusted detection threshold in the case $\mathrm{DNR}_r = 10$ dB when the detection threshold is set for $\mathrm{DNR}_r^\eta = -10$ dB is larger than that required to get $P_{fa}^{(e)} = p_{fa}$, which

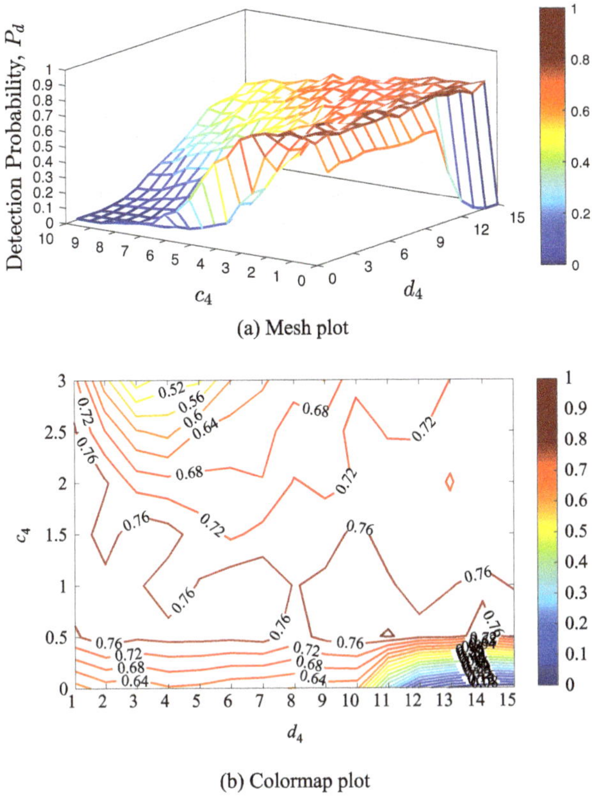

Figure 10.9 *Detection probability as a function of polynomial kernel parameters d_4 and c_4 of the proposed KNR detector when* DNR_r = 20 dB, SNR_s = −14 dB *and* $d_{2,3} = 1$

contributes to the above detection performance limitation. It is also seen that this new threshold setting cannot affect the detection performance improvements of the SIR-LRT and KSIR-LRT detectors by increasing the DNR_r. In this case, the performance gains are about 12 dB and 11 dB for the SIR and KSIR-LRT detectors, respectively. From Figure 10.12, one can further see that around 1.75 dB and 1 dB gains can be obtained by the proposed KSIR-LRT method as compared to that of the SIR-LRT detector for the DNR_r of −10 dB and 10 dB at $P_d = 0.9$, respectively. These gains are equal to 0.8 dB and 3.5 dB comparing the KNR-LRT detector to the NR-LRT test. Furthermore, the NR detector performs better than the SIR-LRT method with the gain of 4.5 dB for DNR_r = −10 dB, while it results in the gain loss of −7.5 dB for DNR_r = 10 dB. To the best of the authors' knowledge, no work has been reported in the literature for this challenging problem of a noisy reference channel-based detector. Thus, for the proposed NR-LRT detector, this shows that it is important to obtain an estimate of the DNR_r when it is assumed that the reference channel is noisy.

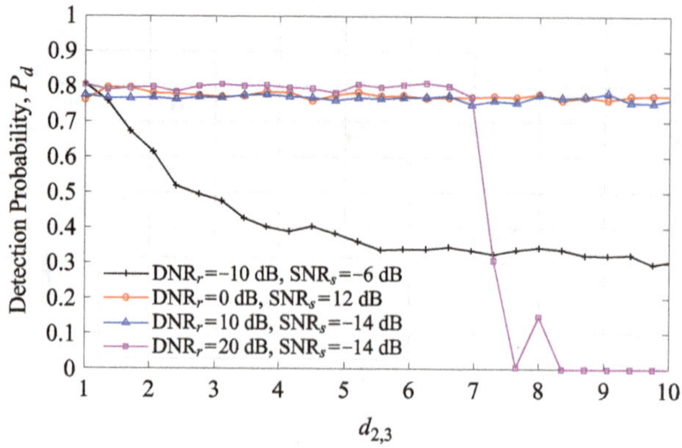

Figure 10.10 *Detection probability as a function of polynomial kernel parameter $d_{2,3}$ of the proposed KNR detector when $d_4 = 3$ and $c_4 = 0.5$ but for different values of* DNR_r *and* SNR_s

In case A2, we consider different ranges of known received DNR_r, described by $[\min\{\mathrm{DNR}_r[\mathrm{dB}]\} + \rho_{\mathrm{dB}}(i-1), \min\{\mathrm{DNR}_r[\mathrm{dB}]\} + \rho_{\mathrm{dB}}i]$ with $\min\{\mathrm{DNR}_r[\mathrm{dB}]\} = -18$ dB and $\rho_{\mathrm{dB}} = 2$ for $i = 1, \ldots, 18$. Here, the detection thresholds are set according to $\mathrm{DNR}_r^{\eta}(i) = \min\{\mathrm{DNR}_r[\mathrm{dB}]\} + \rho_{\mathrm{dB}}(i-1)$ in the ith uncertainty interval (UI) to achieve detectors with fixed levels. The results of this simulation are shown in Figure 10.13. As can be seen, the obtained (empirical) false alarm probability, say $P_{fa}^{(e)}$, is always below the desired false alarm probability $p_{fa} = 10^{-3}$ for the proposed NR-LRT and KNR-LRT detectors. For the DNR_r values resulting in $P_{fa}^{(e)} < p_{fa}$, the adjusted detection thresholds are larger than those required to get $P_{fa}^{(e)} = p_{fa}$. In such cases, it is not surprising that the detection performance of the proposed detectors is degraded, as shown in the following. From Figure 10.13, it is also seen that the KNR-LRT detector has lower

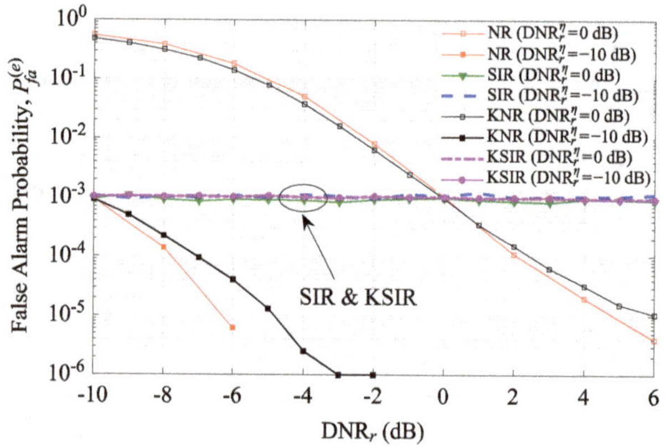

Figure 10.11 *Empirical probability of false alarm as functions of* DNR_r *for two cases of* $\mathrm{DNR}_r^{\eta} = 0$ *dB and* $\mathrm{DNR}_r^{\eta} = -10$ *dB and when* $p_{fa} = 10^{-3}$

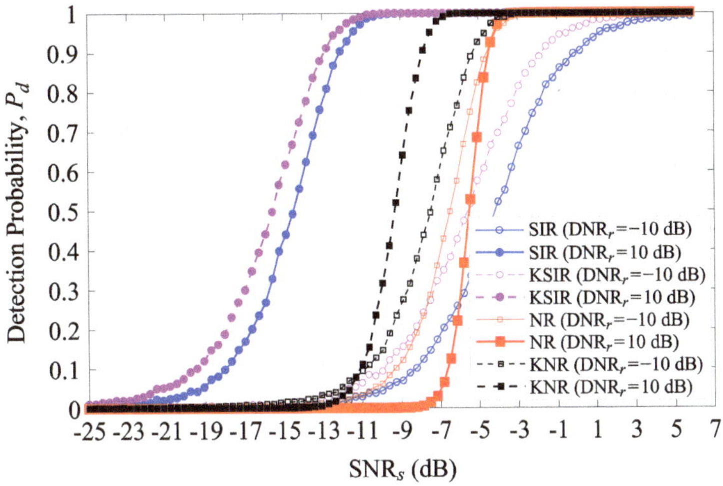

Figure 10.12 *Detection probability versus SNR$_s$ when detection thresholds are set for* $\mathrm{DNR}_r^\eta = -10$ *dB, resulting in* $P_{fa}^{(e)} \leq 10^{-3}$ *for two cases* $\mathrm{DNR}_r = -10$ *dB and* $\mathrm{DNR}_r = 10$ *dB*

variation of false alarm probability as compared to that of the NR-LRT detector; thus, it may result in better detection performance for $\mathrm{DNR}_r(i) = \min\{\mathrm{DNR}_r[\mathrm{dB}]\} + \rho_{\mathrm{dB}} i$ when its detection threshold is set for $\mathrm{DNR}_r^\eta(i) = \min\{\mathrm{DNR}_r[\mathrm{dB}]\} + \rho_{\mathrm{dB}}(i-1)$. Besides, it is observed from Figure 10.13 that for the small values of DNR$_r$ for which

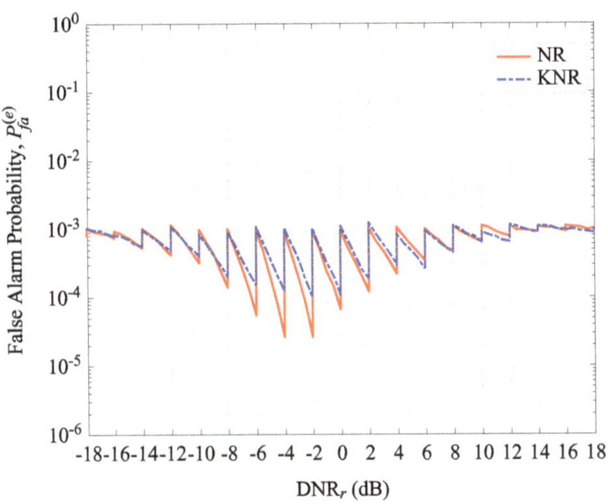

Figure 10.13 *Empirical probability of false alarm for different ranges of* DNR$_r$ *when* $p_{fa} = 10^{-3}$ *and the detection thresholds are set for* $\mathrm{DNR}_r^\eta = \min\{\mathrm{DNR}_r[\mathrm{dB}]\} + \rho_{\mathrm{dB}}(i-1)$ *in the ith interval*

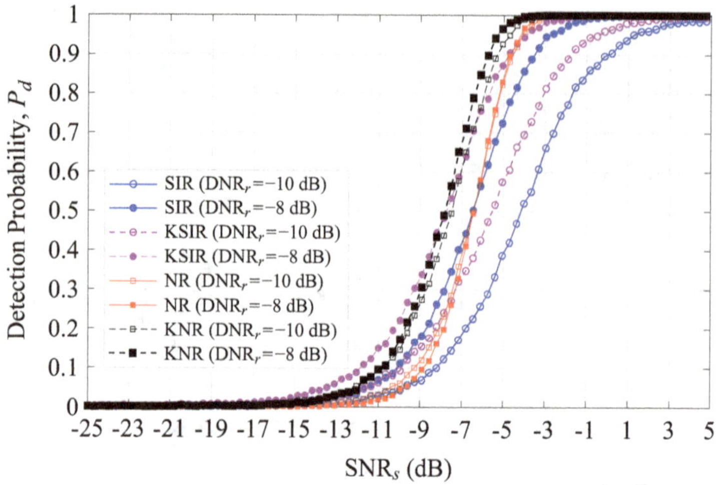

Figure 10.14 *Detection probability versus SNR$_s$ when the detection thresholds of the considered detectors are set for* DNR$_r^\eta$ = *–10 dB, resulting in* $P_{fa}^{(e)} < 10^{-3}$ *for* DNR$_r$ = *–8 dB (i.e., level-*10^{-3}*) and* $P_{fa}^{(e)} = 10^{-3}$ *for* DNR$_r$ = *–10 dB (i.e., size-*10^{-3}*)*

the noise of reference channel dominates the direct signal, we can expect to achieve $P_{fa}^{(e)} = p_{fa} = 10^{-3}$.

In Figures 10.14–10.16, we investigate the impact of different values of DNR$_r$ on the detection performances of the NR, KNR, SIR, and KSIR detectors under case A2 for $i = 1$ (i.e., DNR$_r \in [-10,-8]$ dB), $i = 5$ (i.e., DNR$_r \in [-2, 0]$ dB), and $i = 16$ (i.e., DNR$_r \in [20, 22]$ dB), respectively. It is observed that the detection performance improvements of the proposed NR-LRT and KNR-LRT detectors are insignificant in the uncertainty range of 2 dB by increasing the value of DNR$_r$, while it is improved significantly for the SIR-LRT and KSIR-LRT detectors. However, the proposed detectors generally perform better in the ith UI as compared to the jth UI when $i > j$. In other words, the larger the DNR$_r$, the better the detection performance, especially for the SIR-LRT and KSIR-LRT detectors. From Figure 10.14 (Figure 10.15), one can observe that around 1.75 dB (1 dB) and 1.25 dB (1 dB) gains can be obtained by the proposed KSIR detector as compared with the SIR detector for the DNR$_r$ of –10 dB (–2 dB) and –8 dB (0 dB) at $P_d = 0.9$, respectively. These gains are 0.8 dB and 1 dB for the KNR-LRT and NR-LRT methods in both Figures 10.14 and 10.15. Besides, the NR-LRT detector performs better than the SIR-LRT method with the gain of 4.5 dB for DNR$_r$ = –10 dB, while this is about 1 dB when DNR$_r$ = –8 dB. As can be seen from Figure 10.15, the NR-LRT detector performs better than the SIR-LRT method with the gain of 0.5 dB for DNR$_r$ = –2 dB, while this is around 1 dB for the SIR-LRT detector when DNR$_r$ = 0 dB. Interestingly, it is observed from Figure 10.16 that the performances of the SIR-LRT and NR-LRT detectors, as well as the KSIR-LRT and KNR-LRT methods, cannot be improved as desired, even with increasing DNR$_r$. As

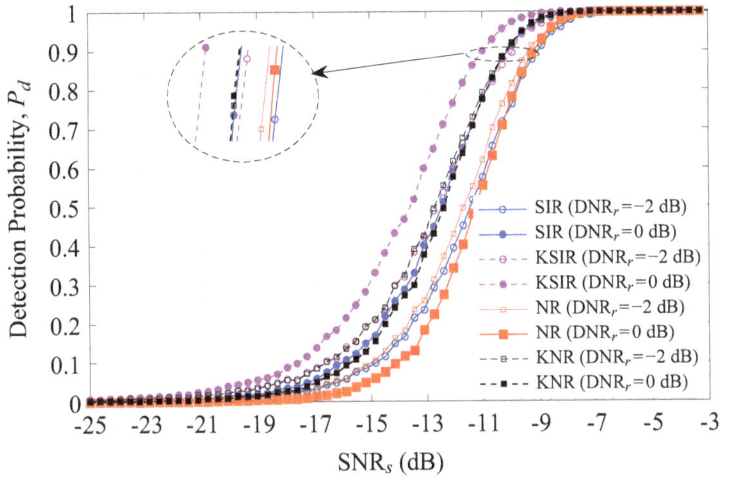

Figure 10.15 *Detection probability versus SNR_s when the detection thresholds of the considered detectors are set for* $\mathrm{DNR}_r^{\eta} = -2$ *dB, resulting in* $P_{fa}^{(e)} < 10^{-3}$ *for* $\mathrm{DNR}_r = 0$ *dB (i.e., level-*10^{-3}*) and* $P_{fa}^{(e)} = 10^{-3}$ *for* $\mathrm{DNR}_r = -2$ *dB (i.e., size-*10^{-3}*)*

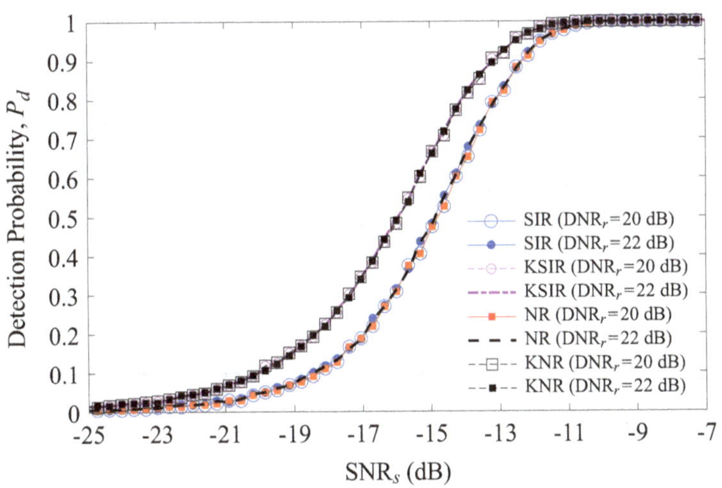

Figure 10.16 *Detection probability versus SNR_s when the detection thresholds of the considered detectors are set for* $\mathrm{DNR}_r^{\eta} = 20$ *dB, resulting in* $P_{fa}^{(e)} < 10^{-3}$ *for* $\mathrm{DNR}_r = 22$ *dB (i.e., level-*10^{-3}*) and* $P_{fa}^{(e)} = 10^{-3}$ *for* $\mathrm{DNR}_r = 20$ *dB (i.e., size-*10^{-3}*)*

always, the proposed kernelized detectors outperform their conventional counterparts with gains of 1 dB.

By comparing the results of Figure 10.12 with those of Figures 10.14–10.16, it is seen that the new proposed threshold-setting strategy provides some detection performance improvements for the noisy reference channel-based detectors as DNR_r increases under the required condition $P_{fa}^{(e)} \leq p_{fa}$. To see better the effect of increasing of DNR_r on the detection performance of the considered detectors, we consider a new case in which $\text{DNR}_r^{\eta} = \text{DNR}_r = \gamma$. The results of this simulation are depicted in Figure 10.17, where there are some points in order. First, the NR-LRT (KNR-LRT) detector significantly outperforms that of the SIR-LRT (KSIR-LRT) detector for $\text{DNR}_r < 0$ dB, while for the other values of DNR_r, the performances of NR-LRT and SIR-LRT (KNR-LRT and KSIR-LRT) detectors are the same. For the NR-LRT detector as compared with SIR-LRT detector, the obtained SNR_s gain is about 8 dB between the case of $\text{DNR}_r = -10$ dB and the case of $\text{DNR}_r = 20$ dB when $P_{fa}^{(e)} = 10^{-3}$ and the detection probability is 0.9. Second, we see that increasing the DNR_r for values greater than 10 dB does not result in improving the detection performance of the proposed counterpart detectors. Particularly, the performance differences for different values of DNR_r become smaller and smaller when DNR_r increases. Third, the same trend can be seen for the proposed kernelized detectors, as shown in Figure 10.17(b). Fourth, it is seen that the kernelized detectors perform better than their counterparts.

The detection performance of the considered detectors is examined in Figure 10.18 for values of the DNR_r greater than 0 dB. Here, we again consider the case of $\text{DNR}_r^{\eta} < \text{DNR}_r$ to attain $P_{fa}^{(e)} \leq p_{fa}$ for different pairs of DNR_r^{η} and DNR_r (or different intervals such as case A2). In this case, we see that the SIR-LRT and KSIR-LRT detectors perform better than the NR-LRT and KNR-LRT detectors, respectively, when 0 dB $< \text{DNR}_r <$ 10 dB. By comparing the results of Figure 10.18(a) and (b) (or that of Figure 10.17(a) and (b)), the SNR_s gains of the KNR-LRT and KSIR-LRT detectors are about 1 dB with respect to the NR-LRT and SIR-LRT detectors, respectively. This also shows the importance of combining the detection theory and kernel theory frameworks to achieve better detection performance.

10.6 Summary

This chapter introduces a new method for setting thresholds in radar applications. Rather than fixing the size of a test, we propose fixing the level of the test for practical uncertainty conditions. We present three new detectors based on detection theory and the combination of detection theory and kernel theory. The DNR_r of the reference channel has a significant impact on the detection thresholds and false alarm regulations of the proposed detectors when using the noisy reference

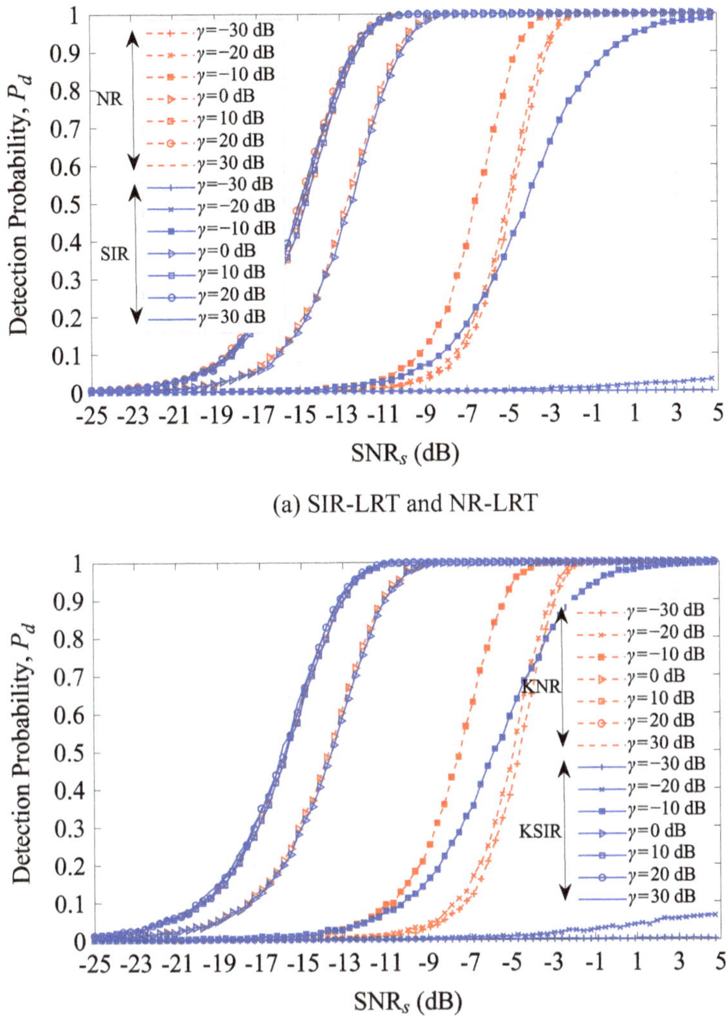

(a) SIR-LRT and NR-LRT

Figure 10.17 *Detection performance comparison of the proposed NR-LRT and SIR-LRT detectors as well as KNR-LRT and KSIR-LRT detectors for different values of* $\gamma = \text{DNR}_r^\eta = \text{DNR}_r$*, resulting in* $P_{fa}^{(e)} = 10^{-3}$

channel. To address this, we suggest setting the detection thresholds of NR-based detectors so that their levels are fixed according to a desired false alarm probability. In our proposed kernel theory-based detectors, we improve detection performance by replacing the inner products of test statistics with appropriate polynomial kernel functions. We use the principle of invariance and maximizing

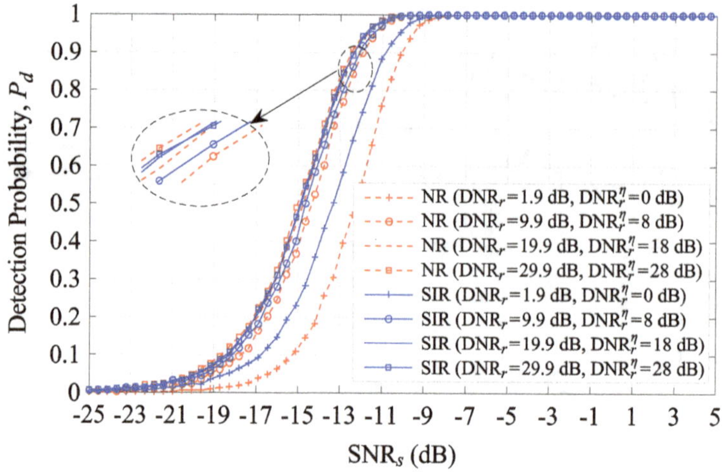

(a) SIR-LRT and NR-LRT

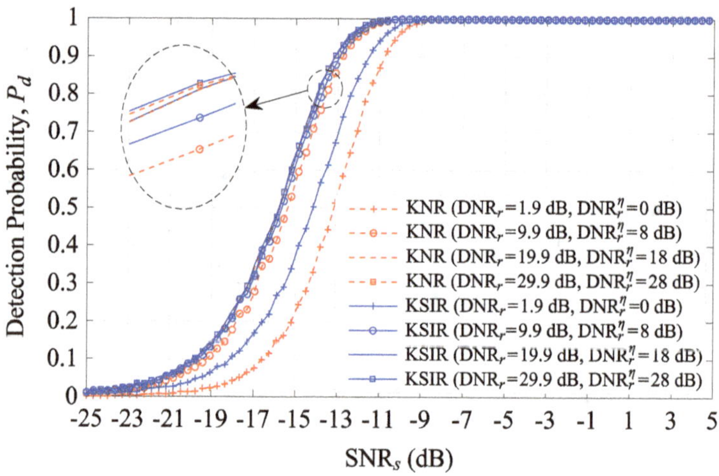

(b) KSIR-LRT and KNR-LRT

Figure 10.18 *Detection performance comparison of the proposed NR-LRT and SIR-LRT detectors, as well as KNR-LRT and KSIR-LRT detectors, for different values of* DNR_r^η *and* DNR_r *with* $\mathrm{DNR}_r > 0$ *dB, resulting in* $P_{fa}^{(e)} \leq 10^{-3}$

detection performance to set the polynomial kernel parameters of the proposed kernelized detectors. Through extensive Monte Carlo simulations, we demonstrate the effectiveness of our threshold-setting strategy and show that the proposed kernelized detectors offer over 1 dB of detection performance gain compared to their counterparts.

References

[1] Hack D.E., Patton L.K., Himed B., Saville M.A. 'Detection in passive MIMO radar networks'. *IEEE Transactions on Signal Processing*. 2014, vol. 62(11), pp. 2999–3012.

[2] Cui G., Liu J., Li H., Himed B. 'Signal detection with noisy reference for passive sensing'. *Signal Processing*. 2015, vol. 108, pp. 389–99.

[3] Liu J., Li H., Himed B. 'Two target detection algorithms for passive multistatic radar'. *IEEE Transactions on Signal Processing*. 2014, vol. 62(22), pp. 5930–39.

[4] Hack D.E., Patton L.K., Himed B., Saville M.A. 'Centralized passive MIMO radar detection without direct-path reference signals'. *IEEE Transactions on Signal Processing*. 2014, vol. 62(11), pp. 3013–23.

[5] Gogineni S., Setlur P., Rangaswamy M., Nadakuditi R.R. 'Passive radar detection with noisy reference channel using principal subspace similarity'. *IEEE Transactions on Aerospace and Electronic Systems*. 2018, vol. 54(1), pp. 18–36.

[6] Santamaria I., Scharf L.L., Via J., Wang H., Wang Y. 'Passive detection of correlated subspace signals in two MIMO channels'. *IEEE Transactions on Signal Processing*. , vol. 65(20), pp. 5266–80.n.d.

[7] Javidan M.H., Zaimbashi A., Liu J. 'Target detection in passive radar under noisy reference channel: a new threshold-setting strategy'. *IEEE Transactions on Aerospace and Electronic Systems*. 2020, vol. 56(6), pp. 4711–22.

[8] Zaimbashi A., Li J. 'Tunable adaptive target detection with kernels in colocated MIMO radar'. *IEEE Transactions on Signal Processing*. 2020, vol. 68, pp. 1500–14.

[9] Casella G., Berger R.L. 'Statistical inference' in Vol. 2. Pacific Grove, CA: Duxbury; 2002.

[10] Schölkopf B., Smola A.J. 'Learning with Kernels' in MIT Press; 2002.

[11] Rojo-Álvarez J.L., Martínez-Ramón M., Muñoz-Marí J., Camps-Valls G. 'Digital signal processing with kernel methods' in Hoboken, NJ, USA: John Wiley and Sons, Inc; p. 111.

[12] Ding G., Wu Q., Yao Y.D., Wang J., Chen Y. 'Kernel-based learning for statistical signal processing in cognitive radio networks: theoretical foundations, example applications, and future directions'. *IEEE Signal Processing Magazine*. 2013, vol. 30(4), pp. 126–36.

[13] Bouboulis P., Theodoridis S. 'The complex Gaussian kernel LMS algorithm'. *International Conference on Artificial Neural Networks*; Springer, Berlin, Heidelberg, 2010.

[14] Salehi A.R., Zaimbashi A. 'Radar target detection with Kernel-based generalized likelihood ratio test'. Presented at *2019 Sixth Iranian Conference on Radar and Surveillance Systems*; Isfahan, Iran. Available from https://ieeexplore.ieee.org/xpl/mostRecentIssue.jsp?punumber=9011554

[15] Salehi A.R., Zaimbashi A., Valkama M. 'Kernelized-likelihood ratio tests for binary phase-shift keying signal detection'. *IEEE Transactions on Cognitive Communications and Networking*. 2021, vol. 7(2), pp. 541–52.

[16] Zhang X., Li H., Himed B. 'Maximum likelihood delay and Doppler estimation for passive sensing'. *IEEE Sensors Journal*. 2019, vol. 19(1), pp. 180–88.

Chapter 11
Noisy RC-based multistatic passive radar target detection

11.1 Introduction

The use of multiple receivers in different locations and multiple transmitters in multistatic passive radar systems enhances the detection of passive radar targets due to the spatial diversity and reduction of scintillation effects in the radar cross section. In comparison to the bistatic or single-input single-output (SISO) configuration, single-input multiple-output (SIMO) or multiple-input multiple-output (MIMO) configurations offer a greater number of observations across various receivers during integration time to achieve longer ranges. This chapter examines a multistatic passive radar with one transmitter and multiple receivers separated spatially, referred to as an SIMO configuration.

There has been limited research on the problem of detecting targets with a two-channel multistatic passive radar. The authors of Reference 1 developed a generalized likelihood ratio test (GLRT) in passive MIMO radar (PMR) networks, termed as PMR-GLRT. This detector uses reference and surveillance signals to exploit correlations in the measurement data and outperforms other detectors that do not utilize all correlations. This study assumes that both the reference channel (RC) and surveillance channel (SC) have equal noise variances of a known value. Although the PMR-GLRT has a simple design, its effectiveness is significantly compromised when the noise variance is uncertain, resulting in the inability to detect targets below a specific signal-to-noise ratio (SNR) threshold, referred to as the *SNR wall*. Moreover, the detector's threshold is strongly influenced by the direct signal-to-noise ratio of the RC, which can result in false alarms or a reduction in detection performance. The authors of Reference 2 explore a new detector for a two-channel passive radar (PR) system with an SISO by using the cross correlation of the main left singular vectors of the received data matrix in the RC and SC. This detector capitalizes on the rank 1 structure of the transmitted signal. Similarly, in Reference 3, the cyclostationary property of the illuminator of opportunity signals is utilized to devise a new detector for an SISO two-channel PR system. However, none of these studies account for the influence of the direct-path signal power-to-noise ratio of the RC, denoted as DNR_r on the detection threshold setting. This chapter aims to explore this topic in more detail.

This chapter focuses on dealing with noisy RC-based SIMO passive radars with calibrated receivers, where the transmitted signal is deterministic but unknown. Two detection problems are considered, one with an RC in each receiver and one without. The former is referred to as a two-channel passive radar, while the latter is known as a one-channel passive radar. Two target detection problems are formulated as composite hypothesis testing problems, and the LRT principle is used to develop two detection algorithms: the P1-LRT and P2-LRT detectors. The P1-LRT detector is found to have an excessive false alarm probability due to uncertainty in the value of the DNR of the available RC. A new threshold-setting strategy is proposed to adjust the level of the P1-LRT detector, and its effectiveness is demonstrated through extensive Monte Carlo (MC) simulation results. The P1-LRT is a fixed-level detector, while the P2-LRT has a constant false alarm rate (CFAR) property, making it a fixed-size test. The P1-LRT detector is shown to outperform the P2-LRT detector significantly. This chapter is based on previous works published in Reference 4.

The current chapter is organized as follows: in section 11.2, the signal model for the SIMO passive radar target detection problem is presented. Two detectors are proposed in section 11.3, one of which is a non-CFAR detector that does not have CFAR property against the averaged value of the direct-to-noise ratio of the RCs. Section 11.4 contains simulation examples to evaluate the detection performance and false alarm regulation of the proposed detectors. To regulate the false alarm rate of the non-CFAR detector, a novel strategy for setting its threshold level rather than its size is proposed. Finally, the chapter is concluded in section 11.5.

11.2 Problem formulation

The scenario being considered is a multistatic radar system that includes a transmitter and N_r receivers that are spatially distributed and each have two channels. This type of radar is known as a two-channel SIMO passive radar. Each receiver is assumed to have an RC that captures the transmitted signal, or direct-path signal, and an SC that senses any potential echoes from a target. The configuration for $N_r = 3$ receivers is illustrated in Figure 11.1. In the single-target detection scenario without any interfering signals, the signals recorded by the RC and SC of the kth receiver, denoted as RC_k and SC_k, respectively, can be expressed as follows:

$$y'_k(t) = \alpha_k s(t - d_k) + n'_k(t) \tag{11.1}$$

$$x'_k(t) = \beta_k s(t - \tau_k)e^{j2\pi f_k t} + w'_k(t) \tag{11.2}$$

where $0 \le t < T$ with T being the integration time; $s(t)$ is the deterministic but unknown transmitted signal; α_k is a deterministic but unknown complex amplitude scaling factor accounting for propagation loss and antenna gain seen at the RC_k; β_k is deterministic but unknown complex coefficient factor considering attenuation due to transmission from transmitted to the SC_k and target reflectivity; (τ_k, f_k) is the known delay-Doppler coordinate of a desired target sensed by the SC_k; $n'_k(t)$ and $w'_k(t)$ are zero-mean additive white Gaussian noise at the RC_k and SC_k with unknown variances σ_r^2 and σ_s^2, respectively. In (11.1), d_k denotes the propagation delay from the transmitter

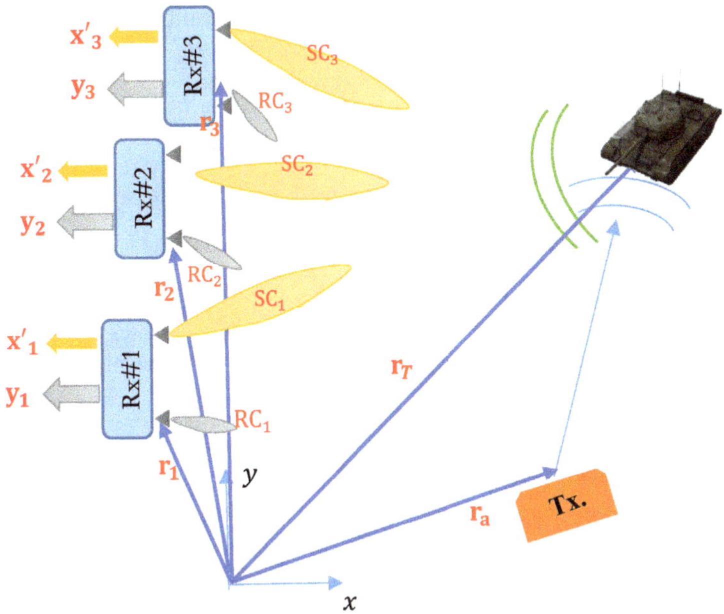

Figure 11.1 *Typical SIMO passive radar system with three two-channel receivers and a single transmitter (Tx.)*

to the RC_k. Since d_k is known *a priori* for stationary transmitters and receivers, we can compensate it to simplify the reference signal models, resulting in

$$y_k(t) = \alpha_k s(t) + n_k(t) \tag{11.3}$$

where $y_k(t) = y_k'(t + d_k)$ and $n_k(t) = n_k'(t + d_k)$.

After sampling observations received in the RC_k and SC_k with proper sampling frequency f_s, we obtain

$$\mathbf{y}_k = \alpha_k\,\mathbf{s} + \mathbf{n}_k \tag{11.4}$$

$$\mathbf{x}_k' = \beta_k\,\Phi(\tau_k,f_k)\mathbf{s} + \mathbf{w}_k' \tag{11.5}$$

where $\mathbf{y}_k, \mathbf{x}_k', \mathbf{s}, \mathbf{n}_k, \mathbf{w}_k \in \mathbb{C}^{N\times 1}$ are column vectors formed by N adjacent samples of $y_k(t)$, $x_k'(t)$, $s(t)$, $n_k(t)$, and $w_k(t)$, respectively. Here, $N = Tf_s$ is the number of samples during the integration time T. By using N-point discrete Fourier transform (DFT) matrix $\mathbf{F}$ with elements $[\mathbf{F}]_{i,l} = e^{-j2\pi(i-1)\frac{\Delta f}{f_S}(l-1)}/\sqrt{N}$, $i, l = 1, ..., N$, where $\Delta f = \frac{1}{T} = \frac{f_S}{N}$ is the frequency bin of N-point DFT, the delay-Doppler compensator matrix $\Phi(\tau_t, f_t)$ can be defined as [4]

$$\Phi(\tau_k,f_k) = \mathbf{D}\left(\frac{f_k}{f_S}\right)\mathbf{F}^H\mathbf{D}(-\tau_k\Delta k)\mathbf{F} \tag{11.6}$$

where $\mathbf{D}(a)$ is a diagonal matrix with the diagonal elements of $[\mathbf{D}(a)]_{ii} = e^{j2\pi(i-1)a}$ for $i = 1, ..., N$.

Let us denote the location of the transmitter, the location of the SC_k (or RC_k), and the hypothesized location-velocity of the target by $\mathbf{r}_a$, $\mathbf{r}_k$, and $(\mathbf{r}_T, \mathbf{v}_T)$, respectively. Now, we can express τ_k and f_k as follows [5]:

$$\tau_k = \frac{||\mathbf{r}_T - \mathbf{r}_a|| + ||\mathbf{r}_T - \mathbf{r}_k||}{c} \tag{11.7}$$

$$f_k = \frac{\mathbf{v}_T^T \left(\frac{(\mathbf{r}_T - \mathbf{r}_a)}{||\mathbf{r}_T - \mathbf{r}_a||} + \frac{(\mathbf{r}_T - \mathbf{r}_k)}{||\mathbf{r}_T - \mathbf{r}_k||} \right)}{\lambda} \tag{11.8}$$

where $\lambda = \frac{c}{f_c}$ is the transmitter's wavelength in which f_c denotes the transmitter's carrier frequency, and c is the speed of light. By using these new parameters, we can respectively rewrite (11.4) and (11.5) as follows:

$$\mathbf{y}_k = \alpha_k \mathbf{s} + \mathbf{n}_k \tag{11.9}$$

$$\mathbf{x}'_k = \beta_k \Phi(\mathbf{r}_T, \mathbf{v}_T)\mathbf{s} + \mathbf{w}'_k \tag{11.10}$$

where $\Phi(\mathbf{r}_T, \mathbf{v}_T)$ can be obtained from (11.6) by replacing τ_k and f_k with those of (11.7) and (11.8), which is a function of $(\mathbf{r}_T, \mathbf{v}_T)$ for stationary transmitters and receivers.

In the following, we consider two target detection scenarios that involve testing for the presence of a target within a specific location-velocity cell. This involves hypothesizing the target's location and velocity as $\mathbf{r}_T$ and $\mathbf{v}_T$, respectively. Both detection scenarios can be defined as binary composite hypothesis tests that involve evaluating null (i.e., $\mathcal{H}_0$) and alternative (i.e., $\mathcal{H}_1$) hypotheses.

Problem 1 (P1): In the case of the two-channel passive radar system, the first detection problem can be formulated as a composite hypothesis testing problem, given by

$$\mathcal{H}_0 : \begin{cases} \mathbf{y}_k = \alpha_k \mathbf{s} + \mathbf{n}_k \\ \mathbf{x}'_k = \mathbf{w}'_k \end{cases} \tag{11.11}$$

$$\mathcal{H}_1 : \begin{cases} \mathbf{y}_k = \alpha_k \mathbf{s} + \mathbf{n}_k \\ \mathbf{x}'_k = \beta_k \Phi(\mathbf{r}_T, \mathbf{v}_T)\mathbf{s} + \mathbf{w}'_k \end{cases} \tag{11.12}$$

Problem 2 (P2): In the case of one-channel passive radar target detection,* we can bypass the RC to obtain

$$\begin{cases} \mathcal{H}_0 : \mathbf{x}'_k = \mathbf{w}'_k \\ \mathcal{H}_1 : \mathbf{x}'_k = \beta_k \Phi(\mathbf{r}_T, \mathbf{v}_T)\mathbf{s} + \mathbf{w}'_k \end{cases} \tag{11.13}$$

*This is the case when there exist obstacles between transmitters and receivers to block the direct-path (transmitter-to-receiver) signals.

In both detection problems P1 and P2, we have $k = 1, \ldots, N_r$. Before proceeding, let us define

$$\mathbf{Z}' = [\mathbf{Y}, \mathbf{X}'] \in \mathbb{C}^{N\times 2N_r} \tag{11.14}$$

where

$$\mathbf{Y} = [\mathbf{y}_1, \ldots, \mathbf{y}_{N_r}] \in \mathbb{C}^{N\times N_r} \tag{11.15}$$
$$\mathbf{X}' = [\mathbf{x}'_1, \ldots, \mathbf{x}'_{N_r}] \in \mathbb{C}^{N\times N_r} \tag{11.16}$$

11.3 LRT-based proposed detectors

In this section, we utilize the LRT principle to create two novel detectors to address the target detection problems P1 and P2. The parameter spaces Θ_0 and Θ_0^c relating to the mentioned detection problems can be succinctly described as

$$\text{P1}: \begin{cases} \Theta_0 = \{\sigma_r^2, \sigma_s^2, \mathbf{s}, \{\alpha_k\}_{k=1}^{N_r}\} \\ \Theta_0^c = \{\sigma_r^2, \sigma_s^2, \mathbf{s}, \{\alpha_k\}_{k=1}^{N_r}, \{\beta_k\}_{k=1}^{N_r}\} \end{cases} \tag{11.17}$$

$$\text{P2}: \begin{cases} \Theta_0 = \{\sigma_s^2\} \\ \Theta_0^c = \{\sigma_s^2, \mathbf{s}, \{\beta_k\}_{k=1}^{N_r}\} \end{cases} \tag{11.18}$$

Appendix 11A demonstrates the necessity of assuming $\sigma_r^2 = \sigma_s^2 = \sigma^2$ to solve detection problem P1. The log-liklihood function (LLF) of the overall hypothesis can be expressed as

$$\begin{aligned} \mathscr{L}(\mathbf{Z}') &= -2NN_r \ln(\pi\sigma^2) \\ &\quad - \frac{1}{\sigma^2} \sum\nolimits_{k=1}^{N_r} \left(\gamma \left\|\mathbf{y}_k - \alpha_k \mathbf{s}\right\|_2^2 + \left\|\mathbf{x}'_k - \beta_k \Phi(\mathbf{r}_T, \mathbf{v}_T)\mathbf{s}\right\|_2^2\right) \end{aligned} \tag{11.19}$$

where γ are, respectively, 1 and 0 for detection problems P1 and P2. Using the fact that unitary transformations preserve lengths of vectors, we can simplify the LLF of (11.19) to write

$$\begin{aligned} \mathscr{L}(\mathbf{Z}) &= -2NN_r \ln(\pi\sigma^2) \\ &\quad - \frac{1}{\sigma^2} \sum_{k=1}^{N_r} \left(\gamma \left\|\mathbf{y}_k - \alpha_k \mathbf{s}\right\|_2^2 + \left\|\mathbf{x}_k - \beta_k \mathbf{s}\right\|_2^2\right) \end{aligned} \tag{11.20}$$

where

$$\mathbf{x}_k = \Phi^H(\mathbf{r}_T, \mathbf{v}_T)\mathbf{x}'_k \tag{11.21}$$

Let us define $\mathbf{Z} = [\mathbf{Y}, \mathbf{X}] \in \mathbb{C}^{N\times 2N_r}$ with $\mathbf{X} = [\mathbf{x}_1, \ldots, \mathbf{x}_k]$, given by

$$\mathbf{X} = \Phi^H(\mathbf{r}_T, \mathbf{v}_T)\mathbf{X}' \tag{11.22}$$

For the detection problems P1 and P2, the test statistics can be generally represented by

$$\Lambda(\mathbf{Z}) = \sup_{\sigma^2,\mathbf{s},\{\alpha_k\}_{k=1}^{N_r}} \{\mathscr{L}(\mathbf{Z})\} - \sup_{\sigma^2,\mathbf{s},\{\alpha_k\}_{k=1}^{N_r},\{\beta_k\}_{k=1}^{N_r}} \{\mathscr{L}(\mathbf{Z})\} \tag{11.23}$$

Let us denote the first term in the right-hand side of (11.23) by $\mathscr{L}_1$, where it can be written as

$$\mathscr{L}_1 = \sup_{\sigma^2,\mathbf{s},\{\alpha_k\}_{k=1}^{N_r}} \left\{-2NN_r \ln(\pi\sigma^2) - \frac{1}{\sigma^2}\sum_{k=1}^{N_r}\left(\gamma \|\mathbf{y}_k - \alpha_k\mathbf{s}\|^2 + \|\mathbf{x}_k\|^2\right)\right\} \tag{11.24}$$

The supremum with respect to (w.r.t.) the unknowns $\{\alpha_k\}_{k=1}^{N_r}$ can be obtained to come up to

$$\mathscr{L}_\alpha = \sup_{\alpha_k} \mathscr{L} = -2NN_r \ln(\pi\sigma^2) - \frac{1}{\sigma^2}\sum_{k=1}^{N_r}\left(\gamma \|\mathbf{y}_k\|^2 - \gamma\frac{\mathbf{s}^H\mathbf{y}_k\mathbf{y}_k^H\mathbf{s}}{\|\mathbf{s}^2\|} + \|\mathbf{x}_k\|^2\right) \tag{11.25}$$

Then, the supremum w.r.t. the unknown vector $\mathbf{s}$ can be computed from

$$\begin{aligned}\mathscr{L}_s = \sup_{\mathbf{s}} \mathscr{L}_\alpha &= -2NN_r \ln(\pi\sigma^2) - \frac{1}{\sigma^2}\sum_{k=1}^{N_r}\|\mathbf{x}_k\|^2 - \gamma\frac{1}{\sigma^2}\sum_{k=1}^{N_r}\|\mathbf{y}_k\|^2 \\ &+\gamma\frac{1}{\sigma^2}\sup_{\mathbf{s}}\left\{\frac{\mathbf{s}^H\sum_{k=1}^{N_r}\mathbf{y}_k\mathbf{y}_k^H\mathbf{s}}{\|\mathbf{s}\|^2}\right\}\end{aligned} \tag{11.26}$$

resulting in

$$\mathscr{L}_s^{(1)} = -2NN_r \ln(\pi\sigma^2) - \frac{1}{\sigma^2}\left(\gamma\sum_{k=1}^{N_r}\|\mathbf{y}_k\|^2 + \sum_{k=1}^{N_r}\|\mathbf{x}_k\|^2 - \gamma\lambda_{\max}(\mathbf{Y}\mathbf{Y}^H)\right) \tag{11.27}$$

where the Rayleigh quotient reaches its maximum value $\lambda_{\max}(\mathbf{Y}\mathbf{Y}^H)$ when $\mathbf{s} = \mathbf{e}_1(\mathbf{Y}\mathbf{Y}^H)$, where $\mathbf{e}_1(\mathbf{Y}\mathbf{Y}^H)$ is the eigenvector associated with the largest eigenvalue of $\mathbf{Y}\mathbf{Y}^H$, say $\lambda_{\max}(\mathbf{Y}\mathbf{Y}^H)$. As a result, the closed-form expression for the first term $\mathscr{L}_1$ can be found as

$$\begin{aligned}\mathscr{L}_1 = \sup_{\sigma^2}\{\mathscr{L}_s^{(1)}\} &= -2NN_r \ln(\pi) - 2NN_r \\ &+ 2NN_r \ln\left(\frac{\gamma\|\mathbf{Y}\|_F^2 + \|\mathbf{X}\|_F^2 - \gamma\lambda_{\max}(\mathbf{Y}^H\mathbf{Y})}{2NN_r}\right)\end{aligned} \tag{11.28}$$

where we have used the fact that $\lambda_{\max}(\mathbf{Y}\mathbf{Y}^H) = \lambda_{\max}(\mathbf{Y}^H\mathbf{Y})$, $\|\mathbf{X}\|_F^2 = \sum_{k=1}^{N_r}\|\mathbf{x}_k\|^2$, and $\|\mathbf{Y}\|_F^2 = \sum_{k=1}^{N_r}\|\mathbf{y}_k\|^2$. In literature, matrix $\mathbf{C} = \mathbf{Y}\mathbf{Y}^H$ is called a covariance matrix, while that of $\mathbf{G} = \mathbf{Y}^H\mathbf{Y}$ is a Gram matrix. Since $N_r << N$, it is more convenient to work with $\mathbf{G} \in \mathcal{C}^{N_r\times N_r}$ rather than $\mathbf{C} \in \mathcal{C}^{N\times N}$ to reduce the computational complexity, as required in practical passive radar target detection scenarios.

The second term on the right-hand side of (11.23), denoted by $\mathscr{L}_2$, can be calculated from

$$\mathscr{L}_2 = \sup_{\sigma^2, \mathbf{s}, \{\alpha_k\}_{k=1}^{N_r}, \{\beta_k\}_{k=1}^{N_r}} \left\{ -2NN_r \ln(\pi\sigma^2) - \frac{1}{\sigma^2} \sum_{k=1}^{N_r} \left(\gamma \left\| \mathbf{y}_k - \alpha_k \mathbf{s} \right\|^2 + \left\| \mathbf{x}_k - \beta_k \mathbf{s} \right\|^2 \right) \right\} \tag{11.29}$$

Following similar steps and derivations of (11.25)–(11.28), we can obtain

$$\begin{aligned} \mathscr{L}_s^{(2)} = \sup_{\mathbf{s}}\{\mathscr{L}_{\alpha,\beta}\} &= -2NN_r \ln(\pi\sigma^2) - \gamma \frac{1}{\sigma^2} \sum_{k=1}^{N_r} \|\mathbf{y}_k\|^2 \\ &+ \frac{1}{\sigma^2} \sup_{\mathbf{s}} \left\{ \frac{\mathbf{s}^H \sum_{k=1}^{N_r} \left(\gamma \mathbf{y}_k \mathbf{y}_k^H + \mathbf{x}_k \mathbf{x}_k^H \right) \mathbf{s}}{\|\mathbf{s}\|^2} \right\} \\ &- \frac{1}{\sigma^2} \sum_{k=1}^{N_r} \|\mathbf{x}_k\|^2 \end{aligned} \tag{11.30}$$

and then this optimization problem can be solved to produce

$$\begin{aligned} \mathscr{L}_s^{(2)} &= -2NN_r \ln(\pi\sigma^2) \\ &- \frac{1}{\sigma^2} \left(\gamma \|\mathbf{Y}\|_F^2 + \|\mathbf{X}\|_F^2 - \lambda_{\max}(\gamma \mathbf{Y}\mathbf{Y}^H + \mathbf{X}\mathbf{X}^H) \right) \end{aligned} \tag{11.31}$$

Finally, the second term on the right-hand side of (11.23) can be rewritten as

$$\begin{aligned} \mathscr{L}_2 = \sup_{\sigma^2}\{\mathscr{L}_s^{(2)}\} = \\ &- 2NN_r \ln(\pi) \\ &- 2NN_r \ln\left(\frac{\gamma \|\mathbf{Y}\|_F^2 + \|\mathbf{X}\|_F^2 - \lambda_{\max}(\gamma \mathbf{Y}\mathbf{Y}^H + \mathbf{X}\mathbf{X}^H)}{2NN_r} \right) \\ &- 2NN_r \end{aligned} \tag{11.32}$$

For the detection problem P1, the first and second terms can be, respectively, simplified to produce

$$\mathscr{L}_1^{(P1)} = -2NN_r \ln(\pi) - 2NN_r \ln\left(\frac{\|\mathbf{Z}\|_F^2 - \|\mathbf{Y}\|_2^2}{2NN_r} \right) - 2NN_r \tag{11.33}$$

$$\mathscr{L}_2^{(P1)} = -2NN_r \ln(\pi) - 2NN_r \ln\left(\frac{\|\mathbf{Z}\|_F^2 - \|\mathbf{Z}\|_2^2}{2NN_r} \right) - 2NN_r \tag{11.34}$$

where we have used the fact that $\|\mathbf{Y}\|_2^2 = \lambda_{\max}(\mathbf{Y}^H\mathbf{Y})$ and replacing the term $\mathbf{Y}\mathbf{Y}^H + \mathbf{X}\mathbf{X}^H$ with $\mathbf{Z}\mathbf{Z}^H$, and then used the fact that $\lambda_{\max}(\mathbf{Z}\mathbf{Z}^H) = \lambda_{\max}(\mathbf{Z}^H\mathbf{Z})$. Inserting (11.33) and (11.34) into (11.23) and after some algebraic manipulations, yields

$$2NN_r \ln\left(\frac{\|\mathbf{Z}\|_F^2 - \|\mathbf{Y}\|_2^2}{\|\mathbf{Z}\|_F^2 - \|\mathbf{Z}\|_2^2} \right) \underset{\mathscr{H}_1}{\overset{\mathscr{H}_0}{\gtrless}} \eta_1' \tag{11.35}$$

or, equivalently,

$$\left(\frac{\|\mathbf{Z}\|_F^2 - \|\mathbf{Y}\|_2^2}{\|\mathbf{Z}\|_F^2 - \|\mathbf{Z}\|_2^2} \right)^{2NN_r} \underset{\mathscr{H}_1}{\overset{\mathscr{H}_0}{\gtrless}} \eta_1 \tag{11.36}$$

where η_1 is the detection threshold of the proposed detector, termed P1-LRT, to be set according to the desired value of the false alarm probability.

For the detection problem P2, the first and second terms can be, respectively, simplified to yield

$$\mathcal{L}_1^{(P_2)} = -2NN_r \ln(\pi) - 2NN_r \ln\left(\frac{\|\mathbf{X}\|_F^2}{2NN_r}\right) - 2NN_r \tag{11.37}$$

$$\mathcal{L}_2^{(P_2)} = -2NN_r \ln(\pi) - 2NN_r \ln\left(\frac{\|\mathbf{X}\|_F^2 - \|\mathbf{X}\|_2^2}{2NN_r}\right) - 2NN_r \tag{11.38}$$

Inserting (11.37) and (11.38) into (11.23) and after some algebraic manipulations, yields

$$2NN_r \ln\left(\frac{\|\mathbf{X}\|_F^2}{\|\mathbf{X}\|_F^2 - \|\mathbf{X}\|_2^2}\right) \underset{\mathcal{H}_1}{\overset{\mathcal{H}_0}{\gtrless}} \eta_2' \tag{11.39}$$

or, equivalently,

$$\left(\frac{\|\mathbf{X}\|_2^2}{\|\mathbf{X}\|_F^2}\right)^{2NN_r} \underset{\mathcal{H}_1}{\overset{\mathcal{H}_0}{\gtrless}} \eta_2 \tag{11.40}$$

where η_2 is the detection threshold of the proposed detector, named P2-LRT, to be adjusted to attain the desired false alarm probability.

11.4 Performance results

In this section, we utilized numerical simulations to assess the effectiveness of the detectors based on P1-LRT and P2-LRT. To accomplish this, we created a vector $\mathbf{s}$ using the expression $\mathbf{s} = \exp\{-j\phi\}$, where the components of vector ϕ were independently and identically distributed over the interval $[0, 2\pi)$. We also determined the coefficients α_k and β_k in order to achieve a specified average signal-to-noise ratio, denoted as SNR_{avg}, in the SCs, and a specified average direct signal-to-noise ratio, denoted as DNR_{avg}, in the RCs. The performance of the detectors was evaluated based on these criteria, given by

$$\text{SNR}_{\text{avg}} = \frac{\frac{1}{N_r}\sum_{k=1}^{N_r} |\beta_k|^2}{\sigma^2} \tag{11.41}$$

$$\text{DNR}_{\text{avg}} = \frac{\frac{1}{N_r}\sum_{k=1}^{N_r} |\alpha_k|^2}{\sigma^2} \tag{11.42}$$

In our situation, it is not a straightforward task to derive the closed-form expressions for both the detection probability and the false alarm probability p_{fa}. Therefore, we rely on

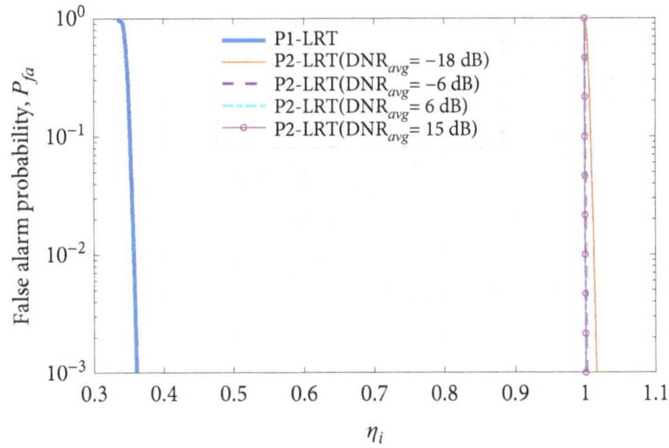

Figure 11.2 *Probability of false alarm as a function of detection thresholds $\{\eta_i\}_{i=1}^2$ of the P1-LRT and P2-LRT detectors for different values of DNR_{avg}*

the results from MC simulations to examine the detection performances and false alarm regulations of our proposed detectors. To ensure the reliability of the MC evaluations, it is generally recommended to carry out at least $\frac{100}{p_{fa}}$ MC runs to obtain a dependable detection threshold. For $p_{fa} = 10^{-3}$, we use 1.2×10^6 MC runs to ensure the accuracy of the results. Additionally, our simulations consider an integration time of $T = 10.24$ ms and a sampling frequency of $f_s = 200$ kHz, which leads to $N = 2{,}048$ samples. Unless stated otherwise, we assume the number of receivers, $N_r = 3$, to be three.

In the context of detection theory, it is common to establish a specific false alarm probability limit and then strive to maximize the detection probability accordingly. Consequently, the detection threshold for the designed test must be selected to align with the desired false alarm probability.. Figure 11.2 depicts the false alarm probabilities versus detection thresholds for the P1-LRT and P2-LRT detectors at different DNR_{avg} values within the range of $[-18, 18]$ dB. It is apparent that the detection threshold $\boldsymbol{\eta}_1$ for the proposed P1-LRT detector with the RC channel is dependent on the DNR_{avg} values, particularly for lower values. Conversely, this is not the case for the P2-LRT detector without any dedicated RC. This implies that when encountering practical situations with low DNR_{avg} values, accurate knowledge of the value is crucial for setting the detection threshold of the proposed P1-LRT detector. Therefore, it is worth considering whether it is possible to determine the detection threshold of the P-LRT in a different form than that suggested in Figure 11.2. The significance of this inquiry lies in the fact that when determining the value of DNR_{avg} in real-life scenarios, there may be some doubt or imprecision, commonly referred to as reference channel power uncertainty (RCPU). Fortunately, there exists a viable and effective approach to establishing detection thresholds using the notion of a test level, as previously explained in the preceding chapter.

To proceed, we will introduce some fresh parameters. Specifically, we will use $\mathrm{DNR}_{\mathrm{avg}}$ to refer to the true $\mathrm{DNR}_{\mathrm{avg}}$, while $\mathrm{DNR}^{\eta}_{\mathrm{avg}}$ will denote the value of $\mathrm{DNR}_{\mathrm{avg}}$ that establishes the detection threshold of the P1-LRT detector. It will become apparent in the subsequent discussion that determining or obtaining the minimum received value of $\mathrm{DNR}_{\mathrm{avg}}$ is essential to create or develop a detector with a fixed level. To gain a better understanding of this novel concept, we will examine two scenarios, as follows:

Case C1: It is assumed that the range of the received $\mathrm{DNR}_{\mathrm{avg}}$ for various radar setups falls between –6 and 6 dB. Additionally, it is assumed that the radar designer or operator is unaware that the P1-LRT detector's detection threshold heavily relies on the received $\mathrm{DNR}_{\mathrm{avg}}$ values and sets the threshold using a fixed $\mathrm{DNR}_{\mathrm{avg}}$ value. In this situation, the focus is on examining how this threshold-setting strategy impacts false alarm regulation and detection performance.

Case C2: Our knowledge regarding the received $\mathrm{DNR}_{\mathrm{avg}}$ includes uncertainties. It is assumed that we have information about the minimum values of the received $\mathrm{DNR}_{\mathrm{avg}}$ as well as the amount of the RCPU, denoted by ρ, which are known based on the accuracy of our knowledge of $\mathrm{DNR}_{\mathrm{avg}}$. This implies that we are aware of the range of the received $\mathrm{DNR}_{\mathrm{avg}}$, which is $[\min \mathrm{DNR}_{\mathrm{avg}}[\mathrm{dB}], \max \mathrm{DNR}_{\mathrm{avg}}[\mathrm{dB}]]$, with $\max \mathrm{DNR}_{\mathrm{avg}}[\mathrm{dB}]$ being equal to $\min \mathrm{DNR}_{\mathrm{avg}}[\mathrm{dB}] + \rho_{\mathrm{dB}}$. Here, $\mathrm{DNR}_{\mathrm{avg}}[\mathrm{dB}] = 10 \log_{10} \mathrm{DNR}_{\mathrm{avg}}$ and $\rho_{\mathrm{dB}} = 10 \log_{10} \rho$. In practical terms, ρ_{dB} represents the level of accuracy of our estimate of $\mathrm{DNR}_{\mathrm{avg}}$. The effectiveness of the proposed threshold-setting method is now being evaluated in two different scenarios, namely C1 and C2. In the first scenario, C1, we assume that the detection thresholds are established based on two different values of $\mathrm{DNR}^{\eta}_{\mathrm{avg}}$, namely, $\mathrm{DNR}^{\eta}_{\mathrm{avg}} = -6$ dB and $\mathrm{DNR}^{\eta}_{\mathrm{avg}} = 6$ dB, while also encountering situations where $\mathrm{DNR}_{\mathrm{avg}} \in [-6, 6]$ dB. The findings from this simulation are presented in Figure 11.3. The proposed P1-LRT detector has a false alarm probability of the desired level ($p_{fa} = 10^{-3}$) when the detection threshold is set based on $\mathrm{DNR}_{\mathrm{avg}} = -6$ dB, resulting in a level-10^{-3}

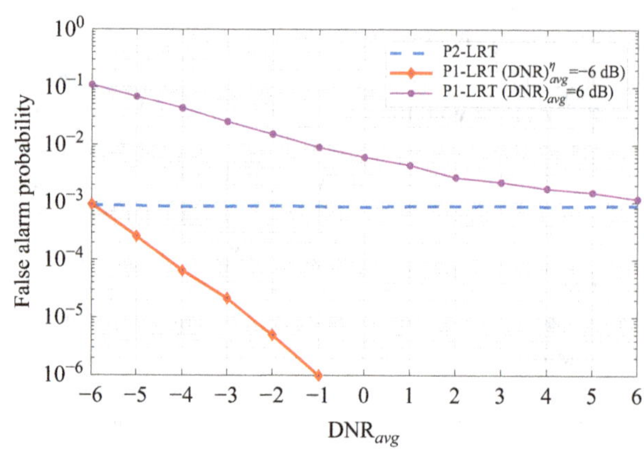

Figure 11.3 Probability of false alarm as functions of DNR_{avg} for two cases of $DNR^{\eta}_{r} = -6$ dB and $DNR^{\eta}_{r} = 6$ dB and when $p_{fa} = 10^{-3}$

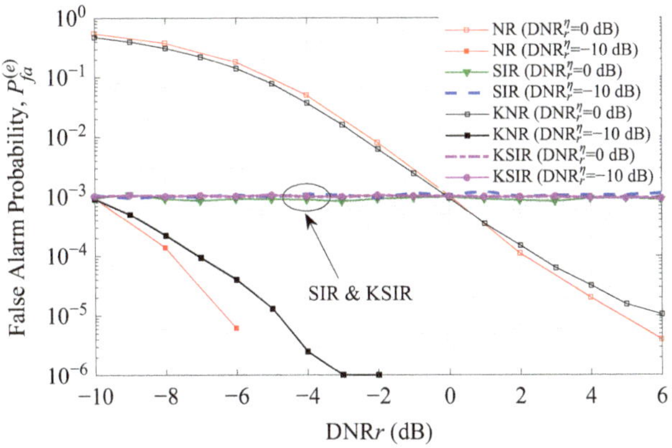

Figure 11.4 *Detection probability versus SNR_{avg} when detection thresholds are set for DNR^{η}_{avg} = –6 dB, resulting in $P_{fa} < 10^{-3}$ for DNR_{avg} = 6 dB (i.e., level-10^{-3}) and $P_{fa} = 10^{-3}$ for DNR_{avg}–6 dB (i.e., size-10^{-3}). Besides, we consider the case of DNR_{avg} = 6 dB with DNR^{η}_{avg} = 6 dB, resulting in size-10^{-3}.*

test. However, setting the detection threshold η_1 based on $\mathrm{DNR}_{\mathrm{avg}}$ = 6 dB leads to excessive false alarm probability, which is not acceptable in practical situations. Thus, it is necessary to determine the minimum value of the received $\mathrm{DNR}_{\mathrm{avg}}$ to achieve a detector with a fixed level ($P_{fa} \leq p_{fa} = 10^{-3}$). The false alarm probability of the P2-LRT detector, without a dedicated RC, is not influenced by the $\mathrm{DNR}_{\mathrm{avg}}$ values.

In practical situations, it may not always be possible to determine the minimum value of the received $\mathrm{DNR}_{\mathrm{avg}}$. In such cases, it is advisable to assume a worst-case scenario for the received $\mathrm{DNR}_{\mathrm{avg}}$ and set the detection threshold accordingly. For instance, in the scenario where $\mathrm{DNR}_{\mathrm{avg}} \in [-6, 6]$ dB, the minimum value of the received $\mathrm{DNR}_{\mathrm{avg}}$ required to achieve a detector with a fixed level is $\mathrm{DNR}^{\eta}_{\mathrm{avg}} = -6$ dB. To compare the performance of the proposed P1-LRT and P2-LRT detectors, we consider two cases of $\mathrm{DNR}_{\mathrm{avg}} = -6$ dB and $\mathrm{DNR}_{\mathrm{avg}}$ = 6 dB under the condition of $\mathrm{DNR}^{\eta}_{\mathrm{avg}} = -6$ dB. Interestingly, increasing the $\mathrm{DNR}_{\mathrm{avg}}$ does not improve the detection probability of the P1-LRT detector. This is because fixing the level of the P1-LRT detector significantly raises the detection threshold for $\mathrm{DNR}_{\mathrm{avg}} > -6$ dB, thereby limiting its detection performance. Additionally, the P2-LRT detector's detection performance is not affected by the received $\mathrm{DNR}_{\mathrm{avg}}$ but is inferior compared with the P1-LRT detector, even for $\mathrm{DNR}_{\mathrm{avg}}$ = 6 dB. In this situation, the increase in SNR is approximately 9.7 and 9.9 dB at $P_d = 0.9$ when $\mathrm{DNR}_{\mathrm{avg}}$ equals –6 and 6 dB, respectively. Additionally, the detection results for the case of $\mathrm{DNR}_{\mathrm{avg}}$ = 6 dB are included in Figure 11.4 when the detection threshold η_1 is set to $\mathrm{DNR}^{\eta}_{\mathrm{avg}}$ = 6 dB. This comparison is only feasible if we have knowledge of the precise value of the received $\mathrm{DNR}_{\mathrm{avg}}$. However, when we compare the cases with $\mathrm{DNR}_{\mathrm{avg}}$ = 6 dB and different values of $\mathrm{DNR}^{\eta}_{\mathrm{avg}}$ set to –6 and 6 dB, we observe a decrease of 4 dB in SNR gain. This reduction in performance comes with the cost of fixing the level of the proposed

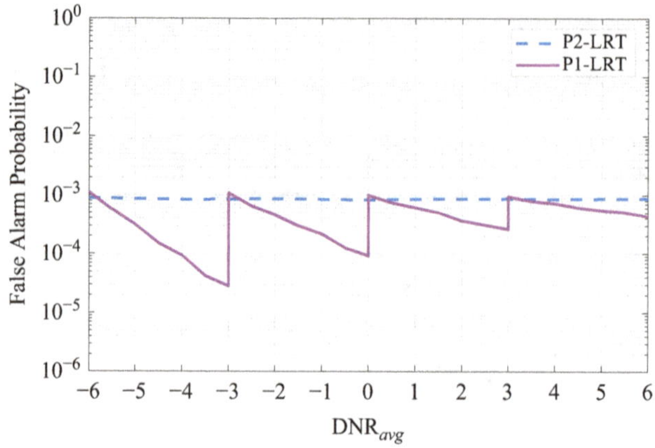

Figure 11.5 *Probability of false alarm as a function of* DNR$_{\text{avg}}$ *when* $p_{fa} = 10^{-3}$ *and the detection thresholds are set for* $DNR_r^{\eta} = \min\{DNR_r[dB]\} + \rho_{\text{dB}}(i-1)$ *in the i-th interval for* $i = 5, \ldots, 8$

P1-LRT detector, but it can be decreased by reducing the uncertainty interval (UI) from 12 dB in Figure 11.4 to a smaller range, as described in the following.

In case C2, we consider different ranges of known received DNR$_{\text{avg}}$, described by $[\min\{\text{DNR}_{\text{avg}}[\text{dB}]\} + \rho_{\text{dB}}(i-1), \min\{\text{DNR}_{\text{avg}}[\text{dB}]\} + \rho_{\text{dB}}i]$ for $i = 1,\ldots,12$ with $\min\{\text{DNR}_{\text{avg}}[\text{dB}]\} = -18$ dB and $\rho_{\text{dB}} = 3$. Here, the detection thresholds are set according to $\text{DNR}^{\eta}_{\text{avg}}(i) = \min\{\text{DNR}_{\text{avg}}[\text{dB}]\} + \rho_{\text{dB}}(i-1)$ in the ith UI to fix the level of the P1-LRT detector. The results of this simulation are shown in Figure 11.5 for $i = 5, \ldots, 8$ to reduce the simulation time. As can be seen, the obtained (empirical) false alarm probability is always below the desired false alarm probability $p_{fa} = 10^{-3}$ for the proposed P1-LRT detector. For the DNR$_{\text{avg}}$ values resulting in $P_{fa} < p_{fa}$, the adjusted detection thresholds are larger than those required to get $P_{fa} = p_{fa}$. In such cases, it is not surprising that the detection performance of the proposed detectors is degraded, as shown in the following.

In Figure 11.6, we investigate the impact of different values of DNR$_{\text{avg}}$ on the detection performances of the P1-LRT and P2-LRT detectors under case C2 for $i = 1$ (i.e., $\text{DNR}_{\text{avg}} \in [-18,-15]$ dB), $i = 5$ (i.e., $\text{DNR}_{\text{avg}} \in [-6,-3]$ dB), $i = 9$ (i.e., $\text{DNR}_{\text{avg}} \in [6, 9]$ dB), and $i = 12$ (i.e., $\text{DNR}_{\text{avg}} \in [15, 18]$ dB), respectively. There are some points in order. First, the proposed P1-LRT detector generally performs better in the ith UI as compared with the jth UI when $i > j$. Second, the P1-LRT significantly outperforms the P2-LRT for both low and high DNR$_{\text{avg}}$ regimes. The detection performance gain is about 13.5 dB for high DNR$_{\text{avg}}$ cases, while it is about 1 dB for $\text{DNR}_{\text{avg}} = -18$ dB. Third, it is seen that the detection performance improvement is insignificant in any UI even with increasing DNR$_{\text{avg}}$ in each UI, which is the cost of fixing the level of P1-LRT detector. Fourth, in contrast to Figure 11.4, here we achieve an SNR gain of 3.6 dB in an interval with a length of 12 dB between the cases of

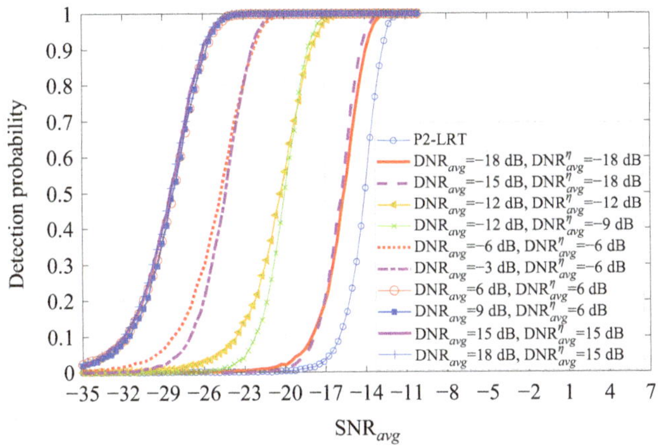

Figure 11.6 *Detection probability versus SNR$_{avg}$ for different values of DNR$_{avg}$ and DNR$^{\eta}_{avg}$. Noting that DNR$^{\eta}_{avg}$ = DNR$_{avg}$ results in size-10^{-3} P1-LRT detector, and DNR$^{\eta}_{avg}$ ≠ DNR$_{avg}$ lead to level-10^{-3} P1-LRT detector.*

$\mathrm{DNR}_{\mathrm{avg}} = -6$ dB and $\mathrm{DNR}_{\mathrm{avg}} = 6$ dB. Fifth, we see that increasing the $\mathrm{DNR}_{\mathrm{avg}}$ for values greater than 6 dB does not result in improving the detection performance of the P1-LRT detector.

Figure 11.7 explores how increasing the number of receivers can enhance the advantages of the proposed detectors, with $\mathrm{SNR}_{\mathrm{avg}}$ being –20 dB. It is evident that the P1-LRT detector outperforms the P2-LRT detector to a significant extent.

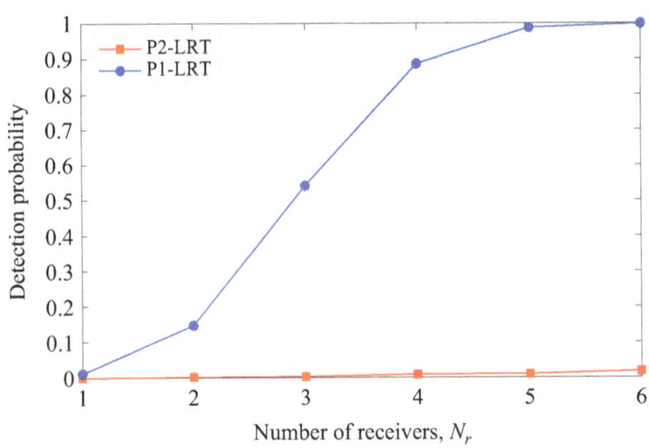

Figure 11.7 *Detection probability as a function of number of receivers with SNR$_{avg}$ = –20 dB, DNR$^{\eta}_{avg}$ = DNR$_{avg}$ = –12 dB, and $P_{fa} = p_{fa} = 10^{-3}$*

11.5 Summary

In this chapter, two problems related to detecting targets in a multistatic setting have been formulated: one involving two channels and another involving a single channel. Two detectors, named P1-LRT and P2-LRT, have been developed to solve these problems using the LRT criterion. It has been observed that the DNR_{avg} of the RC has a significant impact on the detection thresholds and false alarm regulation of the P1-LRT detector. To address this issue, a strategy has been proposed to set the detection threshold of the P1-LRT detector to achieve a fixed level of desired false alarm probability. Despite being a fixed-size test, the performance of the P2-LRT detector is not as good as that of the P1-LRT detector. MC simulation examples have been used to demonstrate the effectiveness of the proposed threshold-setting strategy and to show that the P1-LRT detector outperforms the P2-LRT detector in low and high DNR_{avg} regimes.

References

[1] Hack D.E., Patton L.K., Himed B., Saville M.A. 'Detection in passive MIMO radar networks'. *IEEE Transactions on Signal Processing*. 2014, vol. 62(11), pp. 2999–3012.

[2] Gogineni S., Setlur P., Rangaswamy M., Nadakuditi R.R. 'Passive radar detection with noisy reference channel using principal subspace similarity'. *IEEE Transactions on Aerospace and Electronic Systems*. 2018, vol. 54(1), pp. 18–36.

[3] Horstmann S., Ramrez D., Schreier P.J. 'Two-channel passive detection of cyclostationary signals'. *IEEE Transactions on Signal Processing*. 2020, vol. 68(10), pp. 2340–55.

[4] Zaimbashi A. 'Multistatic passive radar sensing algorithms with calibrated receivers'. *IEEE Sensors Journal*. 2020, vol. 20(14), pp. 7878–85.

[5] Solatzadeh Z., Zaimbashi A. 'Accelerating target detection in passive radar sensors: Delay-Doppler-acceleration estimation'. *IEEE Sensors Journal*. 2018, vol. 18(13), pp. 5445–54.

Appendix 11A
General detection problems P1

Here, we will demonstrate the necessity of assuming that $\sigma_r^2 = \sigma_s^2 = \sigma^2$ in solving detection problem P1. Initially, suppose that these variances are distinct values. Consequently, the test statistic for the LRT can be expressed as

$$\Lambda(\mathbf{Z}) = \sup_{\sigma_s^2, \sigma_r^2, \mathbf{s}, \{\alpha_k\}_{k=1}^{N_r}} \{\mathcal{L}(\mathbf{Z})\} - \sup_{\sigma_s^2, \sigma_r^2, \mathbf{s}, \{\alpha_k\}_{k=1}^{N_r}, \{\beta_k\}_{k=1}^{N_r}} \{\mathcal{L}(\mathbf{Z})\} \tag{11A.1}$$

The first and second terms on the right-hand side of (11A.1) can be given by

$$\mathcal{L}_1 = \sup_{\sigma_s^2, \sigma_r^2, \mathbf{s}, \{\alpha_k\}_{k=1}^{N_r}} \left\{ - NN_r \ln(\pi\sigma_s^2) - NN_r \ln(\pi\sigma_r^2) - \frac{1}{\sigma_r^2} \sum_{k=1}^{N_r} \|\mathbf{y}_k - \alpha_k \mathbf{s}\|^2 - \frac{1}{\sigma_s^2} \sum_{k=1}^{N_r} \|\mathbf{x}_k\|^2 \right\} \tag{11A.2}$$

$$\mathcal{L}_2 = \sup_{\sigma_s^2, \sigma_r^2, \mathbf{s}, \{\alpha_k\}_{k=1}^{N_r}, \{\beta_k\}_{k=1}^{N_r}} \left\{ - NN_r \ln(\pi\sigma_s^2) - NN_r \ln(\pi\sigma_r^2) - \frac{1}{\sigma_r^2} \sum_{k=1}^{N_r} \|\mathbf{y}_k - \alpha_k \mathbf{s}\|^2 - \frac{1}{\sigma_s^2} \sum_{k=1}^{N_r} \|\mathbf{x}_k - \beta_k \mathbf{s}\|^2 \right\} \tag{11A.3}$$

Like the P1-LRT detector derivation, $\mathcal{L}_1$ can be obtained as

$$\mathcal{L}_1 = -NN_r \ln\left(\pi\left(\frac{\|\mathbf{Y}\|_F^2 - \|\mathbf{Y}\|_2^2}{NN_r}\right)\right) - NN_r \ln\left(\pi\left(\frac{\|\mathbf{X}\|_F^2}{NN_r}\right)\right) - 2NN_r \tag{11A.4}$$

which has a closed-form expression without any concern regarding to the assumption of $\sigma_r^2 \neq \sigma_s^2$. However, after many manipulations, $\mathcal{L}_2$ takes the following form

$$\mathcal{L}_2 = -2NN_r \ln\left(\frac{\pi}{NN_r}\right) - NN_r \inf_{\mathbf{g}} \left\{ \ln\left\{ \left(\|\mathbf{Y}\|_F^2 - \mathbf{g}^H \mathbf{Y}\mathbf{Y}^H \mathbf{g}\right)\left(\|\mathbf{X}\|_F^2 - \mathbf{g}^H \mathbf{X}\mathbf{X}^H \mathbf{g}\right) \right\} \right\} \tag{11A.5}$$

where

$$\mathbf{g} = \frac{\mathbf{s}}{\|\mathbf{s}\|} \tag{11A.6}$$

Due to the difficulty in finding a solution to this optimization problem, we have restricted our analysis to the scenario where $\sigma_s^2 = \sigma_s^2 = \sigma^2$. This enables us to derive a closed-form solution for the second term of the P1-LRT statistic.

Part IV

Multistatic passive radar target detection without reference channels

This part deals with detecting passive radar targets in situations where a reliable reference channel is not available due to various factors, such as low signal-to-noise caused by obstructions in the direct path between transmitters and receivers or strong multipath clutter. In such cases, incorporating both the reference and surveillance channels to formulate the target detection problem may lead to inadequate performance of the designed detectors in detecting the targets. Alternatively, one can improve performance by removing the reference channel and utilizing multiple surveillance channels, which allows for more data to be collected and provides spatial diversity for a target's radar cross section. Additionally, interchannel correlations among different receivers can be utilized for target detection and estimating unknown transmitter signals. This approach falls under the second category of passive radar target detection methods, specifically Category 2 (Ca.2). This category itself can be considered a special case within the first category (Ca.1), where only the surveillance channel is included to formulate the target detection problem.

This part consists of two chapters. In Chapter 12, a novel approach for detecting targets in multistatic passive radar is introduced. This chapter delves into the target detection problem in passive multistatic radar setups, where there is a single transmitter and multiple widely spaced receivers. Throughout this chapter, several detection algorithms are developed using different criteria and making various assumptions about the problem's parameter space. This framework assumes that receiver thermal noise is the main interference in detection. However, in practice, the direct-path signal is typically stronger than the target echo and should be taken into account. Chapter 13 presents the development of a target detector based on the Rao statistic, which takes into account both calibrated and uncalibrated receivers in a multiple-input multiple-output (MIMO) setup. In the calibrated scenario, it is assumed that all receivers share the same noise variance, whereas in the uncalibrated scenario, noise variances vary among spatial and frequency receivers. While the effect of the direct-path interference signal is taken into account in Chapter 13, the effect of multipath signal is ignored in both Chapters 12 and 13. This is mainly related to the fact that handling multipath signals as interference signals to solve the detection problem in a closed-form manner may prove impractical. Although we do not incorporate reference channels to formulate detection problem, we can utilize them separately in order to eliminate residuals from multipath signals in some practical situations. To achieve this, we can estimate the transmitted signals using the reference channels and then employ the interference cancellation method introduced in Chapter 7 to remove the multipath signal. The success of this approach hinges on the quality of the reference channels and the estimation methods considered. Further investigation and exploration of this area are required as we are working on providing details for the second edition of the book.

Chapter 12

Multistatic passive radar target detection without direct-path interference

In this chapter, the focus is on detecting targets in a passive multistatic radar that has one transmitter and multiple receivers, which are spatially separated. The configuration used here is the single-input multiple-output (SIMO), and the reference channel (RC) is not used. Different solutions to this problem are developed under the Rao, likelihood ratio test (LRT), and geometrical representation frameworks with different assumptions made about the parameter space. A unified framework for multistatic passive radar target detection is created by formulating three distinct target detection problems as composite hypothesis testing problems. This results in four new detectors. The proposed target detection methods are analytically proven to possess constant false alarm rate (CFAR) behavior, irrespective of noise variance uncertainties across different receivers. Monte Carlo (MC) simulations are used to assess the detection performance of the proposed detectors and demonstrate their capabilities. This chapter is based on the previous work published in Reference 1.

This chapter is structured as follows. Section 12.1 presents the signal model and various target detection problems. In section 12.2, numerous multistatic passive radar target detectors are formulated. The performance of the proposed detectors under different noise variance uncertainties across receivers is analyzed in section 12.3. The results of numerical simulations are provided in section 12.4. Finally, a summary is presented in section 12.5.

12.1 Signal model and problem formulation

Assuming an SIMO configuration with one transmitter and N_r receivers placed at different locations, as depicted in Figure 12.1 for the case of $N_r = 3$, the signal received by the kth receiver's surveillance channel (SC_k) can be expressed as

$$x_k'(t) = \beta_k s(t - \tau_k)e^{j2\pi f_k t} + w_k'(t) \tag{12.1}$$

The transmitted signal and the integration time are, respectively, denoted by $s(t)$ and T, such that $0 \leq t < T$. The complex coefficient factor β_k is used to account for attenuation caused by transmission from the transmitter to the SC_k and the reflectivity of the target. The desired target's delay-Doppler coordinate is represented as (τ_k, f_k) at the SC_k. The additive white Gaussian noise at SC_k is denoted by $w_k'(t)$,

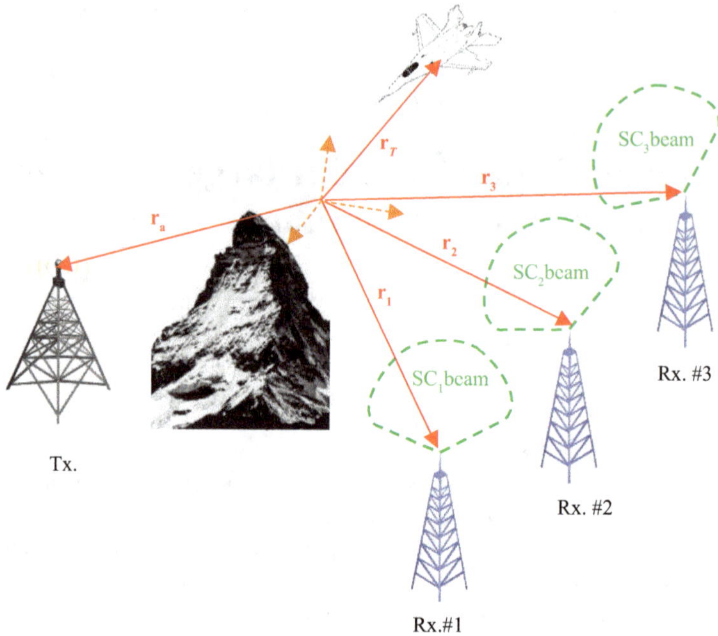

Figure 12.1 *Typical SIMO passive radar system with one transmitter (Tx.) and three surveillance channel (SC) receivers (Rx)*

which has an unknown zero-mean variance of σ_k^2. It is important to note that σ_k^2 may differ from σ_j^2 for $k \neq j$ due to uncertainties in receiver calibration, specifically when receivers are not calibrated.

After sampling observations received in the SC$_k$ with proper sampling frequency f_s, we obtain

$$\mathbf{x}'_k = \beta_k \, \Phi(\tau_k, f_k)\mathbf{s} + \mathbf{w}'_k \tag{12.2}$$

where $\mathbf{x}'_k$, $\mathbf{s}$, and $\mathbf{w}'_k \in \mathbb{C}^{N\times 1}$ are column vectors formed by N adjacent samples of $x'_k(t)$, $s(t)$, and $w'_k(t)$, respectively. Here, $N = Tf_s$ is the number of samples during the integration time T. By using N-point discrete Fourier transform (DFT) matrix $\mathbf{F}$ with elements $[\mathbf{F}]_{i,l} = e^{-j2\pi(i-1)\frac{\Delta f}{f_s}(l-1)}/\sqrt{N}, i, l = 1, ..., N$, where $\Delta f = \frac{1}{T} = \frac{f_s}{N}$ is the frequency bin of N-point DFT, the delay-Doppler compensator matrix $\Phi(\tau_t, f_t)$ can be defined as [1]

$$\Phi(\tau_k, f_k) = \mathbf{D}\left(\frac{f_k}{f_s}\right)\mathbf{F}^H\mathbf{D}(-\tau_k\Delta f)\mathbf{F} \tag{12.3}$$

where $\mathbf{D}(a)$ is a diagonal matrix with the diagonal elements of $[\mathbf{D}(a)]_{ii} = e^{j2\pi(i-1)a}$ for $i = 1, ..., N$.

Let us denote the location of the transmitter, the location of the SC$_k$, and the hypothesized location velocity of the target by $\mathbf{r}_a$, $\mathbf{r}_k$, and $(\mathbf{r}_T, \mathbf{v}_T)$, respectively. Now, we can express τ_k and f_k as follows [2]

$$\tau_k = \frac{\|\mathbf{r}_T - \mathbf{r}_a\| + \|\mathbf{r}_T - \mathbf{r}_k\|}{c} \tag{12.4}$$

$$f_k = \frac{\mathbf{v}_T^T \left(\frac{(\mathbf{r}_T - \mathbf{r}_a)}{\|\mathbf{r}_T - \mathbf{r}_a\|} + \frac{(\mathbf{r}_T - \mathbf{r}_k)}{\|\mathbf{r}_T - \mathbf{r}_k\|} \right)}{\lambda} \tag{12.5}$$

where $\lambda = \frac{c}{f_c}$ is the transmitter's wavelength in which f_c denotes the transmitter's carrier frequency, and c is the speed of light. By these new parameters, we can rewrite (12.2) as follows

$$\mathbf{x}_k' = \beta_k \, \Phi(\mathbf{r}_T, \mathbf{v}_T)\mathbf{s} + \mathbf{w}_k' \tag{12.6}$$

where $\Phi(\mathbf{r}_T, \mathbf{v}_T)$ can be obtained from (12.3) by replacing τ_k and f_k with those of (12.4) and (12.5), which is a function of $(\mathbf{r}_T, \mathbf{v}_T)$ for stationary transmitter and receivers.

In the following, we will examine various target detection scenarios where we are testing for the presence of a target within a particular location and velocity range represented by $(\mathbf{r}_T, \mathbf{v}_T)$. The target's hypothesized location and velocity are denoted by $\mathbf{r}_T$ and $\mathbf{v}_T$, respectively. These detection scenarios can be formulated as a binary composite hypothesis testing between null (i.e., $\mathcal{H}_0$), which indicates the absence of a target, and the alternative hypothesis ($\mathcal{H}_1$), which implies the presence of a target that is

$$\begin{cases} \mathcal{H}_0 : \mathbf{x}_k' = \mathbf{w}_k' \\ \mathcal{H}_1 : \mathbf{x}_k' = \beta_k \, \Phi(\mathbf{r}_T, \mathbf{v}_T)\, \mathbf{s} + \mathbf{w}_k' \end{cases} \tag{12.7}$$

where $k = 1, \ldots, N_r$.

Let us define

$$\mathbf{x}_k = \Phi^H(\mathbf{r}_T, \mathbf{v}_T)\mathbf{x}_k' \tag{12.8}$$

and

$$\mathbf{w}_k = \Phi^H(\mathbf{r}_T, \mathbf{v}_T)\mathbf{w}_k' \tag{12.9}$$

Now, we can rewrite the detection problem (12.7) as

$$\begin{cases} \mathcal{H}_0 : \mathbf{x}_k = \mathbf{w}_k \\ \mathcal{H}_1 : \mathbf{x}_k = \beta_k \, \mathbf{s} + \mathbf{w}_k \end{cases} \tag{12.10}$$

or, equivalently,

$$\begin{cases} \mathcal{H}_0 : \mathbf{X} = \mathbf{W} \\ \mathcal{H}_1 : \mathbf{X} = \mathbf{s}\mathbf{b}^T + \mathbf{W} \end{cases} \tag{12.11}$$

where

$$\mathbf{X} = [\mathbf{x}_1, \ldots, \mathbf{x}_{N_r}] \in \mathbb{C}^{N \times N_r} \tag{12.12}$$

$$\mathbf{W} = [\mathbf{w}_1, \ldots, \mathbf{w}_{N_r}] \in \mathbb{C}^{N \times N_r} \tag{12.13}$$

$$\mathbf{b} = [\beta_1, \ldots, \beta_{N_r}]^T \in \mathbb{C}^{N_r \times 1} \tag{12.14}$$

By defining $\mathbf{Y} = \mathbf{X}^T = [\mathbf{y}_1, ..., \mathbf{y}_N] \in \mathbb{C}^{N_r \times N}$ and $\mathbf{G} = \mathbf{W}^T = [\mathbf{g}_1, ..., \mathbf{g}_N] \in \mathbb{C}^{N_r \times N}$, we obtain

$$\begin{cases} \mathcal{H}_0 : \mathbf{Y} = \mathbf{G} \\ \mathcal{H}_1 : \mathbf{Y} = \mathbf{b}\mathbf{s}^T + \mathbf{G} \end{cases} \tag{12.15}$$

or, equivalently,

$$\begin{cases} \mathcal{H}_0 : \mathbf{y}_n = \mathbf{g}_n \\ \mathcal{H}_1 : \mathbf{y}_n = \mathbf{b}s_n + \mathbf{g}_n \end{cases} \tag{12.16}$$

where $\mathbf{y}_n = [\mathbf{Y}]_{*n} \in \mathbb{C}^{N_r \times 1}$, $\mathbf{g}_n = [\mathbf{G}]_{*n} \in \mathbb{C}^{N_r \times 1}$, and $s_n = [\mathbf{s}]_n$, where $n = 1, ..., N$. The noise $\mathbf{g}_n$ denotes the distributed array thermal noise whose elements are statistically independent and identically distributed (IID) circularly symmetric complex Gaussian, specifically $\mathbf{g}_n \sim \mathcal{CN}(0, \boldsymbol{\Sigma})$ with $\boldsymbol{\Sigma} = \text{diag}(\sigma_1^2, ..., \sigma_{N_r}^2)$. Due to the receiver calibration uncertainties, we assume that the noise variances σ_k^2 differ from σ_j^2 for $k \neq j$. In the following, based on different assumptions on the vectors **s** and **b**, we formulate different detection problems, described as follows:

Problem 1 ($\mathbf{P}_1$): Let us examine the scenario where the transmitted signal has a constant modulus, meaning that $|s_n| = \gamma$ and the value of γ is unknown. Additionally, let's suppose that the vector representing the channel parameters, denoted as **b**, is distributed according to a Gaussian distribution. As such, we have

$$\mathbf{b}_{N_r \times 1} \sim \mathcal{CN}(0, \mathbf{R}_\mathbf{b}) \tag{12.17}$$

As a result, the received signal vector $\mathbf{y}_n$ can be characterized through a circular complex Gaussian distribution with zero mean, i.e., $\mathbf{y}_n \sim \mathcal{CN}(\mathbf{0}, \mathbf{R}_Y)$, where

$$\begin{cases} \mathcal{H}_0 : \mathbf{R}_Y = \boldsymbol{\Sigma} \\ \mathcal{H}_1 : \mathbf{R}_Y = \gamma^2\mathbf{R}_\mathbf{b} + \boldsymbol{\Sigma} \end{cases} \tag{12.18}$$

This problem can be considered as the issue of distinguishing between a diagonal matrix and an arbitrary Hermitian one. This is made possible by taking into account that the temporal noises are both statistically independent from each other and from the signal vector **b**. With this in mind, problem (12.18) can be represented as

$$\begin{cases} \mathcal{H}_0 : \mathbf{R}_Y = \text{diag}(\sigma_1^2, \ldots, \sigma_{N_r}^2) \\ \mathcal{H}_1 : \mathbf{R}_Y \succ \text{diag}(\sigma_1^2, \ldots, \sigma_{N_r}^2) \end{cases} \tag{12.19}$$

In order to address the issue of target detection, we utilize the Rao and LRT criteria to devise new detectors. These new detectors have been named P_1-Rao and P_1-LRT. Additionally, we have introduced a volume-based detector, which is referred to as P_1-Vol, to enable passive target detection.

Problem 2 ($\mathbf{P}_2$): By using multiplexing techniques like DAB and DVB-T signals, certain signal sources that are opportunistic can be accurately represented as a Gaussian process, according to the central limit theorem [3]. In line with this, the values s_n for $n = 1, ..., N$ can be represented as IID circular complex Gaussian random variables, with zero mean and an unknown variance of σ_s^2 (i.e., $s_n \sim \mathcal{CN}(0, \sigma_s^2)$). In this scenario, the channel parameter vector **b** is modeled as an unknown deterministic vector. The received signal vector, denoted as $\mathbf{y}_n$, is characterized by a circular complex Gaussian distribution with zero mean, denoted as $\mathcal{CN}(\mathbf{0}, \mathbf{R}_Y)$, where

$$\begin{cases} \mathcal{H}_0 : \mathbf{R}_Y = \boldsymbol{\Sigma} \\ \mathcal{H}_1 : \mathbf{R}_Y = \sigma_s^2 \mathbf{b}\mathbf{b}^H + \boldsymbol{\Sigma} \end{cases} \tag{12.20}$$

with $\boldsymbol{\Sigma} = \text{diag}(\sigma_1^2, \ldots, \sigma_{N_r}^2)$. Here, the matrix $\mathbf{R}_Y$ is an unknown diagonal matrix under hypothesis $\mathcal{H}_0$, while it is a diagonal plus rank-1 matrix under alternative hypothesis.

To solve this problem, we again resort to the Rao and LRT principles in order to propose four detectors. Three of them are the same as those of the P_1-Rao, P_1-LRT, and P_1-Vol, while a new LRT detector is also devised under a low signal-to-noise ratio (SNR) approximation, named P_2-LRT.

In P_1 and P_2, the log-likelihood function (LLF) of the general hypothesis can be expressed as

$$\mathcal{L}(\mathbf{Y}; \mathbf{R}_Y) = -NN_r \ln \pi - N \ln \left(\left| \mathbf{R}_Y \right| \right) - \text{tr}\{\mathbf{R}_Y^{-1} \mathbf{Y}\mathbf{Y}^H\} \tag{12.21}$$

where $\mathbf{Y} = [\mathbf{y}_1, ..., \mathbf{y}_N]$ is an $N_r \times N$ matrix with columns that are IID as the multivariate complex normal distribution $\mathcal{CN}(0, \mathbf{R}_Y)$, subject to two conditions, namely: (i) $N_r < N$ and (ii) $\mathbf{R}_Y$, is a positive definite matrix.

Problem 3 ($\mathbf{P_3}$): Consider the case where both vectors **s** and **b** can be modeled as unknown but deterministic vectors. This problem is known as detecting an unknown rank-1 signal using uncalibrated distributed passive radar receivers, where their respective thermal noise variances are different and unknown.

This problem has been solved by utilizing the generalized LRT (GLRT) in low-SNR scenarios as presented in Reference 4. In this chapter, we introduce a novel LRT-based detector, also known as the P_2-LRT detector, which can serve as an alternative and more comprehensive representation of the GLRT.

Here, it is readily found that the LLF of the general hypothesis can be expressed as [1]

$$\mathcal{L} = -NN_r \ln \pi - N \ln \left(\prod_{k=1}^{N_r} \sigma_k^2 \right) - \sum_{k=1}^{N_r} \frac{1}{\sigma_k^2} [\mathbf{X}^H \mathbf{X}]_{kk} \tag{12.22}$$

12.2 Detectors design

This section presents various novel detectors for solving the passive multistatic detection problems P_1, P_2, and P_3. The findings of this chapter provide a framework for SIMO passive radar detection without RC.

12.2.1 Proposed detectors for detection problem P_1

For the first detection problem, three detectors based on the Rao, volume-based, and LRT criteria are derived, which are called P_1-Rao, P_1-Vol, and P_1-LRT, respectively.

12.2.1.1 P_1-Rao detector

In this subsection, we resort to the Rao framework to devise a Rao-based detector for problem P_1. Let us first rewrite the detection problem (12.20) to obtain

$$\begin{cases} \mathcal{H}_0 : \boldsymbol{\theta} = \text{vec}\{\mathbf{R}_Y\} \\ \mathcal{H}_1 : \boldsymbol{\theta} \neq \text{vec}\{\mathbf{R}_Y\} \end{cases} \tag{12.23}$$

Let us define the augmented vector corresponding to $\boldsymbol{\theta}$ by $\underline{\boldsymbol{\theta}}$. In our case, $\boldsymbol{\theta}$ = vec $\{\mathbf{R}_Y\}$ has dimension $N_r^2 \times 1$, while its augmented version is $2N_r^2 \times 1$.

Proposition 1. *The Rao test statistic $\Lambda_{P_1-Rao}(\mathbf{Y})$ of the detection problem P_1 can be expressed compactly as*

$$\Lambda_{P_1-\text{Rao}}(\mathbf{Y}) = 2\text{tr}\left\{(\widehat{\mathbf{R}}_Y\widehat{\boldsymbol{\Sigma}}_0^{-1} - \mathbf{I})^2\right\} \tag{12.24}$$

where

$$\widehat{\mathbf{R}}_Y = \frac{1}{N}\mathbf{Y}\mathbf{Y}^H \in \mathbb{C}^{N_r \times N_r} \tag{12.25}$$

and $\widehat{\boldsymbol{\Sigma}}_0$ is referred to as the maximum likelihood estimation (MLE) of $\mathbf{R}_Y$ under the null hypothesis $\mathcal{H}_0$, given by

$$\widehat{\boldsymbol{\Sigma}}_0 = \text{Diag}(\widehat{\mathbf{R}}_Y) \tag{12.26}$$

Proof. The derivation of the test statistic (12.24) is given in Appendix 12A. Here, η_1 is the decision threshold of the proposed Rao-based detector for the detection problem P_1. Here, the hypothesis $\mathcal{H}_1$ is accepted if $\Lambda_{P_1-\text{Rao}}(\mathbf{Y})$ is greater than η_1; otherwise, the null hypothesis $\mathcal{H}_0$ is accepted. Besides, it is shown in Appendix 12A that the proposed Rao test can be equivalently written as

$$\Lambda_{P_1-\text{Rao}}(\widehat{\mathbf{C}}_X) = ||\widehat{\mathbf{C}}_X||_F^2 \tag{12.27}$$

where $\widehat{\mathbf{C}}_X$ is known as distributed spatial coherence matrix, given by

$$\widehat{\mathbf{C}}_X = \widehat{\mathbf{D}}_{\mathbf{X}}^{-1/2}\widehat{\mathbf{R}}_{\mathbf{X}}\widehat{\mathbf{D}}_{\mathbf{X}}^{-1/2} \tag{12.28}$$

where

$$\widehat{\mathbf{R}}_{\mathbf{X}} = \frac{1}{N}\mathbf{X}^H\mathbf{X} \in \mathbb{C}^{N_r \times N_r} \tag{12.29}$$

$$\widehat{\mathbf{D}}_{\mathbf{X}} = \text{Diag}(\widehat{\mathbf{R}}_{\mathbf{X}}) \tag{12.30}$$

It is not difficult to show that the sample covariance matrices (SCMs) $\widehat{\mathbf{R}}_{\mathbf{X}}$ and $\widehat{\mathbf{R}}_{\mathbf{Y}}$ are sufficient statistics for the covariance-based hypothesis testing problem. Besides, $[\widehat{\mathbf{R}}_{\mathbf{X}}]_{kk} = [\widehat{\mathbf{R}}_{\mathbf{Y}}]_{kk} = \frac{1}{N}||\mathbf{x}_k||^2$.

12.2.1.2 P$_1$-Vol detector

Broadly speaking, volume-based methods have been proposed to enhance the reliable spectrum sensing in cognitive radios. Examples of these techniques can be found in References 5 and 6. Now, for the first time, such techniques are being utilized for detecting targets in multistatic passive radar systems. Specifically, the statistic of volume-based detector can be defined based on the received signal SCM, given by

$$\Lambda_{P_1-\mathrm{Vol}}(\widehat{\mathbf{R}}_{\mathbf{Y}}) = \frac{|\mathbf{D}|}{|\widehat{\mathbf{R}}_{\mathbf{Y}}|} \tag{12.31}$$

where

$$\mathbf{D} = \mathrm{diag}\left(\left\|[\widehat{\mathbf{R}}_{\mathbf{Y}}]_{1*}\right\|, \ldots, \left\|[\widehat{\mathbf{R}}_{\mathbf{Y}}]_{N_r*}\right\|\right) \tag{12.32}$$

Invoking Hadamard's inequality [7], we find

$$|\widehat{\mathbf{R}}_{\mathbf{Y}}| \leq \prod_{k=1}^{N_r} \left\|[\widehat{\mathbf{R}}_{\mathbf{Y}}]_{k*}\right\| = |\mathbf{D}| \tag{12.33}$$

resulting in $\Lambda_{P_1-\mathrm{Vol}}(\widehat{\mathbf{R}}_{\mathbf{Y}}) \geq 1$. Under the hypothesis $\mathscr{H}_0$, it can be shown that $\Lambda_{P_1-\mathrm{Vol}}(\widehat{\mathbf{R}}_{\mathbf{Y}})$ asymptotically approaches one as the number of samples tends to infinity. In contrast, $\Lambda_{P_1-\mathrm{Vol}}(\widehat{\mathbf{R}}_{\mathbf{Y}})$ is increased under the hypothesis $\mathscr{H}_1$; thus, it can provide a good indicator for the presence of a target signal. For $N_r = 3$ and under hypothesis $\mathscr{H}_1$, $|\widehat{\mathbf{R}}_{\mathbf{Y}}|$ is the volume of a parallelepiped whose adjacent sides are vectors $[\widehat{\mathbf{R}}_{\mathbf{Y}}]_{1*}$, $[\widehat{\mathbf{R}}_{\mathbf{Y}}]_{2*}$, and $[\widehat{\mathbf{R}}_{\mathbf{Y}}]_{3*}$, while $|\mathbf{D}|$ is the volume of a cube whose adjacent sides are vectors $[\|[\widehat{\mathbf{R}}_{\mathbf{Y}}]_{1*}\|, 0, 0]^{\mathrm{T}}$, $[0, \|[\widehat{\mathbf{R}}_{\mathbf{Y}}]_{2*}\|, 0]^T$, and $[0, 0, \|[\widehat{\mathbf{R}}_{\mathbf{Y}}]_{3*}\|]^T$. This means that the corresponding adjacent sides' vector of parallelepiped and cube has same lengths, resulting in $|\mathbf{D}| > |\widehat{\mathbf{R}}_{\mathbf{Y}}|$. Therefore, we can consider the detection threshold η_2 for the proposed P$_1$-Vol detector, where the hypothesis $\mathscr{H}_1$ is accepted if $\Lambda_{P_1-\mathrm{Vol}}(\widehat{\mathbf{R}}_{\mathbf{Y}})$ is greater than η_2; otherwise, the null hypothesis $\mathscr{H}_0$ is accepted.

12.2.1.3 P$_1$-LRT-based proposed detectors

The LRT principle is utilized to create a detector for problem P$_1$ in the sequel. The parameters associated with the null hypothesis are denoted as Θ_0, while Θ is defined as the set of all possible parameters, consisting of both Θ_0 and Θ_0^c. Here, Θ_0 and Θ_0^c are given as

$$\mathrm{P}_1 : \begin{cases} \Theta_0 = \{\boldsymbol{\Sigma} = \mathrm{diag}(\sigma_1^2, \ldots, \sigma_{N_r}^2)\} \\ \Theta_0^c = \{\mathbf{Q} \succ \mathbf{0}\} \end{cases} \tag{12.34}$$

Proposition 2. *The LRT statistic $\Lambda_{P_1-LRT}(\mathbf{Y})$ of the detection problem P_1 for testing the presence of a target can be expressed compactly as*

$$\Lambda_{P_1-\mathrm{LRT}}(\mathbf{Y}) = \frac{\prod_{i=1}^{N_r} [\widehat{\mathbf{R}}_{\mathbf{Y}}]_{kk}}{|\widehat{\mathbf{R}}_{\mathbf{Y}}|} \tag{12.35}$$

or, equivalently,

$$\Lambda_{P_1-\text{LRT}}(\mathbf{Y}) = |\widehat{\mathbf{C}}_{\mathbf{Y}}|^{-1} \tag{12.36}$$

$$\Lambda_{P_1-\text{LRT}}(\mathbf{X}) = |\widehat{\mathbf{C}}_{\mathbf{X}}|^{-1} \tag{12.37}$$

where the distributed spatial coherence matrix $\widehat{\mathbf{C}}_{\mathbf{Y}}$ *can be defined as*

$$\widehat{\mathbf{C}}_{\mathbf{Y}} = \widehat{\mathbf{D}}_{\mathbf{Y}}^{-1/2}\widehat{\mathbf{R}}_{\mathbf{Y}}\widehat{\mathbf{D}}_{\mathbf{Y}}^{-1/2} \tag{12.38}$$

$$\widehat{\mathbf{D}}_{\mathbf{Y}} = \text{Diag}(\widehat{\mathbf{R}}_{\mathbf{Y}}) \tag{12.39}$$

Both matrices $\widehat{\mathbf{C}}_{X}$ *and* $\widehat{\mathbf{C}}_{\mathbf{Y}}$ *have ones on their diagonals. In addition, it is easy to show that*

$$|\widehat{\mathbf{C}}_{X}| = |\widehat{\mathbf{C}}_{\mathbf{Y}}| = \frac{|\widehat{\mathbf{R}}_{\mathbf{X}}|}{\prod_{k=1}^{N_r}[\widehat{\mathbf{R}}_{\mathbf{X}}]_{kk}} = \frac{|\widehat{\mathbf{R}}_{\mathbf{Y}}|}{\prod_{k=1}^{N_r}[\widehat{\mathbf{R}}_{\mathbf{Y}}]_{kk}} \tag{12.40}$$

Proof. The derivation of the test statistic (12.35) is given in Appendix 12B. Let η_3 be the decision threshold of the proposed LRT-based detection problem P_1. Here, the hypothesis $\mathscr{H}_1$ is accepted if $\Lambda_{P_1-\text{LRT}}$ is greater than η_3; otherwise, the null hypothesis $\mathscr{H}_0$ is accepted.

The P_1-LRT detector's test statistic, also known as the inverse of the Hadamard ratio, can be referred to as the inverse Hadamard ratio (IHR) test. Similar to the volume-based detector, the IHR test experiences a significant decrease in $|\widehat{\mathbf{R}}_{\mathbf{Y}}|$ when a target signal is present, making it a reliable indicator for detecting targets.

12.2.2 Proposed detectors for detection problem P_2

The detection problems P_2 can be approached in two ways:

P_2-Case 1 (P_2-C1): The first approach involves disregarding the rank-1 structure of the desired signal in (12.20), which simplifies the hypothesis testing problem (12.20) as

$$\begin{cases} \mathscr{H}_0 : \mathbf{R}_{\mathbf{Y}} = \text{diag}(\sigma_1^2, \ldots, \sigma_{N_r}^2) \\ \mathscr{H}_1 : \mathbf{R}_{\mathbf{Y}} \succ \text{diag}(\sigma_1^2, \ldots, \sigma_{N_r}^2) \end{cases} \tag{12.41}$$

Proposition 3. *The Rao, LRT, and volume-based detectors for this case have the same forms as those of (12.24), (12.31), and (12.35), respectively.*

P_2-Case 2 (P_2-C2): Alternatively, if the rank-1 structure is taken into consideration, the parameter spaces Θ_0 and Θ_0^c can be defined as

$$P_2 : \begin{cases} \Theta_0 = \{\mathbf{\Sigma} = \text{diag}(\sigma_1^2, \ldots, \sigma_{N_r}^2)\} \\ \Theta_0^c = \{\sigma_s^2, \mathbf{b}, \mathbf{\Sigma} = \text{diag}(\sigma_1^2, \ldots, \sigma_{N_r}^2)\} \end{cases} \tag{12.42}$$

A new algorithm for detecting multistatic passive radar (PR) can be developed using the LRT framework. However, solving this problem for every value of N_r is not feasible. Instead, an exact LRT detector is derived for $N_R = 2$, while a low-SNR approximation of this detector is developed for $N_r > 2$.

Proposition 4. *The low-SNR approximation of the LRT statistic* $\Lambda_{P_2-C_2-\mathrm{LRT}}(\mathbf{Y})$ *of the detection problem* P_2 *can be expressed compactly as*

$$\Lambda_{P_2-C_2-\mathrm{LRT}}(\mathbf{Y}) = \lambda_1(\widehat{\mathbf{C}}_{\mathbf{Y}}) \tag{12.43}$$

In the special case of $N_r = 2$, *the exact* $P_2 - C_2 - \mathrm{LRT}$ *test statistic becomes*

$$\Lambda_{P_2-C_2-\mathrm{LRT}}(\mathbf{Y}) = |\widehat{\mathbf{C}}_{\mathbf{Y}}|^{-1} \tag{12.44}$$

In (12.44), $\widehat{\mathbf{C}}_{\mathbf{Y}} \in \mathbb{C}^{2\times 2}$ *is known as coherence matrix with ones in its diagonal and sample correlation coefficients between different distributed receivers in its off-diagonal.*

Proof. The derivation of the above test statistics is given in Appendix 12C. In the sequel, these proposed LRT-based detectors are called in the general name of P2-C2-LRT.

12.2.3 Proposed detector for detection problem P_3

In the detection problem P_3 the parameter spaces Θ_0 and Θ_0^c can be specified as

$$\mathrm{P}_3: \begin{cases} \Theta_0 = \{\boldsymbol{\Sigma} = \mathrm{diag}(\sigma_1^2, \ldots, \sigma_{N_r}^2)\} \\ \Theta_0^c = \{\mathbf{s}, \mathbf{b}, \boldsymbol{\Sigma} = \mathrm{diag}(\sigma_1^2, \ldots, \sigma_{N_r}^2)\} \end{cases} \tag{12.45}$$

As such, we devise a new LRT-based multistatic PR detection algorithm, called P_3-LRT.

Proposition 5. *The LRT statistic* $\Lambda_{P_3-\mathrm{LRT}}(\mathbf{X})$ *of the detection problem* P_3 *can be expressed compactly as*

$$\Lambda_{P_3-\mathrm{LRT}}(\mathbf{X}) = \rho\lambda_1\left(\widehat{\mathbf{C}}_X\right) = \rho\|\widehat{\mathbf{C}}_X\|_2 \tag{12.46}$$

The value of ρ *is an arbitrary but large scalar that can be absorbed into the detection threshold. As* $\lambda_1(\widehat{\mathbf{C}}_X) = \lambda_1(\widehat{\mathbf{C}}_{\mathbf{Y}})$, *we can deduce that* $\Lambda \mathrm{P}_3 - \mathrm{LRT}(\mathbf{Y}) = \rho\lambda_1(\widehat{\mathbf{C}}_{\mathbf{Y}})$. *Therefore, the tests* P_3*-LRT and* P_2*-C2-LRT are equivalent and have the same performance but are based on different assumptions.*

Proof. The derivation of the test statistic (12.46) is given in Appendix 12D.

12.3 CFAR analysis of proposed detectors

In this section, we utilize the principle of invariance to examine how the proposed detectors may exhibit CFAR behavior when dealing with uncertainties regarding noise variances across different receivers. The goal of the invariance theory is to find transformations that keep the hypothesis testing problem invariant (see, e.g., References 8, 9, and the references therein). The unknown diagonal matrix $\boldsymbol{\Sigma}'$, which contains positive entries, is used for proving CFAR behavior. To demonstrate CFAR behavior, we must illustrate that transforming $\boldsymbol{\Sigma}'^{\frac{1}{2}}\mathbf{y}_n$ maintains the data's distribution family while obtaining $\Lambda_i(\boldsymbol{\Sigma}'^{\frac{1}{2}}\mathbf{Y}) = \Lambda_i(\mathbf{Y})$. Under the $\mathscr{H}_0$ hypothesis, the distribution of the transformed data $\boldsymbol{\Sigma}'^{\frac{1}{2}}\mathbf{y}_n$ follows the $\mathscr{CN}(\mathbf{0}, \boldsymbol{\Sigma}'\boldsymbol{\Sigma})$ format, where the matrix $\boldsymbol{\Sigma}'\boldsymbol{\Sigma}$ is also unknown and diagonal. In addition, we have

$$\begin{aligned}
&\Lambda_{P_1-\text{Rao}}(\boldsymbol{\Sigma}'^{\frac{1}{2}}\mathbf{Y}) \\
&\quad= 2\text{tr}\left\{(\boldsymbol{\Sigma}'^{\frac{1}{2}}\widehat{\mathbf{R}}_{\mathbf{Y}}\boldsymbol{\Sigma}'^{\frac{1}{2}}\text{Diag}^{-1}(\boldsymbol{\Sigma}'^{\frac{1}{2}}\widehat{\mathbf{R}}_{\mathbf{Y}}\boldsymbol{\Sigma}'^{\frac{1}{2}}) - \mathbf{I})^2\right\} \\
&\quad= 2\text{tr}\left\{(\boldsymbol{\Sigma}'^{\frac{1}{2}}\widehat{\mathbf{R}}_{\mathbf{Y}}\boldsymbol{\Sigma}'^{-\frac{1}{2}}\text{Diag}^{-1}(\widehat{\mathbf{R}}_{\mathbf{Y}}) - \mathbf{I})^2\right\} \\
&\quad= 2\text{tr}\left\{(\boldsymbol{\Sigma}'^{\frac{1}{2}}\widehat{\mathbf{R}}_{\mathbf{Y}}\text{Diag}^{-1}(\widehat{\mathbf{R}}_{\mathbf{Y}})\boldsymbol{\Sigma}'^{-\frac{1}{2}} - \mathbf{I})^2\right\} \\
&\quad= 2\text{tr}\left\{(\boldsymbol{\Sigma}'^{\frac{1}{2}}(\widehat{\mathbf{R}}_{\mathbf{Y}}\text{Diag}^{-1}(\widehat{\mathbf{R}}_{\mathbf{Y}}) - \mathbf{I})\boldsymbol{\Sigma}'^{-\frac{1}{2}})^2\right\} \\
&\quad= \Lambda_{\text{P}_1-\text{Rao}}(\mathbf{Y})
\end{aligned} \tag{12.47}$$

Since

$$\begin{aligned}
&\widehat{\mathbf{C}}_{\boldsymbol{\Sigma}'^{\frac{1}{2}}\mathbf{Y}} \\
&\quad= \text{Diag}^{-\frac{1}{2}}(\boldsymbol{\Sigma}'^{\frac{1}{2}}\widehat{\mathbf{R}}_{\mathbf{Y}}\boldsymbol{\Sigma}'^{\frac{1}{2}})\boldsymbol{\Sigma}'^{\frac{1}{2}}\widehat{\mathbf{R}}_{\mathbf{Y}}\boldsymbol{\Sigma}'^{\frac{1}{2}}\text{Diag}^{-\frac{1}{2}}(\boldsymbol{\Sigma}'^{\frac{1}{2}}\widehat{\mathbf{R}}_{\mathbf{Y}}\boldsymbol{\Sigma}'^{\frac{1}{2}}) \\
&\quad= \text{Diag}^{-\frac{1}{2}}(\widehat{\mathbf{R}}_{\mathbf{Y}})\boldsymbol{\Sigma}'^{-\frac{1}{2}}\boldsymbol{\Sigma}'^{\frac{1}{2}}\widehat{\mathbf{R}}_{\mathbf{Y}}\boldsymbol{\Sigma}'^{\frac{1}{2}}\boldsymbol{\Sigma}'^{-\frac{1}{2}}\text{Diag}^{-\frac{1}{2}}(\widehat{\mathbf{R}}_{\mathbf{Y}}) \\
&\quad= \widehat{\mathbf{C}}_{\mathbf{Y}}
\end{aligned} \tag{12.48}$$

we find that

$$\Lambda_{P_1-\text{LRT}}(\boldsymbol{\Sigma}'^{\frac{1}{2}}\mathbf{Y}) = \Lambda_{P_1-\text{LRT}}(\mathbf{Y}) \tag{12.49}$$

$$\Lambda_{P_2-C_2-\text{LRT}}(\boldsymbol{\Sigma}'^{\frac{1}{2}}\mathbf{Y}) = \Lambda_{P_2-C_2-\text{LRT}}(\mathbf{Y}) \tag{12.50}$$

$$\Lambda_{P_3-\text{LRT}}(\boldsymbol{\Sigma}'^{\frac{1}{2}}\mathbf{Y}) = \Lambda_{P_3-\text{LRT}}(\mathbf{Y}) \tag{12.51}$$

For the P_1-Vol detector, it is easy to show that

$$\Lambda_{P_1-\text{Vol}}(\boldsymbol{\Sigma}'^{\frac{1}{2}}\widehat{\mathbf{R}}_{\mathbf{Y}}) \neq \Lambda_{P_1-\text{Vol}}(\widehat{\mathbf{R}}_{\mathbf{Y}}) \tag{12.52}$$

This means that the P_1-Vol detector is not equipped with the ability to handle uncertain levels of noise variance in various receivers, while the other detectors, including the P_1-Rao, P_1-LRT, P_2-C2-LRT, and P_3-LRT, have the capability to maintain constant detection thresholds even when encountering varying noise variances across different receivers. As a result, these detectors demonstrate CFAR behavior in the presence of any noise variance uncertainties among different receivers.

12.4 Performance results

This section utilizes numerical simulations to assess the effectiveness of the suggested detectors, which are outlined in Table 12.1. The evaluation is conducted by generating vector $\mathbf{s}$ through the expression $\mathbf{s} = \exp\{-j\phi\}$, where the elements of vector ϕ are independently and uniformly distributed on the interval $[0, 2\pi)$, unless otherwise specified. The vector $\mathbf{b}$ can be separated into two parts as

Table 12.1 Detectors' specification

Abbreviation	Test statistic	Equation
P_1-Rao	$2\mathrm{tr}\left\{(\widehat{\mathbf{R}}_{\mathbf{Y}}\widehat{\boldsymbol{\Sigma}}_0^{-1} - \mathbf{I})^2\right\}$	(12.24)
P_1-Vol	$\frac{\lvert\mathbf{D}\rvert}{\lvert\widehat{\mathbf{R}}_{\mathbf{Y}}\rvert}$	(12.31)
P_1-LRT	$\frac{\prod_{i=1}^{N_r}[\widehat{\mathbf{R}}_{\mathbf{Y}}]_{kk}}{\lvert\widehat{\mathbf{R}}_{\mathbf{Y}}\rvert}$ or $\lvert\widehat{\mathbf{C}}_{\mathbf{Y}}\rvert^{-1}$	(12.35) and (12.36)
P_2-C1-Rao	$2\mathrm{tr}\left\{(\widehat{\mathbf{R}}_{\mathbf{Y}}\widehat{\boldsymbol{\Sigma}}_0^{-1} - \mathbf{I})^2\right\}$	(12.24)
P_2-C1-Vol	$\frac{\lvert\mathbf{D}\rvert}{\lvert\widehat{\mathbf{R}}_{\mathbf{Y}}\rvert}$	(12.31)
P_2-C1-LRT	$\frac{\prod_{i=1}^{N_r}[\widehat{\mathbf{R}}_{\mathbf{Y}}]_{kk}}{\lvert\widehat{\mathbf{R}}_{\mathbf{Y}}\rvert}$	(12.35)
P_2-C2-LRT	$\lambda_1(\widehat{\mathbf{C}}_{\mathbf{Y}})$	(12.43)
P_2-C2-LRT ($N_r = 2$)	$\lvert\widehat{\mathbf{C}}_{\mathbf{Y}}\rvert^{-1}$	(12.44)
P_3-LRT	$\lVert\widehat{\mathbf{C}}_X\rVert_2$	(12.46)

$\mathbf{b} = \zeta \odot \mathbf{d}$. The coefficient vector $\mathbf{d}$ is drawn from a circular Gaussian distribution (i.e., $\mathcal{CN}(0_{N_r}, \mathbf{I}_{N_r})$) and scaled by entries of vector ζ, denoted by $\{\zeta_k\}$, to achieve a desired averaged SNR in the SCs, denoted by $\mathrm{SNR}_{\mathrm{avg}}$, where

$$\mathrm{SNR}_{\mathrm{avg}} = \frac{1}{N_r}\sum_{k=1}^{N_r}\frac{\lvert\zeta_k\rvert^2}{\sigma_k^2} \tag{12.53}$$

where unequal receiver noise variances are chosen from the vector $[\sigma_1^2, \sigma_2^2, \sigma_3^2, \sigma_4^2, \sigma_5^2, \sigma_6^2] = [1, 1.5, 0.25, 1.2, 2, 0.75]$. In (12.53), $\frac{\lvert\zeta_k\rvert^2}{\sigma_k^2}$ is the input SNR at the kth receiver (i.e., SC_k), denoted by SNR_k.

Due to the complexity of our detectors, we cannot easily derive closed-form formulas for both the detection probability and false alarm probability. Therefore, we are relying on MC simulation results to evaluate the performance of our detectors regarding false alarms and detection accuracy.

12.4.1 False alarm probability evaluations

Figure 12.2 illustrates the assessment of false alarm probabilities versus detection threshold for the proposed detectors using two different sets of noise variances. The first set of noise variances, $[\sigma_1^2, \sigma_2^2]$, is evaluated for $N_r = 2$ and $N = 16$, with values of [1, 1.5] and [0.5, 1.25]. The second set of noise variances, $[\sigma_1^2, \sigma_2^2, \sigma_3^2, \sigma_4^2, \sigma_5^2, \sigma_6^2]$,

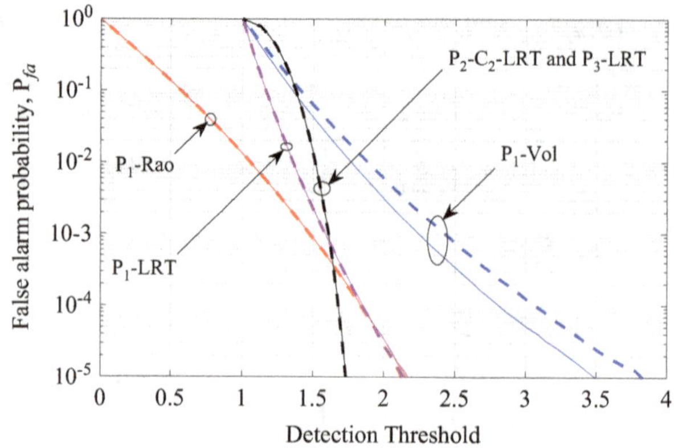

(a) $N_r = 2$ and $N = 16$

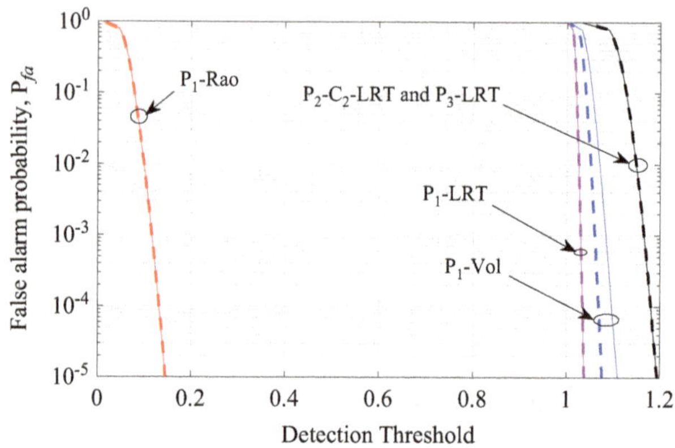

(b) $N_r = 6$ and $N = 1024$

Figure 12.2 *False alarm probability as a function of detection threshold for two different sets of noise variances. The lines denote the results of the cases of* $[\sigma_1^2, \sigma_2^2] = [1, 1.5]$ *and* $[\sigma_1^2, \sigma_2^2, \sigma_3^2, \sigma_4^2, \sigma_5^2, \sigma_6^2] = [1, 1.5, 0.25, 1.2, 2, 0.75]$, *while "- -" stands for* $[\sigma_1^2, \sigma_2^2] = [0.5, 1.25]$ *and* $[\sigma_1^2, \sigma_2^2, \sigma_3^2, \sigma_4^2, \sigma_5^2, \sigma_6^2] = [1, 1, 1, 1, 1, 1]$.

is evaluated for $N_r = 6$ and $N = 1{,}024$, with values of $[1, 1.5, 0.25, 1.2, 2, 0.75]$ and $[1, 1, 1, 1, 1, 1]$. To ensure the accuracy of the results, 10^7 MC simulation runs were conducted to establish the detector thresholds. As depicted in Figure 12.2, all the detectors, except P_1-Vol detector, exhibit the CFAR feature concerning different sets of noise variances across multiple receivers. Notably, P_1-Rao, P_1-LRT, P_1-C2-LRT, and P_3-LRT detectors display constant detection thresholds even when different noise variances across multiple receivers change. These numerical results validate the CFAR findings obtained in section 12.3. In addition, the data in Figure 12.2 can be

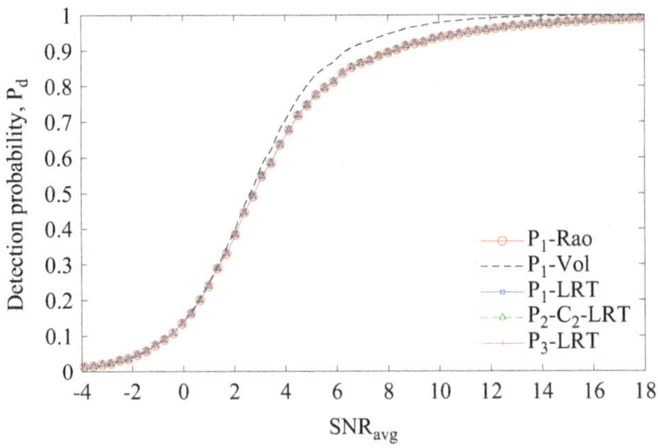

Figure 12.3 Detection probability as a function of SNR$_{avg}$, where the number of receive antennas is equal to 2, the number of time samples is 16, and the desired false alarm probability is set equal to 10^{-3}

utilized to determine the decision thresholds of the proposed detectors to achieve the desired false alarm probability of $p_{fa} = 10^{-3}$.

12.4.2 Detection performance evaluations

In Figures 12.3 and 12.4, we compare the detection performances of the considered detectors for $N = 16$ as the number of receive antennas has varied from 2 to 6, respectively. For $N_r = 2$, as shown in Figure 12.3, it is seen that the P_1-Vol detector performs the best, while the others have the same performance. For $N_r = 6$, it is observed that the proposed P_1-Rao performs the best, while they are followed by P_2-C2-LRT (or P_3-LRT),

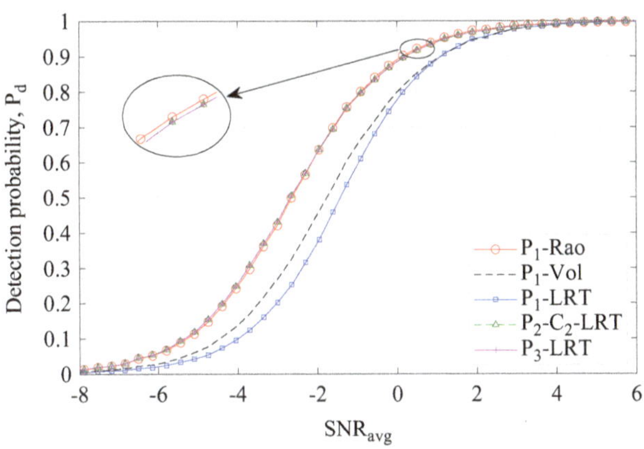

Figure 12.4 Detection probability as a function SNR$_{avg}$, where the number of receive antennas is equal to 6, the number of time samples is 16, and the desired false alarm probability is set equal to 10^{-3}

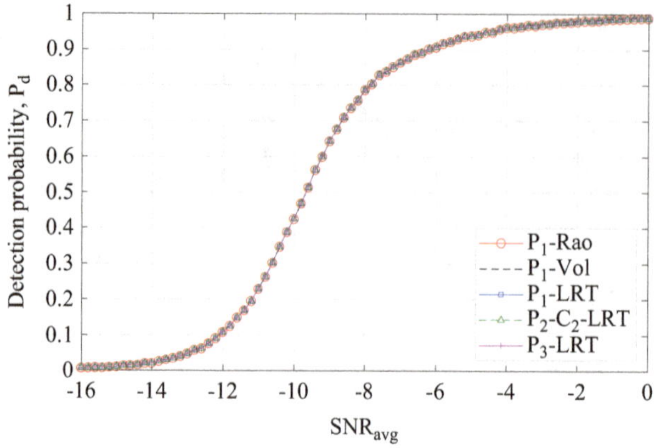

Figure 12.5 *Detection probability as a function of* SNR_{avg}*, where the number of receive antennas is equal to 2, the number of time samples is 1,024, and the desired false alarm probability is set equal to* 10^{-3}

P_1-Vol, and P_1-LRT detectors. The above results for $N = 1{,}024$ are repeated and their results are depicted in Figures 12.5 and 12.6, where all considered detectors perform the same for $N_r = 6$, as shown in Figure 12.5. It is observed from Figure 12.6 that the detection performance of the P_2-C2-LRT (or P_3-LRT) is followed by P_1-Rao, P_1-LRT, and P_1-Vol detectors.

The receiver operating curves (ROCs) of the considered detectors are demonstrated in Figure 12.7, where the number of antennas equals 6, the number of time samples is 16, and the averaged SNR equals 0 dB. Here, it can be observed that the proposed P_1-Rao demonstrates superior performance compared to other detectors that have been

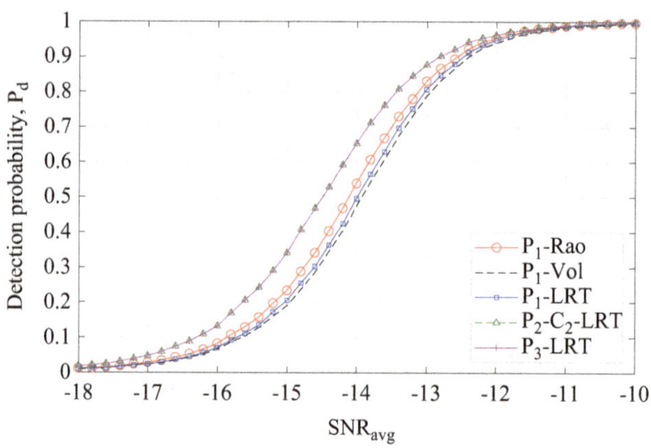

Figure 12.6 *Detection probability as a function of* SNR_{avg}*, where the number of receive antennas is equal to 6, the number of time samples is 1,024, and the desired false alarm probability is set equal to* 10^{-3}

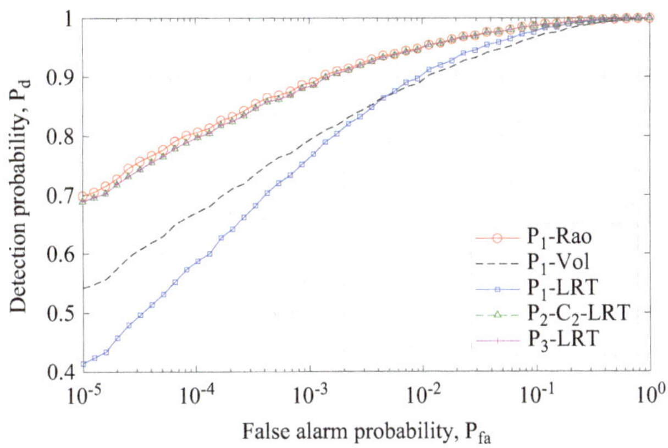

Figure 12.7 *ROCs of the considered detectors, where the number of receive antennas is equal to 6, the number of time samples is 16, and the averaged SNR is set equal to 0 dB*

proposed. Additionally, it can be noted that the P_1-Vol detector is better than P_1-LRT but not as effective as the other proposed detectors. Figure 12.8 depicts the ROCs of the aforementioned detectors under the condition of 1,024 time samples and an average SNR of –13 dB. As illustrated in this figure, the P_3-LRT (or P_2-C2-LRT) detector provides the best detection performance among all proposed algorithms, followed by the P_1-Rao, P_1-LRT, and P_1-Vol detectors.

Finally, we have conducted a simulation where we utilized a circular Gaussian distribution (specifically, $\mathcal{CN}(\mathbf{0}_N, \mathbf{I}_N)$) for the vector $\mathbf{s}$. The results of this simulation, shown in Figures 12.9 and 12.10, demonstrate similar trends to those observed in

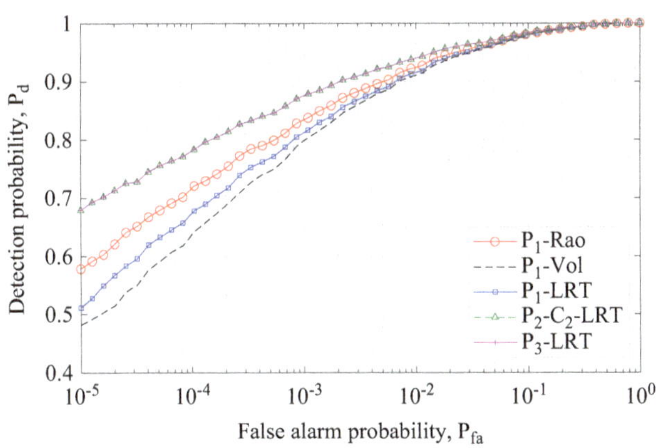

Figure 12.8 *ROCs of the considered detectors, where the number of receive antennas is equal to 6, the number of time samples is 1,024, and the averaged SNR is set equal to –13 dB*

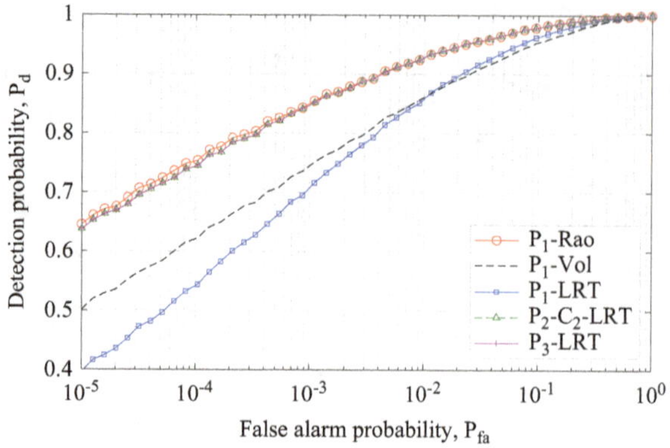

Figure 12.9 *ROCs of the considered detectors, where the number of receive antennas is equal to 6, the number of time samples is 16, and the averaged SNR is set equal to 0 dB. Here, the vector* **s** *is drawn from a circular Gaussian distribution.*

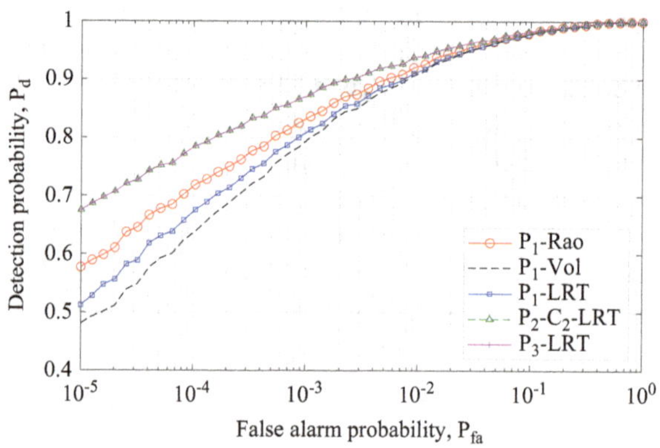

Figure 12.10 *ROCs of the considered detectors, where the number of receive antennas is equal to 6, the number of time samples is 1,024, and the averaged SNR is set equal to –13 dB. Here, the vector* **s** *is drawn from a circular Gaussian distribution.*

Figures 12.7 and 12.8. However, we have observed a detection performance degradation for small spatial–temporal sample sizes, particularly for the P_1-Rao and P_2-C2-LRT (or P_3-LRT) detectors.

12.5 Summary

This chapter has unified the target detection problem in an uncalibrated multistatic passive radar with distributed receivers. To achieve this, we formulated and solved three multistatic target detection problems using the Rao, geometric argument, and LRT principles, resulting in four new detectors. Moreover, we carried out extensive MC simulations to evaluate the performance of the proposed detectors and found that each one has its own unique capabilities. In addition, we utilized the invariance theory to examine the potential CFAR behavior of the detectors against noise variance uncertainties across different receivers. Our analysis showed that, with the exception of the P_1-Vol detector, all the proposed detectors exhibit robustness against noise variance uncertainties in uncalibrated receivers.

References

[1] Zaimbashi A. 'A unified framework for multistatic passive radar target detection under uncalibrated receivers'. *IEEE Transactions on Signal Processing*. 2021, vol. 69(1), pp. 695–708.

[2] Solatzadeh Z., Zaimbashi A. 'Accelerating target detection in passive radar sensors: delay-Doppler-acceleration estimation'. *IEEE Sensors Journal*. 2018, vol. 18(13), pp. 5445–54.

[3] Cui G., Liu J., Li H., Himed B. 'Signal detection with noisy reference for passive sensing'. *Signal Processing*. 2015, vol. 108, pp. 389–99.

[4] Hack D.E., Patton L.K., Himed B. 'Multichannel detection of an unknown rank-one signal with uncalibrated receivers'. *2014 IEEE International Conference on Acoustics, Speech and Signal Processing (ICASSP)*; Florence, 2014. pp. 2987–91.

[5] Huang L., Qian C., Xiao Y., Zhang Q.T. 'Performance analysis of volume-based spectrum sensing for cognitive radio'. *IEEE Transactions on Wireless Communications*. 2014, vol. 14(1), pp. 317–330.

[6] Huang L., So H.C., Qian C. 'Volume-based method for spectrum sensing'. *Digital Signal Processing*. 2014, vol. 28, pp. 48–56.

[7] Seber G.A.F. A matrix handbook for statisticians. Hoboken, NJ, USA: John Wiley; 2008 Nov 7. Available from http://doi.wiley.com/10.1002/9780470226797

[8] Zaimbashi A., Valkama M. 'Impropriety-based multiantenna spectrum sensing with I/Q imbalanced radios'. *IEEE Transactions on Vehicular Technology*. 2019, vol. 68(9), pp. 8693–8706.

[9] Zaimbashi A. 'Multiband FM-based passive bistatic radar: target range resolution improvement'. *IET Radar, Sonar & Navigation*. 2016, vol. 10(1), pp. 174–585.

[10] Kay S., Zhu Z. 'The complex parameter Rao test'. *IEEE Transactions on Signal Processing*. 2016, vol. 64(24), pp. 6580–88.

[11] Liu J., Liu W., Chen B., Liu H., Li H., Hao C. 'Modified Rao test for multichannel adaptive signal detection'. *IEEE Transactions on Signal Processing*. 2015, vol. 64(3), pp. 714–25.

[12] Liu J., Li J. 'Robust detection in MIMO radar with steering vector mismatches'. *IEEE Transactions on Signal Processing*. 2019, vol. 67(20), pp. 5270–80.

[13] Sun M., Liu W., Liu J., Hao C. 'Complex parameter Rao, Wald, gradient, and Durbin tests for multichannel signal detection'. *IEEE Transactions on Signal Processing*. 2021, vol. 70, pp. 117–31.

Appendix 12A
$\mathbf{P_1}$-Rao detector

The complex parameter Rao test statistic for testing the hypothesis testing problem (12.41) may be obtained from

$$\Lambda\left(\mathbf{Y};\underline{\boldsymbol{\theta}}\right)=\left[\frac{\partial\mathscr{L}(\mathbf{Y};\underline{\boldsymbol{\theta}})}{\partial\underline{\boldsymbol{\theta}}^*}\right]^H\underline{\widetilde{\mathbf{I}}}^{-1}\left(\underline{\boldsymbol{\theta}}\right)\left[\frac{\partial\mathscr{L}(\mathbf{Y};\underline{\boldsymbol{\theta}})}{\partial\underline{\boldsymbol{\theta}}^*}\right] \tag{12A.1}$$

where the augmented vector corresponding to $\boldsymbol{\theta}$ can be defined by $\underline{\boldsymbol{\theta}}=[\boldsymbol{\theta}^T,\boldsymbol{\theta}^H]^T$, and the score function $\frac{\partial\mathscr{L}(\mathbf{Y};\underline{\boldsymbol{\theta}})}{\partial\underline{\boldsymbol{\theta}}^*}$ is given by

$$\frac{\partial\mathscr{L}(\mathbf{Y};\underline{\boldsymbol{\theta}})}{\partial\underline{\boldsymbol{\theta}}^*}=\begin{bmatrix}\frac{\partial\mathscr{L}(\mathbf{Y};\boldsymbol{\theta})}{\partial\boldsymbol{\theta}^*}\\ \frac{\partial\mathscr{L}(\mathbf{Y};\boldsymbol{\theta})}{\partial\boldsymbol{\theta}}\end{bmatrix} \tag{12A.2}$$

In (12A.1), $\underline{\widetilde{\mathbf{I}}}\left(\underline{\boldsymbol{\theta}}\right)$ is the Fisher information matrix (FIM), given by

$$\begin{aligned}\underline{\widetilde{\mathbf{I}}}\left(\underline{\boldsymbol{\theta}}\right)&=E\left\{\left[\frac{\partial\mathscr{L}(\mathbf{Y};\underline{\boldsymbol{\theta}})}{\partial\underline{\boldsymbol{\theta}}^*}\right]\left[\frac{\partial\mathscr{L}(\mathbf{Y};\underline{\boldsymbol{\theta}})}{\partial\underline{\boldsymbol{\theta}}^*}\right]^H\right\}\\&=\begin{bmatrix}\widetilde{\mathbf{I}}(\boldsymbol{\theta}) & \widetilde{\mathbf{J}}(\boldsymbol{\theta})\\ \widetilde{\mathbf{J}}^*(\boldsymbol{\theta}) & \widetilde{\mathbf{I}}^*(\boldsymbol{\theta})\end{bmatrix}\end{aligned} \tag{12A.3}$$

The $N_r\times N_r$- dimensional blocks $\widetilde{\mathbf{I}}\left(\theta\right)$ and $\widetilde{\mathbf{J}}\left(\theta\right)$ in the FIM may be obtained as

$$\widetilde{\mathbf{I}}\left(\boldsymbol{\theta}\right)=E\left\{\frac{\partial\mathscr{L}(\mathbf{Y};\underline{\boldsymbol{\theta}})}{\partial\boldsymbol{\theta}^*}\frac{\partial\mathscr{L}(\mathbf{Y};\underline{\boldsymbol{\theta}})}{\partial\boldsymbol{\theta}^*}^H\right\} \tag{12A.4}$$

and

$$\widetilde{\mathbf{J}}\left(\boldsymbol{\theta}\right)=E\left\{\frac{\partial\mathscr{L}(\mathbf{Y};\underline{\boldsymbol{\theta}})}{\partial\boldsymbol{\theta}^*}\frac{\partial\mathscr{L}(\mathbf{Y};\underline{\boldsymbol{\theta}})}{\partial\boldsymbol{\theta}}^H\right\} \tag{12A.5}$$

In the special case of $\widetilde{\mathbf{J}}\left(\boldsymbol{\theta}\right)=0$, the statistic (12A.1) can be simplified as

$$\Lambda\left(\mathbf{Y};\boldsymbol{\theta}\right)=2\left[\frac{\partial\mathscr{L}(\mathbf{Y};\underline{\boldsymbol{\theta}})}{\partial\boldsymbol{\theta}}\right]^T\widetilde{\mathbf{I}}^{-1}\left(\boldsymbol{\theta}\right)\left[\frac{\partial\mathscr{L}(\mathbf{Y};\underline{\boldsymbol{\theta}})}{\partial\boldsymbol{\theta}^*}\right] \tag{12A.6}$$

To obtain the Rao test statistic, denoted as $\Lambda\left(\mathbf{Y};\widehat{\boldsymbol{\theta}}_0\right)$, in the specific case, the MLE $\widehat{\boldsymbol{\theta}}_0$ is used under the hypothesis $\mathscr{H}_0$. In the general case, the MLE $\widehat{\underline{\boldsymbol{\theta}}}_0$ is used for $\underline{\boldsymbol{\theta}}$ instead [10–13]. In order to calculate $\widetilde{\mathbf{I}}\left(\boldsymbol{\theta}\right)$ and $\widetilde{\mathbf{J}}\left(\boldsymbol{\theta}\right)$, we need to obtain

$$\frac{\partial \mathscr{L}(\mathbf{Y};\boldsymbol{\theta})}{\partial \boldsymbol{\theta}} = \frac{\partial \mathscr{L}(\mathbf{Y};\text{vec}\{\mathbf{R}_Y\})}{\partial \text{vec}\{\mathbf{R}_Y\}} = \text{vec}\left\{\frac{\partial \mathscr{L}(\mathbf{Y};\mathbf{R}_Y)}{\partial \mathbf{R}_Y}\right\} \tag{12A.7}$$

It follows from (12.21) that

$$\frac{\partial \mathscr{L}(\mathbf{Y};\mathbf{R}_Y)}{\partial \mathbf{R}_Y} = N\left(\left(\mathbf{R}_Y^{-1}\widehat{\mathbf{R}}_Y - \mathbf{I}\right)\mathbf{R}_Y^{-1}\right)^T \tag{12A.8}$$

where we have used

$$\partial\left|\mathbf{R}_Y\right| = \left|\mathbf{R}_Y\right| \text{tr}\left(\mathbf{R}_Y^{-1}\partial\mathbf{R}_Y\right) \tag{12A.9}$$

$$\partial\mathbf{R}_Y^{-1} = -\mathbf{R}_Y^{-1}\partial\mathbf{R}_Y\mathbf{R}_Y^{-1} \tag{12A.10}$$

Substituting (12A.8) into (12A.7) and using

$$\text{vec}(\mathbf{ABC}) = (\mathbf{C}^{\mathrm{T}} \otimes \mathbf{A})\text{vec}(\mathbf{B}) \tag{12A.11}$$

yields

$$\begin{aligned}\frac{\partial \mathscr{L}(\mathbf{Y};\boldsymbol{\theta})}{\partial \boldsymbol{\theta}} &= N\left(\left(\mathbf{R}_Y^{-1}\otimes\mathbf{R}_Y^{-T}\right)\text{vec}(\widehat{\mathbf{R}}_Y^T) - \text{vec}(\mathbf{R}_Y^{-T})\right)\\ &= N\text{vec}\left\{\mathbf{R}_Y^{-T}\left(\widehat{\mathbf{R}}_Y\mathbf{R}_Y^{-T} - \mathbf{I}\right)\right\}\end{aligned} \tag{12A.12}$$

Similarly, we may obtain

$$\begin{aligned}\frac{\partial \mathscr{L}(\mathbf{Y};\boldsymbol{\theta})}{\partial \boldsymbol{\theta}^*} &= N\left(\left(\mathbf{R}_Y^{-*}\otimes\mathbf{R}_Y^{-H}\right)\text{vec}(\widehat{\mathbf{R}}_Y) - \text{vec}(\mathbf{R}_Y^{-H})\right)\\ &= N\text{vec}\left\{\mathbf{R}_Y^{-H}\left(\widehat{\mathbf{R}}_Y\mathbf{R}_Y^{-H} - \mathbf{I}\right)\right\}\end{aligned} \tag{12A.13}$$

The FIM of (12A.4) can be obtained as

$$\begin{aligned}\frac{\widetilde{\mathbf{I}}(\boldsymbol{\theta})}{N^2} &= \left(\mathbf{R}_Y^{-*}\otimes\mathbf{R}_Y^{-H}\right)E\left\{\text{vec}(\widehat{\mathbf{R}}_Y)\text{vec}^H(\widehat{\mathbf{R}}_Y)\right\}\left(\mathbf{R}_Y^{-T}\otimes\mathbf{R}_Y^{-1}\right)\\ &- \left(\mathbf{R}_Y^{-*}\otimes\mathbf{R}_Y^{-H}\right)E\left\{\text{vec}(\widehat{\mathbf{R}}_Y)\right\}\text{vec}^H\{\mathbf{R}_Y^{-H}\}\\ &- \text{vec}\{\mathbf{R}_Y^{-H}\}E\left\{\text{vec}^H(\widehat{\mathbf{R}}_Y)\right\}\left(\mathbf{R}_Y^{-T}\otimes\mathbf{R}_Y^{-1}\right)\\ &+ \text{vec}\{\mathbf{R}_Y^{-H}\}\text{vec}^H\{\mathbf{R}_Y^{-H}\}\end{aligned} \tag{12A.14}$$

To proceed, let us define $\mathbf{W} = \mathbf{YY}^H$, then $\mathbf{W}$ is said to have a (nonsingular) Wishart distribution with N degrees of freedom, and we write $\boldsymbol{W} \sim \mathscr{W}_{N_r}(N, \mathbf{R}_Y)$. As such, we also have [1, 7]

$$E\{\mathbf{YY}^H\} = N\mathbf{R}_Y \tag{12A.15}$$

$$\text{var}\{\text{vec}(\mathbf{YY}^H)\} = 2NP_{N_r}(\mathbf{R}_Y \otimes \mathbf{R}_Y) \tag{12A.16}$$

where P_{N_r} is the symmetrizer matrix, defined in Reference 7. By using (12A.15) and (12A.16), we find

$$E\left\{\text{vec}(\widehat{\mathbf{R}}_Y)\right\} = \text{vec}(\mathbf{R}_Y) \tag{12A.17}$$

$$E\left\{\text{vec}(\widehat{\mathbf{R}}_Y)\text{vec}^H(\widehat{\mathbf{R}}_Y)\right\} = \frac{2}{N}\mathbf{P}_{N_r}\left(\mathbf{R}_Y\otimes\mathbf{R}_Y\right) + \text{vec}\{\mathbf{R}_Y\}\text{vec}^H\{\mathbf{R}_Y\} \tag{12A.18}$$

By substituting (12A.17)–(12A.18) into (12A.14) and after some algebraic manipulations, we obtain

$$\widetilde{\mathbf{I}}(\boldsymbol{\theta}) = 2NP_{N_r}\left(\mathbf{R}_Y^{-T} \otimes \mathbf{R}_Y^{-*}\right) \tag{12A.19}$$

Thus,

$$\widetilde{\mathbf{I}}^{-1}(\boldsymbol{\theta}) = \frac{1}{2N}\left(\mathbf{R}_Y^{T} \otimes \mathbf{R}_Y^{*}\right)\mathbf{P}_{N_r}^{-1} \tag{12A.20}$$

In a similar way, we can prove that

$$\widetilde{\mathbf{J}}^{*}(\boldsymbol{\theta}) = E\left\{\frac{\partial \mathscr{L}(\mathbf{X};\underline{\boldsymbol{\theta}})}{\partial \boldsymbol{\theta}} \frac{\partial \mathscr{L}(\mathbf{X};\underline{\boldsymbol{\theta}})}{\partial \boldsymbol{\theta}}^{T}\right\} = 0 \tag{12A.21}$$

where we have used the fact that

$$E\left\{\mathrm{vec}(\widehat{\mathbf{R}}_Y^{T})\mathrm{vec}^{T}(\widehat{\mathbf{R}}_Y^{T})\right\} = \mathrm{vec}(\mathbf{R}_Y^{T})\mathrm{vec}^{T}(\mathbf{R}_Y^{T}) \tag{12A.22}$$

The Rao test statistic can be obtained from the special case (12A.6). Plugging (12A.12), (12A.13), and (12A.20) into (12A.6), and after some algebraic manipulations, yields

$$\begin{aligned}
&\Lambda\left(\mathbf{Y};\mathbf{R}_Y\right)\\
&= 2\left[\frac{\partial \mathscr{L}(\mathbf{Y};\mathrm{vec}\{\mathbf{R}_Y\})}{\partial \mathrm{vec}\{\mathbf{R}_Y\}}\right]^{T}\widetilde{\mathbf{I}}^{-1}\left(\mathbf{R}_Y\right)\left[\frac{\partial \mathscr{L}(\mathbf{Y};\mathrm{vec}\{\mathbf{R}_Y\})}{\partial \mathrm{vec}\{\mathbf{R}_Y^{*}\}}\right]\\
&= 2N\mathrm{tr}\left\{\left(\widehat{\mathbf{R}}_Y\mathbf{R}_Y^{-1} - \mathbf{I}\right)\mathbf{R}_Y^{*}\mathbf{R}_Y^{-1}\left(\widehat{\mathbf{R}}_Y\mathbf{R}_Y^{-1} - \mathbf{I}\right)\right\}
\end{aligned} \tag{12A.23}$$

where we have used identity [7]

$$P_{N_r}^{-1}\mathrm{vec}(\mathbf{C}) = \mathrm{vec}(\mathbf{C}) \tag{12A.24}$$

The MLE of $\mathbf{R}_Y$ under hypothesis $\mathscr{H}_0$ may be obtained from

$$\widehat{\boldsymbol{\Sigma}}_0 = \underset{\boldsymbol{\Sigma}}{\mathrm{argmax}}\left\{-NN_r\ln\pi - N\ln\left(|\boldsymbol{\Sigma}|\right) - \mathrm{tr}\left\{\boldsymbol{\Sigma}^{-1}\mathbf{Y}\mathbf{Y}^{H}\right\}\right\} \tag{12A.25}$$

The optimization problem with respect to $\boldsymbol{\Sigma}$ is not convex. However, we can transform it into a convex problem by applying the transformation $\mathbf{E} = g(\boldsymbol{\Sigma}) = \boldsymbol{\Sigma}^{-1}$. With this transformation, we come up with

$$\widehat{\mathbf{E}} = \underset{\mathbf{E}}{\mathrm{argmax}}\left\{\mathscr{L}_0\right\} \tag{12A.26}$$

where

$$\begin{aligned}
\mathscr{L}_0 &= -NN_r\ln\pi + N\ln\left(|\mathbf{E}|\right) - \mathrm{tr}\left\{\mathbf{E}\mathbf{Y}\mathbf{Y}^{H}\right\}\\
&= -NN_r\ln\pi + N\ln\left(\prod_{k=1}^{N_r}[\mathbf{E}]_{kk}\right) - \sum_{k=1}^{N_r}[\mathbf{E}]_{kk}[\mathbf{Y}\mathbf{Y}^{H}]_{kk}
\end{aligned} \tag{12A.27}$$

Now, we must determine which diagonal elements of matrix $\mathbf{E}$ maximize $\mathscr{L}_0$. Taking the gradients of $\mathscr{L}_0$ with respect to the (k, k)th element of the matrix $\mathbf{E}$ yields

$$\frac{\partial \mathcal{L}_0}{\partial [\mathbf{E}]_{kk}} = N \frac{1}{[\mathbf{E}]_{kk}} - [\mathbf{Y}\mathbf{Y}^H]_{kk} \tag{12A.28}$$

Setting the gradient equals to zero, we obtain

$$[\mathbf{E}]_{kk} = \frac{N}{[\mathbf{Y}\mathbf{Y}^H]_{kk}} \tag{12A.29}$$

resulting in

$$\widehat{\boldsymbol{\Sigma}}_0 = \mathrm{Diag}(\widehat{\mathbf{R}}_Y) \tag{12A.30}$$

By substituting (12A.30) into (12A.23) and taking into account the fact that the diagonal elements of $\widehat{\boldsymbol{\Sigma}}_0$ are real, the Rao test statistic may be obtained as

$$\Lambda\left(\mathbf{Y};\mathbf{R}_Y\right) = 2N\mathrm{tr}\left\{(\widehat{\mathbf{R}}_Y\widehat{\boldsymbol{\Sigma}}_0^{-1} - \mathbf{I})^2\right\} \tag{12A.31}$$

where we have used the result of $\mathrm{tr}\left(\mathrm{AB}\right) = \mathrm{vec}^T\left(\mathrm{A}^T\right)\mathrm{vec}\left\{\mathrm{B}\right\}$. Besides, this can be rewritten as

$$\Lambda\left(\mathbf{Y};\mathbf{R}_Y\right) = 2N\mathrm{tr}\left\{\left(\widehat{\boldsymbol{\Sigma}}_0^{\frac{1}{2}}(\widehat{\boldsymbol{\Sigma}}_0^{\frac{-1}{2}}\widehat{\mathbf{R}}_Y\widehat{\boldsymbol{\Sigma}}_0^{\frac{-1}{2}} - \mathbf{I})\widehat{\boldsymbol{\Sigma}}_0^{\frac{-1}{2}}\right)^2\right\} \tag{12A.32}$$

resulting in

$$\Lambda\left(\mathbf{Y};\mathbf{R}_Y\right) = 2N\|\widehat{\mathbf{C}}_X - \mathbf{I}\|_F^2 \tag{12A.33}$$

This completes the proof of Proposition 1.

Appendix 12B

P_1-LRT detector

The test statistic of the P_1-LRT detector can be represented by

$$\Lambda_{p_1}(\mathbf{Y}) = \sup_{\boldsymbol{\Sigma}} \{\mathscr{L}(\mathbf{Y})\} - \sup_{\mathbf{R_Y}} \{\mathscr{L}(\mathbf{Y})\} \tag{12B.1}$$

The first part of the expression for $\Lambda_{p_1}(\mathbf{Y})$, denoted by $\mathscr{L}_{p_1}^{(1)}$, can be written as

$$\mathscr{L}_{p_1}^{(1)} = \sup_{\boldsymbol{\Sigma}} \left\{ -NN_r \ln \pi - N \ln\left(|\boldsymbol{\Sigma}|\right) - N\mathrm{tr}\left\{\boldsymbol{\Sigma}^{-1}\widehat{\mathbf{R}}_{\mathbf{Y}}\right\}\right\} \tag{12B.2}$$

By using (12A.28)–(12A.31), we get

$$\mathscr{L}_{p_1}^{(1)} = -NN_r \ln \pi - N \ln \prod_{k=1}^{N_r} [\widehat{\mathbf{R}}_{\mathbf{Y}}]_{kk} - NN_r \tag{12B.3}$$

The second term of (12B.1), denoted by $\mathscr{L}_{p_1}^{(2)}$, can be written as

$$\mathscr{L}_{p_1}^{(2)} = \sup_{\mathbf{R}_Y} \left\{ -NN_r \ln \pi - N \ln\left(|\mathbf{R}_Y|\right) - N\mathrm{tr}\left\{\mathbf{R}_Y^{-1}\widehat{\mathbf{R}}_{\mathbf{Y}}\right\}\right\} \tag{12B.4}$$

Our task now is to find the matrix $\mathbf{R}_{\mathrm{Y}}$ that can maximize the value of $\mathscr{L}p_1^{(2)}$. To do so, we need to calculate the gradients of the cost function in $\mathscr{L}_{p_1}^{(2)}$ with respect to the matrix $\mathbf{R}_{\mathrm{Y}}$, given by

$$\frac{\partial \mathscr{L}_2}{\partial \mathbf{R}_Y} = -N\mathbf{R}_Y^{-1} + N\mathbf{R}_Y^{-1}\widehat{\mathbf{R}}_{\mathbf{Y}}\mathbf{R}_Y^{-1} \tag{12B.5}$$

where we have used (12A.10) and (12A.11). Setting the gradient equals to zero, the optimum $\mathbf{R}_{\mathrm{Y}}$ is obtained as $\widehat{\mathbf{R}}_Y$, resulting in

$$\mathscr{L}_{p_1}^{(2)} = -NN_r \ln \pi - N \ln(|\widehat{\mathbf{R}}_{\mathbf{Y}}|) - NN_r \tag{12B.6}$$

Substituting (12B.3) and (12B.6) into (12B.1) and after some algebraic manipulations, we obtain

$$\Lambda_{p_1}(\mathbf{Y}) = N \ln\left(\frac{|\widehat{\mathbf{R}}_Y|}{\prod_{k=1}^{N_r} [\widehat{\mathbf{R}}_Y]_{kk}}\right) \tag{12B.7}$$

This completes the proof of Proposition 2.

Appendix 12C
P_2-C2-LRT detector

The test statistic of the P_2-LRT detector may be obtained from

$$\Lambda_{P_2}(\mathbf{Y}) = \sup_{\boldsymbol{\Sigma}}\{\mathscr{L}(\mathbf{Y})\} - \sup_{\sigma_s^2, b, \boldsymbol{\Sigma}}\{\mathscr{L}(\mathbf{Y})\} \tag{12C.1}$$

As before, the first term of (12C.1) is given by (12B.3). The second term of (12C.1), denoted by $\mathscr{L}_{P2}^{(2)}$, can be simplified as

$$\begin{aligned}\mathscr{L}_{P2}^{(2)} = \sup_{g,\boldsymbol{\Sigma}}\{&-NN_r \ln \pi - N\ln(|\mathbf{g}\mathbf{g}^H + \boldsymbol{\Sigma}|)\\ &-N\mathrm{tr}\{(\mathbf{g}\mathbf{g}^H + \boldsymbol{\Sigma})^{-1}\widehat{\mathbf{R}}_{\mathrm{Y}}\}\}\end{aligned} \tag{12C.2}$$

where we have defined $\mathbf{g} = \sigma_s^2\mathbf{b}$. Using

$$|\mathbf{g}\mathbf{g}^H + \boldsymbol{\Sigma}| = \frac{\left(1 + \mathbf{g}^H\boldsymbol{\Sigma}^{-1}\mathbf{g}\right)}{|\boldsymbol{\Sigma}^{-1}|} \tag{12C.3}$$

$$(\mathbf{g}\mathbf{g}^H + \boldsymbol{\Sigma})^{-1} = \boldsymbol{\Sigma}^{-1} - \frac{\left(\boldsymbol{\Sigma}^{-1}\mathbf{g}\right)\left(\boldsymbol{\Sigma}^{-1}\mathbf{g}\right)^H}{1 + \mathbf{g}^H\boldsymbol{\Sigma}^{-1}\mathbf{g}} \tag{12C.4}$$

and after some algebraic manipulations, $\mathscr{L}_{P2}^{(2)}$ can be expressed as

$$\begin{aligned}\mathscr{L}_{P2}^{(2)} = \sup_{g,\boldsymbol{\Sigma}}\Bigg\{&-NN_r\ln\pi - N\ln(1 + \mathbf{g}^H\boldsymbol{\Sigma}^{-1}\mathbf{g})\\ &+N\ln|\boldsymbol{\Sigma}^{-1}| - N\mathrm{tr}\{\boldsymbol{\Sigma}^{-1}\widehat{\mathbf{R}}_{\mathrm{Y}}\} + N\frac{\mathbf{g}^H\boldsymbol{\Sigma}^{-1}\widehat{\mathbf{R}}_{\mathrm{Y}}\boldsymbol{\Sigma}^{-1}\mathbf{g}}{1 + \mathbf{g}^H\boldsymbol{\Sigma}^{-1}\mathbf{g}}\Bigg\}\end{aligned} \tag{12C.5}$$

To take advantage of the fact that

$$\sup_{\mathbf{t}\neq 0}\frac{\mathbf{t}^H\mathbf{A}t}{\mathbf{t}^H\mathbf{B}t} = \lambda_1\left(\mathbf{B}^{-1}\mathbf{A}\right) \tag{12C.6}$$

we need to redefine the unknown parameters such that

$$\rho \triangleq 1 + \mathbf{g}^H\widehat{\mathbf{R}}_{\mathrm{Y}}^{-1}\mathbf{g} \tag{12C.7}$$

$$\mathbf{E} \triangleq \widehat{\mathbf{R}}_{\mathrm{Y}}^{-1} \tag{12C.8}$$

$$\mathbf{t} \triangleq \frac{\mathbf{E}\mathbf{g}}{\sqrt{\mathbf{g}^H\mathbf{E}\mathbf{g}}} \tag{12C.9}$$

Now, (12C.5) can be re-written as

$$\mathcal{L}_{p2}^{(2)} = \sup_{\mathbf{E}}\left\{\sup_{\rho}\left\{-NN_r\ln\pi - N\ln\rho + N\ln|\mathbf{E}| - N\mathrm{tr}\left\{\mathbf{E}\widehat{\mathbf{R}}_{\mathrm{Y}}\right\} + N\tfrac{\rho-1}{\rho}\sup_{\mathbf{t}}\left\{\tfrac{\mathbf{t}^H\mathbf{S}\mathbf{t}}{\mathbf{t}^H\mathbf{E}^{-1}\mathbf{t}}\right\}\right\}\right\}$$
$$= \sup_{\mathbf{E}}\left\{\sup_{\rho}\left\{-NN_r\ln\pi - N\ln\rho + Nln|\mathbf{E}| - N\mathrm{tr}\{\mathbf{ES}\} + N\tfrac{\rho-1}{\rho}\lambda_1(\mathbf{ES})\right\}\right\}$$

Next, the value of ρ to maximize (12C.10) can be obtained from

$$\frac{-N}{\rho} + N\frac{\rho-(\rho-1)}{\rho^2}\lambda_1(\mathbf{E}\widehat{\mathbf{R}}_{\mathrm{Y}}) = 0 \tag{12C.10}$$

resulting in $\rho = \lambda_1(\mathbf{E}\widehat{\mathbf{R}}_{\mathbf{Y}})$. Accordingly, the second term of (12C.10) can be rewritten as

$$\mathcal{L}_{p2}^{(2)} = -NN_r\ln\pi - N\ln|\widehat{\mathbf{R}}_{\mathrm{Y}}| - N + N\mathcal{G}_{p2} \tag{12C.11}$$

where $\mathcal{G}_{p2}$ is defined as

$$\begin{aligned}\mathcal{G}_{p2} &= \sup_{E}\left\{\ln\left(\frac{\prod_{k=1}^{N_r}\lambda_k\left(\mathbf{E}\widehat{\mathbf{R}}_{\mathrm{Y}}\right)}{\lambda_1\left(\mathbf{E}\widehat{\mathbf{R}}_{\mathrm{Y}}\right)}\right) - \left(\sum_{k=1}^{N_r}\lambda_k\left(\mathbf{E}\widehat{\mathbf{R}}_{\mathrm{Y}}\right) - \lambda_1\left(\mathbf{E}\widehat{\mathbf{R}}_{\mathrm{Y}}\right)\right)\right\}\\ &= \sup_{\boldsymbol{\Sigma}}\left\{\ln\left(\frac{\left|\boldsymbol{\Sigma}^{-1}\widehat{\mathbf{R}}_{\mathrm{Y}}\right|}{\lambda_1\left(\boldsymbol{\Sigma}^{-1}\widehat{\mathbf{R}}_{\mathrm{Y}}\right)}\right) - \left(\mathrm{tr}\left(\boldsymbol{\Sigma}^{-1}\widehat{\mathbf{R}}_{\mathrm{Y}}\right) - \lambda_1\left(\boldsymbol{\Sigma}^{-1}\widehat{\mathbf{R}}_{\mathrm{Y}}\right)\right)\right\}\\ &= -\inf_{\boldsymbol{\Sigma}}\left\{\sum_{k=2}^{N_r}\left[\lambda_k\left(\boldsymbol{\Sigma}^{-1}\widehat{\mathbf{R}}_{\mathrm{Y}}\right) - \ln\left(\lambda_k\left(\boldsymbol{\Sigma}^{-1}\widehat{\mathbf{R}}_{\mathrm{Y}}\right)\right) - 1\right]\right\}\\ &\quad -(N_r-1)\end{aligned} \tag{12C.12}$$

In general, this problem cannot be solved, however, for the special case of $N_r = 2$, we have

$$\begin{aligned}\mathcal{G}_{p2} &= \inf_{\boldsymbol{\Sigma}}\left\{\left[\lambda_2(\boldsymbol{\Sigma}^{-1}\widehat{\mathbf{R}}_{\mathrm{Y}}) - \ln\left(\lambda_2(\boldsymbol{\Sigma}^{-1}\widehat{\mathbf{R}}_{\mathrm{Y}}\right) - 1\right]\right\}\\ &\quad -(N_r-1)\end{aligned} \tag{12C.13}$$

It is easy to show that

$$\inf_{\boldsymbol{\Sigma}}\left\{\left[\lambda_2(\boldsymbol{\Sigma}^{-1}\widehat{\mathbf{R}}_{\mathrm{Y}}) - \ln\left(\lambda_2(\boldsymbol{\Sigma}^{-1}\widehat{\mathbf{R}}_{\mathrm{Y}})\right) - 1\right]\right\} = 0 \tag{12C.14}$$

In this special case (i.e., $N_r = 2$), the LRT test can be further derived as

$$NN_r\left(1 + \ln|\widehat{\mathbf{C}}_Y|^{-\frac{1}{N_r}}\right) \underset{H_0}{\overset{H_1}{\gtrless}} \eta_{p_2'} \tag{12C.15}$$

or, equivalently,

$$|\widehat{\mathbf{C}}_Y|^{-1} \underset{H_0}{\overset{H_1}{\gtrless}} \eta_{p2} \tag{12C.16}$$

where η_{p_2} and $\eta_{p_2'}$ are detection thresholds to be adjusted according to a desired false alarm probability.

The LRT detector can be approximated for the general case of $N_r \geq 2$. To this end, we consider a scenario with low-SNR where the SCM $\widehat{\mathbf{R}}_{\mathbf{Y}}$ becomes nearly diagonal. As a result, the maximum value of (12C.12) occurs when $\widehat{\mathbf{\Sigma}} = \text{Diag}(\widehat{\mathbf{R}}_{\mathbf{Y}})$. Substituting this into (12C.11) and constructing the LRT test statistic of (12C.1) yield

$$\begin{aligned}\Lambda_{p_2}(\mathbf{Y}) &= N(N_r - 1) \\ &+ N\left\{\ln\left(\frac{|\widehat{\mathbf{\Sigma}}^{-1}\widehat{\mathbf{R}}_{\mathbf{Y}}|}{\lambda_1(\widehat{\mathbf{\Sigma}}^{-1}\widehat{\mathbf{R}}_{\mathbf{Y}})}\right) - \left(\text{tr}(\widehat{\mathbf{\Sigma}}^{-1}\widehat{\mathbf{R}}_{\mathbf{Y}}) - \lambda_1(\widehat{\mathbf{\Sigma}}^{-1}\widehat{\mathbf{R}}_{\mathbf{Y}})\right)\right\} \\ &- N\ln|\widehat{\mathbf{C}}_Y|\end{aligned} \tag{12C.17}$$

It is easy to show that

$$|\widehat{\mathbf{\Sigma}}^{-1}\widehat{\mathbf{R}}_{\mathbf{Y}} - \lambda\mathbf{I}| = |\widehat{\mathbf{C}}_Y - \lambda\mathbf{I}| \tag{12C.18}$$

This means that the eigenvalues of $\widehat{\mathbf{\Sigma}}^{-1}\widehat{\mathbf{R}}_{\mathbf{Y}}$ and $\widehat{\mathbf{C}}_Y$ are the same. Utilizing this and after carrying out some algebraic manipulations, we arrive at

$$\Lambda(\mathbf{Y}) = N\left(\lambda_1(\widehat{\mathbf{C}}_Y) - \ln(\lambda_1(\widehat{\mathbf{C}}_Y) - 1)\right) \tag{12C.19}$$

The proof is concluded by noting that the low-SNR approximated LRT's test statistic is equal to $\lambda_1(\widehat{\mathbf{C}}_Y)$ since $\Lambda(\mathbf{Y})$ is monotonically increasing with respect to $\lambda_1(\widehat{\mathbf{C}}_Y) \geq 1$.

Appendix 12D
P_3-LRT detector

The test statistic of the P_3-LRT detector may be obtained from

$$\Lambda_{p_3}(\mathbf{X}) = \sup_{\mathbf{\Sigma}}\{\mathscr{L}(\mathbf{X})\} - \sup_{\mathbf{s},\mathbf{b},\mathbf{\Sigma}}\{\mathscr{L}(\mathbf{X})\} \tag{12D.1}$$

The first term on the righthand side of (12D.1), denoted by $\mathscr{L}_{p_3}^{(1)}$, can be given by

$$\mathscr{L}_{p_3}^{(1)} = \sup_{\mathbf{\Sigma}}\left\{ - NN_r \ln \pi - N \ln\left(\prod_{k=1}^{N_r} \sigma_k^2\right) - \sum_{k=1}^{N_r} \frac{1}{\sigma_k^2}[\mathbf{X}^H\mathbf{X}]_{kk}\right\} \tag{12D.2}$$

Similar to (12D.3), we can obtain

$$\mathscr{L}_{p_3}^{(1)} = -NN_r \ln \pi - N \ln \prod_{k=1}^{N_r} \frac{[\mathbf{X}^H\mathbf{X}]_{kk}}{N} - NN_r \tag{12D.3}$$

or

$$\mathscr{L}_{p_3}^{(1)} = -NN_r \ln \pi - N \ln \prod_{k=1}^{N_r} \frac{\|\mathbf{x}_k\|^2}{N} - NN_r \tag{12D.4}$$

The second term on the righthand side of (12D.1), denoted by $\mathscr{L}_{p_3}^{(2)}$, can be given by

$$\mathscr{L}_{p_3}^{(2)} = \sup_{\mathbf{s}}\left\{ \sup_{\{\sigma_k^2\}_{k=1}^{N_r}} \left\{ \sup_{\{\beta_k\}_{k=1}^{N_r}} \left\{ -NN_r \ln \pi - N \ln\left(\prod_{k=1}^{N_r} \sigma_k^2\right) - \sum_{k=1}^{N_r} \frac{1}{\sigma_k^2}\|\mathbf{x}_k - \mathbf{s}\beta_k\|^2 \right\}\right\}\right\} \tag{12D.5}$$

The supremum with respect to the unknown parameters $\{\beta_k\}_{k=1}^{N_r}$ is further obtained as

$$\mathscr{L}_\beta = -NN_r \ln \pi - N \ln\left(\prod_{k=1}^{N_r} \sigma_k^2\right) - \sum_{k=1}^{N_r} \frac{1}{\sigma_k^2}\left\|\mathbf{x}_k - \mathbf{s}\left(\frac{\mathbf{s}^H\mathbf{x}_k}{\|\mathbf{s}\|^2}\right)\right\|^2 \tag{12D.6}$$

This can be further simplified as

$$\mathscr{L}_\beta = -NN_r \ln \pi - N \ln\left(\prod_{k=1}^{N_r} \sigma_k^2\right) - \sum_{k=1}^{N_r} \frac{\|\mathbf{\Pi}_{\mathbf{s}}^{\perp}\mathbf{x}_k\|^2}{\sigma_k^2} \tag{12D.7}$$

where $\mathbf{\Pi}_{\mathbf{s}}^{\perp}$ is the projection matrix into the orthogonal complement of vector $\mathbf{s}$, given by

$$\mathbf{\Pi}_{\mathbf{s}}^{\perp} = \mathbf{I} - \frac{\mathbf{s}\mathbf{s}^H}{\|\mathbf{s}\|^2} \tag{12D.8}$$

Then, the supremum with respect to the unknowns $\{\sigma_k^2\}_{k=1}^{N_r}$ can be computed from

$$\begin{aligned}\mathscr{L}_\sigma &= \sup_{\{\sigma_k^2\}_{k=1}^{N_r}} \left\{ -NN_r \ln \pi - N \ln \left(\prod_{k=1}^{N_r} \sigma_k^2 \right) - \sum_{k=1}^{N_r} \frac{\left\| \mathbf{\Pi}_{\mathbf{s}}^{\perp} \mathbf{x}_k \right\|^2}{\sigma_k^2} \right\} \\ &= -NN_r \ln \pi - N \ln \left(\prod_{k=1}^{N_r} \frac{\left\| \mathbf{\Pi}_{\mathbf{s}}^{\perp} \mathbf{x}_k \right\|^2}{N} \right) - \mathrm{NN_r}\end{aligned} \tag{12D.9}$$

The point at which $\mathscr{L}_\sigma$ attains its maximum is

$$\widehat{\sigma_k^2} = \tfrac{1}{N} \left\| \mathbf{\Pi}_{\mathbf{s}}^{\perp} \mathbf{x}_k \right\|^2 \tag{12D.10}$$

Before proceeding, let us define

$$\mathbf{g} \triangleq \frac{\mathbf{s}}{\|\mathbf{s}\|} \tag{12D.11}$$

where $\|\mathbf{g}\| = 1$. Then, the supremum with respect to the unknown vector $\mathbf{g}$ can be written as

$$\begin{aligned}\mathscr{L}_g &= \sup_g \left\{ -NN_r \ln \pi - N \ln \left(\prod_{k=1}^{N_r} \frac{\mathrm{tr}\left(\left(\mathbf{I} - \mathbf{g}\mathbf{g}^H \right) \mathbf{x}_k \mathbf{x}_k^H \right)}{N} \right) - NN_r \right\} r \\ &= -NN_r \ln \pi - N \ln \prod_{k=1}^{N_r} \frac{\|\mathbf{x}_k\|^2}{N} - N \inf_{\mathrm{g}} \left\{ \ln \left(\prod_{k=1}^{N_r} \left(1 - \frac{\left| \mathbf{g}^H \mathbf{x}_k \right|^2}{\|\mathbf{x}_k\|^2} \right) \right) \right\} - NN_r \\ &= -NN_r \ln \pi - N \ln \prod_{k=1}^{N_r} \frac{\|\mathbf{x}_k\|^2}{N} - N \inf_{\mathrm{g}} \left\{ \sum_{k=1}^{N_r} \ln \left(1 - \frac{\left| \mathbf{g}^H \mathbf{x}_k \right|^2}{\|\mathbf{x}_k\|^2} \right) \right\} - NN_r\end{aligned} \tag{12D.12}$$

Our objective is to find the vector $\mathbf{g}$ that minimizes $\sum_{k=1}^{N_r} \ln \left(1 - \frac{\left| \mathbf{g}^H \mathbf{x}_k \right|^2}{\left\| \mathbf{x}_k \right|^2} \right)$. By applying the Cauchy–Schwartz inequality, we find that

$$\frac{\left| \mathbf{g}^H \mathbf{x}_k \right|^2}{\|\mathbf{x}_k\|^2} \leq \frac{\|\mathbf{g}\|^2 \|\mathbf{x}_k\|^2}{\|\mathbf{x}_k\|^2} = 1 \tag{12D.13}$$

The upper bound is attained if and only if $\mathbf{g} \triangleq \zeta \frac{\mathbf{x}_k}{\|\mathbf{x}_k\|}$, which is not reachable, in general, for every values of k, implying that

$$0 < \prod_{k=1}^{N_r} \left(1 - \frac{\left| \mathbf{g}^H \mathbf{x}_k \right|^2}{\|\mathbf{x}_k\|^2} \right) \leq 1 \tag{12D.14}$$

Consequently, we find

$$-\infty < \sum_{k=1}^{N_r} \ln \left(1 - \frac{\left| \mathbf{g}^H \mathbf{x}_k \right|^2}{\|\mathbf{x}_k\|^2} \right) \leq 0 \tag{12D.15}$$

This means that

$$\inf_{\mathbf{g}} \left\{ \sum_{k=1}^{N_r} \ln \left(1 - \frac{\left|\mathbf{g}^H \mathbf{x}_k\right|^2}{\|\mathbf{x}_k\|^2} \right) \right\} \to -\infty \tag{12D.16}$$

Thus the minimum of $\sum_{k=1}^{N_r} \ln \left(1 - \frac{\left|\mathbf{g}^H \mathbf{x}_k\right|^2}{\|\mathbf{x}_k\|^2} \right)$ is a finite but a large negative value. Using

$$\ln\left(1 + z_k\right) = z - \frac{z_k^2}{2} + \frac{z_k^3}{3} - \frac{z_k^4}{4} + \ldots, \qquad -1 < |z_k| \leq 1 \tag{12D.17}$$

with $1 < z_k = -\frac{\left|\mathbf{g}^H \mathbf{x}_k\right|^2}{\|\mathbf{x}_k\|^2} < 0$, resulting in $\ln\left(1 + z_k\right) < 0$, helps us to approximate

$$\begin{aligned} &\inf_{\mathbf{g}} \left\{ \sum_{k=1}^{N_r} \ln \left(1 - \frac{\left|\mathbf{g}^H \mathbf{x}_k\right|^2}{\|\mathbf{x}_k\|^2} \right) \right\} \\ &\approx \rho \inf_{\mathbf{g}} \left\{ \sum_{k=1}^{N_r} - \frac{\left|\mathbf{g}^H \mathbf{x}_k\right|^2}{\|\mathbf{x}_k\|^2} \right\} \\ &= \rho \ \sup_{\mathbf{g}} \left\{ \mathbf{g}^H \left(\sum_{k=1}^{N_r} \frac{\mathbf{x}_k \mathbf{x}_k^H}{\|\mathbf{x}_k\|^2} \right) \mathbf{g} \right\} \\ &= \rho \ \lambda_1 \left(\sum_{k=1}^{N_r} \frac{\mathbf{x}_k \mathbf{x}_k^H}{\|\mathbf{x}_k\|^2} \right) \\ &= \rho \ \lambda_1 \left(\frac{1}{N} \mathbf{X} \widehat{\mathbf{D}}_{\mathbf{X}}^{-1} \mathbf{X}^H \right) \end{aligned} \tag{12D.18}$$

Observing that the LRT metric is influenced by the minimum $\sum_{k=1}^{N_r} \ln \left(1 - \frac{\left|\mathbf{g}^H \mathbf{x}_k\right|^2}{\|\mathbf{x}_k\|^2} \right)$ value rather than the corresponding $\mathbf{g}$ value that minimizes it. To remedy this approximation, we use a positive large value ρ. Finally, the closed-form expression for $\mathscr{L}_{P3}^{(2)}$ can be obtained as

$$\mathscr{L}_{P3}^{(2)} = -NN_r \ln \pi - N \ln \prod_{k=1}^{N_r} \frac{\|\mathbf{x}_k\|^2}{N} + \rho\lambda_1 \left(\frac{1}{N} \mathbf{X} \widehat{\mathbf{D}}_{\mathbf{X}}^{-1} \mathbf{X}^H \right) - NN_r \tag{12D.19}$$

Plugging (12D.19) and (12D.4) into (12D.1), we obtain the test statistic of the P_3-LRT detector, given by

$$\Lambda_{P3}(\mathbf{X}) = \rho\lambda_1 \left(\frac{1}{N} \mathbf{X} \widehat{\mathbf{D}}_{\mathbf{X}}^{-1} \mathbf{X}^H \right) = \rho \|\widehat{\mathrm{C}}_X\|_2 \tag{12D.20}$$

This completes the proof of Proposition 3.

It should be emphasized that in Reference 4, this detection problem has been addressed in situations where the SNR is low, resulting in identical forms. Despite sharing the same form in their assessments, there are notable distinctions in their MLEs. For instance, to demonstrate this, the authors of Reference 4 have made an approximation of the MLE obtained in (12D.10) by

$$\hat{\sigma}_k^2 \approx \frac{1}{N}\|\mathbf{x}_k\|^2 \tag{12D.21}$$

On the other hand, we utilize exact MLE formulas for $\hat{\sigma}_k^2$ as stated in (12D.10). Then, we approximate $\ln(1+z_k)$ by ρz_k, where ρ is a very large positive value. Ultimately, we use $\inf_{\mathbf{g}}\left\{\sum_{k=1}^{N_r}\ln(1+z_k)\right\} \cong \rho \inf_{\mathbf{g}}\left\{\sum_{k=1}^{N_r} z_k\right\}$. Our findings demonstrate that the minimum of $\inf_{\mathbf{g}}\left\{\sum_{k=1}^{N_r} z_k\right\}$ is not negative infinity but rather an extremely large negative value. In order to obtain the precise value of $\inf_{\mathbf{g}}\left\{\sum_{k=1}^{N_r}\ln(1+z_k)\right\}$, which is necessary in the P_3-LRT statistic and may not be achieved by $\inf_{\mathbf{g}}\left\{\sum_{k=1}^{N_r} z_k\right\}$, we can increase ρ to a positive value. It is worth noting that (12D.20) reveals an interesting fact that the decision threshold of the P_2-LRT detector can absorb the variable ρ, making it unnecessary in the proposed detector. This supports our goal of developing a more informative expression for the GLRT detector, resulting in an LRT-based detector that can be considered a precise solution for the detection problem P_3.

Chapter 13

MIMO passive radar target detection with direct-path interference

13.1 Introduction

In this chapter, we explore the problem of detecting targets in passive MIMO radar with calibrated or uncalibrated receivers, and direct-path interference. Our setup consists of multiple transmitters that do not overlap in frequency but share a common bandwidth and multiple receivers that are spatially separated and do not use reference channels. When the receivers are calibrated, we assume that they all have the same level of noise variance, but for uncalibrated receivers, the noise variance can vary across both frequency and space. To address the target detection problem in the presence of direct-path interference, we formulate it as a composite hypothesis testing problem. The liklihood ratio test (LRT) test, as described in references 1 and 2, lacks a closed-form expression. Consequently, we employ the Rao test as an alternative approach to bypass the need for calculating the Maximum Likelihood Estimation (MLE) of the unknown parameters under the alternative hypothesis. This test only requires the MLE to be calculated under the null hypothesis and involves the computation of the Fisher information matrix (FIM), enabling closed-form Rao statistics for both calibrated and uncalibrated problems. To demonstrate the effectiveness of the proposed Rao detectors, Monte-Carlo (MC) simulations are provided. The material of this chapter is based on our previous work published in Reference 3. This method belongs to the Ca.2 category of passive radar techniques for detecting targets.

The rest of this chapter is structured in the following way: section 13.2 focuses on the signal model and problem formulation of a multistatic passive radar with uncalibrated receivers and direct-path interference. The closed-form Rao test for both uncalibrated and calibrated scenarios is explained thoroughly in section 13.3. We present and analyze our MC simulation results in section 13.4. Finally, section 13.5 provides summarizing remarks for this chapter.

13.2 Detection problem formulation

The task at hand is to detect a target in an MIMO passive radar system, which consists of multiple transmitters and receivers as depicted in Figure 13.1. The system

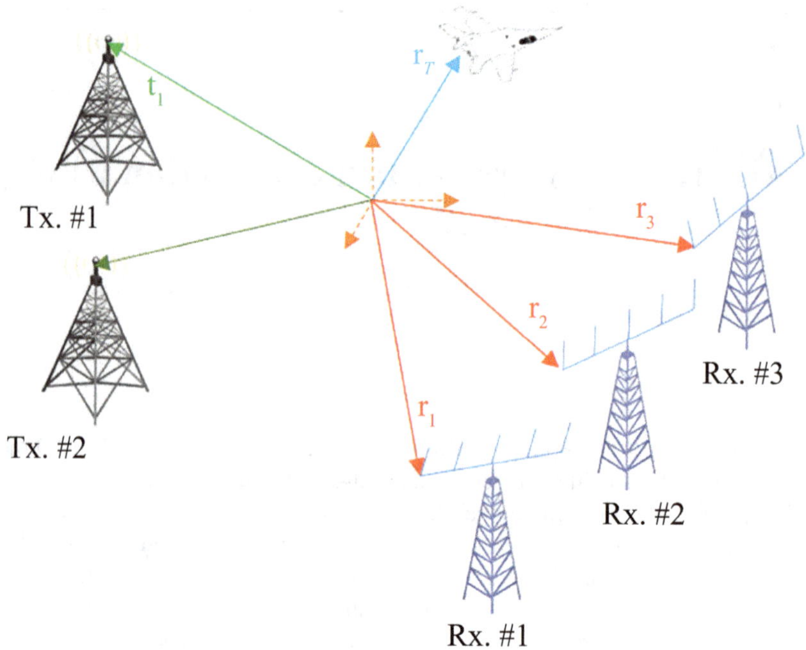

Figure 13.1 *A conceptual MIMO radar system operates in a passive mode with three receivers (Rx) located at positions* $\mathbf{r}_1$, $\mathbf{r}_2$, *and* $\mathbf{r}_3$, *and two transmitters (Tx) positioned at* $\mathbf{t}_1$ *and* $\mathbf{t}_2$. *The aim of the system is to detect a target present at the location* $\mathbf{r}_T$.

comprises N_t transmitters with respective locations denoted by $\{\mathbf{t}_j\}_{j=1}^{N_t}$ and a scene that contains a target with position–velocity coordinates of $(\mathbf{r}_T, \mathbf{v}_T)$. The positions of the N_r receivers are denoted by $\mathbf{r}_k$, where $k = 1, ..., N_r$. The assumption is that all the transmitters and receivers are stationary and their positions are known. The direct-path and the target-path propagation delays can be given by

$$d_{jk} = \frac{\|\mathbf{r}_k - \mathbf{t}_j\|}{c} \tag{13.1}$$

$$p_{jk} = \frac{\|\mathbf{r}_T - \mathbf{t}_j\| + \|\mathbf{r}_T - \mathbf{r}_k\|}{c} \tag{13.2}$$

where c is the speed of light. Similarly, target-path Doppler shifts in hertz can be obtained as

$$f_{jk} = \frac{\mathbf{v}_T^T \left(\frac{\mathbf{r}_T - \mathbf{t}_j}{\|\mathbf{r}_T - \mathbf{t}_j\|} + \frac{\mathbf{r}_T - \mathbf{r}_k}{\|\mathbf{r}_T - \mathbf{r}_k\|} \right)}{\lambda_j} \tag{13.3}$$

The jth transmitter wavelength in meters is given by $\lambda_j = \frac{c}{f_j^{(c)}}$, where $f_j^{(c)}$ is the jth transmitter carrier frequency in hertz. It is assumed that the transmitted signals do not overlap in frequency and have a common bandwidth of B hertz. The received signal at each receiver is downconverted to baseband and then sampled at a rate of f_s. This results in sampled signals of length L, where $L = Tf_s$ and T is the integration time. Let $\mathbf{x}_{jk}$ denotes the sampled complex baseband signal collected by the kth receiver on the jth frequency channel, expressed as

$$\mathbf{x}_{jk} = \beta_{jk}\,\mathbf{D}_{d_{jk}}(0)\mathbf{s}_j + \alpha_{jk}\,\mathbf{D}_{p_{jk}}(f_{jk})\mathbf{s}_j + \mathbf{n}_{jk} \tag{13.4}$$

where

- $\mathbf{s}_j \in \mathcal{C}^{L\times 1}$ is the time domain and unknown signal vector transmitted by the jth transmitter,
- β_{jk} is a complex scaling coefficient, including the antenna scaling factor and the channel propagation effects from the jth transmitter to the kth receiver,
- $\mathbf{D}_\tau(f) \in \mathcal{C}^{L\times L}$ is a unitary linear operator to perform time delay τ and Doppler shift f, i.e., $\mathbf{D}_\tau(f)^H\mathbf{D}_\tau(f) = \mathbf{D}_\tau(f)\mathbf{D}_\tau(f)^H = \mathbf{I}_L$, represented by

$$\mathbf{D}_\tau(f) = \mathbf{B}\left(\frac{f}{f_s}\right)\mathbf{F}^H\mathbf{B}(-\tau\Delta f)\mathbf{F} \tag{13.5}$$

  where $\mathbf{B}(a) \in \mathcal{C}^{L\times L}$ is a diagonal matrix with diagonal elements $[\mathbf{B}(a)]_{ii} = e^{-j2\pi(i-1)a}$ for $i = 1, ..., N$; $\mathbf{F} \in \mathcal{C}^{L\times L}$ is the N-point discrete Fourier transform (DFT) matrix with elements $[\mathbf{F}]_{i,l} = e^{-j2\pi(i-1)\frac{\Delta f}{f_S}(l-1)}/\sqrt{N}$, $i, l = 1, ..., N$, where $\Delta f = \frac{f_S}{N}$ is the frequency bin of N-point DFT [4]. In (13.4), e.g., $\mathbf{D}_{p_{jk}}(f_{jk})$ denotes the unitary linear operator with time delay p_{jk} and Doppler shift f_{jk}.
- α_{jk} is a complex scaling coefficient accounting for the antenna gain factor, the channel propagation effects in the target path from the jth transmitter to the kth receiver, and the target reflectivity.
- $\mathbf{n}_{jk} \in \mathcal{C}^{L\times 1}$ is formed by the thermal noise samples of the jth frequency channel at the kth receiver, modeled as zero-mean white Gaussian noise with unknown variance σ_{jk}^2. The unknown noise variance σ_{jk}^2 can in practice differ in each frequency/transmitter channel for $k = 1, \ldots, N_r$ and $j = 1, \ldots, N_t$, then $\sigma_{jk}^2 \neq \sigma_{j'k'}^2$ for $k \neq k'$ and $j \neq j'$. Thus, we consider fully uncalibrated receivers.

Let the observation from N_r receivers due to jth transmitter be vectorized as $\mathbf{x}_j = [\mathbf{x}_{j1}^T, \ldots, \mathbf{x}_{jN_r}^T]^T$ and define $\boldsymbol{\alpha}_j = [\boldsymbol{\alpha}_{j1}, \ldots, \boldsymbol{\alpha}_{jN_r}]^T$, $\boldsymbol{\beta}_j = [\boldsymbol{\beta}_{j1}, \ldots, \boldsymbol{\beta}_{jN_r}]^T$, $\mathbf{n}_j = [\mathbf{n}_{j1}^T, \ldots, \mathbf{n}_{jN_r}^T]^T$, $\mathcal{D}_{d_j} = \mathrm{diag}\{\mathbf{D}_{d_{j1}}, \ldots, \mathbf{D}_{d_{jN_r}}\}$, and $\mathcal{D}_{p_j} = \mathrm{diag} = \{\mathbf{D}_{p_{j1}}, \ldots, \mathbf{D}_{p_{jN_r}}\}$. Now, we can rewrite (13.4) compactly as

$$\mathbf{x}_j = \mathcal{D}_{d_j}(\boldsymbol{\beta}_j \otimes \mathbf{s}_j) + \mathcal{D}_{p_j}(\boldsymbol{\alpha}_j \otimes \mathbf{s}_j) + \mathbf{n}_j \tag{13.6}$$

13.3 Detectors design

Our focus is on conducting hypothesis testing to determine whether a target is present within a hypothesized location-velocity represented by $(\mathbf{r}_H, \mathbf{v}_H)$, using the signal model described in (13.4). This can be expressed as a binary hypothesis test, where we compare the null and alternative hypotheses $\mathcal{H}_0$ and $\mathcal{H}_1$, respectively, i.e.,

$$\begin{cases} \mathcal{H}_0\colon \mathbf{x}_j = \mathcal{D}_{dj}(\boldsymbol{\beta}_j \otimes \mathbf{s}_j) + \mathbf{n}_j \\ \mathcal{H}_1\colon \mathbf{x}_j = \mathcal{D}_{dj}(\boldsymbol{\beta}_j \otimes \mathbf{s}_j) + \mathcal{D}_{Pj}(\boldsymbol{\alpha}_j \otimes \mathbf{s}_j) + \mathbf{n}_j \end{cases} \tag{13.7}$$

We can divide the vector of unknown parameters $\boldsymbol{\theta}$ into two parts as $\boldsymbol{\theta} = [\boldsymbol{\theta}_r^T, \boldsymbol{\theta}_s^T]^T$. The relevant parameter vector $\boldsymbol{\theta}_r$ is represented by $\boldsymbol{\alpha} \in \mathcal{C}^{N_rN_t\times 1}$, where $\boldsymbol{\alpha} = [\boldsymbol{\alpha}_1^T, ..., \boldsymbol{\alpha}_{N_t}^T]^T$. The nuisance parameter vector is denoted as $\boldsymbol{\theta}_s = [\boldsymbol{\beta}^T, \mathbf{s}^T, \sigma_1^T, ..., \sigma_{N_t}^T]^T \in \mathcal{C}^{(N_r+L+N_r)N_t\times 1}$, where $\boldsymbol{\beta} = [\boldsymbol{\beta}_1^T, ..., \boldsymbol{\beta}_{N_t}^T]^T$, $\mathbf{s} = [\mathbf{s}_1^T, ..., \mathbf{s}_{N_t}^T]^T$, and $\boldsymbol{\sigma}_j = [\sigma_{j1}^T, ..., \sigma_{jN_r}^T]^T$. By doing this, we can express the hypothesis testing problem in (13.7) as

$$\begin{cases} \mathcal{H}_0 : \boldsymbol{\alpha} = \mathbf{0}_{N_tN_r}, & \boldsymbol{\theta}_s \\ \mathcal{H}_1 : \boldsymbol{\alpha} \neq \mathbf{0}_{N_tN_r}, & \boldsymbol{\theta}_s \end{cases} \tag{13.8}$$

In order to distinguish between the two alternatives, there are various decision criteria that can be utilized, such as LRT, Rao test, and Wald test [5–7]. We have opted to use Rao test, which yields decision statistics in closed forms for calibrated and uncalibrated receivers.

Proposition 1. *In the uncalibrated case, the Rao test statistic can be succinctly represented as*

$$\Lambda_{\text{Unc}}(\mathbf{x}) = 2\sum_{j=1}^{N_t} \|\mathbf{g}_j\|^2 - 2\mathbf{g}^H\mathbf{R}(\mathbf{K}+\mathbf{R}^H\mathbf{R})^{-1}\mathbf{R}^H\mathbf{g} \tag{13.9}$$

where

$$\mathbf{g} = \begin{bmatrix} \mathbf{g}_1 \\ \mathbf{g}_2 \\ \vdots \\ \mathbf{g}_{N_t} \end{bmatrix} \triangleq \mathbf{E}^{-\frac{1}{2}} \begin{bmatrix} \mathbf{u}_1 \\ \mathbf{u}_2 \\ \vdots \\ \mathbf{u}_{N_t} \end{bmatrix} \tag{13.10}$$

$$\mathbf{E} = \text{Diag}\left(\mathbf{E}^{(1)}, \ldots, \mathbf{E}^{(N_t)}\right) \tag{13.11}$$

$$\mathbf{E}^{(j)} = \|\hat{\mathbf{s}}_j\|^2 \text{diag}\left(\frac{1}{\widehat{\sigma_{j1}^2}}, \ldots, \frac{1}{\widehat{\sigma_{jN_r}^2}}\right) \tag{13.12}$$

$$\mathbf{u}_j \triangleq \text{diag}\left(\frac{1}{\widehat{\sigma_{j1}^2}}, \ldots, \frac{1}{\widehat{\sigma_{jN_r}^2}}\right) \begin{bmatrix} \left(\mathbf{D}_{Pj1}\hat{\mathbf{s}}_j\right)^H \boldsymbol{\Pi}^{\perp}_{\mathbf{D}_{dj1}\hat{\mathbf{s}}_j} \mathbf{x}_{j1} \\ \vdots \\ \left(\mathbf{D}_{PjN_r}\hat{\mathbf{s}}_j\right)^H \boldsymbol{\Pi}^{\perp}_{\mathbf{D}_{djN_r}\hat{\mathbf{s}}_j} \mathbf{x}_{jN_r} \end{bmatrix} \tag{13.13}$$

$$\mathbf{R}=\mathbf{E}^{-\frac{1}{2}}\left[\begin{array}{ll}\mathbf{F} & \mathbf{G}\end{array}\right] \tag{13.14}$$

$$\mathbf{F}=\operatorname{Diag}\left(\mathbf{F}^{(1)}, \ldots, \mathbf{F}^{(N_t)}\right) \tag{13.15}$$

$$\mathbf{F}^{(j)}=\operatorname{diag}\left(\frac{1}{\widehat{\sigma}_{j1}^{2}}, \ldots, \frac{1}{\widehat{\sigma}_{jNr}^{2}}\right) \operatorname{diag}\left(\hat{\mathbf{s}}_{j}^{H} \mathbf{D}_{p_{j1}}^{H} \mathbf{D}_{d_{j1}} \hat{\mathbf{s}}_{j}, \ldots, \hat{\mathbf{s}}_{j}^{H} \mathbf{D}_{p_{jN_r}}^{H} \mathbf{D}_{d_{jN_r}} \hat{\mathbf{s}}_{j}\right) \tag{13.16}$$

$$\mathbf{G}=\operatorname{Diag}\left(\mathbf{G}^{(1)}, \ldots, \mathbf{G}^{(N_t)}\right) \tag{13.17}$$

$$\mathbf{G}^{(j)}=\operatorname{diag}\left(\frac{1}{\widehat{\sigma_{j1}^{2}}}, \ldots, \frac{1}{\widehat{\sigma_{jN_r}^{2}}}\right)\left[\begin{array}{c}\widehat{\beta}_{j1} \hat{\mathbf{s}}_{j}^{H} \mathbf{D}_{p_{j1}}^{H} \mathbf{D}_{d_{j1}} \\ \vdots \\ \widehat{\beta}_{jN_r} \hat{\mathbf{s}}_{j}^{H} \mathbf{D}_{p_{jN_r}}^{H} \mathbf{D}_{d_{jN_r}}\end{array}\right] \tag{13.18}$$

$$\mathbf{K}=-\left[\begin{array}{cc}\mathbf{E} & \mathbf{P} \\ \mathbf{P}^{H} & \mathbf{Q}\end{array}\right] \tag{13.19}$$

$$\mathbf{P}=\operatorname{Diag}\left(\mathbf{P}^{(1)}, \ldots, \mathbf{P}^{(N_t)}\right) \tag{13.20}$$

$$\mathbf{P}^{(j)}=\operatorname{diag}\left(\frac{1}{\widehat{\sigma_{j1}^{2}}}, \ldots, \frac{1}{\widehat{\sigma_{jN_r}^{2}}}\right) \widehat{\beta}_{j} \hat{\mathbf{s}}_{j}^{H} \tag{13.21}$$

$$\mathbf{Q}=\operatorname{Diag}\left(\mathbf{Q}^{(1)}, \ldots, \mathbf{Q}^{(N_t)}\right) \tag{13.22}$$

$$\mathbf{Q}^{(j)}=\widehat{\boldsymbol{\beta}}_{j}^{H} \operatorname{diag}\left(\frac{1}{\widehat{\sigma_{j1}^{2}}}, \ldots, \frac{1}{\widehat{\sigma_{jN_r}^{2}}}\right) \widehat{\boldsymbol{\beta}}_{j} \mathbf{I}_{L} \tag{13.23}$$

$$\left\{\begin{array}{r}\hat{\mathbf{s}}_{j}=\mathbf{e}_{1}\left(\boldsymbol{\Phi}_{j} \boldsymbol{\Sigma}^{-1} \boldsymbol{\Phi}_{j}^{H}\right) \\ \boldsymbol{\Phi}_{j}=\left[\mathbf{D}_{d_{j1}}^{H} \mathbf{x}_{j1}, \ldots, \mathbf{D}_{d_{jN_r}}^{H} \mathbf{x}_{jN_r}\right] \\ \boldsymbol{\Sigma}=\operatorname{diag}\left(\left\|\mathbf{x}_{jk}\right\|^{2}, \ldots,\left\|\mathbf{x}_{jN_r}\right\|^{2}\right) \\ \widehat{\sigma_{jk}^{2}}=\frac{1}{L}\left\|\boldsymbol{\Pi}_{\mathbf{D}_{d_{jk}} \hat{\mathbf{s}}_{j}}^{\perp} \mathbf{x}_{jk}\right\|^{2} \\ \boldsymbol{\Pi}_{\mathbf{D}_{d_{jk}} \hat{\mathbf{s}}_{j}}^{\perp} \triangleq \mathbf{I}_{L}-\frac{\mathbf{D}_{d_{jk}} \hat{\mathbf{s}}_{j} \hat{\mathbf{s}}_{j}^{H} \mathbf{D}_{d_{jk}}^{H}}{\hat{\mathbf{s}}_{j}^{H} \hat{\mathbf{s}}_{j}} \\ \widehat{\beta}_{jk}=\frac{\hat{\mathbf{s}}_{j}^{H} \mathbf{D}_{d_{jk}}^{H} \mathbf{x}_{jk}}{\hat{\mathbf{s}}_{j}^{H} \hat{\mathbf{s}}_{j}}\end{array}\right. \tag{13.24}$$

Proof. The process for obtaining the test statistic (13.9) is explained in Appendix 13A. By taking into account the block diagonal structure of matrices **E**, **F**, **G**, **P**, and **Q**, the Unc-Rao test statistic can be restated as

$$\Lambda_{\text{Unc}}(\mathbf{x})=2 \sum_{j=1}^{N_t}\left(\left\|\mathbf{g}_{j}\right\|^{2}-\mathbf{g}_{j}^{H} \mathbf{R}_{j}\left(\mathbf{K}_{j}+\mathbf{R}_{j}^{H} \mathbf{R}_{j}\right)^{-1} \mathbf{R}_{j}^{H} \mathbf{g}_{j}\right) \tag{13.25}$$

where

$$\mathbf{R}_{j}=\mathbf{E}^{(j)-\frac{1}{2}}\left[\begin{array}{ll}\mathbf{F}^{(j)} & \mathbf{G}^{(j)}\end{array}\right] \tag{13.26}$$

$$\mathbf{K}_j = -\begin{bmatrix} \mathbf{E}^{(j)} & \mathbf{P}^{(j)} \\ \mathbf{P}^{(j)H} & \mathbf{Q}^{(j)} \end{bmatrix} \tag{13.27}$$

Proposition 2. *If the receivers are calibrated, the Rao test statistic can be acquired by*

$$\Lambda_C(\mathbf{x}) = \frac{2}{\widehat{\sigma^2}} \sum_{j=1}^{N_t} \left(\|g_j\|^2 - g_j^H \mathcal{R}_j \left(\mathcal{K}_j + \mathcal{R}_j^H \mathcal{R}_j\right)^{-1} \mathcal{R}_j^H g_j \right) \tag{13.28}$$

where

$$[g_j]_k = \left(\mathbf{D}_{Pjk} \frac{\hat{\mathbf{s}}_j}{\|\hat{\mathbf{s}}_j\|}\right)^H \mathbf{\Pi}_{\mathbf{D}_{djk}^{\perp} \hat{\mathbf{s}}_j} \mathbf{x}_{jk} \tag{13.29}$$

$$\mathcal{R}_j = \mathcal{E}^{(j)-\frac{1}{2}} [\mathcal{F}^{(j)}, \mathcal{G}^{(j)}], \tag{13.30}$$

$$\mathcal{E}^{(j)} = \|\hat{\mathbf{s}}_j\|^2 \mathbf{I}_{N_r}, \tag{13.31}$$

$$\mathcal{F}^{(i)} = \operatorname{diag}\left(\hat{\mathbf{s}}_j^H \mathbf{D}_{P1}^H \mathbf{D}_{dj1} \hat{\mathbf{s}}_j, \ldots, \hat{\mathbf{s}}_j^H \mathbf{D}_{PN_r}^H \mathbf{D}_{djN_r} \hat{\mathbf{s}}_j\right) \tag{13.32}$$

$$\mathcal{G}^{(j)} = \operatorname{diag}(\widehat{\rho}_j) \begin{bmatrix} \hat{\mathbf{s}}_j^H & \mathbf{D}_{Pj1}^H & \mathbf{D}_{dj1} \\ \vdots & \vdots & \vdots \\ \hat{\mathbf{s}}_j^H & \mathbf{D}_{PjN_r}^H & \mathbf{D}_{djN_r} \end{bmatrix} \tag{13.33}$$

$$\widehat{\rho}_j = [\hat{\rho}_{j1}, \ldots, \hat{\rho}_{jN_r}]^T \tag{13.34}$$

$$\mathcal{K}_j = -\begin{bmatrix} \mathcal{E}^{(j)} & \mathcal{P}^{(j)} \\ \mathcal{P}^{(j)H} & \mathcal{O}^{(j)} \end{bmatrix} \tag{13.35}$$

$$\mathcal{P}^{(j)} = \widehat{\rho}_j \hat{\mathbf{s}}_j^H \tag{13.36}$$

$$\mathcal{O}^{(j)} = \|\widehat{\rho}_j\|^2 \mathbf{I}_L, \tag{13.37}$$

$$\hat{\mathbf{s}}_j = \mathbf{e}_1\left(\mathcal{Q}_j \mathcal{Q}_j^H\right) \tag{13.38}$$

$$\widehat{\sigma^2} = \frac{1}{LN_r N_t} \sum_{j=1}^{N_t} \left(\sum_{k=1}^{N_r} \|\mathbf{x}_{jk}\|^2 - \frac{\hat{\mathbf{s}}_j^H \mathcal{Q}_j \mathcal{Q}_j^H \hat{\mathbf{s}}_j}{\hat{\mathbf{s}}_j^H \hat{\mathbf{s}}_j} \right) \tag{13.39}$$

$$\mathcal{Q}_j = \left[\mathbf{D}_{dj1}^H \mathbf{x}_{j1}, \ldots, \mathbf{D}_{djN_r}^H \mathbf{x}_{jN_r}\right] \tag{13.40}$$

$$\hat{\rho}_{jk} = \frac{\hat{\mathbf{s}}_j^H \mathbf{D}_{djk}^H \mathbf{x}_{jk}}{\hat{\mathbf{s}}_j^H \hat{\mathbf{s}}_j} \tag{13.41}$$

In contrast to Reference 8, this chapter utilizes no approximations in deriving the Rao detector for calibrated receivers.

13.4 Performance results

In this section, the numerical simulation outcomes are used to evaluate the efficiency of the proposed detectors. To make a comparison, we analyze the uncalibrated single-input multiple-output(SIMO)-based P1-Rao detector that was introduced in Chapter 12. It is important to note that this detector was formulated without the presence of a direct-path interference signal, and it is referred to as the P1-Rao detector in Chapter 12 but called NoDPI-Unc-Rao in this section.

In our simulations, we generate $\mathbf{s}_j$ according to $\mathbf{s}_j = \exp\{-\mathrm{j}\boldsymbol{\phi}_j\}$, where the vectors $\phi_j \in \mathscr{R}^{L\times 1}$ are independent with independent and identically distributed elements uniformly distributed on $[0, 2\pi)$. The average signal-to-noise ratio ($\mathrm{SNR_{avg}}$) is defined as

$$\mathrm{SNR_{avg}} = \frac{1}{N_r N_t} \sum_{j=1}^{N_t} \sum_{k=1}^{N_r} \frac{|\alpha_{jk}|^2}{\sigma_{jk}^2} \tag{13.42}$$

and the average direct-path interference-to-noise ratio, denoted as $\mathrm{DNR_{avg}}$, is given by

$$\mathrm{DNR_{avg}} = \frac{1}{N_r N_t} \sum_{j=1}^{N_t} \sum_{k=1}^{N_r} \frac{|\beta_{jk}|^2}{\sigma_{jk}^2} \tag{13.43}$$

Obtaining precise closed-form expressions for the detection probability and false alarm probability of our detectors is challenging. Therefore, we rely on MC simulations to evaluate the false alarm regulation and detection performance of our proposed calibrated and uncalibrated detectors. To ensure reliable MC evaluations, we follow the commonly agreed practice of conducting at least $\frac{100}{p_{fa}}$ and 10^4 independent MC trials to compute the empirical false alarm and detection probabilities, respectively. In this section, the empirical and the desired nominal false alarm probabilities are denoted by P_{fa} and p_{fa}, respectively. We first analyze the performance of the considered detectors in two multistatic configurations, namely SIMO and MIMO. Subsequently, we demonstrate the effectiveness of the Rao test statistic for detecting a target in a five-element MIMO scenario.

13.4.1 Performance of SIMO systems

To begin with, we examine an SIMO passive radar setup in which there is a single transmitter, or $N_t = 1$, and four dispersed receivers, or $N_r = 4$. It is assumed that the receivers lack calibration, with $\boldsymbol{\sigma} = \sigma_0^2[1, 0.5, 0.75, 1.3]$, unless otherwise stated. We compare the efficacy of the proposed Unc-Rao and NoDPI-Unc-Rao detectors, where the latter is intended for uncalibrated receivers without direct-path signal interference. The empirical false alarm probabilities, denoted as P_{fa}, as a function of detection threshold are plotted in Figure 13.2 for received signal length $L = 128$ and $\mathrm{DNR}^{\eta}_{\mathrm{avg}}$ of −10 dB and 10 dB. The detector thresholds are established via 10^6 MC simulation runs to ensure the accuracy of the findings. In this case, it is not possible to use the C-Rao detector that was created for the calibrated scenario.

In Figure 13.2, it is evident that the threshold of the proposed Unc-Rao and NoDPI-Unc-Rao detectors depends on the values of $\mathrm{DNR_{avg}}$ as they require estimation of the unknown transmitted signal. A pertinent query that arises in this context is whether it is feasible to establish the detection threshold for the mentioned

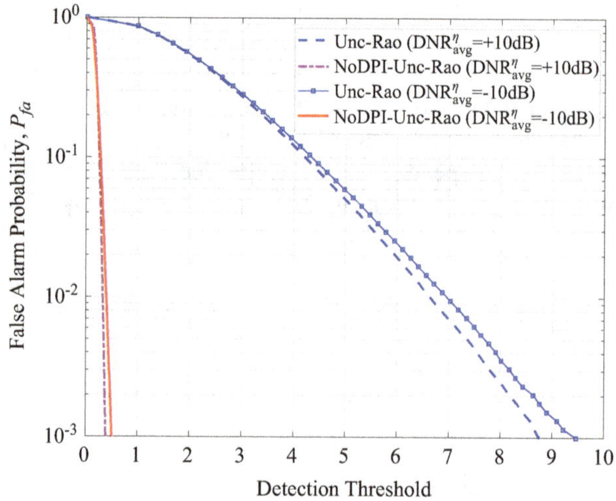

Figure 13.2 *Empirical false alarm probability of the NoDPI-Unc-Rao and Unc-Rao detectors versus detection threshold for an SIMO configuration*

detectors as the exact value of DNR_{avg} is unknown. However, in practical scenarios, the range of variation of DNR_{avg} may be available. In such cases, an effective and practical method to set the detection threshold may be through the concept of the level of a test, as explained in earlier chapters. The ensuing discussion introduces a new approach to setting the threshold in target detection problems when practical uncertainties are a concern. To apply this idea, we can establish DNR^{η}_{avg} as the DNR_{avg} values exploited to set the detection thresholds of the detectors under examination. In order to gain a deeper understanding of this new concept, the variation of P_{fa} versus DNR_{avg} is depicted in Figure 13.3 for different values of DNR^{η}_{avg}, given by −10 dB and +10 dB. When DNR^{η}_{avg} is at +10 dB, the NoDPI-Unc-Rao detector produces an excessive false alarm rate, especially for DNR_{avg} values below +10 dB. However, the proposed Unc-Rao detector shows robustness, performing almost like a constant false alarm rate (CFAR) against DNR_{avg}. In the case of DNR^{η}_{avg} = −10 dB, both detectors lead to a detector with a fixed level of 10^{-3} for all values of $DNR_{avg} \geq -10$ dB.

By utilizing the concept of *level-of-test*, it becomes feasible to modify the detection thresholds of the detectors being considered. To achieve this, the minimum received DNR_{avg} value is first determined in order to obtain detectors with fixed levels of P_{fa}. Subsequently, the detection threshold of the NoDPI-Unc-Rao detector can be set based on this minimum value to facilitate the handling of the direct-path interference signal with a fixed level. Conversely, a sufficiently high value of DNR^{η}_{avg} is required to establish the threshold for the proposed Unc-Rao detector, which results in a detector with a fixed size. Nevertheless, utilizing a test with a constant level, instead of a constant size of P_{fa}, may lead to a reduction in the detection

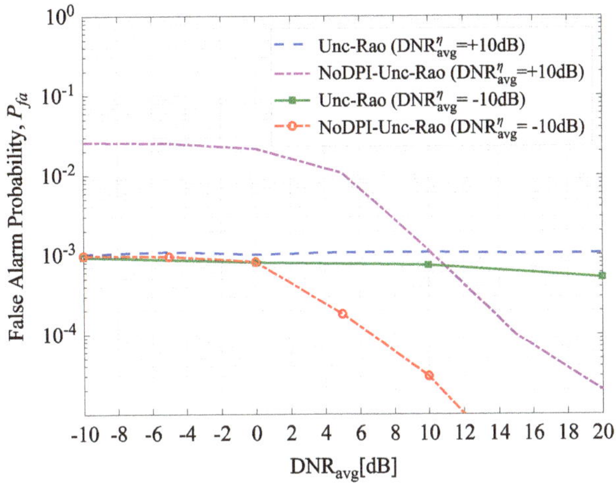

Figure 13.3 *False alarm regulation of the NoDPI-Unc-Rao and Unc-Rao detectors versus DNR_{avg} for an SIMO configuration (i.e., $N_t = 1$, $N_r = 4$), $DNR^{\eta}_{avg} = \{-10, +10\}$ dB, and received signal length* $L = 128$

performance of the detectors since a lower probability of false alarm typically means a lower probability of detection.

Figure 13.4 presents a comparison of the detection performance between the Unc-Rao and NoDPI-Unc-Rao detectors. The Unc-Rao detector uses $\text{DNR}^{\eta}_{\text{avg}} = +10$ dB, while the NoDPI-Unc-Rao detector uses $\text{DNR}^{\eta}_{\text{avg}} = -10$ dB to ensure $P_{fa} \leq p_{fa} = 10^{-3}$. The Unc-Rao detector achieves $P_{fa} = p_{fa} = 10^{-3}$. For higher values of SNR_{avg} at $\text{DNR}^{\eta}_{\text{avg}} = -10$ dB, the NoDPI-Unc-Rao detector outperforms the proposed detector, but its performance significantly deteriorates as DNR_{avg} increases. At $\text{DNR}_{\text{avg}} = 0$ dB, the proposed detector achieves a 10 dB SNR_{avg} gain over the NoDPI-Unc-Rao detector.

The detection probability as a function of DNR_{avg} is shown in Figure 13.5, using the same threshold setting strategy as Figure 13.4 and with SNR_{avg} set to –6 dB. The NoDPI-Unc-Rao detector experiences a significant decline in target detection performance with increasing DNR_{avg}, whereas the proposed Unc-Rao detector's performance improves. Additionally, the proposed detector is capable of not only reducing direct-path signal interference but also exploiting it for passive detection.

13.4.2 Performance of MIMO systems

In the case of MIMO configuration, we consider $N_r = 2$ and $N_r = 4$. For comparison purposes, we consider two scenarios as follows:

- Fully calibrated case with $\boldsymbol{\sigma}_1 = \sigma_0^2[1, 1, 1, 1]$ and $\sigma_2^2 = \sigma_0^2[1, 1, 1, 1]$
- Uncalibrated case with $\boldsymbol{\sigma}_1 = \sigma_0^2[1, 0.25, 0.25, 0.25]$ and $\boldsymbol{\sigma}_2 = 1.15 \times \sigma_0^2[1, 0.25, 0.25, 0.25]$

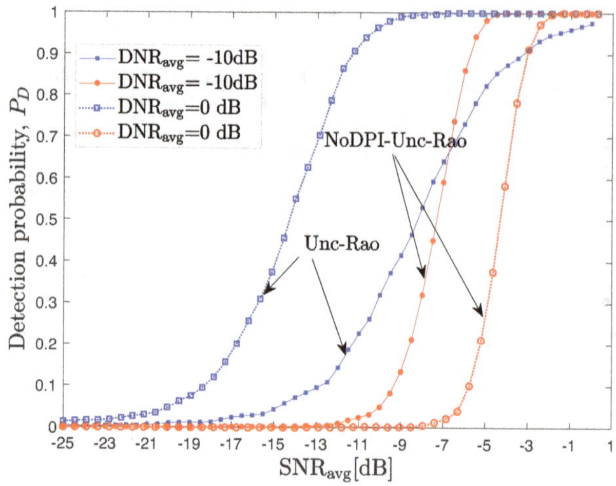

Figure 13.4 *Detection probability versus SNR_{avg} when we use DNR^{η}_{avg} = +10 dB for threshold setting of the proposed Unc-Rao detector, while it is DNR^{η}_{avg} = –10 dB for the NoDPI-Unc-Rao one. Here, we have an SIMO configuration (i.e., $N_t = 1$, $N_r = 4$), $L = 128$, and the desired false alarm probability is set equal to* 10^{-3}.

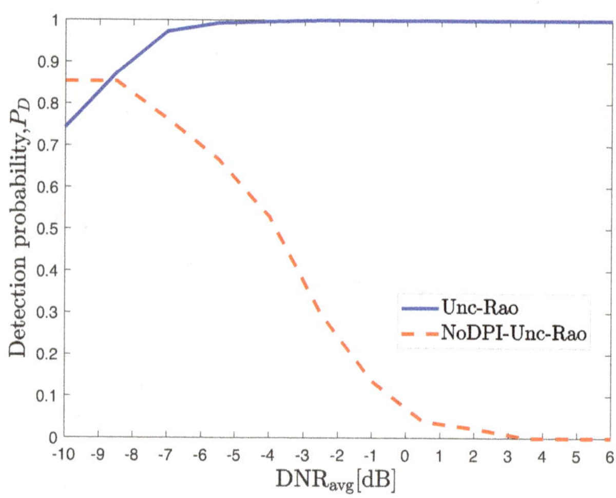

Figure 13.5 *Detection probability versus DNR_{avg} when we use DNR^{η}_{avg} = +10 dB for threshold setting of the proposed Unc-Rao detector, while it is DNR^{η}_{avg} = –10 dB for the NoDPI-Unc-Rao one. Here, we have an SIMO configuration (i.e., $N_t = 1$, $N_r = 4$), $L = 128$, SNR_{avg} = –6 dB, and the desired false alarm probability is set equal to* 10^{-3}.

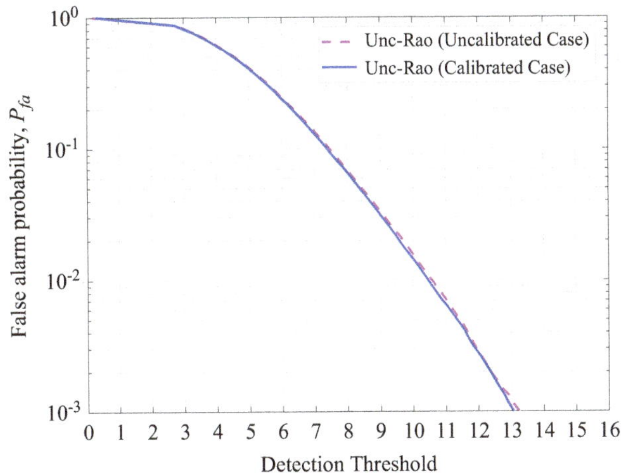

Figure 13.6 *Empirical false alarm probability of the Unc-Rao detector versus detection threshold for a MIMO configuration (i.e., $N_t = 2$, $N_r = 4$), $L = 128$, $DNR^{\eta}_{avg} = +10$ dB, and two calibrated and uncalibrated cases are considered*

For both scenarios, a noise variance of approximately −121 dB is obtained by utilizing $\sigma_0^2 = 8.28 \times 10^{-13}$. This value is practical and reasonable for the particular radar systems in question. The false alarm probability as a function of detection threshold for both Unc-Rao and C-Rao detectors is depicted in Figures 13.6 and 13.7, respectively. In both cases, the detection thresholds are set based on $\text{DNR}^{\eta}_{\text{avg}} = +10$ dB for the detectors.

The results presented in Figure 13.6 demonstrate that the proposed Unc-Rao detector effectively exhibits the CFAR property in the presence of varying noise variances in different frequency and spatial channels. In contrast, as depicted in Figure 13.7, C-Rao detection thresholds are not constant when switching from a calibrated to an uncalibrated case, indicating that it is not a CFAR detector for uncalibrated passive radar receivers. Surprisingly, for this particular example, switching from the uncalibrated to calibrated cases increases C-Rao detection threshold. This suggests that if the detection threshold for C-Rao is established based on this calibrated example and then applied to the uncalibrated one, the resulting false alarm probability will be lower than the desired value. However, this conclusion cannot be extended to other examples, and we only use this example to evaluate the detection performance fairly.

In order to study the effect of the $\text{DNR}^{\eta}_{\text{avg}}$ values on the threshold settings of the MIMO configuration Unc-Rao and C-Rao detectors, we examined the calibrated case and plotted the curve of false alarm probability as a function of DNR_{avg} for two different values of $\text{DNR}^{\eta}_{\text{avg}}$, namely −10 dB and +10 dB. The results of this simulation can be seen in Figure 13.8. While it was previously claimed in Reference 8 that the C-Rao detector had CFAR properties against $\text{DNR}^{\eta}_{\text{avg}}$ due to its similarity with LRT detectors in References 1 and 2 for high values of L, this is not actually the case

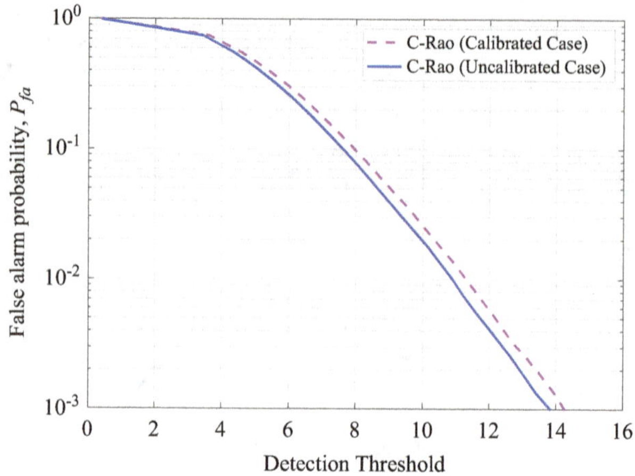

Figure 13.7 *Empirical false alarm probability of the C-Rao detector as a function of detection threshold for an MIMO configuration (i.e., $N_t = 2$, $N_r = 4$), $L = 128$, $DNR^{\eta}_{avg} = +10$ dB, and two calibrated and uncalibrated cases are considered*

for smaller values of $\text{DNR}^{\eta}_{\text{avg}}$, as shown in Figure 13.8. However, we observed that both detectors with $\text{DNR}^{\eta}_{\text{avg}} = 10$ dB were fairly robust with respect to changes in DNR_{avg} and maintained a fixed-level false alarm probability of 10^{-3} (i.e., $P_{fa} \leq 10^{-3}$)

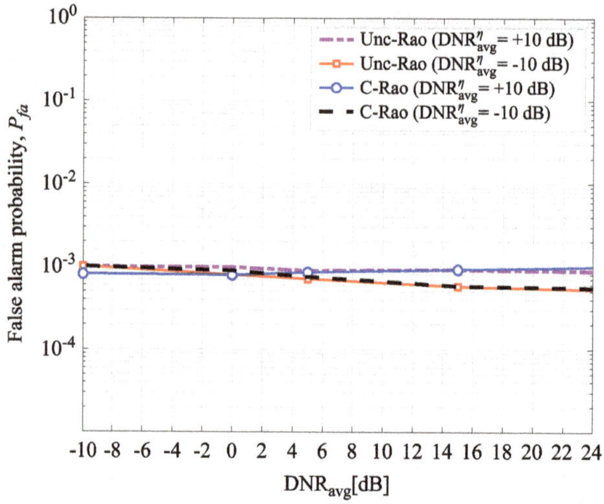

Figure 13.8 *False alarm regulation of the Unc-Rao and C-Rao detectors versus of DNR_{avg} for an MIMO configuration (i.e., $N_t = 2$, $N_r = 4$), calibrated case, $DNR^{\eta}_{avg} = \{-10, +10\}$ dB, and received signal length $L = 128$*

over the entire range of tested $\mathrm{DNR_{avg}}$ values, even though they were not perfectly CFAR.

In order to assess the efficacy of the Unc-Rao and C-Rao detectors, we used the above specific example and adjusted the detection threshold of the C-Rao detector according to the calibrated receiver, ensuring that it achieved a P_{fa} of no more than p_{fa} when applied to the uncalibrated receiver. The results of this simulation for the calibrated receiver are illustrated in Figure 13.9, while the results for the uncalibrated receiver are reported in Figure 13.10. It is worth noting that the Unc-Rao detector can be used for either calibrated or uncalibrated receivers due to its CFAR property. As expected, the C-Rao detector demonstrated superior performance compared to the Unc-Rao detector in the calibrated case, with a $\mathrm{SNR_{avg}}$ gain of approximately 0.8 dB for $P_D = 0.9$. However, in the case of the uncalibrated receiver, the Unc-Rao detector outperforms the C-Rao detector. Typically, the C-Rao method is not advised for uncalibrated receivers due to its high susceptibility to threshold settings and inability to regulate its level or size for varying values of $\boldsymbol{\sigma}$, where the scenario of Figure 13.10 is a special case discovered after conducting thorough and extensive searches. The impact of utilizing low values of $\mathrm{DNR}^{\eta}_{\mathrm{avg}}$ to set thresholds on the detection performance is illustrated in Figure 13.10. A comparison between two scenarios, where $\mathrm{DNR}^{\eta}_{\mathrm{avg}}$ is either −10 dB or +10 dB, reveals that fixed-level detectors are obtained in the former case, while fixed-size detectors are obtained in the latter case. It is possible to adjust the detection threshold of the Unc-Rao detector to obtain either fixed-level or fixed-size detectors. However, it should be noted that the former scenario leads to smaller losses in $\mathrm{SNR_{avg}}$ compared to the latter scenario.

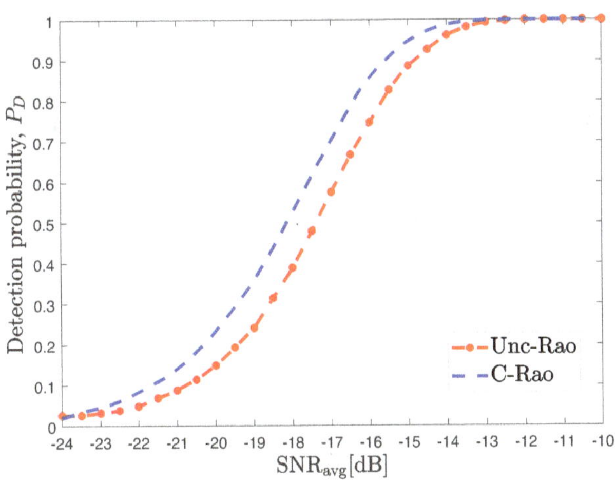

Figure 13.9 *Detection probability versus $\mathrm{SNR_{avg}}$ for calibrated receivers. Here, we have an MIMO configuration (i.e., $N_t = 2$, $N_r = 4$), $L = 128$, and the desired false alarm probability is set equal to 10^{-3}. Here, we obtain $P_{fa} = p_{fa} = 10^{-3}$ for both detectors.*

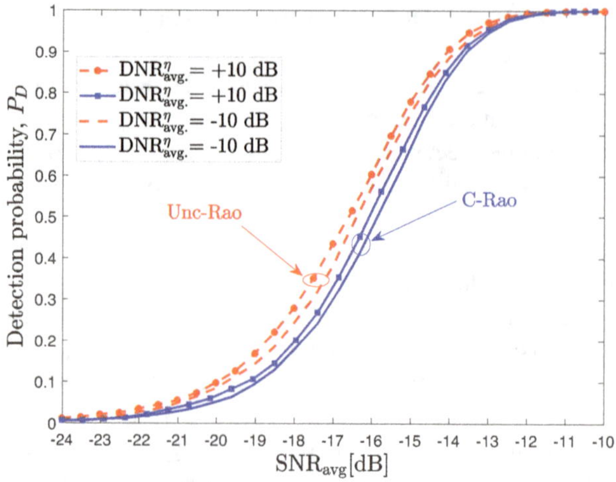

Figure 13.10 *Detection probability versus SNR_{avg} for uncalibrated receivers. Here, we have an MIMO configuration (i.e., $N_t = 2$, $N_r = 4$), $L = 128$, and the desired false alarm probability is set equal to 10^{-3}. Here, we have $P_{fa} = p_{fa} = 10^{-3}$ for the proposed Unc-Rao detector, while it is $P_{fa} < p_{fa} = 10^{-3}$ for the C-Rao detector.*

13.4.3 Rao test statistic for target localization

In this section, the effectiveness of the Rao test statistic is demonstrated in a passive MIMO radar situation that involves three receivers situated at distances of $\mathbf{r}_1 = [1, 30]^T$ km, $\mathbf{r}_2 = [5, 50]^T$ km, and $\mathbf{r}_3 = [12, 80]^T$ km. There are two transmitters located at $\mathbf{t}_1 = [20, 20]^T$ km and $\mathbf{t}_2 = [50, 10]^T$ km, and a target located at $\mathbf{r}_T = [35, 45]^T$ km. Additionally, each multichannel receiver has a five-element uniform linear array (ULA) to direct the beam toward the desired target area. By applying beamforming at each multichannel receiver and summing up the signals from individual elements with specific weights, the signal model can be represented as explained in (13.4). In Figures 13.11 and 13.12, the Rao test statistic as a function of hypothesized x–y coordinate is plotted. Here, it is assumed that the target hypothesized velocity is matched to that of the true one, and we use $L = 512$, $\text{DNR}_{\text{avg}} = +10$ dB, and $\text{SNR}_{\text{avg}} = -10$ dB. To calculate SNR_{avg}, we take into account the differences in α_{jk} values resulting from different bistatic geometries, with their magnitudes provided as

$$|\alpha_{jk}| = \frac{\lambda \sigma_{\text{rcs}}}{4\pi \left\| \mathbf{r}_T - \mathbf{t}_j \right\| \left\| \mathbf{r}_T - \mathbf{r}_k \right\|} \sqrt{\frac{p_j(\mathbf{r}_T) G_a^{(k)}(\mathbf{r}_T)}{4\pi}} \tag{13.44}$$

The isotropic power emitted by the jth transmitter toward the target location $\mathbf{r}_T$ is represented by $p_j(\mathbf{r}_T)$, while the total gain of the kth receiver antenna toward the target location is denoted by $G_a^{(k)}(\mathbf{r}_T)$. It is important to note that the values of SNR_{avg} can be modified using σ_{rcs}. As shown in Figures 13.11 and 13.12, the Rao test statistic peaks at the true target position. More precisely, an ellipsoid can be seen

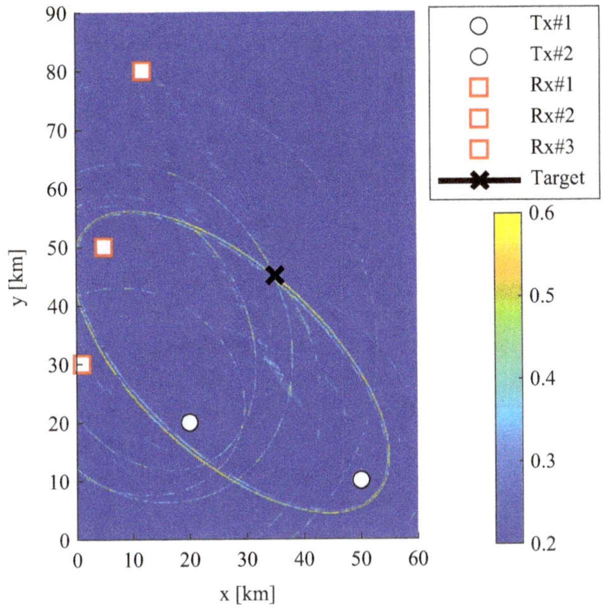

Figure 13.11 Image of the Rao test statistic as a function of hypothesized x–y coordinate (i.e., $\mathbf{r}_H = [x, y]^T$) when $L = 512$, $DNR_{avg} = +10$ dB, $SNR_{avg} = -10$ dB, and the five-element ULA multichannel receivers

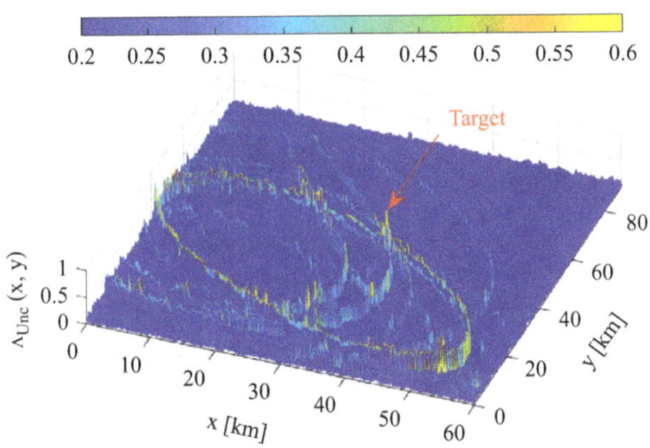

Figure 13.12 The Rao test statistic as a function of hypothesized $x - y$ coordinate (i.e., $\mathbf{r}_H = [x, y]^T$) when $L = 512$, $DNR_{avg} = +10$ dB, $SNR_{avg} = -10$ dB, and the five-element ULA multichannel receivers

for each of the transmitter–receiver pairs for which a target is detected. As such, each ellipsoid contains the target location; thus, by intersecting different ellipsoids, the position of the target can be obtained. All of these are done through the Rao test statistic, with a large peak at the actual target position.

13.5 Summary

This chapter has provided details to develop the Rao test for detecting targets in calibrated/uncalibrated passive radar systems. These detectors are advantageous compared to traditional LRT approaches because they are in closed form, eliminating the need for iterative algorithms of LRT-based detectors. This makes them more practical. The chapter provides several MC simulation results to demonstrate the effectiveness of the proposed Rao-based detectors in comparison to the detector discussed in Chapter 12. The results indicate that the proposed detectors outperform their counterparts for various multistatic configurations, particularly in terms of detection probability and false alarm regulation.

References

[1] Zhang X., Li H., Himed B. 'Multistatic detection for passive radar with direct-path interference'. *IEEE Transactions on Aerospace and Electronic Systems*. 2017, vol. 53(2), pp. 915–25.

[2] Zhang X., Li H., Himed B. 'A direct-path interference resistant passive detector'. *IEEE Signal Processing Letters*. 2017, vol. 24(6), pp. 818–22.

[3] Zaimbashi A., Greco M.S. 'Multistatic passive radar target detection under uncalibrated receivers with direct-path interference'. *IEEE Transactions on Aerospace and Electronic Systems*. 2022, vol. 58(11), pp. 5443–55.

[4] Hack D.E., Patton L.K., Himed B., Saville M.A. 'Detection in passive MIMO radar networks'. *IEEE Transactions on Signal Processing*. 2014, vol. 62(11), pp. 2999–3012.

[5] Liu J., Zhou W., Zaimbashi A., Li H. 'Persymmetric adaptive array detection of spread spectrum signals'. *IEEE Transactions on Information Theory*. 2020, vol. 66(12), pp. 7828–34.

[6] Sun M., Liu W., Liu J., Hao C. 'Complex parameter Rao, Wald, gradient, and Durbin tests for multichannel signal detection'. *IEEE Transactions on Signal Processing*. , vol. 70, pp. 117–31.n.d.

[7] Ciuonzo D., Carotenuto V., De Maio A. 'On multiple covariance equality testing with application to SAR change detection'. *IEEE Transactions on Signal Processing*. , vol. 65(19), pp. 5078–91.n.d.

[8] Fazlollahpoor M., Derakhtian M., Khorshidi S. 'Rao detector for passive MIMO radar with direct-path interference'. *IEEE Transactions on Aerospace and Electronic Systems*. 2020, vol. 56(4), pp. 2999–3009.

Appendix 13A
Unc-Rao test statistic derivation

The decision statistic for the Rao test is provided as

$$\Lambda(\mathbf{x}) = 2\left[\frac{\partial \mathfrak{T}(\boldsymbol{\theta}_r, \boldsymbol{\theta}_s; \mathbf{x})}{\partial \boldsymbol{\theta}_r}\right]^T \left[\mathcal{I}_{rr}^{-1}(\boldsymbol{\theta})\right]_{rr} \left[\frac{\partial \mathfrak{T}(\boldsymbol{\theta}_r, \boldsymbol{\theta}_s; \mathbf{x})}{\partial \boldsymbol{\theta}_r^H}\right]^T_{\boldsymbol{\theta}=(\alpha=0, \hat{\boldsymbol{\theta}}_{s,0})} \quad (13A.1)$$

The test data vector, denoted as $\mathbf{x}$, is a complex vector of dimensions $LN_rN_t \times 1$ consisting of multiple vectors $\mathbf{x}_1, ..., \mathbf{x}_{N_t}$, i.e., $\mathbf{x} = [\mathbf{x}_1^T, ..., \mathbf{x}_{N_t}^T]^T \in \mathcal{C}^{LN_rN_t\times 1}$. The joint probability density function of this vector can be expressed as

$$\begin{aligned}\mathfrak{T}(\boldsymbol{\theta}_r, \boldsymbol{\theta}_s; \mathbf{x}) &= \prod_{j=1}^{N_t} \mathcal{L}(\boldsymbol{\theta}_r, \boldsymbol{\theta}_s; \mathbf{x}_j) = \prod_{j=1}^{N_t}\prod_{k=1}^{N_r} \mathcal{L}(\boldsymbol{\theta}_r, \boldsymbol{\theta}_s; \mathbf{x}_{jk}) \\ &= \left(\frac{1}{\pi^{N_tN_r}\prod_{j=1}^{N_t}\prod_{k=1}^{N_r}\sigma_{jk}^2}\right)^L \\ &\times \exp\left(-\sum_{j=1}^{N_t}\sum_{k=1}^{N_r}\frac{\left\|\mathbf{x}_{jk} - \beta_{jk}\mathbf{D}_{d_{jk}}\mathbf{s}_j - \alpha_{jk}\mathbf{D}_{p_{jk}}\mathbf{s}_j\right\|^2}{\sigma_{jk}^2}\right)\end{aligned} \quad (13A.2)$$

In (13A.1), $\frac{\partial}{\partial \boldsymbol{\theta}_r}$ denotes the gradient operator with respect to $\boldsymbol{\theta}_r$, $\hat{\boldsymbol{\theta}}_{s,0}$ is the MLE of $\boldsymbol{\theta}_s$ under $\mathcal{H}_0$. Let us define the FIM partitioned as

$$\begin{aligned}\mathcal{I}(\boldsymbol{\theta}) &= \begin{bmatrix}\mathcal{I}_{rr}(\boldsymbol{\theta}) & \mathcal{I}_{rs}(\boldsymbol{\theta}) \\ \mathcal{I}_{sr}(\boldsymbol{\theta}) & \mathcal{I}_{ss}(\boldsymbol{\theta})\end{bmatrix} \\ &:: \begin{bmatrix}N_rN_t\times N_rN_t & N_rN_t\times QN_t \\ QN_t\times N_rN_t & QN_t\times QN_t\end{bmatrix}_{(N_r+Q)N_t\times(N_r+Q)N_t}\end{aligned} \quad (13A.3)$$

where $Q = 2N_r + L$. In (13A.1), $[\mathcal{I}^{-1}(\boldsymbol{\theta})]_{rr}$ is the $N_rN_t \times N_rN_t$ matrix formed by the N_rN_t rows and N_rN_t columns of $\mathcal{I}^{-1}(\boldsymbol{\theta})$, obtained as

$$[\mathcal{I}^{-1}(\boldsymbol{\theta})]_{rr} = \left(\mathcal{I}_{rr}(\boldsymbol{\theta}) - \mathcal{I}_{rs}(\boldsymbol{\theta})\mathcal{I}_{ss}^{-1}(\boldsymbol{\theta})\mathcal{I}_{rs}^H(\boldsymbol{\theta})\right)^{-1} \quad (13A.4)$$

For our problem, using (13A.2), we have

$$\left[\frac{\partial \mathcal{L}(\boldsymbol{\theta}_r, \boldsymbol{\theta}_s; \mathbf{x}_j)}{\partial \boldsymbol{\alpha}_j^H}\right]_k = \frac{1}{\sigma_{jk}^2}\left(\mathbf{s}_{p_{jk}}^H\mathbf{x}_{jk} - \boldsymbol{\alpha}_{jk}\mathbf{s}_{p_{jk}}^H\mathbf{s}_{p_{jk}}\right), \quad (13A.5)$$

$$\left[\frac{\partial \mathscr{L}(\boldsymbol{\theta}_r, \boldsymbol{\theta}_s; \mathbf{x}_j)}{\partial \boldsymbol{\beta}_j^H}\right]_k = \frac{1}{\sigma_{jk}^2}\left(\mathbf{s}_{d_{jk}}^H \mathbf{u}_{jk} - \boldsymbol{\beta}_{jk}\mathbf{s}_{d_{jk}}^H \mathbf{s}_{d_{jk}}\right), \tag{13A.6}$$

$$\frac{\partial \mathscr{L}(\boldsymbol{\theta}_r, \boldsymbol{\theta}_s; \mathbf{x}_j)}{\partial \sigma_{jm}^2} = -\frac{L}{\sigma_{jm}^2} + \frac{\|\boldsymbol{\Pi}_{\mathbf{D}_{d_{jm}}\hat{\mathbf{s}}_j}^{\perp} \mathbf{x}_{jm}\|^2}{\sigma_{jm}^4}, \tag{13A.7}$$

$$\left[\frac{\partial \mathscr{L}(\boldsymbol{\theta}_r, \boldsymbol{\theta}_s; \mathbf{x}_j)}{\partial \mathbf{s}_j}\right]^T = \sum_{k=1}^{N_r} \frac{\beta_{jk}\mathbf{x}_{jk}^H \mathbf{D}_{d_{jk}} - |\beta_{jk}|^2 \mathbf{s}_j^H \mathbf{D}_{d_{jk}}^H \mathbf{D}_{d_{jk}} + \mathbf{p}^H(\boldsymbol{\alpha}_{jk})}{\sigma_{jk}^2}. \tag{13A.8}$$

where

$$\mathbf{s}_{h_{jk}} \triangleq \mathbf{D}_{h_{jk}}\mathbf{s}_j; \quad h \in \{\mathrm{p}, \mathrm{g}\} \tag{13A.9}$$

$$\mathbf{x}_{jk} \triangleq \mathbf{x}_{jm} - \beta_{jk}\mathbf{D}_{d_{jk}}\mathbf{s}_j \tag{13A.10}$$

$$\mathbf{u}_{jm} \triangleq \mathbf{x}_{jm} - \alpha_{jk}\mathbf{D}_{p_{jk}}\mathbf{s}_j \tag{13A.11}$$

The vector $\mathbf{p}$ in (13A.8) is influenced by α_{jk}, which becomes zero when the MLE of the unknown parameters is substituted under the hypothesis $\mathscr{H}_0$.

The MLE of $\mathbf{s}_j$ can be acquired through

$$\hat{\mathbf{s}}_j = \operatorname{argsup}\left\{-LN_r \ln\pi - L\prod_{k=1}^{N_r}\frac{1}{L}\|\boldsymbol{\Pi}_{\mathbf{s}_{\mathbf{d}_{jk}}}^{\perp}\mathbf{x}_{jk}\|^2 - LN_r\right\} \tag{13A.12}$$

where

$$\boldsymbol{\Pi}_{\mathbf{s}_{d_{jk}}}^{\perp} = \mathbf{I} - \frac{\mathbf{s}_{d_{jk}}\mathbf{s}_{d_{jk}}^H}{\|\mathbf{s}_{d_{jk}}\|^2} \tag{13A.13}$$

This can be rewritten as

$$\mathbf{s}_{d_j} = \underset{\mathbf{s}_{d_j}}{\operatorname{arginf}}\left\{\ln\left(\prod_{k=1}^{N_r}\frac{\operatorname{tr}\left(\left(\mathbf{I} - \frac{\mathbf{s}_{d_{jk}}\mathbf{s}_{d_{jk}}^H}{\|\mathbf{s}_{d_{jk}}\|^2}\right)\mathbf{x}_{jk}\mathbf{x}_{jk}^H\right)}{L}\right)\right\} \tag{13A.14}$$

By defining

$$\mathbf{g}_j \triangleq [\mathbf{g}_{j1}^{\mathrm{T}}, \ldots, \mathbf{g}_{jN_r}^{\mathrm{T}}]^{\mathrm{T}} \tag{13A.15}$$

where

$$\mathbf{g}_{jk} \triangleq \frac{\mathbf{s}_{d_{jk}}}{\|\mathbf{s}_{d_{jk}}\|} \tag{13A.16}$$

The optimization problem (13A.14) can be equivalently written as

$$\hat{\mathbf{g}}_j = \underset{g_j}{\operatorname{arginf}}\left\{\ln\left(\prod_{k=1}^{N_r}\left(1 - \frac{|\mathbf{g}_{jk}^H\mathbf{x}_{jk}|^2}{\|\mathbf{x}_{jk}\|^2}\right)\right)\right\} \tag{13A.17}$$

Invoking the Cauchy–Schwartz inequality, we get

$$\frac{\left|\mathbf{g}_{jk}{}^{H}\mathbf{x}_{jk}\right|^{2}}{\left\|\mathbf{x}_{jk}\right\|^{2}} \leq \frac{\left\|\mathbf{g}_{jk}\right\|^{2}\left\|\mathbf{x}_{jk}\right\|^{2}}{\left\|\mathbf{x}_{jk}\right\|^{2}} = 1 \tag{13A.18}$$

The upper bound is attained if and only if $\mathbf{g}_{jk} \triangleq \zeta \frac{\mathbf{x}_{jk}}{\|\mathbf{x}_{jk}\|}$. However, this condition cannot be met for all values of j and k, implying that

$$0 < \prod_{k=1}^{N_r}\left(1 - \frac{\left|\mathbf{g}_{jk}^{H}\mathbf{x}_{jk}\right|^{2}}{\left\|\mathbf{x}_{jk}\right\|^{2}}\right) \leq 1 \tag{13A.19}$$

Consequently, we find

$$-\infty < \sum_{k=1}^{N_r} \ln\left(1 - \frac{\left|\mathbf{g}_{jk}^{H}\mathbf{x}_{jk}\right|^{2}}{\left\|\mathbf{x}_{jk}\right\|^{2}}\right) \leq 0 \tag{13A.20}$$

This means that

$$\inf_{\mathbf{g}_j}\left\{\sum_{k=1}^{N_r} \ln\left(1 - \frac{\left|\mathbf{g}_{jk}^{H}\mathbf{x}_{jk}\right|^{2}}{\left\|\mathbf{x}_{jk}\right\|^{2}}\right)\right\} \rightarrow -\infty \tag{13A.21}$$

Thus, the minimum of $\sum_{k=1}^{N_r} \ln\left(1 - \frac{\left|\mathbf{g}_{jk}^{H}\mathbf{x}_{jk}\right|^{2}}{\left\|\mathbf{x}_{jk}\right\|^{2}}\right)$ is a negative value that is finite but considerably high. Using

$$\ln\left(1 + z_{jk}\right) = z - \frac{z_{jk}^{2}}{2} + \frac{z_{jk}^{3}}{3} - \frac{z_{jk}^{4}}{4} + \cdots \quad -1 < \left|z_{jk}\right| \leq 1 \tag{13A.22}$$

with $1 < z_{jk} = -\frac{\left|\mathbf{g}_{jk}{}^{H}\mathbf{x}_{jk}\right|^{2}}{\left\|\mathbf{x}_{jk}\right\|^{2}} < 0$, resulting in $\ln\left(1 + z_{jk}\right) < 0$, helps us to approximate

$$\inf_{\mathbf{g}_j}\left\{\ln\left(\prod_{k=1}^{N_r}\left(1 - \frac{\left|\mathbf{g}_{jk}{}^{H}\mathbf{x}_{jk}\right|^{2}}{\left\|\mathbf{x}_{jk}\right\|^{2}}\right)\right)\right\} \approx \inf_{\mathbf{g}_j}\left\{\sum_{k=1}^{N_r} -\frac{\left|\mathbf{g}_{jk}{}^{H}\mathbf{x}_{jk}\right|^{2}}{\left\|\mathbf{x}_{jk}\right\|^{2}}\right\} \tag{13A.23}$$

Thus, the problem reduces to

$$\begin{aligned}\sup_{\mathbf{g}_j}\left\{\sum_{k=1}^{N_r} \frac{\left|\mathbf{g}_{jk}{}^{H}\mathbf{x}_{jk}\right|^{2}}{\left\|\mathbf{x}_{jk}\right\|^{2}}\right\} &= \sup_{\mathbf{s}_j}\left\{\frac{\mathbf{s}_j{}^{H}\left(\sum_{k=1}^{N_r} \frac{\mathbf{D}_{d_{jk}}^{H}\mathbf{x}_{jk}\mathbf{x}_{jk}^{H}\mathbf{D}_{d_{jk}}}{\left\|\mathbf{x}_{jk}\right\|^{2}}\right)\mathbf{s}_j}{\mathbf{s}_j{}^{H}\mathbf{s}_j}\right\} \\ &= \mathbf{e}_1\left(\sum_{k=1}^{N_r} \frac{\mathbf{D}_{d_{jk}}^{H}\mathbf{x}_{jk}\mathbf{x}_{jk}^{H}\mathbf{D}_{d_{jk}}}{\left\|\mathbf{x}_{jk}\right\|^{2}}\right)\end{aligned} \tag{13A.24}$$

where $\mathbf{e}_1(\mathbf{A})$ is eigenvector associated to the largest eigenvalue of matrix $\mathbf{A}$. We further define

$$\mathbf{\Phi}_j = \left[\mathbf{D}_{dj1}^H \mathbf{x}_{j1}, \ldots, \mathbf{D}_{djN_r}^H \mathbf{x}_{jN_r} \right] \tag{13A.25}$$

$$\mathbf{\Sigma} = \mathrm{diag}\left(\|\mathbf{x}_{jk}\|^2, \ldots, \|\mathbf{x}_{jN_r}\|^2 \right) \tag{13A.26}$$

Then (13A.12) can be written as

$$\hat{\mathbf{s}}_j = \mathbf{e}_1 \left(\mathbf{\Phi}_j \mathbf{\Sigma}^{-1} \mathbf{\Phi}_j^H \right) \tag{13A.27}$$

Using (13A.5), (13A.6), and (13A.7), respectively, the MLE of the other unknown parameters can be obtained as

$$\widehat{\alpha}_{jk} = \frac{\hat{\mathbf{s}}_j^H \mathbf{D}_{Pjk}^H \mathbf{x}_{jk}}{\hat{\mathbf{s}}_j^H \hat{\mathbf{s}}_j} \tag{13A.28}$$

$$\widehat{\beta}_{jk} = \frac{\hat{\mathbf{s}}_j^H \mathbf{D}_{djk}^H \mathbf{x}_{jk}}{\hat{\mathbf{s}}_j^H \hat{\mathbf{s}}_j} \tag{13A.29}$$

$$\widehat{\sigma_{jk}^2} = \frac{1}{L} \| \mathbf{\Pi}^{\perp}_{\mathbf{D}_{djk} \hat{\mathbf{s}}_j} \mathbf{x}_{jk} \|^2 \tag{13A.30}$$

In what follows, we aim to obtain the blocks of FIM, i.e., $\mathcal{I}_{rr}(\boldsymbol{\theta})$, $\mathcal{I}_{rs}(\boldsymbol{\theta})$, and $\mathcal{I}_{ss}(\boldsymbol{\theta})$. To compute $\mathcal{I}_{rr}(\boldsymbol{\theta})$, we have

$$\mathcal{I}_{rr}(\boldsymbol{\theta}) = \mathrm{Diag}\left(\mathcal{I}_{rr}^{(1)}(\boldsymbol{\theta}), \ldots, \mathcal{I}_{rr}^{(N_t)}(\boldsymbol{\theta}) \right) \tag{13A.31}$$

where

$$\left[\mathcal{I}_{rr}^{(j)}(\boldsymbol{\theta}) \right]_{mn} = E\left\{ \left[\frac{\partial \mathcal{L}(\boldsymbol{\theta}_r, \boldsymbol{\theta}_s; \mathbf{x}_j)}{\partial \alpha_j^*} \right]_m \left[\frac{\partial \mathcal{L}(\boldsymbol{\theta}_r, \boldsymbol{\theta}_s; \mathbf{x}_j)}{\partial \alpha_j^*} \right]_n^H \right\} \tag{13A.32}$$

Applying (13A.5) and taking expectation, we obtain

$$\begin{aligned} \sigma_{jm}^2 \sigma_{jn}^2 \left[\mathcal{I}_{\mathbf{rr}}^{(j)}(\boldsymbol{\theta}) \right]_{mn} = {} & \mathbf{s}_{P\,jm}^H E\left\{ \mathbf{x}_{jm} \mathbf{x}_{jn}^H \right\} \mathbf{s}_{\mathbf{P}jn} - \alpha_{jn}^* \mathbf{s}_{P\,jm}^H E\left\{ \mathbf{x}_{jm} \right\} \mathbf{s}_{P\,jn}^H \mathbf{s}_{Pjn} \\ & - \alpha_{jm} \mathbf{s}_{Pjm} \mathbf{s}_{P\,jm}^H E\left\{ \mathbf{x}_{jn}^H \right\} \mathbf{s}_{Pjn} + \alpha_{jm} \alpha_{jn}^* \mathbf{s}_{P\,jm}^H \mathbf{s}_{Pjm} \mathbf{s}_{P\,jn}^H \mathbf{s}_{Pjn} \end{aligned} \tag{13A.33}$$

Using $\mathbf{x}_{jk} \sim \mathcal{CN}(\alpha_{jk} \mathbf{D}_{Pjk} \mathbf{s}_j, \sigma_{jk}^2 \mathbf{I}_L)$, the elements of matrix $\mathcal{I}_{\mathbf{rr}}^{(j)}(\boldsymbol{\theta})$ can be obtained as

$$\left[\mathcal{I}_{\mathbf{rr}}^{(j)}(\boldsymbol{\theta}) \right]_{mn} = \begin{cases} \dfrac{1}{\sigma_{jm}^2} \| \mathbf{D}_{Pjm} \mathbf{s}_j \|^2 + \zeta(\alpha_{jm}, \alpha_{jm}) & \mathrm{n} = \mathrm{m} \\ \rho(\alpha_{jm}, \alpha_{jn}) & \mathrm{n} \neq \mathrm{m} \end{cases} \tag{13A.34}$$

By substituting the MLE of unknown parameters under $\mathcal{H}_0$ hypothesis, we obtain

$$\mathcal{I}_{\mathbf{rr}}^{(j)}\left(\widehat{\boldsymbol{\theta}}_0 \right) = \|\hat{\mathbf{s}}_j\|^2 \mathrm{diag}\left(\frac{1}{\widehat{\sigma_{j1}^2}}, \ldots, \frac{1}{\widehat{\sigma_{jN_r}^2}} \right) \tag{13A.35}$$

To compute $\mathcal{I}_{ss}(\theta)$, we have

$$\left[\mathcal{I}_{ss}(\boldsymbol{\theta})\right]_{mn} = E\left\{\left[\frac{\partial \mathscr{L}(\boldsymbol{\theta}_r, \boldsymbol{\theta}_s; \mathbf{x}_j)}{\partial \boldsymbol{\theta}_s^*}\right]_m \left[\frac{\partial \mathscr{L}(\boldsymbol{\theta}_r, \boldsymbol{\theta}_s; \mathbf{x}_j)}{\partial \boldsymbol{\theta}_s^*}\right]_n^H\right\} \tag{13A.36}$$

where $\boldsymbol{\theta}_s = \left[\boldsymbol{\beta}^T, \mathbf{s}^T, \boldsymbol{\sigma}^T\right]^T$ with $\boldsymbol{\sigma} = [\sigma_1, \ldots, \sigma_{N_t}]^T$. Then $\mathcal{I}_{ss}(\boldsymbol{\theta})$ can be written in the form of a block matrix

$$\begin{aligned}\mathcal{I}_{ss}(\boldsymbol{\theta}) &= \begin{bmatrix} \mathcal{I}_{\beta\beta}(\boldsymbol{\theta}) & \mathcal{I}_{\beta \mathrm{s}}(\boldsymbol{\theta}) & \mathcal{I}_{\beta\sigma}(\boldsymbol{\theta}) \\ \mathcal{I}_{\mathrm{s}\beta}(\boldsymbol{\theta}) & \mathcal{I}_{\mathrm{ss}}(\boldsymbol{\theta}) & \mathcal{I}_{\mathrm{s}\sigma}(\boldsymbol{\theta}) \\ \mathcal{I}_{\sigma\beta}(\boldsymbol{\theta}) & \mathcal{I}_{\sigma \mathrm{s}}(\boldsymbol{\theta}) & \mathcal{I}_{\sigma\sigma}(\boldsymbol{\theta}) \end{bmatrix} \\ &:: \begin{bmatrix} N_rN_t \times N_rN_t & N_rN_t \times LN_t & N_rN_t \times N_rN_t \\ LN_t \times N_rN_t & LN_t \times LN_t & LN_t \times N_rN_t \\ N_rN_t \times N_rN_t & N_rN_t \times LN_t & N_rN_t \times N_rN_t \end{bmatrix}\end{aligned} \tag{13A.37}$$

Thus, we need to compute blocks $\mathcal{I}_{\beta\beta}(\boldsymbol{\theta})$, $\mathcal{I}_{\beta \mathrm{s}}(\boldsymbol{\theta})$, $\mathcal{I}_{\beta\sigma}(\boldsymbol{\theta})$, $\mathcal{I}_{\mathrm{ss}}(\boldsymbol{\theta})$, $\mathcal{I}_{\sigma \mathrm{s}}(\boldsymbol{\theta})$, and $\mathcal{I}_{\sigma\sigma}(\boldsymbol{\theta})$. Matrix $\mathcal{I}_{\beta\beta}(\boldsymbol{\theta})$ can be represented as

$$\mathcal{I}_{\beta\beta}(\boldsymbol{\theta}) = \mathrm{Diag}\left(\mathcal{I}_{\beta\beta}^{(1)}(\boldsymbol{\theta}), \ldots, \mathcal{I}_{\beta\beta}^{(N_t)}(\boldsymbol{\theta})\right) \tag{13A.38}$$

where

$$\left[\mathcal{I}_{\beta\beta}^{(j)}(\boldsymbol{\theta})\right]_{mn} = E\left\{\left[\frac{\partial \mathscr{L}(\boldsymbol{\theta}_r, \boldsymbol{\theta}_s; \mathbf{x}_j)}{\partial \beta_j^*}\right]_m \left[\frac{\partial \mathscr{L}(\boldsymbol{\theta}_r, \boldsymbol{\theta}_s; \mathbf{x}_j)}{\partial \beta_j^*}\right]_n^H\right\} \tag{13A.39}$$

Using $\mathbf{u}_{jk} \sim \mathscr{CN}(\beta_{jk}\, \mathbf{D}_{d_{jk}}\mathbf{s}_j, \sigma_{jk}^2\mathbf{I})$, we get

$$\left[\mathcal{I}_{\beta\beta}^{(j)}(\boldsymbol{\theta})\right]_{mn} = \begin{cases} \dfrac{1}{\sigma_{jm}^2}\|\mathbf{s}_j\|^2 & n = m \\ 0 & n \neq m \end{cases} \tag{13A.40}$$

Thus,

$$\mathcal{I}_{\beta\beta}^{(j)}\left(\widehat{\boldsymbol{\theta}}_0\right) = \|\hat{\mathbf{s}}_j\|^2\, \mathrm{diag}\left(\frac{1}{\widehat{\sigma_{j1}^2}}, \ldots, \frac{1}{\widehat{\sigma_{jN_r}^2}}\right) \tag{13A.41}$$

Matrix $\mathcal{I}_{\beta \mathrm{s}}(\boldsymbol{\theta})$ can be written as

$$\mathcal{I}_{\beta \mathrm{s}}(\boldsymbol{\theta}) = \mathrm{Diag}\left(\mathcal{I}_{\beta \mathrm{s}}^{(1)}(\boldsymbol{\theta}), \ldots, \mathcal{I}_{\beta \mathrm{s}}^{(N_t)}(\boldsymbol{\theta})\right) \tag{13A.42}$$

The mth row of matrix $\mathcal{I}_{\beta \mathrm{s}}^{(j)}(\boldsymbol{\theta})$ can be represented as

$$\left[\mathcal{I}_{\beta \mathrm{s}}^{(j)}(\boldsymbol{\theta})\right]_{m*} = E\left\{\left[\frac{\partial \mathscr{L}(\boldsymbol{\theta}_r, \boldsymbol{\theta}_s; \mathbf{x}_j)}{\partial \beta_j^*}\right]_m \left[\left[\frac{\partial \mathscr{L}(\boldsymbol{\theta}_r, \boldsymbol{\theta}_s; \mathbf{x}_j)}{\partial \mathbf{s}_j}\right]^T\right]_*\right\} \tag{13A.43}$$

Substituting (13A.6) and (13A.8) into (13A.43) produces

$$\left[\mathcal{I}^{(j)}_{\beta s}(\boldsymbol{\theta})\right]_{m*} = \frac{1}{\sigma^2_{jm}}\beta_{jm}\mathbf{s}_j^H + \mathbf{r}_\alpha^T + \mathbf{g}_\alpha^T \tag{13A.44}$$

The functions $\mathbf{r}_\alpha$ and $\mathbf{g}_\alpha$ are dependent on α. By substituting the MLEs of the unknown parameters under $\mathscr{H}_0$ hypothesis, both functions equate to zero, and we come up with

$$\mathcal{I}^{(j)}_{\beta s}\left(\widehat{\boldsymbol{\theta}}_0\right) = \operatorname{diag}\left(\frac{1}{\widehat{\sigma}^2_{j1}}, \ldots, \frac{1}{\widehat{\sigma}^2_{jN_r}}\right)\widehat{\beta}_j\hat{\mathbf{s}}_j^H \tag{13A.45}$$

Matrix $\mathcal{I}_{\beta\sigma}(\boldsymbol{\theta})$ can be represented as

$$\mathcal{I}_{\beta\sigma}(\boldsymbol{\theta}) = \operatorname{Diag}\left(\mathcal{I}^{(1)}_{\beta\sigma}(\boldsymbol{\theta}), \ldots, \mathcal{I}^{(N_t)}_{\beta\sigma}(\boldsymbol{\theta})\right) \tag{13A.46}$$

where

$$\left[\mathcal{I}^{(j)}_{\beta\sigma}(\boldsymbol{\theta})\right]_{mn} = E\left\{\left[\frac{\partial\mathscr{L}(\boldsymbol{\theta}_r, \boldsymbol{\theta}_s; \mathbf{x}_j)}{\partial\beta_j^*}\right]_m\left[\frac{\partial\mathscr{L}(\boldsymbol{\theta}_r, \boldsymbol{\theta}_s; \mathbf{x}_j)}{\partial\boldsymbol{\sigma}}\right]_n\right\} \tag{13A.47}$$

Applying (13A.6) and (13A.9) into (13A.47) produces

$$\left[\mathcal{I}^{(j)}_{\beta\sigma}(\boldsymbol{\theta})\right]_{mn} = \frac{1}{\sigma^2_{jm}}E\left\{\left(\mathbf{s}^H_{d\,jm}\mathbf{u}_{jm} - \beta_{jm}\mathbf{s}^H_{d\,jm}\mathbf{s}_{djm}\right)\frac{\left\|\mathbf{u}_{jn} - \beta_{jn}\mathbf{s}_{djm}\right\|^2}{\left(\sigma^2_{jn}\right)^2}\right\} \tag{13A.48}$$

Using $\mathbf{u}_{jn} \sim \mathscr{CN}(\beta_{jn}\,\mathbf{D}_{d_{jn}}\mathbf{s}_j, \sigma^2_{jn}\mathbf{I})$, it follows that

$$\mathcal{I}^{(j)}_{\beta\sigma}(\boldsymbol{\theta}) = \mathbf{0}_{N_r\times N_r} \tag{13A.49}$$

Matrix $\mathcal{I}_{ss}(\boldsymbol{\theta})$ can be written as

$$\mathcal{I}_{ss}(\boldsymbol{\theta}) = \operatorname{Diag}\left(\mathcal{I}^{(1)}_{ss}(\boldsymbol{\theta}), \ldots, \mathcal{I}^{(N_t)}_{ss}(\boldsymbol{\theta})\right) \tag{13A.50}$$

where

$$\mathcal{I}^{(j)}_{ss}(\boldsymbol{\theta}) = E\left\{\frac{\partial\mathscr{L}(\boldsymbol{\theta}_r, \boldsymbol{\theta}_s; \mathbf{x}_j)}{\partial\mathbf{s}_j^*}\left[\frac{\partial\mathscr{L}(\boldsymbol{\theta}_r, \boldsymbol{\theta}_s; \mathbf{x}_j)}{\partial\mathbf{s}_j^*}\right]^T\right\} \tag{13A.51}$$

Using (13A.8), followed by some algebra, yields

$$\mathcal{I}^{(j)}_{ss}(\boldsymbol{\theta}) = \sum_{k=1}^{N_r}\frac{\left|\beta_{jk}\right|^2}{\sigma^2_{jk}}\mathbf{I}_L + \mathbf{H}_\alpha \tag{13A.52}$$

The matrix $\mathbf{H}_\alpha$ is influenced by $\boldsymbol{\alpha}$. By replacing the MLE of parameters based on the $\mathscr{H}_0$ hypothesis, $\mathbf{H}_\alpha$ becomes zero. As a result, we obtain

$$\mathcal{I}^{(j)}_{ss}\left(\widehat{\boldsymbol{\theta}}_0\right) = \widehat{\beta}_j^H \operatorname{diag}\left(\frac{1}{\widehat{\sigma}^2_{j1}}, \ldots \frac{1}{\widehat{\sigma}^2_{jN_r}}\right)\widehat{\beta}_j\mathbf{I}_L \tag{13A.53}$$

Matrix $\mathcal{I}_{\sigma\sigma}(\boldsymbol{\theta})$ is defined as

$$\mathcal{I}_{\sigma\sigma}(\boldsymbol{\theta}) = \text{Diag}\left(\mathcal{I}_{\sigma\sigma}^{(1)}(\boldsymbol{\theta}), \ldots, \mathcal{I}_{\sigma\sigma}^{(N_t)}(\boldsymbol{\theta})\right) \tag{13A.54}$$

where

$$\left[\mathcal{I}_{\sigma\sigma}^{(j)}(\boldsymbol{\theta})\right]_{mn} = E\left\{\left[\frac{\partial \mathcal{L}(\boldsymbol{\theta}_r, \boldsymbol{\theta}_s; \mathbf{x}_j)}{\partial \sigma^2}\right]_m \left[\frac{\partial \mathcal{L}(\boldsymbol{\theta}_r, \boldsymbol{\theta}_s; \mathbf{x}_j)}{\partial \sigma^2}\right]_n^H\right\} \tag{13A.55}$$

The diagonal elements of this matrix can be represented as

$$\left[\mathcal{I}_{\sigma\sigma}^{(j)}(\boldsymbol{\theta})\right]_{kk} = -\frac{L}{\sigma_{jk}^2}\frac{L}{\left(\sigma_{jk}^2\right)} + \frac{1}{\left(\sigma_{jk}^2\right)^2\left(\sigma_{jk}^2\right)^2} E\left\{\left\|\mathbf{n}_{jk}\right\|^4\right\} \tag{13A.56}$$

Since $E\left\{\left\|\mathbf{n}_{jk}\right\|^4\right\} = \sigma_{jk}^4 L^2 + 2\sigma_{jk}^4 L$, we obtain

$$\left[\mathcal{I}_{\sigma\sigma}^{(j)}(\boldsymbol{\theta})\right]_{kk} = \frac{2L}{\sigma_{jk}^4} \tag{13A.57}$$

Following the same steps, one can obtain that the off-diagonal elements of matrix $\mathcal{I}_{\sigma\sigma}^{(j)}(\boldsymbol{\theta})$ are all zero, resulting in

$$\mathcal{I}_{\sigma\sigma}^{(j)}\left(\widehat{\boldsymbol{\theta}}_0\right) = \text{diag}\left(\frac{2L}{\widehat{\sigma}_{j1}^4}, \ldots, \frac{2L}{\widehat{\sigma}_{jN_r}^4}\right) \tag{13A.58}$$

It can be shown that matrix $\mathcal{I}_{\sigma\mathrm{s}}(\boldsymbol{\theta})$ can be defined as

$$\mathcal{I}_{\sigma\mathrm{s}}(\boldsymbol{\theta}) = \text{Diag}\left(\mathcal{I}_{\sigma\mathrm{s}}^{(1)}(\boldsymbol{\theta}), \ldots, \mathcal{I}_{\sigma\mathrm{s}}^{(N_t)}(\boldsymbol{\theta})\right) \tag{13A.59}$$

where

$$\left[\mathcal{I}_{\sigma\mathrm{s}}^{(j)}(\boldsymbol{\theta})\right]_{m*} = E\left\{\left[\frac{\partial \mathcal{L}(\boldsymbol{\theta}_r, \boldsymbol{\theta}_s; \mathbf{x}_j)}{\partial \sigma^2}\right]_m \left[\frac{\partial \mathcal{L}(\boldsymbol{\theta}_r, \boldsymbol{\theta}_s; \mathbf{x}_j)}{\partial \mathbf{s}_j}\right]^T\right\} \tag{13A.60}$$

Plugging (13A.7) and (13A.8) into (13A.60) followed by some algebra, it yields

$$\mathcal{I}_{\sigma\mathrm{s}}^{(j)}(\boldsymbol{\theta}) = \mathbf{0}_{N_r\times \mathrm{L}} \tag{13A.61}$$

Another block of matrix $\mathcal{I}(\boldsymbol{\theta})$, i.e., $\mathcal{I}_{rs}(\boldsymbol{\theta})$, is defined as

$$\mathcal{I}_{rs}(\boldsymbol{\theta}) = \left[\begin{array}{ccc}\mathcal{I}_{\alpha\beta}(\boldsymbol{\theta}) & \mathcal{I}_{\alpha\mathrm{s}}(\boldsymbol{\theta}) & \mathcal{I}_{\alpha\sigma}(\boldsymbol{\theta})\end{array}\right] \tag{13A.62}$$

where

$$\mathcal{I}_{\alpha\beta}(\boldsymbol{\theta}) = \text{Diag}\left(\mathcal{I}_{\alpha\beta}^{(1)}(\boldsymbol{\theta}), \ldots, \mathcal{I}_{\alpha\beta}^{(N_t)}(\boldsymbol{\theta})\right) \tag{13A.63}$$

$$\mathcal{I}_{\alpha\mathrm{s}}(\boldsymbol{\theta}) = \text{Diag}\left(\mathcal{I}_{\alpha\mathrm{s}}^{(1)}(\boldsymbol{\theta}), \ldots, \mathcal{I}_{\alpha\mathrm{s}}^{(N_t)}(\boldsymbol{\theta})\right) \tag{13A.64}$$

$$\mathcal{I}_{\alpha\sigma}(\boldsymbol{\theta}) = \text{Diag}\left(\mathcal{I}_{\alpha\sigma}^{(1)}(\boldsymbol{\theta}), \ldots, \mathcal{I}_{\alpha\sigma}^{(N_t)}(\boldsymbol{\theta})\right) \tag{13A.65}$$

Similar to the our previous computations, it can be derived that

$$\mathcal{I}_{\alpha\beta}^{(j)}\left(\widehat{\boldsymbol{\theta}}_0\right) = \text{diag}\left(\frac{1}{\widehat{\sigma_{j1}^2}}\hat{\mathbf{s}}_j^H\,\mathbf{D}_{Pj1}^H\,\mathbf{D}_{dj1}\hat{\mathbf{s}}_j, \ldots, \frac{1}{\widehat{\sigma_{jN_r}^2}}\hat{\mathbf{s}}_j^H\,\mathbf{D}_{PjN_r}^H\,\mathbf{D}_{djN_r}\hat{\mathbf{s}}_j\right) \tag{13A.66}$$

$$\mathcal{I}_{\alpha s}^{(j)}\left(\widehat{\boldsymbol{\theta}}_0\right) = \text{diag}\left(\frac{1}{\widehat{\sigma_{j1}^2}}, \ldots, \frac{1}{\widehat{\sigma_{jN_r}^2}}\right)\begin{bmatrix} \widehat{\beta}_{j1}\hat{\mathbf{s}}_j^H\,\mathbf{D}_{Pj1}^H\,\mathbf{D}_{dj1} \\ \vdots \\ \widehat{\beta}_{jN_r}\hat{\mathbf{s}}_j^H\,\mathbf{D}_{PjN_r}^H\,\mathbf{D}_{djN_r} \end{bmatrix} \tag{13A.67}$$

and

$$\mathcal{I}_{\alpha\sigma}^{(j)}\left(\widehat{\boldsymbol{\theta}}_0\right) = \mathbf{0}_{N_r\times N_r} \tag{13A.68}$$

In order to compute the Unc-Rao test statistic, we need to compute

$$\frac{\partial\mathcal{L}\left(\boldsymbol{\theta}_r, \boldsymbol{\theta}_s; \mathbf{x}\right)}{\partial\boldsymbol{\theta}_r^*} = \begin{bmatrix} \dfrac{\partial\mathcal{L}\left(\boldsymbol{\theta}_r, \boldsymbol{\theta}_s; \mathbf{x}_1\right)}{\partial\boldsymbol{\alpha}_1^*} \\ \vdots \\ \dfrac{\partial\mathcal{L}\left(\boldsymbol{\theta}_r, \boldsymbol{\theta}_s; \mathbf{x}_{N_t}\right)}{\partial\boldsymbol{\alpha}_{N_t}^*} \end{bmatrix} \tag{13A.69}$$

Plunging (13A.5) into (13A.69), followed by some algebra, results in

$$\frac{\partial\mathcal{L}\left(\boldsymbol{\theta}_r, \boldsymbol{\theta}_s; \mathbf{x}_j\right)}{\partial\boldsymbol{\alpha}_j^*} = \text{diag}\left(\frac{1}{\widehat{\sigma_{j1}^2}}, \ldots, \frac{1}{\widehat{\sigma_{jN_r}^2}}\right)\begin{bmatrix} (\mathbf{D}_{Pj1}\hat{\mathbf{s}}_j)^H\boldsymbol{\Pi}^{\perp}_{\mathbf{D}_{dj1}\hat{\mathbf{s}}_j}\mathbf{x}_{j1} \\ \vdots \\ (\mathbf{D}_{PjN_r}\hat{\mathbf{s}}_j)^H\boldsymbol{\Pi}^{\perp}_{\mathbf{D}_{djN}\hat{\mathbf{s}}_j}\mathbf{x}_{jN_r} \end{bmatrix} \tag{13A.70}$$

This has been obtained after substituting the MLE of unknowns under $\mathcal{H}_0$ hypothesis. Finally, inserting the obtained results into (13A.1), after some calculations, we come up with the Unc-Rao test statistic of (13.9).

Index

Accelerating target detection problem
  multilayer GLR-based detection algorithm 253–254
  problem formulation 252–253
  signal modeling 248–252
  successive and recursive (SaR) detection algorithm 256–257
  3D-MLa-GLRT detection algorithm
    chirp-FFT implementation 255–256
  2D-SaR detection algorithm
    detection performance 257–262
  3D-SaR detection algorithm 257
    performance evaluation 262–268
Adaptive filter-based interference removal approaches 46–51
  least mean squares (LMS) algorithms 47–49
  recursive least squares (RLS) algorithms 48–51
Analog television-based passive radar systems 225
  CZT-MLa-GLR detection algorithm
    recursive implementation 230–232
  detection problem formulation 226–227
  multilayer GLR-based detection algorithm 228–229
    chirp z-transform implementation 229–230
  robust S-CZT-MLa-GLR detection algorithm 233–234
    performance analysis 234–244
  S-CZT-MLa-GLR detection algorithm 231, 232
    performance analysis 234–244
Analog television signal characteristic 19–23
Analytical performance analysis 87–89
Antenna radiation pattern loss 33
ASGLR 85
Autoambiguity function (AAF) 10, 14, 15, 21–23

Binary hypothesis GLRT-based detector 202–205, 202–206
Bistatic radar maximum range equation 34
Bistatic range-Doppler function 53–58
Bistatic velocity 24, 25
Bit randomization 12
B-ITPA. *See* Broadband imperative target positioning algorithm (B-ITPA)
Broadband FM-based passive bistatic radar (PBR) system 172–174
  analytical performance evaluation 180–181
  broadband UMPI test
    design 174
    first stage 174–177
    performance quality improvement 185–187
    range resolution improvement 181–185
    second stage 177–180
    simulation 187–194
Broadband imperative target positioning algorithm (B-ITPA) 177–180

Cartesian bistatic velocity resolution 29, 31, 32
Cartesian coordinate 28
Cascaded integrator-comb (CIC) filter 85

Cassini ovals 34
Cauchy Schwartz inequality 80
Cell under test (CUT) 51
Constant false alarm rate (CFAR)
  behavior, detectors 325–326
  test 94
  2D-CA-CFAR detector 51, 100, 102
Channel coding 12–13
Clutter power spectral density (PSD) 91
Complex amplitude mismatch (CAM) 153–154, 160
Conventional passive radar systems
  detection performance evaluation 64–67
  interference removal approaches 44
    adaptive filter-based interference removal 46–51
    subspace-based interference removal 44–46
  multipath removal capability
    moving multipath scatterers 59–64
    stationary multipath scatterers 53–59
  reference channel (RC) 39–44
  signal processing steps 52
  surveillance channel (SC) 39–44
  target detection approaches 51–52
  traditional signal processing techniques 52–67
C-Rao detectors 357–359
CZT-MLa-GLR detection algorithm, recursive implementation 230–232

Delay-Doppler map 229
Detection loss (DL) 89, 98–100, 122–123, 148, 153, 209, 210, 229
Detection performance evaluation 94–102
Direct signal-to-noise ratio (DNR) 33, 42, 43, 53, 234, 235, 308–311, 353
Discrete chirp-fast Fourier transform (DCFT) 255, 256
Discrete Fourier transform 150
Doppler frequency migrations (DFMs) 250
Doppler resolution 7
DVB-T signal characteristic 11–19
Dynamic range 34

Equivalent monostatic range (EMR) 34, 36
Exponential power spectrum density (PSD) 59–60
Extending cross-ambiguity function (ECAF) 247
Extensive cancellation algorithm (ECA) 43–46
Extensive cancellation algorithm and cell-averaging (ECA-CA) detection algorithm 100–102, 221, 222

False alarm regulation evaluation 92–94
Fast Fourier transform (FFT) 85, 118, 207, 282
First-order Taylor approximation 24
Fisher information matrix (FIM) 347
FM radio signal characteristic 6–11
FM-based passive bistatic radar (FBPBR) systems 107
  broadband target detection algorithm (*see* Broadband FM-based passive bistatic radar (PBR) system)
FM-based single-band PR
  detection performance 104–106
  GLRT-based detector 76–81
  high-SNR RC-based problem formulation 74–87
  target detection algorithm 89–91
    false alarm regulation evaluation 92–94
    performance evaluation 94–102
  two-stage GLRT-based detector
    first stage 81–84
    second stage 85–87
FR-BH-GLR detector 206–207

Generalized likelihood ratio test (GLRT) 73, 248, 321
GLRT-based detector 76–81

High resolution imperative target positioning (HRITP) algorithm 145–147, 154, 157
High-quality performance (HQP) detector 171
High-range resolution (HRR) detector 171
High-resolution UMPI test
  design 142–143
  first stage 143–145
  second stage 145–147
High-SNR RC-based problem formulation 74–76
  GLRT-based detector 76–81
  two-stage GLRT-based detector
    first stage 81–84
    second stage 85–87
Hitchhiking-based passive bistatic radar 4–5

Imperative target positioning (ITP) algorithm 87, 94, 177
Integrated sidelobe level ratio (ISLR) 6, 8, 9, 123, 124, 171, 181, 185, 186
Interference cancellation (IC) parameter 53
Interfering target matrix (ITM) 116, 117, 145
Inverse fast Fourier transform (IFFT) 282
Inverse Hadamard ratio (IHR) test 324

Kernel trick 279
Kernelization 279
Kernelized noisy reference (KNR)-LRT detector 282–290, 292, 294, 295, 296
Kernelized simplified ideal reference (KSIR)-LRT detector 282–290, 292, 294, 295, 296

Least mean squares (LMS) algorithms 47–49, 61–63
Least squares (LS) estimation method 45
Likelihood ratio test (LRT) test 273, 347
Log-liklihood function (LLF) 275, 276, 303

Manastash Ridge Radar 5
M-ary hypothesis testing-based detection approach 200–201
  analytical performance evaluation 209–210
  binary hypothesis GLRT-based detector 202–205
    forward and recursive implementation 205–206
  FR-BH-GLR detector 206–207
  PI-FR-BH-GLR detector
    active radar detection 211–212
    comparative analysis 217–222
    FFT-based implementation 207–209
    passive radar detection 211–212
    performance 210, 212–217
Maximal invariant (MI) statistic 113–116, 168–169
Maximum likelihood estimation (MLE) 347
MIMO passive radar target detection
  C-Rao detectors 357–359
  detectors design 350–352
  NoDPI-Unc-Rao detector 353–356
  performance 352–353
    MIMO configuration 355–360
    Rao test statistic 360–362
    SIMO configuration 353–356

problem formulation 347–349
Unc-Rao detectors 353–360
Modified chirp fast Fourier transform (MCFFT) 248
Monte Carlo (MC) simulation 96, 126, 157, 236, 262, 273, 274, 296
Moving Pictures Experts Group (MPEG)-2 video data stream 12
M-tap transversal filter 46
Multiband FM-based passive bistatic radar 107
analytical performance analysis 120–121, 147–148
SW1-C target model 122–123
SW1-I target model 121
high-resolution UMPI test
design 142–143
first stage 143–145
second stage 145–147
maximal invariant statistic 153–154, 168–169
range resolution improvement 148–153
signal modeling 108–111, 138–142
simulation 154–163
UMPI-based detector
design 111–120
first stage 111–116
second stage 116–120
Multiband imperative target positioning (MITP) algorithm 116, 118–120
Multiband uniformly most powerful invariant (UMPI) test 107
Multilayer GLR-based detection algorithm 228–229, 253–254
chirp z-transform implementation 229–230
CZT-MLa-GLR detection algorithm
recursive implementation 230–232
Multistage processing and cell-averaging (MP-CA) detection algorithm 218–221
Multitarget scenario (MTS) 266, 267

National Television System Committee 19
Neyman-Pearson criterion 76
NoDPI-Unc-Rao detector 353–356
Noise temperature 33
Noisy RC-based bistatic passive target detection 273–274
Kernel-based detectors 278
Kernelized LRT-based proposed detectors 280–282
preliminaries 278–280
Kernelized NR (KNR)-LRT detector 282–290, 292, 294, 295, 296
Kernelized SIR (KSIR)-LRT detector 282–290, 292, 294, 295, 296
LRT-based proposed detector 275–278
noisy reference channel-LRT based (NR-LRT) detector 277, 281–283, 287–292, 294, 295, 296
problem formulation 274–275
simplified ideal reference LRT-based detector 278, 281, 283–285, 287, 289, 292, 294–296
Noisy RC-based SIMO passive radars 299–300
LRT-based proposed detectors 303–306
performance 306–311
P1-LRT detector 300, 306–308
P2-LRT detector 300, 306–308
problem formulation 300–303
Noisy reference channel-LRT based (NR-LRT) detector 277, 281–283, 287–292, 294, 295, 296

Orthogonal frequency division multiplexing (OFDM) transmission 12–14
Orthogonal frequency division multiplexing (OFDM)-based waveform 211

P_1-LRTdetector 300, 306–308, 323–324, 339

P_1-Rao detector 320–322, 326, 328–331, 335–338
P_1-Vol detector 320–323, 326, 328–331
P_2 detectors 324–325
P_2-C2-LRT detector 325, 327–329, 332, 340–342
P_2-LRT detector 300, 306–308
P_3-LRT detector 325, 343–346
Partitioned matrix inversion Lemma 78, 86
Passive bistatic radar velocity resolution 29
Passive radar (PR), *See specific types*
  advantages 3–4
  bistatic 22–23
  coverage 31–36
  geometry 22–32
  illuminators 5–6
    analog television (ATV) 19–23
    digital video broadcasting-terrestrial (DVB-T) 11–19
    frequency modulation (FM) radio 6–11
  multiple-input multiple-output (MIMO) 23
  multistatic 23
  power budget 31–36
Peak-to-sidelobe level ratio (PSLR) 6, 8, 9, 123, 124, 151, 152, 181
Phase alternating line (PAL) 19
PI-FR-BH-GLR detector
  active radar detection 211–212
  comparative analysis 217–222
  FFT-based implementation 207–209
  passive radar detection 211–212
  performance 210, 212–217
Positioning algorithm 85–87
Power spectrum density (PSD) 41

Radio data system (RDS) 8
Range migrations (RMs) 250
Range resolution (RR) 6, 8, 9, 123, 124, 148–153
Rao test statistic 360–362
Recursive least squares-cell averaging RLS-CA method 65, 66
Recursive least squares (RLS) algorithms 48–51, 63
Reference channel (RC) 26, 39–44, 274
Reference channel power uncertainty (RCPU) 307
Robust S-CZT-MLa-GLR detection algorithm 234–244

Satellite-based passive bistatic 4
S-CZT-MLa-GLR detection algorithm 231, 232, 234–244
Second-order real polynomial (SORP) kernel 279
Sequential cancellation algorithm (SCA) 43–46
Sequential couleurs avec memoire 19
Signal conditioning 42
Signal processing gain 34
Signal-to-noise power ratio 33, 34
Signal-to-noise ratio (SNR) 52, 53, 95–96, 156, 263, 188, 353
SIMO passive radar system 317
  detectors
    CFAR behavior 325–326
    design 321
    false alarm probability *vs.* threshold 327–329
    P_2 detectors 324–325
    P_3 detectors 325
    performance 325–327, 329–332
    P_1-LRT-based proposed detectors 323–324
    P_1-Rao detector 322
    P_1-Vol detector 323
    specification 327
  problem formulation 317–321
  signal model 317–321
Simplified ideal reference LRT-based detector 278, 281, 283–285, 287, 289, 292, 294–296, 295
Single-channel PR target detection approaches 42–43

Singular eigenvalue decomposition (SVD) 143
SISO passive radar system 274
SNR loss (SNRL) 160, 161
Stereo transmission 7
Subsidiary Carrier Authorization (SCA) 8
Subspace-based interference removal 44–46
Surveillance channel (SC) 39–44, 274
SW1-C target model 122–123, 135–136
SW1-I target model 121, 135–136

Target range resolution (TRR) 181
3D-MLa-GLRT detection algorithm 254
  chirp-FFT implementation 255–256
3D-SaR detection algorithm 257, 262–268
Transmission parameter signaling (TPS) 13
Two-channel PR target detection techniques 42, 43
2D-SaR detection algorithm 257–262
Two-stage GLRT-based detector
  first stage 81–84
  second stage 85–87

UMPI-based detector 111
  first stage 111–116
  second stage 116–120
Unc-Rao detectors 353–360
Uniformly most powerful invariant (UMPI) criterion 171, 174–180

Very high-frequency (VHF) FM-based passive bistatic radar 5

Zoom-FFT method 229

www.ingramcontent.com/pod-product-compliance
Lightning Source LLC
LaVergne TN
LVHW011834200926
848484LV00002B/355